Preface

Due to increasing penetration by the telephone, together with rapid growth in telex service, the inland telegram has lost its traditional leading position in the telecommunication field although there is still considerable growth in overseas telegrams. Telegraph techniques are of fundamental importance since they provide the bases for instruments and transmission not only in general telegraphy, but for telex and low-speed data communication.

Space does not permit full detailed treatment of the many teleprinters available today, so I have described general characteristics of those in current use in the United Kingdom (and elsewhere) and concentrated detail on unique design features of each. References are given to detailed descriptions to be found elsewhere.

Maintenance procedures are dealt with incidentally, but both this subject and telegraph power plants are covered in detail in *The International Telex Service*.

I am indebted to the Post Office for permission to publish official information and photographs, and pleased to acknowledge frequent reference to the Post Office Electrical Engineers' Journal. I

should like to thank G. H. E. Collins and F. J. Vallance for their excellent illustrations.

I wish to acknowledge permission to publish technical information and photographs generously supplied by industrial organizations — ITT Creed Ltd, The Marconi Co. Ltd, Muirhead & Co. Ltd, The Plessey Co. Ltd, Standard Telephones & Cables Ltd, The Telephone Manufacturing Co. Ltd, Hasler AG (Switzerland), Kokusai Denshin Denwa Co. Ltd (Japan), Philips Telecommunication Industry (Holland), Sagem SA (France), Siemens and Halske AG (Federal German Republic).

Except where otherwise stated, references to Recommendations in footnotes are those of the CCITT:

Volume IIb:	*Telegraph Operation*,	Series F
Volume III:	*Line Transmission*,	Series G and H
Volume IV:	*Maintenance of International Lines*,	Series M
Volume VII:	*Telegraph Technique*,	Series R, S, T, U
Volume VIII:	*Data Transmission*,	Series V

R.N.R.

Contents

1 Codes and Alphabets

Apart from specialized needs, such as facsimile telegraphy or character recognition, information to be sent by telegraphy is converted into some form of code for ease of transmission and recognition. The information to be transmitted is printed language — letters, figures, symbols, punctuation and mathematical signs, denoted collectively by the term *characters*. The introduction of printing telegraphy with page-type machines necessitated sending additional information for controlling movements of a remote paper carriage. Development of further facilities has increased the scope to include station-calling and identification signals, while more recently the process controls of automation have expanded the number of 'characters' which may need to be signalled.

1.1 Codes

Information codes are constructed from two or more basic states or *significant conditions* — represented, for example, by current-on and current-off. For transmission, these significant conditions are allotted a specific duration to form the unit element from which a code can be constructed. According to the range of information to be signalled, the basic code is made up from grouping a number of elements — rarely less than five — selected from one or other of these two (or more) electrical conditions. The duration coupled with the significant condition define the characteristics of any telegraph-signal element.

Each group of elements or character signal is signalled for reception as a whole — usually but not necessarily on a sequential-time basis — to convey a discrete piece of information, for example a letter or a figure. A telegraph alphabet is built up from the set of possible code-combinations (or from a selection of these) using a unique pattern to represent each character.

Codes may be classified in various ways: earlier codes were of non-uniform length (duration), but constant-length codes show advantages for mechanized telegraphy. The two most desirable features in any code are: (1) maximum efficiency in conveying a given amount of information with the least number of code elements, i.e. at the maximum commercial speed (characters/s) which will result from a given element-modulation rate; and (2) freedom from undetectable errors in the presence of signal distortion or of mutilation due to transmission conditions. These two factors are in mutual conflict.

1.2 Binary signals

It has been found that binary, or two-condition, signals, in which any element of the code must take either of two conditions, are adequate and convenient for transmitting information: most codes used in telegraphy are in binary form. Codes having three or more conditions are inferior to two-condition codes when transmission bandwidth and signal/noise ratios are considered.

For identification the two discrete conditions of a code have distinguishing names. In early forms of signal-recording telegraphy indelible marks were recorded on a paper tape to indicate an *active* signal condition, the spaces separating these marks indicating the contrary signal condition. In the course of evolution, the conditions of the binary code have continued to be identified by the terms *mark* and *space*, even though without their original significance and indeed in fields outside telegraphy.

More recently, to avoid possible confusion in transmission over international circuits, particularly when operated by radio-teleprinter, the distinctive terms *start* condition and *stop* condition of the start–stop principle have been adopted, together with their diagrammatic nomenclature A and Z respectively.

Table 1.1 Binary coding equivalence between notation symbols and significant conditions

Equipment	Significant conditions	
General telegraphy	Space S	Mark M
Start—stop telegraphy	Start	Stop
Diagrammatic symbol	*A*	*Z*
Binary notation	0	1
Perforated tape	No hole	Hole
Single-current telegraphy*	No current	(Positive) current
Double-current telegraphy*	Negative current	Positive current
Amplitude modulation	Tone-off	Tone-on
Frequency-shift modulation	Upper frequency	Lower frequency
Phase modulation (with reference phase)	Opposite phase to the reference phase	Reference phase
Differential phase modulation	Inversion of phase	No phase inversion

*Note. These conventions refer specifically to association with telegraph alphabet No. 2. They also relate specifically to the potential on an *international* circuit: since very few circuits operate nowadays by direct currents over the international circuit itself, this convention has little practical significance. Most telegraph-operating administrations observe this convention (− start, + stop if double-current operation) within their national teleprinter networks. In the United Kingdom and in certain other networks overseas the traditional *national* standard for teleprinter operation is the opposite of this:

positive potential = start element (space)
negative potential = stop element (mark)

All international teleprinter circuits connecting the United Kingdom are operated by a.c. (voice-frequency) methods: consequently this apparent discrepancy is entirely without significance.

From electronic technology, particularly in switching and computers, the digits 0 and 1 have been used to designate the OFF and ON conditions of a binary circuit or device. The need for equivalence between the various notation symbols used for the significant conditions of a binary code has been met by a Recommendation* which is summarized in Table 1.1 There is no fundamental difference in the significance of these various terms used to denote the two binary states.

The number of possible arrangements or code-combinations available with a binary code is readily determined. For example, if the code consisted of two units only, the first element could be either the mark M condition or the space S condition, giving two possibilities for this position. Associated with each of these, the second code element could also take either of the two conditions M or S, giving a total of 2 x 2 = 4 combinations, as shown in the column opposite.

More generally, the number of possible code-combinations when using binary signals is 2^n, where n is the number of elements in the code.

1.3 Unequal-length codes

These are codes in which the duration of a code-character is not constant over the range of symbols coded. Such codes are largely falling into obsolescence because of advantages from adopting equal-length codes to mechanized and electronic methods.

Combination	*1st element*	*2nd element*
1	M	M
2	M	S
3	S	M
4	S	S

The most widely known example of unequal-length code is the Morse Code and its derivatives developed mainly for more economic working over submarine-cable d.c. circuits. There is no limit to the number of character-combinations available from an unequal-length code.

*Recommendation V1.

1.4 Morse code

This is an unequal-length binary code. The two significant conditions, mark and space, each have more than one choice of basic duration: Morse code characters are constructed from widely varying numbers of marks and spaces, resulting in character-signals of many differing durations.

Mark signals may be either short (the *dot*) or long (the *dash*). The dot is the elementary or unit signal in the Morse code; the duration of the dash is three times that of the dot, and the alphabet is constructed from combinations of dots and dashes. Space signals are essential to separate mark elements in the code: between dot or dash elements within one character, the duration of a space is equal to that of one dot. Since neither the dot nor the dash is identified without its separating space, it may alternatively be considered that a dot-plus-space occupies two unit elements, and a dash-plus-space four elements, the overall time to transmit a dash being twice that of the dot. Between characters the space has a duration equal to that of three dots, while between words the space duration is seven dots. These timings are illustrated in Fig 1.1 by a time-chart for the word THEY.

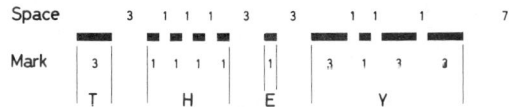

Fig 1.1 Morse Code for THEY

The efficiency of the Morse code is low. There is considerable difference in the number of elements required to signal the various symbols — in fact a 19 : 1 (or 20 : 2) ratio for the longest (zero) and shortest (E) characters in the alphabet. In comparing commercial speeds of various codes the average Morse character is regarded as consisting of about 8·5 units for plain English text, a value which would increase if messages included a high proportion of figures.

The Morse code is highly vulnerable to interference which would result in transposition between dots and dashes: for example, a split dash may produce two dots, or two dots may run together to form one dash — faults which usually result in the change of one letter into another. A characteristic of the Morse code is that while transpositions between dots and dashes are likely to cause undetected errors, those between marks and spaces are unlikely to cause undetectable errors. The Morse code has no fixed modulation rate — its speed can be widely varied to suit prevailing conditions.

Morse code was originally designed for manual operation using a simple telegraph key for forming the signals, working to a visual or audible type of receiver. If received as a tone in a headgear receiver, a trained telegraphist can accurately decipher weak Morse signals in the presence of considerable noise and interference.

Morse code was later adapted by Wheatstone to high-speed telegraphy using machine-driven transmitters controlled from paper tape perforated according to the code. (Owing to the nature of the code, transverse perforation of a Morse character on tape is not possible — there cannot be more than one dot per centre-hole.) At the receiving station, a perforated tape which is a duplicate of that at the sending station can be reconstituted from the signals received. Printing receivers, operated from the perforated tape reproduced over high-speed Wheatstone—Morse systems are still in use in some parts of the world, but Morse systems as a whole, including cable-code Morse, are rapidly becoming obsolescent. This has evolved mainly from the need to standardize on 5-unit teleprinter systems to facilitate worldwide interconnection of circuits. For well-developed fixed networks carrying heavy traffic, use of the Morse code is now obsolete. It is still used for communication between stations either or both of which are mobile — ships, aircraft or cars — if separated by distances too great for successful low-power radio-telephony. The reason is that at sending and receiving stations the Morse equipment and its power supplies can be of simple type, requiring little in the way of accommodation, maintenence or technical skill. For the same reason, Morse systems still have their place in remote areas carrying light traffic over fixed installations.

1.5 Constant-length codes

Although printing-telegraph receivers using unequal-length codes have been used, a code using a constant and minimum number of elements is the preferred arrangement for a straightforward design of telegraph-printing machine. Modern telegraph systems use constant-length codes.

In a constant-length code all combinations are formed from equal numbers of elements each having an equal duration which becomes the basic time unit for the system. In practical systems the number of units may range according to requirements between five and seven, or more. A 4-unit binary code, for example, will give only $2^4 = 16$ combinations, insufficient for any alphabet, yet adequate for the numerals 0–9 together with a few control signals.

1.6 5-unit Code
The 5-unit code is an equal-length binary code using five elements of equal duration for each character. It is in universal use for start–stop (teleprinter) systems and also widely used for synchronous telegraph systems.

Each of the five elements can take either of two conditions and the maximum number of code-combinations is $2^5 = 32$. With a basic need for the 26 letters of the Roman alphabet together with 10 numerals, this number of possible code-combinations is insufficient: by allocating two from the 32 combinations for case-shift functions, and using primary and secondary printing cases at the receiver, preselected by these two shift combinations, an alphabet approaching 64 characters becomes available.

The 5-unit code is highly efficient in signalling information with the minimum number of code elements. For the same reason it is particularly vulnerable to error in transmission over circuits liable to interruption, since any inadvertent change in the significant condition of one or more elements will result in reception of a code-combination allocated to a different character.

1.7 6-unit code
The maximum number of code-combinations possible with a 6-unit code is $2^6 = 64$. With this increased availability compared with 5-unit code, a 6-unit code has been used to a slight extent for a variety of purposes. One use is to provide the full set of letters, figures and other symbols of the 5-unit alphabet but without the need for a case-shift. For messages which consist of approximately equal

proportions of letters and figures, use of a 6-unit code can result in economy of line and operating time due to the absence of case-shift signals: otherwise it is less efficient than a 5-unit code, but with the same vulnerability to error. In isolated cases it is used for such purposes as stock-market quotations or teletypesetting.

The 6-unit code is used on synchronous time-division multiplex systems operated over intercontinental submarine cables. For switched services, e.g. telex, it is necessary to transmit supervisory ('call' and 'clear') signals which consist of long-duration mark and space conditions. As these signals would be indistinguishable from the 5-mark and 5-space combinations, and spare code-combinations are not available for this purpose within the 5-unit alphabet, a 6-unit code has to be used to make available the extra signal combinations. This code is less efficient than the 5-unit code, but it is less liable to undetected errors since it contains a large number of unallocated or redundant code-combinations: certain forms of signal mutilation during transmission may result in reception of an unallocated combination, a form of error which it is possible to detect at the receiver.

Another form of 6-unit code results from the addition of a single checking element to the 5-unit code to form a protected code.

1.8 7-unit code
The number of possible code-combinations available from a 7-unit binary code is $2^7 = 128$.

International agreement between telegraph-operating administrations and agencies, and also with ISO, serving the interests of users of data-processing machines needing data interchange over telecommunication circuits, resulted in the adoption of 7-unit code which provides expanded facilities, such as capital letters as well as lower case, numerals, and a variety of printed symbols and control signals to an aggregate of 128 without need for a case-shift. It cannot readily be used over the telex system, which is designed for a 5-unit code, but is used over leased telegraph circuits. It is probable that a new exchange network designed to carry 7-unit signals will ultimately be established.

The composition of combinations available from

a 7-unit code follows a pattern of proportion shown in Table 1.2

Table 1.2 Mark/space ratios available in 7-unit code

Mark/space ratio	Number of combinations
0 : 7	1
1 : 6	7
2 : 5	21
3 : 4	35
4 : 3	35
5 : 2	21
6 : 1	7
7 : 0	1

Selection of either the 3 : 4 or the 4 : 3 mark/space ratio provides 35 combinations having this unique characteristic. The 3 : 4 ratio is chosen as the basis of an error-detecting code in which received signals are checked for correct 3 : 4 ratio. The 35 available combinations conveniently provide equivalence for the 32 combinations of the 5-unit input and output signals of the alphabet No. 2, together with two combinations for the long-mark and long-space supervisory signals, and also a 35th combination which is used to initiate an automatic request ('ARQ') for repetition of a mutilated character. The efficiency of this 7-unit ARQ code is 5/7 or 71·4 per cent. Since only 35 from the available 128 combinations are used there is a redundancy of (128 − 35) = 93 combinations. This 7-unit code is used universally for error-protected HF radio-teleprinter circuits.

Patterns similar to that of Table 1.2 exist for other codes, and some use has been made of the unique-ratio combinations — for example, figure security within a special alphabet using the 2 : 3 ratio of 5-unit code.

ALPHABETS

A telegraph alphabet is a table of correspondence between an agreed set of characters and the signals which represent them. They have been designed for various applications and the more important are described below.

1.9 Morse international alphabet

The Morse alphabet was designed to be memorized by the sending and receiving operators, although extensively used later in high-speed mechanized

systems. It is based upon the allocation of the shortest and simplest signals for letters of most frequent occurrence in the English language. Table 1.3 shows the arrangement of the international Morse alphabet, although variants exist for national uses.

Table 1.3 International Morse alphabet

LETTERS

a	· —	i	· ·	r	· — ·
b	— · · ·	j	· — — —	s	· · ·
c	— · — ·	k	— · —	t	—
d	— · ·	l	· — · ·	u	· · —
e	·	m	— —	v	· · · —
accented e	· · — · ·	n	— ·	w	· — —
f	· · — ·	o	— — —	x	— · · —
g	— — ·	p	· — — ·	y	— · — —
h	· · · ·	q	— — · —	z	— — · ·

FIGURES

1	· — — — —	6	— · · · ·
2	· · — — —	7	— — · · ·
3	· · · — —	8	— — — · ·
4	· · · · —	9	— — — — ·
5	· · · · ·	0	— — — — —

FIGURES—SHORTENED VERSION

1	· —	6	— · · · ·
2	· · —	7	— · · ·
3	· · · —	8	— · ·
4	· · · · —	9	— ·
5	· · · · ·	0	—

PUNCTUATION MARKS AND MISCELLANEOUS SIGNS

Full stop (period)	.	· — · — · —
Comma	,	— — · · — —
Colon or division sign	:	— — — · · ·
Question mark	?	· · — — · ·
Apostrophe	'	· — — — — ·
Hyphen or dash or subtraction sign	—	— · · · · —
Fraction bar or division sign	/	— · · — ·
Left-hand bracket (parenthesis)	(	— · — — ·
Right-hand bracket (parenthesis)	)	— · — — · —
Inverted commas (quotation marks)	" "	· — · · — ·
Double hyphen (equals)	=	— · · · —
Understood		· · · — ·
Error		· · · · · · · ·
Cross or addition sign	+	· — · — ·
Invitation to transmit		— · —
Wait		· — · · ·
End of work		· · · — · —
Starting signal (to precede every transmission)		— · — · —
Multiplication sign	x	— · · —

1.10 ITU 5-unit alphabet No. 2

This alphabet, the teleprinter alphabet, shown in Table 1.4, is standardized for worldwide use over the telex network, for the gentex service and for general use in start–stop apparatus operating at 50 or 75 bauds.

With few exceptions, each code-combination is allocated to two purposes preselected by one of the case-shift keys, so virtually doubling the number of printed symbols in relation to code-combinations. The layout of the keyboard is not specified by the table; the usual aim is to follow typewriter-keyboard practice.

In particular the numerals 1, 2, 3 . . . 8, 9, 0 are placed in order along the back row. As the related letters QWE . . . IOP (English keyboards) share the same code-combinations with the numerals they must share the same keys unless the numerals appear on a separate and exclusive row of keys in a 4-row keyboard.

The CARRIAGE RETURN signal is to ensure that the printing position for the next character following the end of a printed line will be at the beginning of a new line, i.e. adjacent to the margin; the LINE FEED makes available a new printing line. Transmitting either of the shift combinations selects in the receiver either the primary or secondary character (or function) which relate to a given 5-unit combination. To transmit a group of letters, the operator must precede this by sending the LETTERS SHIFT combination: similarly, any figures must be preceded by the FIGURES SHIFT combination. The receiver continues to print in the letters (or figures) case until the contrary shift combination is received. With national alphabets which comprise more than 26 letters, some of these have to be placed in the secondary case and, to avoid excessive use of the shift keys, an automatic case-shift mechanism may be included in such keyboards.

The SPACE, i.e. character-space combination, causes the insertion of one space of character width for each transmission of the SPACE combination. The spacing or feeding action which a page-type receiver normally applies when printing any symbol is suppressed on receipt of CARRIAGE RETURN and LINE FEED combinations; it is also suppressed should combination 32 be received.

Secondary combinations related to letters F, G and H are reserved for national use; in the United Kingdom they are used for %, @ and £ respectively. In the systems of foreign administrations they are used for other purposes, for example accented

letters or letters outside the 26-letter Roman alphabet. On a keyboard the keytops for letters F, G and H are marked in a distinguishing colour and operators are advised that the secondary functions of these keys should not be used on international connections

Table 1.4 ITU 5-unit teleprinter alphabet No. 2

No. of signal	Letter case	Figure case	No. of element				
			1	2	3	4	5
1	A	—	o	o			
2	B	?	o			o	o
3	C	:		o	o	o	
4	D	WRU	o			o	
5	E	3	o				
6	F	%	o			o	o
7	G	@		o		o	o
8	H	£			o		o
9	I	8		o	o		
10	J	*Bell*	o	o		o	
11	K	(	o	o	o	o	
12	L	)		o			o
13	M	.			o	o	o
14	N	,			o	o	
15	O	9				o	o
16	P	0		o	o		o
17	Q	1	o	o	o		o
18	R	4		o		o	
19	S	'	o		o		
20	T	5					o
21	U	7	o	o	o		
22	V	=		o	o	o	o
23	W	2	o	o			o
24	X	/	o		o	o	o
25	Y	6	o		o		o
26	Z	+	o				o
27	*Carriage return*					o	
28	*Line feed*			o			
29	*Letters shift*		o	o	o	o	o
30	*Figures shift*		o	o		o	o
31	*Space*				o		
32	*Not used*						

0 = mark

Note. When it is desired to have a printed record of the receipt of certain functional code signals, the following standardized* symbols may be used:

☒ = WHO ARE YOU? (WRU)
☐ = BELL
< = CARRIAGE RETURN
≡ = LINE FEED

*Recommendations S4 and S5.

as their use is liable to result in printing different symbols at the distant station.

The secondary of combination 4 (D in the letters case) is the WRU (WHO ARE YOU?) signal used for operating the station-identification device (the 'answer-back' unit) of a distant teleprinter. The secondary of combination 10 (J in the letters case) is used to operate a call alarm at the distant station. The receipt of either of these inhibits paper spacing on page-type receivers.

The special symbols shown at the foot of Table 1.4 may be printed to indicate receipt of code-combinations relating to the functional signals — CARRIAGE RETURN, LINE FEED, WHO ARE YOU and BELL; this may be needed when printing on perforated tape.

It is necessary to permit operation of the carriage-return, line-feed, letters-shift, figures-shift and character-space functions irrespective of whether the receiving teleprinter has been put into the letters or figures case: consequently there are no separate primary and secondary allocations for code-combinations 27, 28, 29, 30, 31 or 32.

In perforated-tape systems, combination 29, the LETTERS SHIFT consisting of all mark elements, may be used to obliterate a combination which has been perforated in error: this it does by converting the faulty combination (after moving the tape to the appropriate position) into five punched holes. This has no effect on reception except to record on the tape a LETTERS SHIFT combination which would change the shift should an associated teleprinter have been at that moment in the figures case; this facility would, however, normally be used in preparing a perforated tape or in some situation in which appropriate shift action can be taken.

Combination 32, comprising all five space elements, is not allocated to a specific purpose; it is liable to be simulated by faults in the line and apparatus which result in a prolonged space condition and, on a receiving teleprinter, this fault is registered as code-combination 32. When using perforated tape, combination 32 does not produce any perforation so that its presence would be indistinguishable from unused tape. A further objection to using an all-space combination is that if it is transmitted repeatedly at maximum operating rate it may simulate the clearing signal used on switched systems and so break down a connection

prematurely. For certain specialized applications, use of combination 32 has been sanctioned for national purposes only.

'Baudot Code'. In certain areas this term is used, wrongly, to describe alphabet No. 2. The Baudot multiplex system was the first telegraph system to make wide use of 5-unit code and it was practically alone in this field for some forty years. For most of this time the transmitter consisted of five piano-type keys on which the operator set up combinations in 5-unit code (which had to be committed to memory) using two fingers of the left hand and three fingers of the right. Although the receiver used a mechanical printer from the outset, it was only later that a typewriter-like keyboard was added to the trans-mitter. The Baudot alphabet was designed primarily for its ease of memorization and manipulation and although more than one version was in existence for national uses, an international version was standardized as the ITU Telegraph Alphabet No. 1, now obsolete; it bears little resemblance to alphabet No. 2.

'Murray Code'. This is another term employed, wrongly, in some areas to describe alphabet No. 2. The alphabet designed by Murray was for machine operation using perforated tape — there was neither intention nor need to memorize the alphabet. It was arranged with letters of most frequent occurrence in the English language represented by fewest holes in perforated tape; the objects were to keep the paper tape as strong as possible by having a minimum number of holes and to impose the least wear on a perforator-punch mechanism. Another new feature of this alphabet was to allocate the 5-mark combination to the ERASE signal so that any unwanted character in a perforated tape could be overpunched with five holes before transmission, to produce this ERASE signal — a facility commonly used today in perforated-tape systems. An important innovation was re-arrangement of numerals in relation to those letters which share the same code-combinations, so that numerals 1, 2, 3 . . . 8, 9, 0 were associated with letters QWE . . . IOP (English keyboards); this enabled an inexpensive compact 3-row keyboard to be designed with numerals occupying positions in sequence in the top row of the keyboard. In the Baudot system, which was designed before the introduction of the typewriter, numerals appeared in haphazard positions when later assembled with corresponding letters in a keyboard layout. For various requirements — teleprinter or synchronous systems — more than one version of the Murray alphabet came into use. The more general form provided a character-space function in association with both the LETTERS SHIFT and FIGURES

bet, from which it differs in the following respects: (1) the allocation of code-combinations to punctuation and other signs is changed; (2) the erase sign is abolished; (3) it uses a separate code-combination for character-space. As a consequence, receivers must be designed so that no paper-spacing occurs on receipt of either letters-shift or figures-shift combinations.

Table 1.5 CCITT 7-unit alphabet No. 3 used for ARQ radio teleprinter systems

No. of alphabet No. 2 signal	Letters and figures cases		5-unit alphabet No. 2	7-unit alphabet No. 3
1	A	—	ZZAAA	AAZZAZA
2	B	?	ZAAZZ	AAZZAAZ
3	C	:	AZZZA	ZAAZZAA
4	D	WRU	ZAAZA	AAZZZAA
5	E	3	ZAAAA	AZZZAAA
6	F		ZAZZA	AAZAAZZ
7	G		AZAZZ	ZZAAAAZ
8	H		AAZAZ	ZAZAAZA
9	I	8	AZZAA	ZZZAAAA
10	J	BELL	ZZAZA	AZAAAZZ
11	K	(	ZZZZA	AAAZAZZ
12	L	)	AZAAZ	ZZAAAZA
13	M	.	AAZZZ	ZAZAAAZ
14	N	,	AAZZA	ZAZAZAA
15	O	9	AAAZZ	ZAAAZZA
16	P	0	AZZAZ	ZAAZAZA
17	Q	1	ZZZAZ	AAAZZAZ
18	R	4	AZAZA	ZZAAZAA
19	S	'	ZAZAA	AZAZAZA
20	T	5	AAAAZ	ZAAZAZ
21	U	7	ZZZAA	AZZAAZA
22	V	=	AZZZZ	ZAAZAAZ
23	W	2	ZZAAZ	AZAAZAZ
24	X	/	ZAZZZ	AAZAZZA
25	Y	6	ZAZAZ	AAZAZAZ
26	Z	+	ZAAAZ	AZZAAAZ
27	carriage return		AAAZA	ZAAAAZZ
28	line feed		AZAAA	ZAZZAAA
29	letters shift		ZZZZZ	AAAZZZA
30	figures shift		ZZAZZ	AZAAZZA
31	space		AAZAA	ZZAZAAA
32	(not used)		AAAAA	AAAAZZZ
	repetition signal		—	AZZAZAA
	signal α		(long A condition)	AZAZAAZ
	signal β		(long Z condition)	AZAZZAA

SHIFT, a requirement which was popular at the time with European users since it relieved the need for a separate character-space key; nevertheless, an alternative version was available which allowed for a separate character-space key together with non-spacing shift keys as now used in alphabet No. 2. The Murray alphabet, long obsolete, was not accorded international recognition.

Alphabet No. 2 was based on the Murray alpha-

1.11 CCITT 7-unit alphabet No. 3 (the error-detecting ARQ alphabet)*

To enable reliable 5-unit teleprinter transmission (using alphabet No. 2) to take place over HF radio circuits, a 7-unit error-detecting alphabet has been constructed employing a constant-ratio 3 : 4 (mark : space) code. This is used in the radio path only, with

*Recommendation S13.

appropriate 5-unit/7-unit code conversion at the line-radio interfaces.

The 3 : 4 code provides 35 unique combinations for allocation to the 32-character alphabet No. 2, together with three additional character signals needed for supervisory purposes. These three are: (1) an RQ signal (known as signal I) − used to request automatic repetition of any transmitted character detected at the receiving station as being in error; (2) a signal termed alpha (α) equivalent to an all-space condition at the input and output − the clearing signal for a switched connection, and also the condition of a disengaged circuit; (3) a signal termed beta (β) which at the input and output of the system is equivalent to an all-mark condition; it is the calling signal for a switched system, and also the 'idle' signal or interval signal applied on a connection at any time when other characters are not being signalled − in effect an extension of the stop signal.

Table 1.5 shows the arrangement of CCITT alphabet No. 3 and indicates its relation to alphabet No. 2. The table also gives an idea of the problem of encoding and decoding the 5-unit/7-unit combinations, the latter containing always a total of three mark and four space elements in each character.

1.12 CCITT 6-unit alphabet No. 4 (cable synchronous systems)*

This alphabet is shown in Table 1.6 and is used for operating teleprinter circuits over synchronous-multiplex cable systems; it was designed to provide two additional combinations for supervisory signals, equivalent to all-space and all-mark signals (corresponding to *alpha* and *beta* in alphabet No. 3) in addition to all the characters of alphabet No. 2 which has no redundancy available for extra signals. Alphabet No. 4 was not designed for error detection though its high redundancy $(2^6 − (2^5 + 2)) = 30$ does in fact afford some protection.

The alphabet is formed simply by pre-fixing combinations Nos 1−31 of alphabet No. 2 with an A-element: combination 32 is prefixed with a Z-element in order to provide a transition. The alpha and beta combinations consist of six A-elements and six Z-elements respectively. One of the redundant

*Recommendation R44.

combinations (ZZAAZZ) is in fact used as a phasing signal, but it should not be regarded as part of the alphabet as it is not available to users of the circuit.

Table 1.6 CCITT 6-unit alphabet No. 4 for cable synchronous multiplex systems

Character	5-unit Alphabet No. 2	6-unit Alphabet No. 4
A	ZZAAA	AZZAAA
B	ZAAZZ	AZAAZZ
C	AZZZA	AAZZZA
D	ZAAZA	AZAAZA
E	ZAAAA	AZAAAA
F	ZAZZA	AZAZZA
G	AZAZZ	AAZAZZ
H	AAZAZ	AAAZAZ
I	AZZAA	AAZZAA
J	ZZAZA	AZZAZA
K	ZZZZA	AZZZZA
L	AZAAZ	AAZAAZ
M	AAZZZ	AAAZZZ
N	AAZZA	AAAZZA
O	AAAZZ	AAAAZZ
P	AZZAZ	AAZZAZ
Q	ZZZAZ	AZZZAZ
R	AZAZA	AAZAZA
S	ZAZAA	AZAZAA
T	AAAAZ	AAAAAZ
U	ZZZAA	AZZZAA
V	AZZZZ	AAZZZZ
W	ZZAAZ	AZZAAZ
X	ZAZZZ	AZAZZZ
Y	ZAZAZ	AZAZAZ
Z	ZAAAZ	AZAAAZ
Carriage return	AAAZA	AAAAZA
Line feed	AZAAA	AAZAAA
Letters	ZZZZZ	AZZZZZ
Figures	ZZAZZ	AZZAZZ
Space	AAZAA	AAAZAA
Character No. 32	AAAAA	ZAAAAA
Continuous A condition (alpha)	−	AAAAAA
Continuous Z condition (beta)	−	ZZZZZZ
Phasing signal	−	ZZAAZZ

1.13 ISO/ASCII/CCITT 7-unit alphabet No. 5 for data communication and telegraphy*

This is an expanded alphabet and the construction of the version used in the United Kingdom is shown in Table 1.7.

The presentation in this table differs from that of the earlier telegraph alphabets to illustrate some flexibility in the alphabet. The seven code elements or *bits* are represented by $b_1, b_2, b_3, b_4, b_5, b_6, b_7$ (reading in binary order of increasing magnitude

*Recommendation V2.

Table 1.7 ISO/ASCII/CCITT alphabet No. 5 for data communication and telegraphy

Bits				Column →	0	1	2	3	4	5	6	7
				b7	0	0	0	0	1	1	1	1
				b6	0	0	1	1	0	0	1	1
				b5	0	1	0	1	0	1	0	1
b4	b3	b2	b1	Row ↓								
0	0	0	0	0	NUL	[TC₇] DLE	SP	0	@	P	`	p
0	0	0	1	1	SOH [TC₁]	[DC₁]	!	1	A	Q	a	q
0	0	1	0	2	STX [TC₂]	[DC₂]	"	2	B	R	b	r
0	0	1	1	3	ETX [TC₃]	[DC₃]	£	3	C	S	c	s
0	1	0	0	4	EOT [TC₄]	[DC₄]	$	4	D	T	d	t
0	1	0	1	5	ENQ [TC₅]	NAK [TC₈]	%	5	E	U	e	u
0	1	1	0	6	ACK [TC₆]	SYN [TC₉]	&	6	F	V	f	v
0	1	1	1	7	BEL	ETB [TC₁₀]	'	7	G	W	g	w
1	0	0	0	8	BS [FE₀]	CAN	(	8	H	X	h	x
1	0	0	1	9	HT [FE₁]	EM	)	9	I	Y	i	y
1	0	1	0	10	LF [FE₂]	SUB	*	:	J	Z	j	z
1	0	1	1	11	VT [FE₃]	ESC	+	;	K	[	k	{
1	1	0	0	12	FF [FE₄]	FS [IS₄]	,	<	L	\	l	|
1	1	0	1	13	CR [FE₅]	GS [IS₃]	-	=	M	]	m	}
1	1	1	0	14	SO	RS [IS₂]	.	>	N	^	n	~
1	1	1	1	15	SI	US [IS₁]	/	?	O	_	o	DEL

from right to left (i.e. b_1 corresponds to the lowest order and b_7 to the highest)). The first four bits (b_1, b_2, b_3, b_4) appear in the left-hand vertical columns of the table; the last three (b_5, b_6, b_7) appear, following the arrows, in the top three horizontal rows. The symbols or functions of the alphabet are displayed in the eight main vertical columns numbered 0–7 inclusive. Each column contains 16 symbols appearing in horizontal rows numbered 0–15 inclusive; the total number of symbols is 2^7, shown in eight columns each of 16 rows, totalling 128 code-combinations.

The alphabet may be considered in four sub-sets:

1. Control (columns 0 and 1);
2. Numerals and signs (columns 2 and 3);
3. Capital letters (columns 4 and 5);
4. Lower-case letters (columns 6 and 7).

Diacritical signs are provided which, in conjunction with BACKSPACE, can be placed as accents over the last character printed; otherwise they represent special signs.

The DELETE character in column 7 row 15 (DEL

Table 1.8 CCITT alphabet No. 5 — transmission-control characters

Character no.	Abbreviated code	Purpose
TC_1	SOH	Start of heading
TC_2	STX	Start of text
TC_3	ETX	End of text
TC_4	EOT	End of transmission
TC_5	ENQ	Enquiry—broadly equivalent to WHO ARE YOU in ITA2
TC_6	ACK	Acknowledge—an affirmative in response to a sender
TC_7	DLE	Data link escape. This character is used in conjunction with others exclusively to provide supplementary data-control functions
TC_8	NAK	Negative acknowledgement—negative response to a sender
TC_9	SYN	Synchronous idle—provided for use in synchronous transmission systems so that, in the absence of any other character, a signal is provided from which synchronism may be achieved or maintained between terminal equipments
TC_{10}	ETB	End-of-transmission block

Table 1.9 CCITT alphabet No. 5 — format effectors

Character no.	Abbreviated code	Purpose
FE_0	BS	Backspace—controls the printing position one printing space backwards on the same printing line: it does not backspace a tape-reperforating mechanism
FE_1	HT	Horizontal tabulation—controls the movement of the printing position to the next of a series of predetermined positions along the printed line
FE_2	LF	Line feed—controls the movement of the printing position to the next printing line
FE_3	VT	Vertical tabulation—controls the movement of the printing position to the next in a series of predetermined printing lines
FE_4	FF	Form feed—controls the movement of the printing position to the first predetermined printing line on the next form
FE_5	CR	Carriage return—controls the movement of the printing position to the first printing position on the same printing line

Although provision is made for case-shift (SI and SO, meaning SHIFT IN and SHIFT OUT), these are for exceptional use. The table shows that code-combinations for a capital letter and for the corresponding small letter differ only in bit 6; discrimination between capitals and small type can be made by a key which changes bit 6 from 0 to 1 and vice versa, there being no need to *transmit* a case-shift signal.

at position 7/15) can be used to obliterate errors in perforated tape by overpunching all seven code positions.

Seven of the code positions are available for national options (these are 4/0, 5/11, 5/12, 5/13, 7/11, 7/12 and 7/13).

The transmission controls TC are functional characters intended to control or facilitate transmission of information over telecommunication networks (see Table 1.8).

Format effectors FE are functional characters which control layout or positioning of input or output information (see Table 1.9).

Device controls DC are functional characters which can be used to control ancillary devices such as tape readers or punches.

Information separators are used to separate and control information in a logical sense: IS1—IS4 refer to Unit, Record, Group and File separators respectively.

There are also eight miscellaneous control characters including BEL (bell), CAN (cancel), SUB (substitute). The SHIFT OUT (SO) provides for printing an alternative set of graphics, up to a maximum of 95, in place of those shown in columns 2—7 in the main table, while retaining the ability to communicate in the standard alphabet; reversion is by use of the SHIFT IN (SI) character. The 32 control characters and DELETE are not affected by SHIFT OUT.

An escape sequence consists of the character ESC (escape) followed by one, two or three characters to obtain additional control functions which may provide, among other things, graphics or graphic sets outside the standard set, although certain prohibitions apply.

The sole purpose of NUL (null) is to accomplish media-fill or time-fill. EM (end of medium) is a control character which may be used to identify the end of a particular portion of information.

The elements (bits) are transmitted in sequence starting with b_1 and ending with b_7, i.e. the lowest binary order is sent first. Fuller details of this alphabet are given elsewhere.*

1.14 Code converters

For interconnecting telegraph systems which use incompatible codes, converters have been designed and used for automatic transformation from one code to another, for example Morse code into 5-unit code and vice versa. Such needs have arisen where the circuits of two operating administrations are to be interconnected or where intercontinental submarine telegraph cables (d.c.) have to be extended over landline systems, or during a transitional stage in the replacement of one system by another.

Apart from the straight code-conversion between corresponding characters, the problem arises of the conversion of characters which exist in one only of the two alphabets concerned -- for example, the 5-unit case-shift signals which are not used with Morse.

With the spread of standardized 5-unit systems throughout the world, the need for code-converters has become much reduced. On the other hand, use of error-detecting and error-correcting codes necessitates conversion between the unprotected and protected codes. Information on encoding and decoding is included in descriptions of special systems, such as the ARQ error-correcting system.

Reference

Chesterman, D. A., 'International Telegraph Alphabet No. 5', *POEE Journal*, **62**, p. 89 (1969).

Data Telecommunication, R. N. Renton (Pitman).

2 Distortion

2.1 Telegraph modulation

Telegraph signals are formed by making changes in the electrical condition of a circuit. In binary systems, these *significant conditions* (designated mark/space, Z/A or 1/0) may be brought about by switching on and off a steady d.c. potential (single-current system); or reversing the direction of a steady d.c. potential (double-current system); or, in systems of binary or higher valency, switching on and off an a.c. potential at one or more voice frequencies. In the United Kingdom, the standard method of using a teleprinter is the double-current system at ±80 V; this will be used to illustrate telegraph modulation.

The change applied to the electrical condition at certain *significant instants,* according to the telegraph signals to be transmitted, is known as *telegraph modulation*; the significance of the telegraph signals so formed is fixed by the time intervals between successive transitions from one significant condition to another. At the receiver, the recognition of these significant instants associated with the significant condition enables the overall telegraph signal – e.g., a character-combination – to be reconstituted.

An ideal telegraph signal is represented in Fig 2.1, a 5-unit start–stop signal (letter Y of alphabet No. 2) from a teleprinter operating at 50 bauds. The term *telegraph signal* embraces the set of transitions which together represent one practical unit of information – e.g. a character-combination.

At a transmitter, successive transitions from one significant condition to another take place abruptly (ignoring transit time) at the instants which characterize the change of condition. These instants – *a, b, c, d, e* and *f* in Fig 2.1 – are the significant instants of the modulation; the time interval between two consecutive ones is known as a *significant interval.* Each of the parts *ab, bc, cd, de* . . . which together constitute a telegraph signal and are distinguished from one another by their nature, magnitude, duration and relative position, is known as a *signal element.* Elements such as *ab, bc, cd, de* and *ef*, which are the shortest signal elements permitted by the code, are *minimum intervals.* The *unit interval* in a telegraph system using an equal-length code is defined as *the interval of time such that the theoretical durations of the significant intervals of a modulation (or restitution) are whole multiples of this interval.*

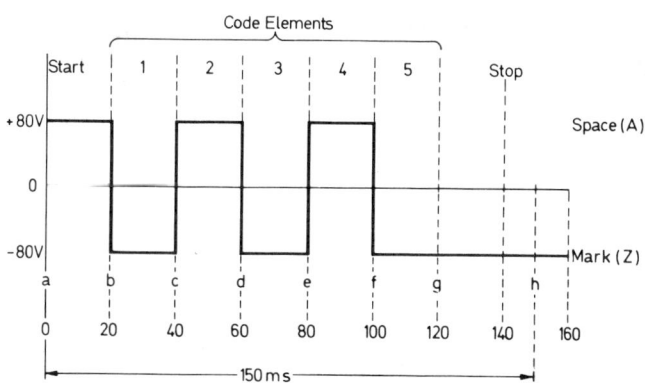

Fig 2.1 Start–stop 5-unit 50-baud double-current teleprinter signal (letter Y of alphabet No. 2)

In most systems, the signal elements have a duration which is an integral multiple of the unit interval. In start—stop telegraph systems the duration of the stop signal may not conform to this relationship, and is 1·0, 1·42, 1·5 (more commonly) or 2·0 times the unit element.

The unit elements in an equal-length code from whose arrangements the telegraph alphabet is formed are known as *code elements* — e.g. in alphabet No. 2 the 5-unit elements which occur between the start and the stop signals are code elements.

In a telegraph receiver, the series of significant conditions is reconstituted from the telegraph modulation by the appropriate device — e.g. the tongue of a relay, the armature of an electromagnet, or a transistor — the process being referred to as *restitution.*

The term *start—stop modulation* (or *restitution*) is applied when it is necessary to distinguish this from isochronous* modulation (or restitution) of synchronous systems: in the latter, the significant intervals are theoretically always equal to the unit element or to an integral multiple thereof.

2.2 Modulation rate

The maximum rate at which information may be transmitted over a channel is inversely proportional to the duration of the shortest element — the unit element of modulation. This factor is of primary importance in considering the technical and economic problems related to telegraph transmission. The modulation rate C is defined as the *reciprocal of the duration* (*expressed in seconds*) *of the unit interval.* The unit of modulation rate corresponds to one unit element of modulation per second and is called the *baud* after Emile Baudot, whose name is associated with the first multiplex printing telegraph system.

Referring to the characteristics of the teleprinter used for telex (Fig 2.1), the unit interval in this case is 20 ms† or 0·02 s. The modulation rate is $C = 1/0·02 = 50$ bauds. In general, if T is the unit

interval in seconds, the modulation rate $C = 1/T$ bauds.

Teleprinters are available which operate at 50, 75, 100 and 110 bauds. The duration of the unit interval for various modulation rates is shown for reference in Table 2.1.

For commercial comparison between various systems it is convenient to know the number of characters which can be transmitted in one second, which depends not only upon modulation rate, but also upon the number of units — including start, stop, and any parity or check bits — which constitute one character.

In a start—stop system, the information signals are phased by prefixing them with a start signal and terminating them by a stop signal. For a 5-unit code with unit interval T (see Fig 2.1), the start element is equal to T but the stop element is nominally $3T/2$, giving a total character-signal duration of $(T + 5T + 3T/2) = 7·5T$ or $7·5/C$. The maximum number of characters which can be signalled in one second is $1/(7·5T)$ or $C/7·5$ for this class of teleprinter.

For a 50-baud teleprinter with a 5-unit code, the start and code elements are each $1/50 = 0·020$ s, but the stop element is $3/(2 \times 50) = 0·030$ s. The minimum time for a character cycle is $7·5T = (7·5 \times 0·020) = 0·150$ s; the maximum number of characters which can be sent in one second is $1/7·5T = C/7·5 = 50/7·5 = 6·6$ char/s.

For a 5-unit teleprinter operating at 75 bauds (with a stop element equal to $3T'/2$), the maximum character-printing rate is $C'/7·5 = 75/7·5 = 10$ char/s; a similar machine operating at 100 bauds would have a printing rate of $100/7·5 = 13·3$ char/s.

In a start—stop system using a 7-unit code (e.g. alphabet No. 5) and a parity bit, each character comprises, in order of transmission,

start signal	1 unit
information bits	7 units
parity bit	1 unit
stop signal	2 units

making a total of 11 units. If the system operates at a modulation rate of 110 bauds, the unit interval is 0·009 s and the maximum printing rate $110/11 = 10$ char/s.

Comparison is also made based upon the number of words per minute of which a system is capable. For this purpose the 'telegraph word' is taken as an

*Isochronous = equal time — e.g. equal speed of revolution and equal phase.
Synchronous = same time — e.g. equal speed and constant phase-difference, which may or may not be zero.
† ms = millisecond = 0·001 second.

arbitrary 5-letter word together with one letter-space, making six characters in all. Conversion from characters/s to words/min is made by multiplying the characters/s by 60/6 = 10. For the 7·5-unit/50-baud teleprinter at 6·6 char/s, the speed may also be expressed as 6·6 x 10 = 66 words/min (wpm). The maximum printing or commercial speed is important to the customer who needs to know

transmitter may be a matter of some difficulty. It is usual at the receiver to restore the abrupt nature of the changes by some form of relay which should have the property of producing instantaneous electrical or mechanical changes at instants corresponding to those at which the amplitude of the operating current is equal to some pre-determined value — for example, in a double-

Table 2.1 Modulation, character and word rates of teleprinters

Modulation rate (bauds)	50	50	75	75	100	110
Unit interval (ms)	20	20	13·3	13·3	10	9
Unit elements per character	7·5	11	7·5	11	7·5	11
Character period (s)	0·150	0·220	0·100	0·146	0·075	0·099
Character rate (per s)	6·6	4·5	10	6·8	13·3	10
Word rate (per min)	66	45	100	68	133	100

how fast his traffic or information can be cleared, and also to the administration or carrier which provides circuits in order to know at what rate a circuit can clear traffic. Some typical values of commercial speeds are included in Table 2.1.

The time taken for a complete signal-reversal cycle (Fig 2.1) is $2T$. The maximum frequency with which this can take place is $f = 1/2T = C/2$. Telegraph binary signals of unit interval T can be regarded as having a fundamental frequency f Hz which is equal to $C/2$, i.e. to one-half the modulation rate in bauds; for example, teleprinter signals at a modulation rate of 50 bauds are equivalent to a fundamental modulating frequency of 25 Hz.

2.3 Telegraph-signal distortion

A modulation (or restitution) suffers from telegraph-signal distortion when all the significant intervals do not have exactly their theoretical durations. Although the changes in significant conditions at the transmitter may be effectively instantaneous, they lose their abrupt character during propagation through the transmission medium. On account of waveform distortion, the recognition at the receiver of the significant intervals between the original transitions or modulations at the

current system, when the current is passing through zero.

The process of reproducing the significant instants of modulation at the receiver by some form of relay is termed the *restitution of modulation*. The time interval between a significant instant of modulation and the corresponding significant instant of restitution is called the *restitution delay*.

If all the restitution delays resulting from a transmission were constant for all significant instants, then all corresponding significant intervals at transmitter and receiver would be equal and no telegraph distortion would arise. In general, due to a number of causes, the restitution delays vary from instant to instant, and telegraph distortion — time distortion of the significant instants — is present in the received signals.

Some effects of transient conditions are shown in Chapters 3 and 4 where the causes of delay in build-up of the received current in d.c. and a.c. circuits are discussed. The degree of imperfection of the received signals is to a large extent dependent upon the relation between the duration of the transient effects and the duration of the unit element of modulation. If the duration of the transient or build-up time necessary to reach the steady state is less than the duration of the shortest element of modulation – i.e. the unit element – the shape of the received signals will be unaffected by the presence of previous signals, as the received current will always reach the steady state before commencement of the next element. When this is so (and in the absence of any external disturbing influences) the response to a modulation will always be the same, and the instantaneous changes produced by a perfect receiving relay will always occur with a *constant delay* after the corresponding operation at the transmitter. In these circumstances, received signals would be in exact accordance with the operation of the transmitter and the system would be telegraphically distortionless despite the presence of waveform distortion. This is a somewhat ideal conception and in most practical cases significant values of distortion will arise.

Restitution delays may vary appreciably from one instant to another; with excessive deviation from constancy, the correct interpretation of the received signals is either impossible or else involves unduly meticulous adjustments. The accuracy and stability of a telegraph circuit is dependent on the extent of these deviations and their value is used to express quantitatively the quality of transmission. A knowledge of the degree of telegraph distortion existing or to be expected is of fundamental importance in determining the reliability of telegraph circuits, judging the quality of the component sections, assessing maintenance tolerances and planning teleprinter networks. The quality of a modulation as expressed by telegraph distortion is independent of the modulation rate since its measurement is based upon the unit interval.

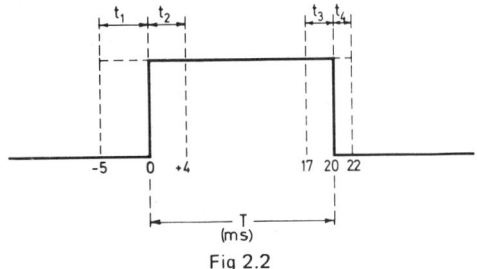

Fig 2.2

Telegraph distortion is essentially a time distortion and is expressed as the percentage of the unit interval by which a transition deviates from the significant instant at which it should theoretically occur. In Fig 2.2 the full line shows a unit element

of theoretical duration, T ms, bounded by two transitions. If either of these deviates by t ms from its theoretical position in time, the distortion δ is expressed by $\delta = (t/T) \times 100\%$. A transition may occur earlier than its theoretical position – by an amount such as t_1 or t_3 – or later than its theoretical position – by an amount such as t_2 or t_4. In a train of signals, the amount of distortion on successive transitions is usually distributed over a range of values, peak values being infrequent. In assessing, it is always the *maximum* value to occur during a given period of observation which is quoted as the distortion value. In Fig 2.2 (where $t_1 > t_2 > t_3 > t_4$) the early distortion quoted would be $\delta = t_1/T \times 100\% = (5/20) \times 100 = 25\%$ early; the late distortion quoted would be $\delta' = t_2/T \times 100\% = (4/20) \times 100 = 20\%$ late. The smaller values of distortion, those at t_3 and t_4, are ignored, being contained in the greater values quoted. Frequently, early and late values are not separately quoted, in which case the largest value only (25% in the example) would be quoted. Transitions which occur later than the theoretical significant instant are regarded as having positive distortion; those which occur earlier are regarded as having negative distortion; this applies equally to space/mark or mark/space transitions.

Distortion resulting from the displacement of either of the significant instants of Fig 2.2 from its ideal position is referred to as *individual distortion.*

A receiving device examines the significant state of an element (e.g. mark or space) at the theoretical centre of the unit interval. If the amount of deviation of a significant instant from the theoretical instant exceeds 50% of the unit interval, an incorrect significant condition might be recorded. The reception of an element having an incorrect condition – e.g. space instead of mark, or vice versa – is considered as an *error*, and such occurrences are measured by *error rate*. Telegraph distortion is concerned, not with changes of the binary state, but with the actual timing of the significant instants, and it is consequently limited to values not exceeding ±50% of the unit interval. A transition with more than 50% distortion would be indistinguishable from an adjacent transition, and correct translation of such a signal into a printed character is impossible.

For synchronous systems the distortion criterion

cannot directly apply since signal regeneration is inherently provided by the receiver; the criterion of signal quality is based upon error rate.

Under unstable HF radio conditions, additional transitions are apt to arise, causing *extras* (*hits*) and *splits* and *misses* which cannot be described in terms of displacements; these signals are often re-generated before being fed to a telegraph receiver. Again, in such cases the assessment of performance cannot be measured by telegraph-signal distortion and is expressed instead by the error rate of the system.

Distortion measurements are commonly made using some form of stroboscopic instrument. If readings are to be observed by the eye viewing a cathode-ray oscilloscope, the period of observa-tion should not be excessively long or the eye will become fatigued and the occasional high value of distortion may be missed. On the other hand, if the period is too short, the likelihood is increased that the true distortion is greater than the observed value. In view of the range in values which occur during a train of signals, the observed value will vary according to the period of observation, and uncertainty follows as to whether peak values have been included or not.

If a precise statement of the degree of distortion is required, it is usual to accompany the quoted value with a statement regarding the probability with which this value might be exceeded. For convenience, a shortened terminology is used, e.g. $P(3)$ indicates a probability (P) that the quoted value of distortion may be exceeded by 1 transition in 1000, the (3) resulting from $\log_{10} 1000 = 3$. Similarly, $P(2)$, $P(4)$ or $P(5)$ values may be quoted for probabilities of 1 in 10^2, 10^4 or 10^5.

The CCITT has recommended* that for distortion measurements made during normal maintenance testing the period of observation should correspond to the examination of at least 800 significant instants, whatever the type of distortion meter used, isochronous or start–stop. At a modulation rate of 50 bauds this results in an observation period of about 30 s; at other rates, the observa-tion should last about 20 s. It is also recommended that the observation time should be divided into two approximately equal parts — one part during

*Recommendation R5.

which the significant instants in advance of their theoretical position could be observed, and the other during which the significant instants later than their theoretical position could be observed.

Distortion measurements can be made separately on transmitters, d.c. lines and VF channels; or over a complete connection between two stations, including the transmitting and receiving equipment. Distortion measured on a channel or equipment when the source of test signals is free from dis-tortion (so that all the measured distortion is that introduced by the channel) is termed the *inherent distortion,* defined as *the degree of distortion of the restituted signal when the modulation is effected without distortion.*

2.4 Isochronous distortion

This mode of distortion measurement can be made only on a continuing stream of signals in which the ideal significant instants occur at times which are whole-number multiples of the unit interval. Unlike start–stop systems, in this case the ideal significant instants cannot be defined, actual early and late distortions cannot be determined and only the range within which the transitions deviate can be measured.

Isochronous distortion is defined as *the ratio to the unit interval of the maximum measured differ-ence (irrespective of sign) between the actual and theoretical intervals separating any two signifi-cant instants of modulation (or restitution), these instants being not necessarily consecutive.* The degree of distortion is expressed as a percentage of the unit interval. The period of observation should preferably be stated (e.g. by quoting a $P(X)$) value.

Consider the restituted signal-train shown by the full line in Fig 2.3, where equally-spaced dotted lines show the ideal or theoretical significant unit intervals T ms. If the transition at X has the maximum late displacement (x ms) and that at Y has the maximum early displacement (y ms), the isochronous distor-tion δ_I is, by definition:

$$\delta_I = \frac{(t_1 - t_2) \times 100\%}{T}$$

But $(t_1 - t_2) = (x + y)$; hence, when measured over the period of observation, the isochronous distor-tion can be expressed as the sum of the time differences (from the ideal instants) of the maximum early and late displacements, expressed as a percentage of the unit interval:

$$\delta_I = \frac{(x + y) \times 100\%}{T}$$

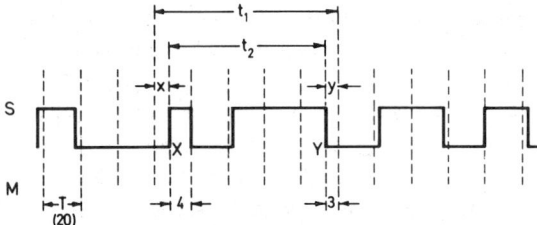

Fig 2.3 Basis of isochronous measurement
Note: 4 should correspond with x

Taking the values illustrated in Fig 2.3, the isochronous distortion for this 50-baud signal ($T = 20$ ms) is

$$\delta_I = \frac{(4 + 3) \times 100}{20} = 35\%$$

Isochronous distortion is alternatively defined as *the algebraic difference between the highest and lowest value of individual distortion affecting the significant instants of an isochronous modulation.* (This difference is independent of the choice of an ideal reference instant.)

2.5 Start—stop distortion

In start—stop systems, the transition at the commencement of the start signal serves as a reference point against which the remaining transitions (a maximum of five for start—stop 5-unit code) of the start—stop signal are measured.

The degree of start—stop distortion is defined as *the ratio to the unit interval of the maximum measured difference (irrespective of sign) between the actual and theoretical intervals separating any significant instant of modulation (or restitution) from the significant instant of the start element immediately preceding it.* This is expressed as a percentage.

In the example illustrated in Fig 2.4, in which the full line shows the distorted significant instants of restitution against the theoretical significant instants (dotted lines), the restitution delays are the differences $(t_1' - t_1)$, $(t_2' - t_2)$, $(t_4' - t_4)$ and $(t_6' - t_6)$. The greatest of these, in absolute value, is $d = (t_2' - t_2)$. The value of start—stop distortion δ_{SS} is the ratio of the difference d to the unit interval T, expressed as a percentage: $\delta_{SS} = (d/T) \times 100\%$.

The maximum distortion shown $(t_2' - t_2)$ is *early* distortion (having a negative value), the significant instant of restitution occurring earlier than the ideal significant instant. Other deviations, such as $(t_4' - t_4)$ and $(t_6' - t_6)$, having positive values, are *late* distortions, with significant instants later than the ideal instants. It is sometimes desirable to quote distortion more precisely by giving both maximum early and maximum late values. From the timings shown, the larger of the early distortions is that shown at E. Measured (in absolute value) from the reference point — the commencement of the start signal — the displacement $E = (t_2' - t_2) = 36 - 40 = -4$ ms. The early distortion is $-4/20 \times 100 = -20\%$ (early). The larger of the late displacements is shown at L, and $L = (t_6' - t_6) = 123 - 120 = 3$ ms. The late

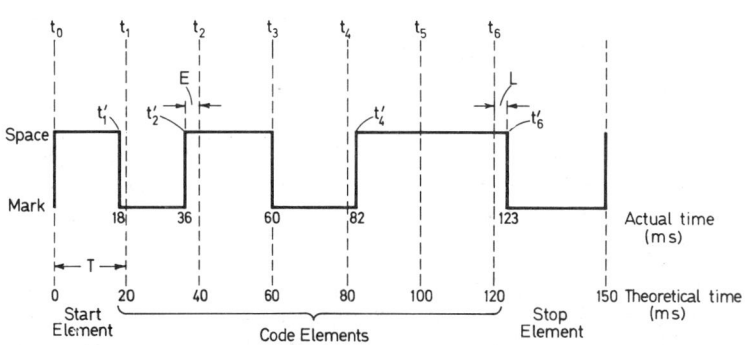

Fig 2.4 Basis of start—stop measurement

distortion is 3/20 x 100 = +15% (late). The distortion to be quoted would be the greater value, 20%, but more specifically it should be quoted as −20% (early), +15% (late).

In expressing start–stop distortion, a single value often quoted refers to the maximum individual distortion of a transition without reference to the sense − early or late. To afford a more accurate picture, two values may be stated, the maximum positive and negative values of distortion. For example, a start–stop distortion of 4% would imply that all transitions lying in the range +4% to −4% could occur. On the other hand, a transmitter, for example, which showed distortion of +4% and −1% might be quoted simply as 4%, yet the quotation (+4% and −1%) is more informative, showing the presence of bias and indicating a means of improvement in performance.

According to the *actual* value taken for the unit interval *T,* a distinction is drawn between *gross start–stop distortion* and *synchronous start–stop distortion.*

Gross start–stop distortion. A distortion-measuring instrument makes its measurements by comparison against a time-base running at the nominal modulation rate. The degree of distortion registered includes the influence of the mean speed deviation of the time-base generator. The resulting measurement is termed the *gross start–stop distortion,* defined as *the degree of distortion determined when the unit interval and the theoretical intervals assumed are exactly those appropriate to the standardized modulation rate.*

Synchronous start–stop distortion. Referring to the deviation in time-base accuracy mentioned above, it may be desirable to eliminate this speed differential by adapting the speed of the distortion-measuring instrument to the momentary mean modulation rate of the signals being measured. The result from a measurement made on this basis is termed the degree of *synchronous start–stop distortion,* defined as *the degree of distortion determined when the unit interval and the theoretical intervals assumed are those appropriate to the actual mean rate of modulation (or restitution).* In

arriving at the mean modulation rate, account is taken only of those significant instants which correspond to a change of condition in the same sense as the commencement of the start element.

These two cases (gross and synchronous start–stop distortion) have no counterpart in isochronous distortion.

2.6 Types and causes of distortion

Three distinct sources of distortion have been recognized: *characteristic, bias* and *fortuitous.* Though each can be present in a line or an equipment individually, a combination of all three is normally found. It is possible to analyse the value of each type of distortion component from the total distortion.

Bias distortion. This is the consistent lengthening of either mark or space elements. Lengthening of mark elements is known as *marking bias*; of space elements is termed *spacing bias.* Bias distortion is caused by asymmetry in transmitting or receiving equipment $\frac{1}{m}$ for example: due to lack of neutrality in adjustment of transmitter or relay contacts; to inequality in voltage of the +80 and −80 V supplies; to the presence of earth currents, or to hysteresis effects.

If a is the mean restitution delay for one type of transition (e.g. space–mark) and b is the mean value of the deviation for the other type of transition (e.g. mark–space), then the bias distortion δ_B is, by definition,

$$\delta_B = \frac{(a-b)}{T} \times 100\%$$

Fortuitous distortion. This is distortion which arises from any random influences $\frac{1}{m}$ for example: from mechanical imperfections of a transmitter, or from electrical interference from power systems or from other telephone or telegraph circuits, including intermodulation interference from adjacent telegraph channels in VFT multiplex systems.

If the distortion on a certain number of significant instants is observed, the difference between the greatest and smallest displacements (as a percentage of the unit interval) will give the fortuitous distortion *for the transitions observed*; the number of transitions observed should be specified (e.g. by the $P(X)$ value).

The conception of fortuitous distortion is closely related to the probability of occurrence of extreme values of displacement. If the greatest and least displacements observed are r_1 and r_2 then, if the observation were carried out over a much longer period, they could be classified into three groups, namely those greater than r_1, those smaller than r_2, and those whose values lie between r_1 and r_2. As a first approximation, the first two groups may be excluded as being of more rare occurrence, and the fortuitous distortion is:

$$\delta_F = \frac{(r_1 - r_2)}{T} \times 100\%$$

This value must be taken in conjunction with a knowledge of what values have been neglected, i.e. the probability (p) that other values of r will occur. In an observation some values could escape being recorded, either because they escaped the attention of the observer or because the period of observation was insufficiently long for these extreme values to occur — in both cases because these values occurred so rarely. For a full statement of fortuitous distortion it is necessary to express the probability that the value of distortion quoted will be exceeded. The normal probability law applies to the occurrence of fortuitous distortion, which implies that small displacements from a mean value of distortion are more likely to occur than large ones, and that positive and negative displacements with respect to the mean are equally likely. The fortuitous frequency/distortion characteristic produces a bell-shaped normal distribution curve, as shown in Fig 2.5. In general, the measured distortion will increase with the time of observation owing to the increased chance of the occurrence of extreme peak values of distortion.

For practical purposes, a distortion/cumulative-frequency characteristic is of more value if it is plotted on arithmetic probability paper. Provided that the distribution is normal, this results in a straight-line graph due to the special scale used for the x-ordinate (see Fig 2.26, p.39).

Characteristic Distortion. This is distortion due to transients which are present in the transmission channel as a result of modulation and arises from the inherent electrical characteristics of the transmission equipment $\frac{1}{m}$ for example it may be caused by the automatic gain control in an amplitude-modulated VFT system. This distortion occurs consistently with any given combination of signal elements and is the one which would be measured at the output of a system when perfect signals are applied to the input, the system is free from bias and there is no interference from other sources. Measurement of characteristic distortion is not associated with the probability factor.

The combination of characteristic and bias distortion is sometimes referred to as the 'system distortion'; that due to all three causes is the 'total' distortion.

The term *cyclic distortion* may be applied to that which occurs with a definite periodicity; it may be due to speed deviations or to some constant irregularity in the cycle of a transmitter, or an interfering alternating current which 'beats' with the telegraph signal. The term is not much used in practice, such causes being treated as fortuitous distortion.

The term *service distortion* is used to express the value of distortion measured on an equipment or channel when it is in service.

2.7 Addition of distortion on multi-link circuits
Telegraph circuits between two offices often comprise a number of component telegraph circuits joined in tandem (i.e. in series), either connected on a quasi-random basis by automatic switching at telex or TAS exchanges, or forming semi-permanent circuits such as those for leased circuits. It is important to be able to estimate the overall distortion which is likely to occur in the complete circuit. The cumulative effect of signal

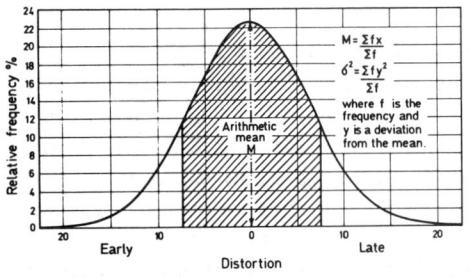

Fig 2.5 Normal distribution curve

distortion in transmission over successive links depends upon the type of distortion concerned.

Characteristic Distortion. If several links, each having the same type of characteristic distortion, are connected in tandem, the overall distortion will in most cases be additive. As an example, if distortion is of the type where short mark signals following long space signals are shortened due to build-up time exceeding the signal duration, then when the signal is repeated into the next link the short signal already reduced in length will suffer even greater shortening, and after several such repetitions may be completely lost. Characteristic distortion must be kept very small on individual channels which are to form part of a switching network.

Bias Distortion. If bias distortion is present in each of a number of tandem-connected channels, the effect will be additive only if the bias is of the same sign in each link, and there will be a net reduction if the bias distortion in one or more of the links is of opposite sign to the others. For a circuit made up of four links in tandem, the chance of bias in all links being of the same sign is 1 in 8, assuming equal probability of marking or spacing bias; in most cases the resultant total bias distortion will be only about twice that for a single link. While the cumulative effect of bias distortion is not so serious as that of characteristic distortion, it must — and usually can — be kept to a low value because bias distortion affects every signal element; it is the main limiting factor for good performance on multi-link switched connections.

Fortuitous Distortion. Since fortuitous distortion is quite irregular and only a small proportion of signal elements suffer maximum distortion, and furthermore a transient may be either advanced or retarded, the chance of any particular signal suffering maximum fortuitous distortion of the same kind in each link is very small. For example, if four links each giving a maximum of 10% fortuitous distortion are connected in tandem, the resulting total is rarely likely to exceed 20%.

The cumulative effect of this is of less importance in multi-link connections than that of either characteristic or bias distortion.

For purposes such as network planning, the CCITT has recommended* the use of a formula for estimating the approximate aggregate distortion to be expected when several VFT channels are connected in tandem. The overall distortion produced in a multi-link path is found to lie between the arithmetic sum and the r.m.s. value of the distortions on the individual links. For VFT channels only:

$$\delta_{INH} = \sum_1^n \delta_C + \sqrt{\sum_1^n (\delta_B)^2 + \sum_1^n (\delta_F)^2}$$

For the combination of a transmitter and a multi-link circuit:

$$\delta_{TEXT} = \sum_1^n \delta_C + \sqrt{\delta_T^2 + \delta_V^2 + \sum_1^n (\delta_B)^2 + \sum_1^n (\delta_F)^2}$$

where:

δ_{INH} = the probable inherent start—stop distortion on standardized text

δ_{TEXT} = the probable gross start—stop distortion in service.

δ_C = the characteristic start—stop distortion of a single channel

δ_B = the bias distortion of a single channel

δ_F = the fortuitous distortion of a single channel

δ_T = the synchronous start—stop distortion of the transmitter

δ_V = the start—stop distortion due solely to the difference between the mean transmitter speed and the standardized speed. (In 5-unit systems the difference to be considered is equal to six times the mean difference for one element.)

n = number of channels connected in tandem.

For example, if the distortion on each channel due to characteristic distortion were 3%, bias distortion 2% and fortuitous distortion 5%, the

*Recommendation R11.

total inherent distortion for a multi-link channel would be:

1 link: $(1 \times 3) + \sqrt{(1 \times 2^2) + (1 \times 5^2)}$

$\qquad = 3 + \sqrt{4 + 25} = 3 + \sqrt{29}$

$\qquad = 3 + 5.4 = 8.4\%.$

2 links: $(2 \times 3) + \sqrt{(2 \times 2^2) + (2 \times 5^2)}$

$\qquad = 6 + \sqrt{8 + 50} = 6 + \sqrt{58}$

$\qquad = 6 + 7.6 = 13.6\%.$

3 links: $(3 \times 3) + \sqrt{(3 \times 2^2) + (3 \times 5^2)}$

$\qquad = 9 + \sqrt{12 + 75} = 9 + \sqrt{87}$

$\qquad = 9 + 9.3 = 18.3\%.$

4 links: $(4 \times 3) + \sqrt{(4 \times 2^2) + (4 \times 5^2)}$

$\qquad = 12 + \sqrt{16 + 100} = 12 + \sqrt{116}$

$\qquad = 12 + 10.7 = 22.7\%.$

5 links: $(5 \times 3) + \sqrt{(5 \times 2^2) + (5 \times 5^2)}$

$\qquad = 15 + \sqrt{20 + 125} = 15 + \sqrt{145}$

$\qquad = 15 + 12 = 27\%.$

The CCITT has also published* limiting values of distortion to be expected on multi-link circuits comprising VFT channels operating at 50 bauds — see Table 2.2.

Table 2.2 Standard limits of transmission quality to be applied in planning international point-to-point telegraph communications and switched networks by means of start—stop apparatus at 50 bauds

Number of channels in tandem	Limit of Distortion on 2:2 reversals at 50 bauds	Limit of Isochronous Distortion on standard-ized text	Limit of inherent start—stop service distortion on standard-ized text
(1)	(%) (2)	(%) (3)	(%) (4)
1	4	10	8
2	7	18	13
3	10	24	17
4	12	28	21
5	—	—	25

Note. The above distortion values do not correspond to conventional degrees of distortion but to routine measurements.

Analysis of distortion. The CCITT has recommended† a method for analysing the total distortion measured on a channel into its components of characteristic, bias and fortuitous distortion, as follows:

(1) Measure the overall distortion Δ on Q9S text at the actual mean modulation rate.

(2) Measure the distortion Δ_1 on 2:2 reversals at the above modulation rate. The reading Δ_1 is the sum of the bias and fortuitous distortions.

By adjusting a compensator fitted to the distortion-measuring equipment (e.g. a compensation winding on the input relay), reduce this distortion to its minimum value, δ, by removing the bias distortion. For practical purposes, δ is the fortuitous and $(\Delta_1 - \delta)$ the bias distortion.

(3) Keeping the distortion-measuring set adjusted

as for the measurement of δ, measure the distortion Δ^1 on Q9S signals at the above mean modulation rate. For all practical purposes $(\Delta^1 - \delta)$ is the characteristic distortion.

The method can be applied using either an isochronous or a start—stop distortion-measuring set. This method gives useful *approximate* results for the equation:

$$\Delta^1 + \Delta_1 - \delta = \Delta$$

2.8 Standard test signal for distortion measurement
To provide a universal start—stop, 5-unit, 50-baud test signal which is suitable for distortion measurements over both national and international circuits, and one which can be used equally well for start—stop and isochronous distortion measurements the CCITT has recommended* use of the test signal known as Q9S — the form in which it is printed on a monitoring teleprinter.

To permit measurements with both isochronous and start—stop distortion-measuring equipment, the test signals must be start—stop ones capable of continuous transmission over the period of observation — with a 1·0 unit stop signal. The latter requirement is necessary so that on a distortion-measuring set designed to read isochronous distortion, using a cathode-ray oscilloscope which traces a circular beam

*Recommendation R57.
†Recommendation R4.

*Recommendation R51.

during each unit interval, a display of distorted signals will appear repeatedly in the same position on the scale of the measuring set. If a stop signal of 1·5 units were used with the test signal, distortion indications of start—stop characters would appear alternately 180° (half an element) out of phase and would be difficult to read.

(start—stop distortion) or a synchronous mode (isochronous distortion). If the distortion of a given signal-train — e.g. the Q9S test signal — is measured for start—stop and also (under the same conditions) for isochronous distortion, two different values may result.

The isochronous-measuring mode is used for

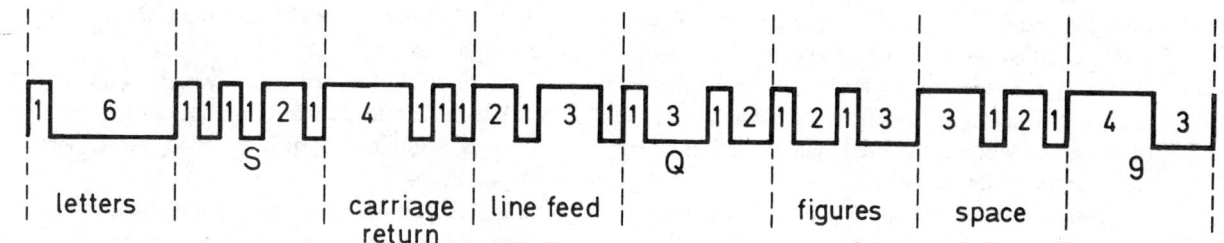

Fig 2.6 Q9S test signal

The composition of the Q9S signal is shown in Fig 2.6. This form of text signal is chosen to give a highly irregular modulation pattern and generate an amount of distortion closely resembling that produced in service with normal message text. It has equal numbers of mark and space elements to facilitate detection of bias distortion and includes combinations of elements most likely to be subject to distortion. One cycle of this 8-character test signal provides 30 transitions. If this signal is repeated for the 30-s observation period recommended, the total number of transitions during this period (200 characters at 400 opm) is 30 × 200/8 = 750 transitions. Since $\log_{10} 750 = 2·875$, the probability coefficient for this test may be expressed as $P(2·9)$ or approximately as $P(3)$.

In addition to the use of the Q9S text, the recommendation also includes the use of continuous mark—space reversals with element proportions 1:1, 2:2 and 1:6. The duration of each element is a multiple of the unit interval corresponding to the modulation rate. It is also permissible to reverse the polarity of the emitted test signals so that, while the 1:1 and 2:2 test signals remain unchanged, two others, 6:1 and 'Q9S reversed', become available.

2.9 Comparison of start—stop and isochronous distortion measurements

The method used for measuring distortion depends upon whether the signals originate in a start—stop

checking and lining-up VFT channels. For end-to-end checking of complete circuits the start—stop mode is preferable; measurements are carried out under actual service conditions and continuous supervision of a circuit is possible.

If the signals of a teleprinter transmitter have a start—stop distortion of $\pm x\%$, the same signals indicated on an isochronous measuring set will read $2x\%$. On the other hand, a start—stop distortion of $\pm x\%$ may be indicated at the receiving end of a channel if only fortuitous isochronous distortion of $x\%$ is present, with the start-signal transition straying within the same range as the remaining transitions. The ratio of isochronous to start—stop distortion may thus vary between 1·0 and 2·0. It can be shown that for 10^4 transitions, and assuming that there are four instants of modulation per start—stop character, a value of about 1·5 is appropriate. In Table 2.2, a comparison of columns 3 and 4 reveals a ratio of about 1·4.

Conventional Distortion*. The distortion of a train of telegraph signals is the maximum value of distortion of all the individual transitions. There is uncertainty whether peak distortion values have been included in measurements which extend over a comparatively short period of observation. It is not advisable to include in any calculation a maximum value which is so infrequent that it does not significantly affect

*Recommendations R54, R55

the error rate. Agreement* has been reached to regard as the *conventional start–stop distortion* that value of distortion which, during a prolonged period of observation, is exceeded by a certain though very small probability. The assigned value for this probability is 1 in 10^5, i.e. one occurrence of distortion exceeding the conventional degree of distortion per 10^5 significant instants of modulation. This value, which corresponds to a $P(5)$ value, is related to an error rate of two erroneous translations by the receiving apparatus per 10^5 alphabetic telegraph signals, resulting from excessive values of distortion when considered in conjunction with the 'margin' of the receiver.

From a study of this subject, formulae have been tentatively derived for estimating by how much the results measured during 30-s observations in the isochronous and start–stop modes should be increased so that they may represent the conventional distortion δ_{CON}.

For isochronous distortion δ_I:

$$\delta_{CON} = 1 \cdot 14(1 \cdot 15 \delta_I - 0 \cdot 15 \delta_B)$$

For start–stop distortion, δ_{SS}:

$$\delta_{CON} = \delta_E + \delta_L / 3 \ldots \text{if } \delta_E > \delta_L, \text{ or}$$

$$\delta_{CON} = \delta_L + \delta_E / 3 \ldots \text{if } \delta_L > \delta_E$$

(Note: δ_B, δ_E and δ_L refer to bias, early and late distortion respectively.)

The reason for two start–stop equations is that the larger of the two values includes the whole of the bias distortion, whereas the smaller value comprises only the characteristic and fortuitous distortions which are to be read at only two-thirds of their peak values.

2.10 Receive margin

A signal transition, after transmission, rarely occurs at the theoretical significant instant so the receiving device must be designed to accept signals with a considerable degree of distortion. The extent of this property of a receiver to translate correctly from signals which suffer distortion is termed the *receiving margin*, defined† as *the maximum*

*Recommendation R56.
†Recommendation S1.

distortion compatible with correct translation, when signals are presented to a receiver under the most unfavourable conditions so far as composition of the signals and of distortion are concerned. The maximum distortion which results in incorrect translation applies without special reference to the form of distortion affecting signals; it is the maximum tolerable value of distortion causing incorrect translation which determines the value of the margin.

The *theoretical margin* is that which could be calculated from the construction data of the apparatus, assuming that it is operating under perfect conditions.

The *effective margin* of an apparatus considered individually is that which could be measured on the apparatus under operating conditions.

The *nominal margin* of a receiver represents the minimum value set for the effective margin of the apparatus under standard operating and adjustment conditions.

For start–stop apparatus, the margin is the maximum start–stop distortion of the modulation which it is possible to apply to the apparatus compatible with correct translation of all signals which it should be able to receive, whether signals composing the modulation are transmitted separately or whether they follow one another with the maximum rapidity corresponding to the modulation rate. In particular, it is convenient to consider (1) the *net margin* which is measured when the modulation rate applied to the apparatus is exactly equal to the standard theoretical rate, and (2) the *synchronous margin* which is measured when the mean unit interval of the modulation applied to the apparatus is equal to that which would result from transmission by the apparatus under examination, assuming it to include a transmitter.

The margin of a *synchronous* receiver is the margin (as defined above) when the distortion taken into account is *isochronous* distortion.

Line distortions of less than 5% are regarded as negligible; values of 10–20% or more on multilink connections are common, 30–40% very high. The higher the value of receiving margin, the better is the receiver. For point-to-point or switched connections over international circuits, the CCITT has recommended* that the limit of the gross start–

*Recommendation R57.

stop distortion which can be present in signals at the input of the local section of the circuit (e.g. at the ultimate telex exchange on a switched connection) is to be 30%.

The CCITT has also recommended* that the effective net margin for start–stop apparatus operating at 50 or 75 bauds, measured from the input of the local end (the *telemargin*) – e.g. including the receiving subscriber's telex exchange line – should be not less than 35%, and for that which operates at 100 bauds the corresponding margin should be not less than 30%. (The reason for the inclusion of the local end is that some start–stop receivers – e.g. single-current apparatus – cannot be separated during operation from their supply and repeater devices). The quoted values are given on the supposition that the local end contributes negligible distortion to the connection and that signals are sent by a transmitter having a nominal cycle equal to or greater than 7·0 units (7·2 for a 100-baud machine).

The CCITT has further recommended that margin should be measured using the prescribed test sentence (THE QUICK BROWN FOX JUMPS OVER THE LAZY DOG), making a first test with identical distortion value on all transitions of the signal-train, obtained by artificially lengthening the start signal, followed by a second test with the start signal shortened. The margin is numerically equal to the distortion which results in less than one error per test sentence. The margin is the *lesser* of the two values of distortion obtained from the two measurements. In making measurements, the signal-generator transmitter should have a 7·5-unit cycle.

Modern mechanical teleprinters usually have a margin in excess of ±40%; an electronic teleprinter could have one approaching ±50%.

Values quoted for circuit distortion and for receiving margin are not strictly directly comparable. A quoted distortion value is the maximum recorded during an observation and this value could well occur, for example, on a transition which is not the same as the one which determines the margin.

In Fig 2.7 the shaded block bounded by *XY* is the interval required by the receiver to read the polarity of the element; during this period the signal *must* have its correct polarity if the character is to be correctly translated. Because of mechanical design

*Recommendation S3.

considerations, manufacturing tolerances and wear, this period cannot be reduced; it is arranged to occur about the centre of the theoretical interval. The signal will be read correctly provided that transition *AC* is not delayed by more than *x*% and transition *DB* is not advanced by more than *y*%. For a modern mechanical teleprinter *x* + *y* exceeds 80% of the nominal element, leaving less than 20% for reading

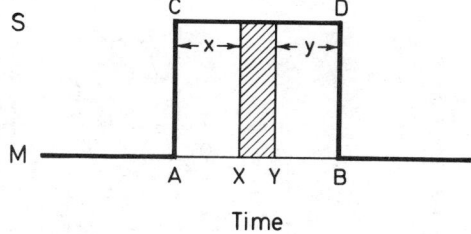

Fig 2.7 Receive margin

the signal. If *XY* could be made negligible so that *x* + *y* = 100%, and the early and late margins were equal, they would each be 50% – the theoretical maximum margin, – a value which could be closely approached by an electronic teleprinter.

Some conditions relating signal distortion to margin of a 50-baud teleprinter are illustrated in the timing diagram of Fig 2.8. The charts relate to:

(*a*) a correctly adjusted teleprinter with balanced receiving margin of ±40%;

(*b*) an incorrectly adjusted teleprinter which results in all the selecting periods being too late (with respect to the commencement of the start signal);

(*c*) a receiving teleprinter running about 6% fast, so that the selecting periods become progressively early with each element;

(*d*) an undistorted received signal (letter R) applied to the teleprinter receiver;

(*e*) received signals with 20% marking bias; the space–mark transitions are advanced by 20% with respect to the mark–space transitions which, being of the same type as the start signal, remain unchanged with respect to the start signal;

(*f*) a signal distorted by 40%. The second code element will not be correctly selected and a printing error will occur (CARRIAGE-RETURN for R).

Fig 2.8 (*c*) illustrates the effect of speed difference between sending and receiving teleprinters. If the speed of the teleprinters on a switched network is within ±1% of the correct value, the maximum speed difference between any two connected teleprinters will be 2%. Such a speed difference is equivalent to a loss of margin of approximately 11% on the fifth code element and 9% on the receiving device. As a consequence, any distortion of the start signal is accepted at the expense of the receive margin of the code elements — for example, if the start signal were distorted by ±50%, examination of the code elements would occur at the theoretical beginning or end of each code element instead of at the mid-point; as a result the margin on the code elements would be zero. Since, in a start—

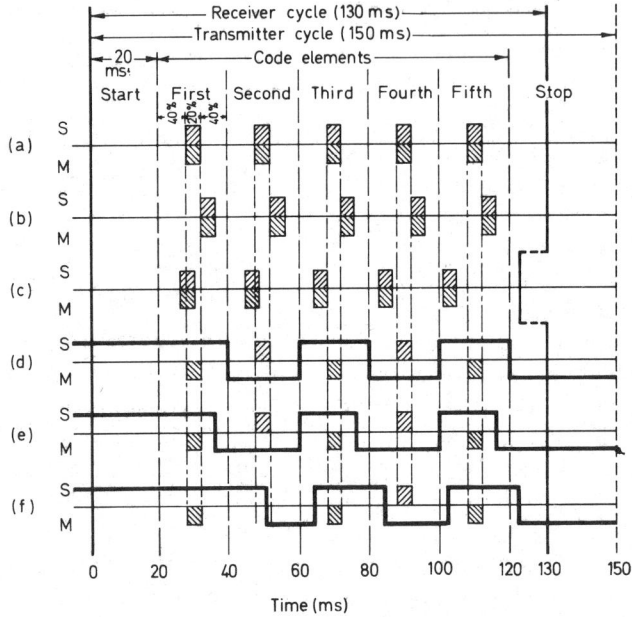

Fig 2.8 Teleprinter margin and signal distortion

fourth code element. In making a calculation relating to speed error, the accumulated error on the fifth code element must be considered for the maximum ill-effect.

Start—stop distortion is equally likely to be early or late. The effective margin of a receiver is determined by whichever of its two margins — early or late — is the less. Since the sum of the early and late margins for a given machine is fixed by the design and adjustment, the greatest effective margin is achieved when these are equal, i.e. balanced. Most teleprinters are equipped with an orientation device or range-finder which enables this balanced adjustment to be obtained.

Start—stop signals are measured from the leading edge of the start signal, and it is this instant which determines the signal examining period of the receiving device. As a consequence, any distortion of stop system, the receiving margin of the code elements is shared with the start signal, the receiving margin of a start—stop system is only one-half that of a synchronous system, which has no start element and as a consequence the signal examination point is always kept central. It follows that a synchronous system is superior to a start—stop system insofar as margin is concerned.

Margin Testers. Equipment for measuring teleprinter receive margin consists mainly of a signal generator which will transmit signals corresponding to the test sentence (THE QUICK BROWN FOX . . .) into the teleprinter at maximum character rate, with an adjustable amount of distortion obtained by shortening and lengthening the start signal so that transitions occur earlier and later than the theoretical significant instants; the amount of injected distortion is read directly from a scale.

For practical reasons, each line of text is terminated not only by (undistorted) characters – CARRIAGE RETURN, LINE FEED and LETTERS SHIFT – but also by a 150-ms mark period prior to the end-of-line signals to allow the teleprinter under test to regain phase should this have been lost due to excessively distorted signals.

Code keys enable any 5-unit (or other) signal, such as R or Y, to be set up for repeated transmission. In testing, the distortion is gradually increased, first positively and then negatively, until failure of correct registration occurs. The standard signal-shaping network is applied to the teleprinter electromagnet circuit. Provision is made for checking neutrality of the electromagnet by measuring its sensitivity to current in either direction.

Margin measurement may be carried out iso-chronously – with test signals generated at the exact nominal modulation rate of the system – or synchronously – with signals transmitted at a modulation rate equivalent to the mean rate of registration of signal elements by the receiving mechanism. Isochronous margin is a measure of the normal service margin of a receiver; synchronous margin is a measure of the efficiency of the mechanical condition of the receiver if the effect of speed-error is ignored.

At telex and TAS out-stations, teleprinter margin is checked by dialling the routing code to gain access to the exchange signal generator which transmits the test sentence (THE QUICK BROWN FOX . . .) with specified and graded amounts of early and late start-signal distortion. Permissible limits are quoted, according to the nature of the connection to the zone exchange from which the test signals are emitted; alternatively, portable testers can be used at out-stations.

2.11 The PO high-speed (400-baud) isochronous TDMS (Fig 2.9)

This TDMS was developed for distortion measurements on high-speed synchronous telegraph systems such as the 7-unit ARQ systems which operate at 96 bauds (2-channel) or 192 bauds (4-channel).

An electronic transmitter, using dekatron tubes, provides undistorted test signals (1:1, 2:2, 6:1, Q9S and 7:7) at 30 or 80 V, single or double current, with both normal and reversed polarity. Seven keys enable any 7-unit combination to be set up for repeated transmission. The output signals are fed

Fig 2.9 PO high-speed (400-baud) isochronous TDMS (*Courtesy of The Post Office*)

via a polarized relay, but an electronic output (limited to 20 mA) is available for use at modulation rates above 200 bauds, at which value the transmit relay begins to introduce excessive distortion. A centre-zero meter enables the output current and voltage to be checked.

Distortion is indicated as bright spots on the circular time-base of a cathode-ray oscilloscope, the period of one revolution being equal to the nominal duration of the unit signal element. Transmitter and receiver are driven by a common oscillator which is adjustable for modulation rates between 40 and 400 bauds. Adequate screening and radio-interference suppression have been included in the design to enable the tester to be used in radio stations.

The RC oscillator provides a basic frequency range from 40 to 100 Hz; this can be multiplied by 1, 2 or 4. A fine frequency control provides an adjustment of ±0·05 to ±0·02% at the upper and lower ends of the main tuning control. The oscillator frequency in Hz and the modulation rate in bauds are numerically equal and the main frequency control is calibrated in bauds.

The receiver has a high impedance. Signals being measured are fed to a pulse-generating transformer and rectified so that either mark–space or space–mark transitions will brighten the cathode-ray trace at the instants of restitution; either direction of transition can be rendered ineffective by a key for comparative measurements. The circular time-base is produced from a quadrature phase-splitter connected to the X and Y deflector plates of the tube; the usual focus and brightness controls are provided for adjustment. Distortion can be read direct off the oscilloscope scale to an accuracy better than 1%.

2.12 The PO start-stop (50-baud) TDMS (TG1157)

This distortion meter automatically phases itself on receipt of any 5-unit start–stop signal to display the timing of each restituted transition on a cathode-ray tube, which has a 6-inch screen fitted with a 4-inch scale calibrated over the range ±40% distortion. Having a high impedance input, the tester can be used for monitoring signals in service and it is fitted to test-desk positions in TAS and telex

Fig 2.10 PO 50-baud start–stop TDMS (TG1157)
(*Courtesy of The Post Office*)

exchanges. A front view of the TDMS is shown in Fig 2.10.

The cathode-ray display is controlled from a time-base having a period equal to the nominal time of a unit signal element (20 ms for a modulation rate of 50 bauds); the cathode-ray traverse is repeated for each of the six possible transitions for start–stop 5-unit signals. To permit distortion of individual instants of restitution to be read, the six traverses are mutually displaced by a vertical deflection circuit which gives a trace comprising a column of six horizontal lines (Fig 2.11). The repetition rate of this time-base is controlled by an oscillator which is automatically started by the receipt of a start signal and stopped when the correct number of beam traverses has occurred.

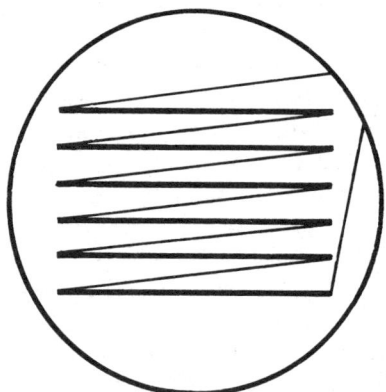

Fig 2.11 Cathode-ray tube ladder trace

For distortion measurement, the trace of the cathode-ray beam is brightened at the instants of restitution, the background trace being suppressed to leave only the bright spots visible to indicate instants of restitution on the time-scale; reading is assisted by using a screen with a moderately long afterglow. To facilitate reading distortion of the instants of restitution, which may be either advanced or retarded relative to their correct timings with respect to the start signal, the commencement of the time-base cycle is delayed by a period of half a unit element (10 ms at 50 bauds) so that undistorted signals are indicated at the centre of each traverse. A distortion indication read on this type of meter is shown in Fig 2.12 with the accompanying timing chart, showing a start–stop distortion of 10% early/15% late.

Although elements 1 and 3 are exactly 20 ms

in duration, in the start—stop mode of measurement in which timing is referred to commencement of the start element, the associated transitions have a distortion of 10%*E* and 15%*L* respectively. At the same

conditions to be set up. When an overall distortion measurement only is required, separation of the individual element traces may be eliminated by operating the GROUP SIGNALS key: distortion is

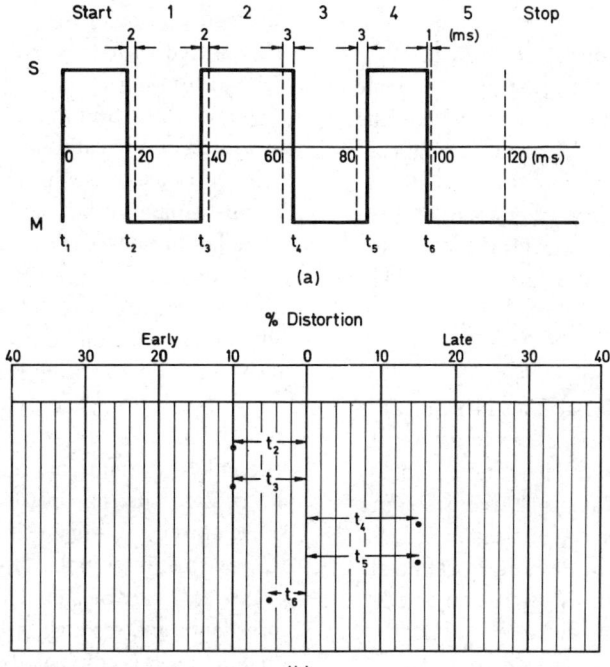

(a)

(b)

Fig 2.12 PO TDMS (TG1157) — reading start—stop distortion

time, although element 2 is lengthened by 25% and element 4 is shortened by 20%, the start—stop distortion indicated is 15%.

The tester is equipped with focus and brilliance as well as gain and shift controls for the *X* and *Y* deflections, and a set of keys enables various test

then read with the spots collapsed upon a single horizontal trace.

Transit times and bounce of transmitter contacts may be measured by throwing the TRANSIT TIME key. In this condition, the spot-pulse generator is inoperative and the trace shows the signal waveform. The duration of the transit time and bounce can be measured against the calibrated scale (see Fig 2.13). With the tester correctly adjusted for 50-baud operation, each division (2%) corresponds to a period of 0·4 ms.

With the SPEED TEST key operated, measurement of modulation rate can be made (if the transmission is at maximum character rate — e.g. from a teleprinter sending 'plugged' signals), by eliminating separation of the element traces and also the automatic stop action so that the primary time-base runs continuously. Indications of instants of restitution will move to left or right across the screen at a rate proportional to speed error. This can be measured by varying time-base frequency

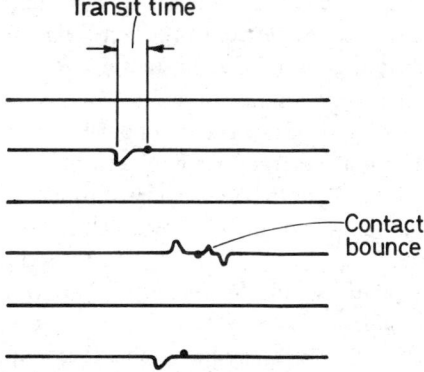

Fig 2.13 PO TDMS (TG1157) — display showing transit time and bounce

with a calibrated control to attain synchronism, shown by bringing the spots to rest. The percentage speed-error is read direct from the scale of the calibrated control-knob.

2.13 The Plessey isochronous/start—stop TDMS

This telegraph-distortion measuring set is available in two versions — the original valve model and the newer product using semiconductors. In each case, the tester comprises two separate units which can be used individually. One instrument is basically a signal generator, the other is the TDMS receiver. Designed in portable form, they are also suitable for rack mounting.

The TDMS Model 5 (Fig 2.14). This instrument is essentially a telegraph test-signal generator but it incorporates a $2\frac{3}{4}$-in. (69-mm) diameter cathode-ray tube for monitoring test signals and providing limited measurements of speed and distortion.

The signal generator provides undistorted test signals — 1:1 and 2:2 reversals, repetition of any selected 5-unit combination and a test message

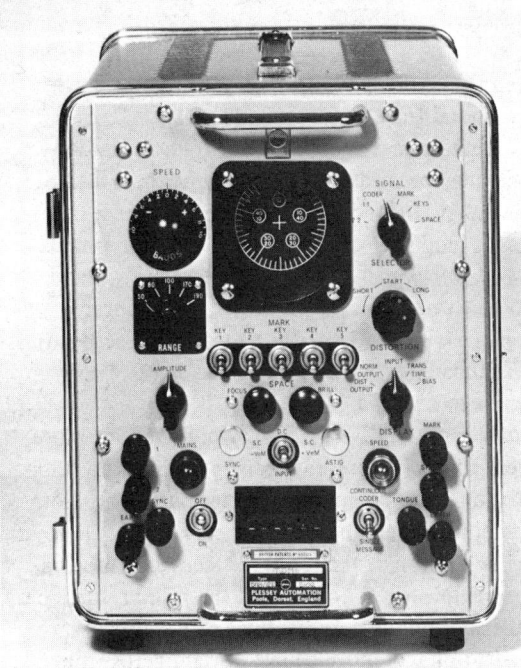

Fig 2.14 Plessey signal generator/TDMS Model 5 (*Courtesy of The Plessey Co.*)

of 100 characters from alphabet No. 2, repeated continuously if required. These, and also long mark and space signals, are selected by the switch shown, the repeated character being set up as desired on the five switches at the centre of the panel. As an alternative to being distortionless, signals can be transmitted with any value of start-signal distortion continuously adjustable up to ±50%; the degree of distortion in the signals is displayed and measured on the cathode-ray oscilloscope. This provision of signals with known distortion enables the instrument to be used to measure margin. Modulation rates available are 50 bauds and any four others in the range 50—190 bauds, and each rate is adjustable within ±10 bauds. With these adjustments, the instrument can be used to measure the modulation rate of an incoming signal with a degree of accuracy within ±0·1 bauds at 50 bauds and ±0·4 bauds at 200 bauds. By means of a link, the stop signal can be made equal to 1·0, 1·5 or 2·0 units. The instrument can be used on single-current or double-current circuits; in the former case, mark polarity can be either positive or negative. It may be adjusted for high or low input impedance and for high or low sensitivity depending on whether it is connected in series or shunt, or as a termination for the circuit being tested.

The instrument reads isochronous distortion on a circular trace controlled from a half-unit time-base, so that 360° displacement corresponds to 50% distortion. The calibrated graticule gives direct reading with 50 divisions each of 1% distortion (see Fig 2.16). Early displacement is denoted by clockwise readings and late displacement by anti-clockwise readings. Controls are provided on the front panel for amplitude, focus and brilliance of the display together with pre-set controls on the rear panel for initial setting-up and maintenance.

The modulation rate of input signals is measured by adjusting the SPEED control until the display — which may be drifting around the trace in one direction or the other — becomes stationary. The modulation rate of the incoming signal is read directly from the SPEED range selector switch (50—190 bauds) and the fine SPEED control (±10 bauds).

Polarized relays may be tested either by plugging a Carpenter relay directly into the jack on the face panel, or using this jack for a relay-base adapter for other types of relay. With the DISPLAY switch set to BIAS and the SIGNAL SELECTOR to 1:1 or 2:2, the relay will vibrate and signals from the relay contacts are displayed on the oscilloscope trace. For a neutral relay, bright spots corresponding to mark and space signals should coincide to produce a single spot. If necessary the relay can be adjusted *in situ* and retested.

If the SELECTOR switch is moved from BIAS to TRANSIT TIME, the relay being driven by 1:1 or 2:2 reversals, transit time will be displayed as a percentage of modulation rate by a brightened arc on the oscilloscope trace (see Fig 2.16). The transit time can be readily calculated, e.g. at 50 bauds the time-base is 100 Hz and 360° represents 10 ms. Any contact bounce will be revealed by additional bright spots at the end of the transit-time arc.

The 'coder' character-generator is a novel device, only $5 \times 2\frac{1}{2} \times 1\frac{1}{2}$ in., mounted to a Carpenter relay-base. The coder consists of a flush, printed circular board bonded to an aluminium backing piece. The disc is etched to give five concentric tracks corresponding to elements of the 5-unit code, and also three tracks for switching functions.

Copper tracks are rhodium plated and connections to the tracks are made by beryllium-copper wipers. Coding and switching contacts take the form of a contact-wiper assembly attached to a moulded contact-bridge so that wiper tips bear lightly on the metallic or insulated tracks of the printed disc. Metallic segments of coding tracks are connected together and through a coding earth-return wiper. The five coding contacts are sampled sequentially, and at the fifth element the disc is stepped from character to character by means of a drive coil, armature, pawl and 100-tooth index wheel; it will generate signals at modulation rates up to 200 bauds. The coder provides the standard test sentence (THE QUICK BROWN FOX . . .) together with numerals, shifts, spaces, carriage-return and line-feed to a total of 100 characters. Other forms of test message can be supplied by a change of printed disc.

The Model 5 TDMS has been superseded by a new design of telegraph signal generator, the TSG40.

The TDMS Model 6 (Fig 2.15). This instrument measures isochronous and start–stop distortion on a $2\frac{3}{4}$-in. diameter cathode-ray tube with a calibrated graticule. The scale differs from that of the Model 5 instrument: in the Model 6, the 360° scale corresponds to 100% distortion, in 50 divisions each of 2%, i.e. 50% early and 50% late when measuring start–stop distortion. It is also possible to measure modulation rate of an incoming signal and to measure bias and transit time of a polarized relay.

The instrument covers modulation rates of 50 bauds and any other four rates within the range 40–190 bauds ± 10 bauds. It will operate to signals in either single-current or double-current mode – either to CCITT or British Post Office polarity conventions. The instrument can be set for high or low input impedance and used for either series or shunt monitoring or as a termination. With double-current systems, shunt monitoring enables distortion to be measured without affecting a circuit which is in service. Focus and brilliance controls, together with pre-set controls for installation and maintenance, are provided.

For measuring isochronous distortion (or modulation rate) operation of the CONTINUOUS switch

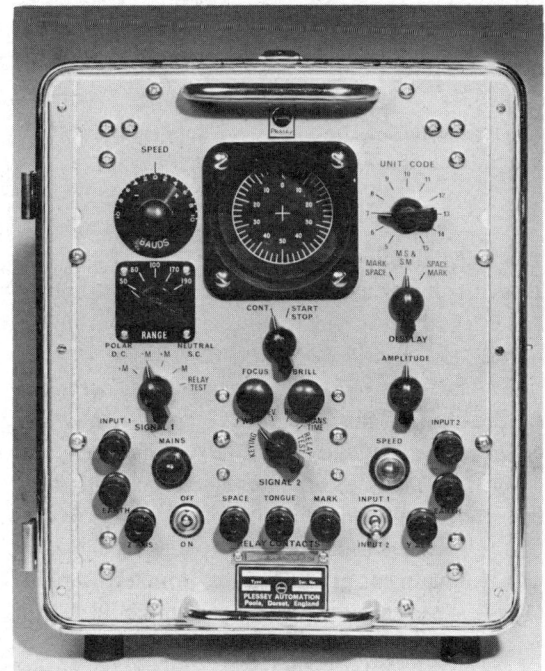

Fig 2.15 Plessey TDMS Model 6 (*Courtesy of The Plessey Co.*)

provides a circular trace on the cathode-ray tube; a typical display showing this is seen in Fig 2.16. If the display rotates slowly, the difference in modulation rate can be adjusted and measured by using the ±10 bauds fine adjustment to bring the display to rest.

must be driven from an external source. The M, T and S contacts are connected to three terminals provided for this purpose on the lower face panel, and selector switches operated to the appropriate positions. The relay may be tested using either circular or spiral time-base trace; in the latter case,

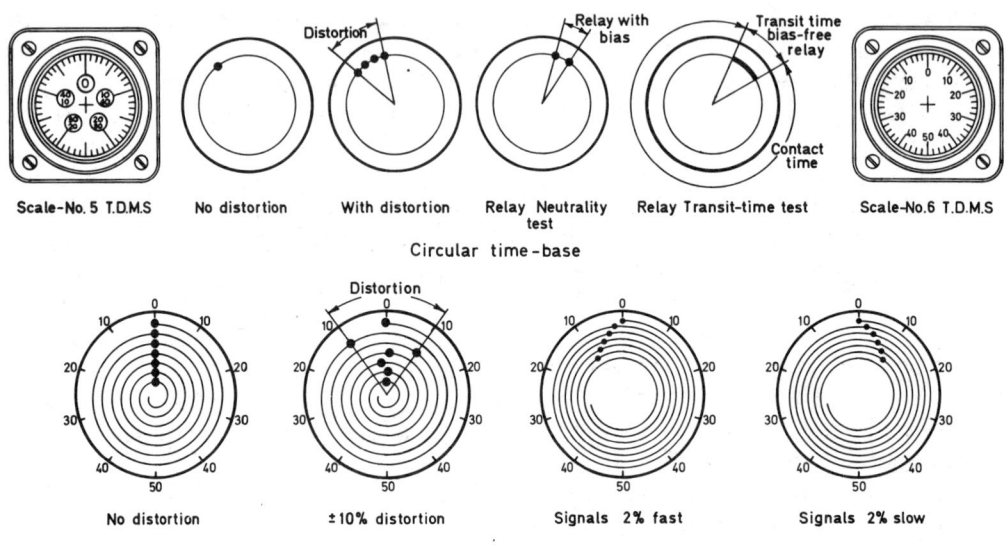

Circular time-base

Scale–No. 5 T.D.M.S   No distortion   With distortion   Relay Neutrality test   Relay Transit-time test   Scale–No.6 T.D.M.S

Spiral time-base

No distortion   ±10% distortion   Signals 2% fast   Signals 2% slow

Fig 2.16 Plessey TDMS.s — typical displays

For measuring start–stop distortion (or modulation rate) operation of the switch to START–STOP sets up a spiral trace on the oscilloscope. The number of turns – one per unit in the spiral can be adjusted for a range of 5 to 15 units, according to the code of the incoming signals. The 7-unit position is used for the start–stop 5-unit code with a 1·5-unit stop element. Mark–space and space–mark transitions can be displayed separately if required. Any difference in modulation rate between incoming signals and the setting of the TDMS time-base will be indicated by a regular displacement of the bright spots from vertical (if no other distortion is present), the spot indicating the last transition of a character being most displaced (see Fig 2.16). The modulation rate of incoming signals can be measured by settings of the SPEED controls necessary to bring the spots to the correct vertical line.

The bias and transit time of a relay (or teleprinter transmitter) can be measured as described for the Model 5 instrument; in this case the relay

2:2 signals will produce four bright spots on the display.

The Model 6 TDMS has been superseded by a new model, the TDMS80.

The Telegraph Signal Generator TSG10 (Fig 2.17). This instrument uses semiconductors throughout and is suitable for rack mounting or as a portable tester.

The facilities follow in principle those described for the Model 5 instrument. At left and right will be seen the coder and output relay respectively; the latter is normally a Carpenter electromagnetic relay but alternatively an electronic plug-in relay, designed on the switched-oscillator principle, is available. The five main control-knobs are, from left to right: (1) the speed selector for three fixed modulation rates – normally 45·5 (U.S.A. rate), 50 and 75 bauds; (2) the selector for the length of the stop signal – 1·0, 1·5 or 2·0 units; (3) the test signal selector — 1:1, coder for 100-character test message, repeated character set up on the 5-unit keys immediately below the knob, or long mark or space signals. By using the five code-keys and a 2-unit stop element, 2:2 reversals may be generated;

and generates signals of 5, 6, 7 or 8 units, including stop signal — the latter being 0·5, 1·0, 1·5 or 2·0 units at choice. A coder is not included, the test signals being 1:1, 2:2 or any characters set up on the eight keys at the left-hand side.

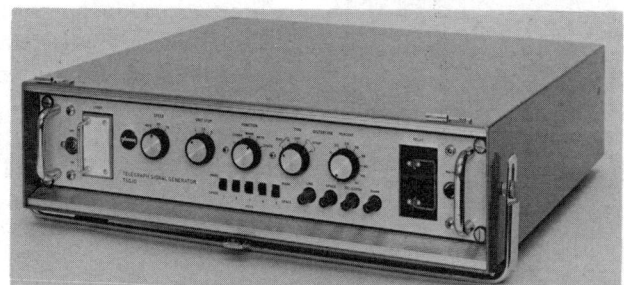

Fig 2.17 Plessey signal generator TSG10 (*Courtesy of The Plessey Co.*)

(4) selector for distortionless signals, or signals having mark or space bias, or early or late start-signal distortion; (5) control of start-signal distortion, with fixed increments up to ±50%. Four terminals are provided to connect the line under test, earth potential and station battery supply (e.g. ±80 V).

Unlike the Model 5 instrument, the TSG10 does not include a monitoring oscilloscope, nor facilities for relay testing.

The Telegraph Signal Generator TSG20. This unit (Fig 2.18) follows the design of the TSG10 and gives similar facilities but with the following differences. It covers the modulation-rate range of 30—300 bauds

The Telegraph Signal Generator TSG40 (Fig 2.19). This unit (which supersedes the Model 5) uses integrated semiconductor circuits and has an isolated electronic output relay.

The facilities which can be selected by control switches include:

5-unit or 7-unit signals (for alphabets Nos 2 or 5)
Stop-signal duration of 1·0, 1·5 or 2·0 units
Modulation rate 25—330 bauds (continuously adjustable, with accuracy ±1%, resolution 0·03 bauds)
Output signals — steady mark or space; 1:1 reversals; 96-character test message (THE QUICK BROWN FOX . . .); 32-character 'RY' or 'U' test; 8-character Q9S; automatically

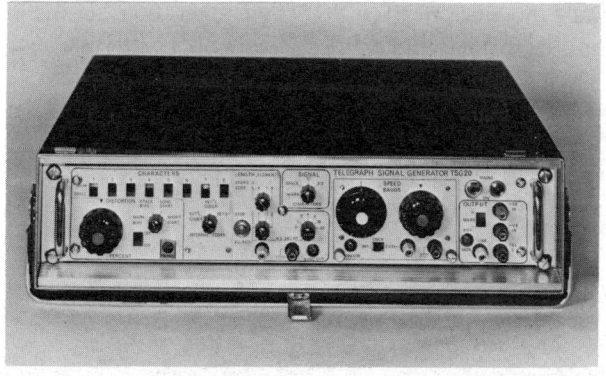

Fig 2.18 Plessey signal generator TSG20 (*Courtesy of The Plessey Co.*)

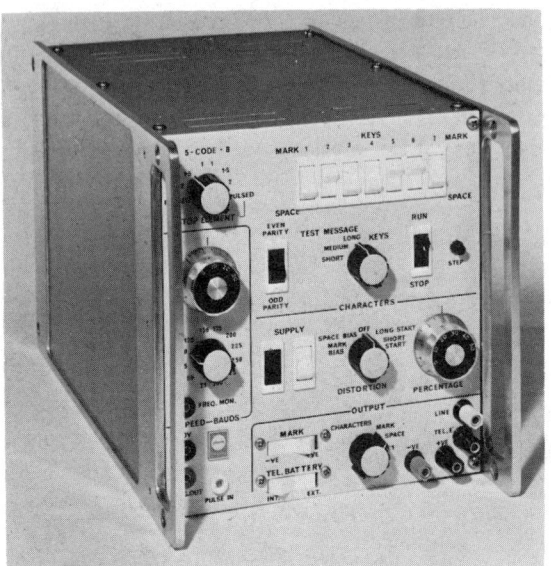

Fig 2.19 Plessey signal generator TSG40 (*Courtesy of The Plessey Co.*)

vides a circular trace for isochronous-distortion measurements and a spiral trace for start—stop distortion. The six control-knobs (Nos 1, 2 and 6 from the left are dual-concentric controls) provide: (1) selection of input conditions — series or shunt, single or double current, ±6, 0/6 or ±80 V, positive or negative mark polarity; also relay tests for bias and transit time; (2) selection of (*a*) the number of units in the code, from 7 to 14; (*b*) the circular or spiral trace; (3) selection of mark—space, space—mark, or all transitions; (4) amplitude control of the trace; (5) fine control (1-baud steps) of modulation rate; (6) selection of modulation rate — 30—330 bauds.

repeated character set up on 5 or 7 keys. (All the foregoing are available in alphabet No. 2 or 5.) Output signals are also available in parallel form at ±6 V

Lengthened/shortened start signal, continuously adjustable over the range 0—50%, with accuracy ±1% overall, and resolution ±0.05%

The equipment is portable, but equally suited for rack mounting.

The TDMS 70 (Fig 2.20). This TDMS uses semiconductors. Designed for rack mounting, it can also form a portable instrument.

The facilities follow broadly those described for the Model 6 instrument. The 3-in. diameter tube pro-

The TDMS80 (Fig 2.21). Incoming signal transitions are displayed upon a raster on a 100 x 60 mm tube, which produces a line or trace for each code element selected, enabling individual elements to be measured on the scale; this is calibrated for early and late start—stop distortion measurement, to an accuracy of 1%, over the range 0—50% in 1% divisions. A x5 switch permits the range to be set to 0—10% in 0.2% divisions. A continuous or free-run single trace is also provided for measurement of modulation rate and assessment of isochronous distortion.

This TDMS will operate over a range of modulation rates 25—330 bauds, continuously variable, with a resolution of 0.03 bauds and accuracy ±1% overall. The display can be set for any code in the range 7- to 13-unit in 1-unit increments.

2.14 Tester TG2402 (Fig 2.22)
This portable TDMS, newly developed by Trend Communications for the British Post Office, follows the patterns already described but with

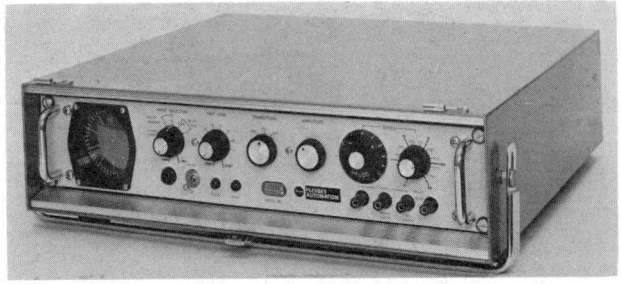

Fig 2.20 Plessey TDMS70 (*Courtesy of The Plessey Co.*)

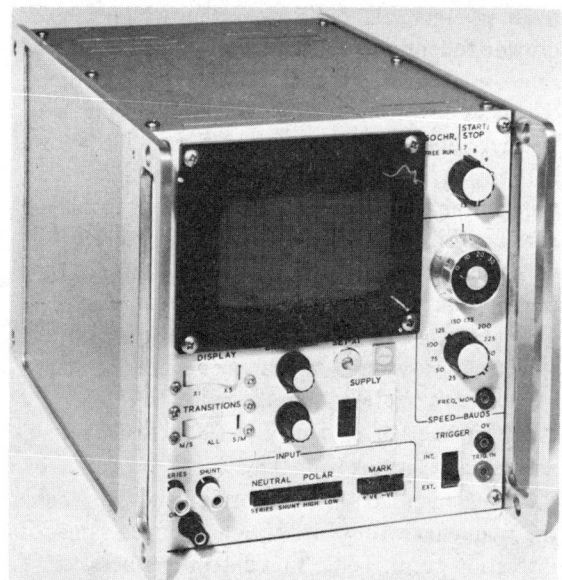

Fig 2.21 Plessey TDMS80 (*Courtesy of The Plessey Co.*)

many additional features and high accuracy. The output relay is electronic.

The Signal Generator. Provides Q9S and QUICK BROWN FOX test messages (from an integrated circuit) in alphabet Nos 2 and 5 with applied bias or start distortion up to ±49% in 1% steps. The 8-unit signals include an odd or even parity bit and consist of three lines, upper case, lower case and signs, transmitted individually or in sequence. Test messages may be released continuously, or one message or character at a time. Each line of text is preceded by (undistorted) CARRIAGE RETURN, LINE FEED, and 200-ms mark, the latter to ensure the stop condition after distorted signals and to allow time for the teleprinter carriage to return at high modulation rates.

A set of eight keys allows any 5-, 6-, 7- or 8-unit code-combination to be set up, with stop signal selected at 1·0, 1·5 or 2·0 units.

Modulation rates (derived from a crystal oscillator) are continuously adjustable in the range 37—1512 bauds, with 2% accuracy and stability better than 1% during 30 s. Fixed rates of 50, 75, 100, 110, 150, 200, 300, 400, 600, 900 and 1200 bauds are provided to an accuracy of ±0·1% and stability better than 0·1% during 30 s.

Other spot rates (e.g. $82\frac{2}{7}$ or $123\frac{3}{7}$ bauds) can be set up by using the fixed and variable controls.

Reversals at 1:1, 2:2, 1:3, 3:1, 1:6 and 6:1 are also available.

The Receiver. The display is similar to that of Fig 2.21, but reads to ±50% with 1% scale dots, brightened at every tenth dot and blanked out at 0%. The display rows may be superimposed (e.g. for peak distortion) or separated (e.g. for characteristic distortion).

For start—stop distortion a row of calibrated dots appears for each possible transition. The single-line display can, for greater accuracy, have the X-scan expanded to cover the range 20%E—20%L; a transition outside this range brings in a warning lamp.

When measuring isochronous distortion a single row of calibration dots is displayed continuously. Single M/S or S/M transitions appear as short vertical lines above or below this line. The isochronous distortion is measured as the total displacement between outermost transition dots. This

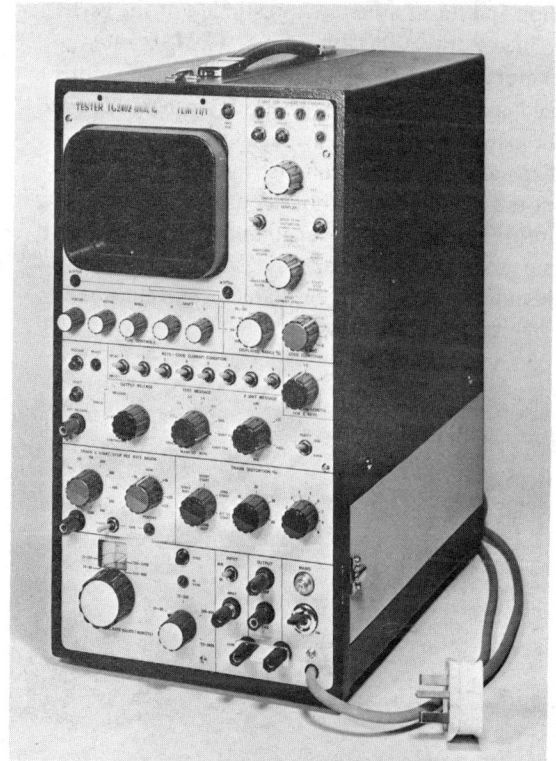

Fig 2.22 Tester TG2402 (*Courtesy of The Post Office*)

display may not appear centrally: any difference between incoming modulation rate and tester time-base would result in the display drifting to left or right.

A HOLD PEAK facility is provided on the single-line start—stop display and, optionally, on the isochronous display: this gives a continuous indication of peak early and late reading, until reset, in addition to the display for each transition, without need for continuous close observation.

The stop—signal duration can be measured and reconstituted waveforms displayed to measure transit time or bounce.

On the 5-unit QUICK BROWN FOX message, character errors may be counted (up to 15), showing inverted elements and characters with distortion in excess of a pre-set limit.

Short start signals can be rejected.

2.15 The telegraph-distortion analyser

When using a direct-indicating type of distortion meter, the observer notes maximum values of distortion which occur during the period of observation and forms a *subjective* estimate of the general value of distortion; this type of TDMS is satisfactory for the day-to-day checking of apparatus and channels in service. The quasi-integrating type of instrument gives an averaged reading which depends upon distortion distribution and the law of the instrument, without conveying much information about dispersion of the values, whilst a peak-indicating meter may give an inordinately high reading due to the isolated occurrence of a high value of distortion.

The statistical study of telegraph distortion

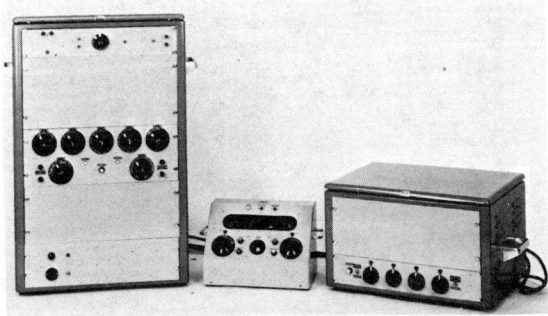

Fig 2.23 Telegraph-distortion analyser (*Courtesy of The Post Office*)

requires a more elaborate instrument which will provide fuller data of distribution of distortion values on restituted signals. For this purpose, the telegraph-distortion analyser illustrated in Fig 2.23 was developed in the Post Office Research Station. Although a laboratory-type instrument, it is transportable for field use when required to collect data on distortion measurements under service conditions in telex exchanges or VFT stations. Its primary use is for measurement on start—stop 5-unit systems, but it is suitable also for use on synchronous systems. Measurements may be made at modulation rates within the range 25—500 bauds (in steps of 25 bauds) to 1000 bauds, with an accuracy varying from 0·5% at the lowest rates to 2% at the high rates. The distortion analyser records the frequency with which various grouped values of distortion occur. To do so it determines between which of a number of pairs of limits the distortion of a transition lies and then records the reception of this transition on a counting meter associated with the range of distortion bounded by this particular pair of limits. The equipment may be set to select mark—space or space—mark transitions, or to respond to both. Space—mark transitions are often chosen because, being of opposite polarity to that of the commencement of the start signal, they are directly affected by bias, which is found to make a dominant contribution to the limit of circuit performance. In the start—stop mode, the cycle may include from seven to eleven elements, and all or any of the transitions in a character may be included in the measurements. In the synchronous mode, facilities are provided for synchronizing to an external timing source but not to the signals themselves.

Results of measurements may be either shown on dekatron 10-digit display tubes or automatically printed out on a teleprinter, together with a punched tape prepared at the same time on a reperforator if required. The operator may exercise control over each step in the test or the equipment may be allowed to run automatically for a continuous series of measurements with all results printed out.

The analyser has an accurate time-scale which commences from reception of the first significant instant of a character signal and is divided into suitable intervals, such as 1% of a unit element. A series of counting meters, each of which is sensitized in turn during a portion of the time-scale, records the occurrence of any significant instants during the allotted period of each meter.

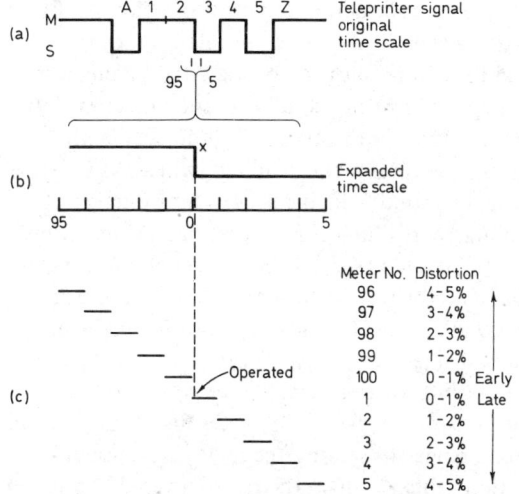

Fig 2.24 Principle of the telegraph-distortion analyser

In Fig 2.24 (a) is shown the waveform of a typical start—stop 5-unit signal, at (b) one of the significant instants is shown against the proposed time-scale. The basic conception of the analyser is that the time-scale of 100 intervals is repeated for each theoretical signal-element period in the character, and during each timing interval a different counting meter is presented to the circuit to detect the possible occurrence of a significant instant, as shown at (c). This arrangement implies the use of 100 counting meters, one for each 1% of the distortion range, but a set of 12 such meters is found to be adequate provided their ranges can be adjusted. Each meter is associated with a given range of distortion which can be adjusted to the requirements of a particular test. Distortion distribution is assessed by considering the number of transitions recorded by each meter, together with the distortion ranges covered by them.

Referring to Fig 2.23, the main unit contains all circuits and controls concerned with the reception of signals and classifying and metering the data derived from them. The control unit contains the circuits and controls by which the main unit may be made to receive or supply information, start and stop either process and display information supplied

by the main unit. The print-out receives information from the control unit and translates it into corresponding characters in alphabet No 2; this information is supplied to the teleprinter and reperforator.

The time-base, controlled from a crystal oscillator, is in the form of a distributor which completes one cycle in a unit interval. This cycle commences half an element before the nominal mean time at which incoming signal elements commence; at the time when the incoming signal is most likely to have a transition, the distributor is halfway through its cycle. Associated with the outputs of the distributor are the 12 counting meters.

When a transition occurs, the distributor is stopped and a metering cycle is started, during which reception of the transition is registered by the counting meter associated with the period of time in which the transition occurred. At the end of the metering cycle, which has a fixed duration of 1 ms, the distributor remains stationary until the normal starting instant in the next element time-scale.

Once a test has started, transitions are recorded in the appropriate counting meters until a predetermined number has been received. For ease of calculation, a test will comprise 100, 1000, 2000, 5000 or 10 000 transitions as selected, on completion of which the test cycle is automatically terminated. At this stage the meter readings are stored on a ferrite-core matrix. When the information is required, each meter reading is passed via a binary counter to a dekatron counter which gives a visible digital display; the same information is printed on the teleprinter, together with the automatic insertion of a preamble concerning the test conditions, carriage-return and line-feed signals being automatically inserted. In each case, either individual or the cumulative totals of the 12 meter readings may be displayed in turn or printed out. The digital display is given on the set of four digital indicator tubes in the control unit, the lower showing which meter is being read. The operator must press a READOUT button once for each meter reading when the displayed readings are being taken.

The arrangement adopted for the meters is for Nos 2—11 to cover equal distortion ranges, which may be 1, 2, 4 or 8%. If the ranges are changed, the instant at which meter No 7 becomes effective remains the same, so that although these 10 meters may be switched to cover between them 10, 20, 40 or 80% of the element period, the mean position of their combined distortion remains constant. A separate control enables this mean position to be varied in steps of 5%. Meters 1–12 then cover the extreme early and late values of distortion. It is found advisable to adjust the meter range and mean-distortion controls so that no transitions are recorded in meters 1 and 12.

In use, with the power supply and the channel or machine being tested connected to the analyser, a test is carried out after first setting the controls to the required positions as regards modulation rate, amplitude of incoming signal, synchronous or start–stop mode, duration of start–stop receive cycle, transitions to be measured within the character, type of transition (mark–space or space–mark), distortion range per meter, mean distortion position, total number of transitions, mode and form of recorded information – individual, cumulative etc., manual or automatic read out, initial test serial-number. Use of the START button initiates the required measurements. A lamp glow indicates completion of measuring the selected

number of transitions. Under automatic conditions results are then printed out – a process taking about 45 s – and a further set of tests follows unless the STOP button is pressed. Finally the test is cleared by pressing the METER RESET button which clears all meters and print-out circuits. By setting a switch, the print-out teleprinter may be used for communication over a circuit being tested.

Typical print-out results are shown below; IND and CUM refer respectively to individual and cumulative total meter readings, whilst ALL – the normal method – includes both individual and cumulative readings. In the first line of each print-out, 0036 and 0037 are the test serial numbers; 1000 and 0000 refer to the number of transitions (0000 indicates 10^4 since four printed digits only are available, zero is always printed as 0); –15 is the mean-distortion setting. The setting of the mean-distortion switch indicates the mean position of the range covered by the meters, i.e. the distortion value at which the range covered by meter 6 ends and that covered by meter 7 begins. The range of meter 1 begins 50% earlier than the setting of the mean-distortion value. The figure 2 is the distortion range covered by each meter.

```
IND

0036 1000  ⌐15     2
   0    0    36   145   297   368   139    15     0     0     0     0

0037 0000  ⌐15     2
   0    2   476  1185  3296  3604  1310   126     1     0     0     0


CUM

0036 1000  ⌐15     2
   0    0    36   181   478   846   985  1000  1000  1000  1000  1000

0037 0000  ⌐15     2
   0    2   478  1663  4956  8563  9873  9999  0000  0000  0000  0000


ALL

0036 1000  ⌐15     2
   0    0    36   145   297   368   139    15     0     0     0     0
   0    0    36   181   478   846   985  1000  1000  1000  1000  1000

0037 0000  ⌐15     2
   0    2   476  1185  3296  3604  1310   126     1     0     0     0
   0    2   478  1663  4956  8563  9873  9999  0000  0000  0000  0000


REPERFORATOR

0036 1000 ⌐15 2
0 0 36  145 297 368 139 15 0 0 0 0

0037 0000 ⌐15 2
0 2 476 1185 3296 3604 1310 126 1 0 0 0
```

Cumulative totals of meter readings are more convenient to apply when the distribution curve is plotted on arithmetic-probability paper — a scale which simplifies drawing conclusions from the tests.

In the print-out examples given, both tests have the same distortion ranges on individual meters, i.e. M = −15% and R = 2%; meter 4, for example, is associated with the range −21 to −19%. The figures 145, 1185, 181 and 1663, representing individual readings or cumulative totals of meter 4, now have the following significance:

Test No. 0036.   14·5% of transitions had distortion values in the range −21 to −19%.

Test No. 0037.   11·85% of transitions had distortion values in the range −21 to −19%.

Test No. 0036.   18·1% of transitions were earlier than −19%.

Test No. 0037.   16·63% of transitions were earlier than −19%.

Individual meter readings have been used to draw the distribution curve shown in Fig 2.25.

The distribution curve plotted in Fig 2.26 is drawn on arithmetic-probability paper from

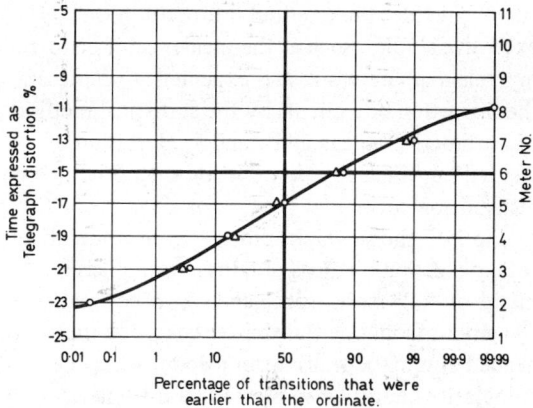

Fig 2.26 Cumulative-frequency graph for telegraph-line distortion

cumulative totals of meter readings. The cumulative count, when plotted on arithmetic-probability graph paper, produces a graph which, over its major portion, is linear with its ends asymptotic to some limiting values. The slope of the graph will increase or decrease if higher or lower values of distortion are recorded. Typical limits are represented by the 0·1 (or 99·9) ordinates − $P(3)$ values; or by the 0·01 (or 99·99) ordinates − $P(4)$ values. Corresponding distortion values will be exceeded by not more than 1 transition in 10^3 ($P(3)$) or 1 in 10^4 ($P(4)$).

For many purposes, production of distribution

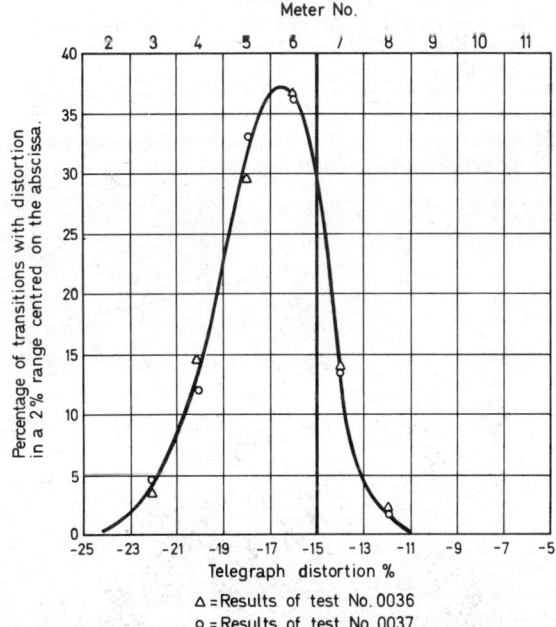

Fig 2.25 Normal distribution graph from TDA measurements

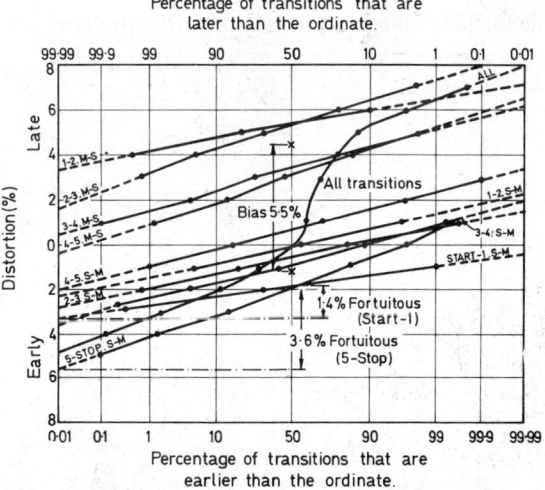

Fig 2.27 Distortion measurement on automatic transmitter No. 3A

curves provides the required information. For more extensive studies, such as the incidence of distortion in a telegraph network, it is expedient to represent the distortion of a circuit by the statistical method of arithmetic — mean and standard deviation of distortion, which together operate on a fundamental distribution curve.

Fig 2.27 shows graphs plotted from cumulative total-distortion readings obtained from measurements on four mark—space and six space—mark transitions from one transmitter head of a multiple-headed automatic transmitter (Model 3A). The cumulative count, when plotted on arithmetic-probability graph paper, produces a graph which is linear over the major portion showing it to be a normal statistical distribution — with the ends asymptotic to some limiting values. The lower (upper) abscissa indicates the percentage of transitions that are earlier (later) than the ordinate. The 1% ordinate indicates that 1 in every 100 transitions has distortion which exceeds the value at which the graph cuts this ordinate — the $P(2)$ values. Similarly, the 0·1 and 99·9 ordinates give the $P(3)$ values, and the 0·01 and 99·99 ordinates give the $P(4)$ values. The Model 3A transmitter has six sets of transmitting contacts — one for each code element and one for the start and stop elements. The slope of each graph demonstrates the fortuitous distortion about the median value. The graphs show also that mark elements are all longer than space elements, since mark—space transitions are all late relative to space—mark transitions. This bias-distortion component of the total distortion, which

is responsible for the double bend in the graph for the distribution of all transitions, is about 5·5% distortion, when considered as the difference between the mean values of the mark—space and space—mark distribution medians.

References

1 Nyquist, H., Schank, R. B. and Cory, S. I. 'Measurement of Telegraph Transmission', *AIEE Transactions*, **46**, p. 367 (1927).
2 Wheeler, L. K. and Tissington, R. S. 'An Electronic TDMS for Start—stop Telegraph Signals', *POEE Journal*, **43**, p. 18 (1950).
3 Dain, G. T. 'An Improved TDMS', *ATE Journal*, **9**, p. 20 (1953).
4 Wheeler, L. K. and Frost, A. C. 'A Telegraph Distortion Analyser', *POEE Journal*, **47**, p. 5 (1954).
5 Wheele, D. H. E. and Collier, E. G. 'Telegraph Distortion on Physical (DC) Lines and Telegraph Machines', *ibid.* **52**, p. 61 (1959).
6 Wheeler, L. K. and Frost, A. C. 'Telegraph Distortion in the Trunk Network of the TAS System', *ibid.* **52**, p. 103 (1959).
7 Carter, R. O. and Wheeler, L. K. 'Some Applications of Electronic Methods to Telegraph Apparatus', *IPOEE Printed Paper No. 199* (1949).
8 Roquet, R. 'Théorie et Technique de la Transmission Télégraphique' (1954), Eyrolles.
9 Wüsteney, H. H. 'Telegraph Distortion and Distortion Measuring', *AIEE Transactions Paper No. 60/61* (Jan. 1960).
10 –, 'Teleprinter Margin and Margin Measurement', *AIEE Transactions Paper No. 62/495* (Jan. 1962).
11 Willington, D. J. 'A New Telegraph Distortion Measuring Set', *POEE Journal*, **68**, p. 29 (1975).

3 D.C. Transmission

3.1 Field of d.c. transmission

For the first hundred years of telegraphy, all circuits – irrespective of length of route – were operated by direct currents. In well-developed networks this method is now superseded, for economic and stability reasons, by a.c. transmission at voice-frequencies for all purposes other than relatively short routes. Even long submarine d.c. telegraph cables, for which the classic propagation theories were developed, are giving way to the economic advantages of a.c. transmission, either by using undersea repeaters or else changing to the medium of the communication satellite.

Telegraph receivers – the teleprinter and reper-forator – are controlled by direct current. For long-distance transmission, a.c. methods are used, so that at some point, at each end of a circuit, there will be a d.c./a.c. boundary, or interface. For multiplex telegraphy, where a telephone-type 'bearer' circuit is shared for a large number of telegraph transmissions, it is economical and convenient to provide the multiplexing equipment at a central office. Here the d.c./a.c. interface will then be found: transmission over the few miles of 'local end' between the user's office and the central office is by direct current. In a dense network it is rarely economical to extend d.c. transmission beyond about 20 miles, though any particular application may be influenced by factors of topography, the existing lineplant layout, convenience or expediency.

Telegraph signals are generated in the form of abrupt changes in potential; the characteristics of the line $\frac{1}{M}$ the distributed resistance and capacitance $\frac{1}{M}$ prevent the transmission of these abrupt changes, and complex transient conditions arise before the final steady-state value of the current is reached. If the time required for the current to build up exceeds the duration of the unit interval of the code the steady-state value will not be attained.

3.2 Signal-shaping network

In a circuit possessing inductance, a direct current does not attain its Ohm's Law value instantaneously; in building up to its full value the current is subjected to a delay which depends upon the ratio of resistance R to inductance L present in the circuit.

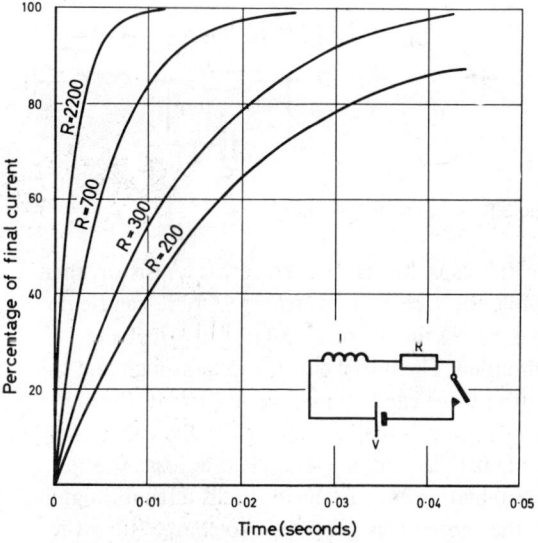

Fig 3.1 Effect of inductance/resistance ratio on growth of current

In telegraph circuits, relays and electromagnets are required to respond to currents of short duration, and in the design of such circuits attention has to be paid to the gradual rise of current and consequent delay in response of the inductive relay or electromagnet. The series of graphs in Fig 3.1 demonstrates the increase in rate of growth of current – or decrease of time-constant – by increasing the total circuit resistance to reduce the ratio L/R. The final value of current will become less as resistance is added, and the ordinate is scaled in percentage of final value. If R and V are increased in the same ratio, the final values of current remain constant. Series resistance is added in telegraph circuits to ensure quick response of relays and electromagnets.

The time required to charge a capacitor depends not only upon its capacitance, C farads, and upon the applied p.d., V volts, but also upon the value of any series resistance, R ohms, present in the circuit.

At the receiving end of a telegraph line the usual termination is an inductive circuit — a relay or an electromagnet — which delays a rapid build-up of current. A signal-shaping network (Fig 3.2) comprising a resistor shunted by a capacitor, the combination being placed in series with the inductance can effect a considerable improvement towards restoring the abrupt nature which is characteristic of telegraph-signal transitions.

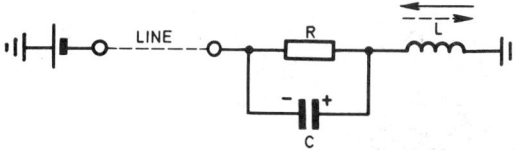

Fig 3.2 Signal-shaping network

If a capacitor is shunted across a resistor, then while the capacitor is uncharged there can be no p.d. across the resistor, and it is in effect short-circuited. The initial current value is high and this current, the capacitor-charging current, flows in series through the inductance. As the capacitor becomes charged, a p.d. appears across the shunt combination and the current falls until ultimately, if the capacitor is given time to charge fully, the Ohm's Law, or steady-state, current is attained.

The effects of this signal-shaping device are shown in Fig 3.3. Curve (1) is for current growth in the inductance with neither added resistance nor capacitance in circuit; the voltage is reduced to give the same steady-state value as in the other curves. Curves (2)–(6) all have a resistor of 4000 Ω in series with the inductance of 6 H, which is that of the electromagnet of a teleprinter. With curve (2) there is no capacitance in circuit, the improvement being entirely due to the reduction of time-constant. Curves (3)–(6) are obtained by adding increasing values of shunting capacitance from 0·2 to 2·0 μF. The initial current can greatly exceed the steady-state value, which is 20 mA derived

from an applied voltage of 80 V, the total circuit resistance including the electromagnet coils being slightly in excess of 4000 Ω. At a certain value of capacitance an oscillatory current results from the interaction of the inductance and the capacitance, and if this oscillation continues too long [curves (5) and (6)] it may interfere with the succeeding signal (which at 50 bauds could arrive after 20 ms).

A sensitive relay has a lightweight moving system

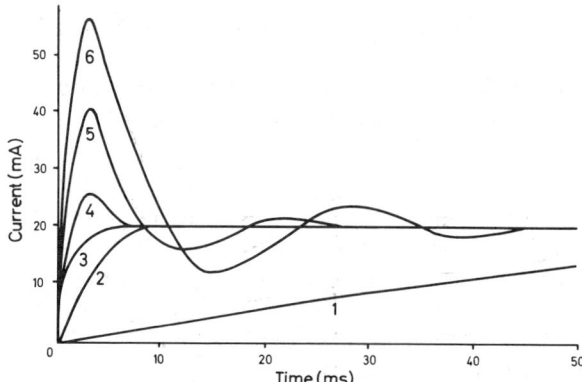

Fig 3.3 Effect of shunted capacitor

and excessive current may cause violent operation of the armature and produce contact bounce. For such a purpose, a termination giving a curve between those of (2) and (4) would be used. On the other hand, the electromagnet of a teleprinter carries a mechanical load, and the overshoot obtained from a termination which gives a curve lying between those of (4) and (5) is beneficial in supplying a high initial energy to obtain fast operation of the armature.

The optimum values of resistance and capacitance depend mainly on the values of steady-state current and inductance, and to some extent on variations of inductance with current, telegraph modulation rate and mechanical design of the receiving device. The best arrangement for a particular case is found by experiment and then standardized: typical values used are 4000 Ω + 4 μF with 80-V operation. The reason for the high value of capacitance is that the electromagnet is itself shunted by a 2-μF capacitor to suppress currents from the e.m.f. generated by the armature moving in the magnetic field; this e.m.f. would otherwise pass into the line and cause interference with telephone circuits working in the same cable. This 2-μF capacitor slows down the rise in current in the electromagnet coils which it shunts.

The shunted capacitor is also effective in assisting a rapid cessation of current when the signal voltage is removed; the tendency for the collapsing field of an inductor to prolong the current is opposed by the capacitor discharge current which is in the opposite direction.

The properties of the shunted capacitor may also be used to improve the waveform of signals received over a long d.c. line. The square-wave telegraph signal which is transmitted is equivalent to a number of sine waves comprising a fundamental frequency together with all *odd* multiple, or harmonic, frequencies. The transmission line attenuates currents in proportion to their frequency, the higher harmonic components are reduced in relative amplitude and waveform distortion results. The shunted capacitor has an *impedance*

$$Z = \frac{R}{\sqrt{1 + \omega^2 C^2 R^2}}$$

which *falls* with rising frequency; due to its presence the relative amplitudes of harmonic components are increased and the waveform is improved. For correcting line distortion, the shunted capacitor could equally well be inserted at the sending end.

As an alternative to using a parallel resistance—capacitance combination in which impedance falls as frequency rises, a resistor in series with an inductor could be shunted across the receiving electromagnet or relay; with this arrangement the impedance of the resistance—inductance circuit *rises* with frequency. This shunt path absorbs a considerable proportion of the low-frequency components, but at higher frequencies a larger proportion of received current passes through the receiving device. This type of correcting circuit is known as an *inductive shunt,* sometimes as a *magnetic shunt.*

3.3 The d.c. transmission line

A transmission line possesses four primary coefficients, resistance R, capacitance C, inductance L and leakance G. With the high-insulation resistance of underground cables the d.c. leakance is small, and so is the natural inductance of the line. Both resistance and capacitance have important influence over propagation of d.c. energy along the line.

Telegraph signals are applied to a d.c. transmission line as abrupt changes in potential. If the transient time, that is the time required for current to build up to the steady value at the receiving end, is less than the duration of the shortest signal element there will be no telegraph distortion due to the line, and the problem is merely one of providing sufficient current to secure reliable operation of the receiving relay; if, however, the build-up time is longer than this there may be some distortion. The stage at which distortion arises is mainly dependent upon the ratio of the operating current of the receiving relay to the steady current, but it would usually be reached when the build-up time approaches 1·5 times the duration of the shortest signal element. When the line conditions become of importance in determining the quality of transmission, the signals are almost wholly composed of *transients.*

Transients arise in a transmission line when, mainly because of capacitance effects, all frequencies do not suffer the same attenuation (i.e. reduction in amplitude) nor the same change in phase during transmission: under these conditions the line does not immediately respond to any change in electrical conditions at the sending end but undergoes a transitory condition while passing from one steady state to another.

Direct-current telegraph-transmission problems are almost entirely concerned with transient conditions and, because the mathematical solution of transient phenomena is very complex, no easily applicable formulae are available for practical use. The transmission technique is largely based upon practical experience supplemented by a knowledge of electrical phenomena derived from mathematical and experimental studies.

In an actual line, resistance and capacitance are distributed, normally uniformly, along it. As a simplification, a length of underground line up to about 20 miles could be represented by the diagram of Fig 3.4(*a*), where the total capacitance, C farads, is depicted as though it were concentrated at the mid-point of the line resistance R. This configuration is called a T-network, from the shape of the graphical representation. It is assumed that a voltage, V, is applied at the sending end and that the receiving end is short-circuited. While the capacitance is uncharged it has no p.d. and so the distant section of the line is short-circuited. The initial current leaving the sending end is $i_S = V \div R/2$; the current at the receiving end is $i_R = 0$. As the capacitance charges, the value of i_S will diminish while i_R will increase until finally, when the capacitance is fully charged and its p.d. becomes $V/2$, then $i_S = i_R = V/R$. The equations governing instantaneous values of currents during the build-up time are:

$$i_S = \frac{V}{R}(1 + e^{-4t/CR})$$

$$i_R = \frac{V}{R}(1 - e^{-4t/CR})$$

The graphs to these equations are shown in Fig 3.4(b). The time-constant is $RC/4$ and the time required for the received current to build up to a given fraction of the final value is proportional to the product of resistance and capacitance of the line.

If the line is now short-circuited at the sending end to terminate a signal element, the capacitance discharges equally into each termination since the circuit is symmetrical, the currents now being

$i_S = i_R = \frac{V}{R}e^{-4t/CR}$. During discharge, current direc-

tion at the sending end is reversed but the received current remains in the same direction. All these conditions are plotted in Fig 3.4(b).

If the voltage is applied to the line through a resistance of $s\ \Omega$ and the line is terminated by a resistance of $r\ \Omega$, the formulae for the received current becomes

$$i_R = \frac{V}{R + s + r}$$

$$\left[1 - e^{\left(\frac{-4\,(R + s + r)}{CR\,(R + 2\,(s + r) + 4sr/R)}\right)\,t}\right]$$

and the time constant is

$$T = \frac{CR}{4}\left(1 + \frac{s + r + 4sr/R}{R + s + r}\right)$$

The time-constant is increased by the presence of resistance at either sending or receiving end of the line; inclusion of resistance will in each case increase the time taken for current to reach the steady value. As the equation is symmetrical in s and r the effect of a given resistor is the same whether it is at the receiving or sending end.

The Arrival Curve. A long d.c. line could be represented by a sufficiently large number of T-sections of the form shown in Fig 3.4(a) joined in series. Though the sending voltage is suddenly applied to the first T-section, the p.d. across this capacitance builds up but slowly to reach its final value. In effect, it is this slowly-developing p.d. across the first capacitance which is applied across the second T-section: as a result the final p.d. across the second capacitance is built up,

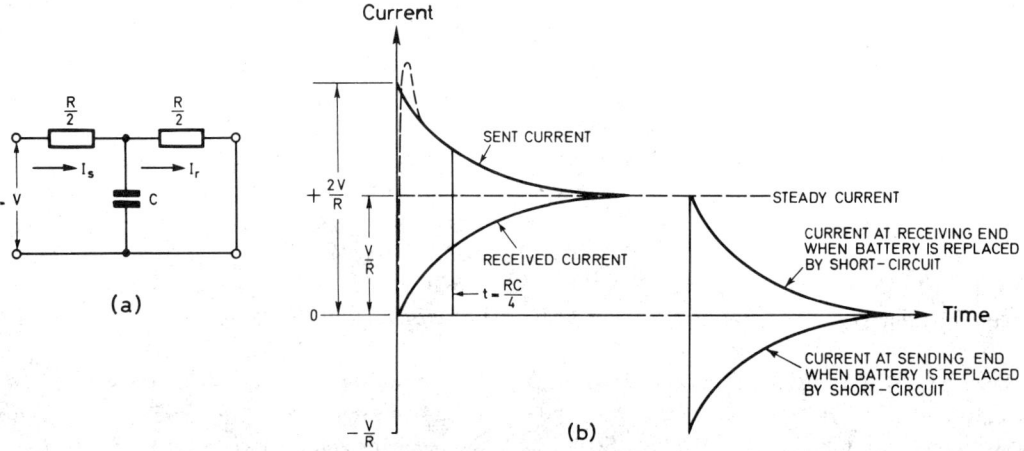

Fig 3.4 Short artificial d.c. line: (a) configuration; (b) sent and received currents

starting from zero, even more slowly to reach its final value. This charging process may be regarded as continuing progressively through section after section of the line. Consequently there is a finite delay – the *silent interval* – while energy is propagated along the line before any significant value arrives at the receiving termination. The graph showing the build up of current i_R at the receiving end when a voltage is suddenly applied to a long d.c. line, previously in a state of rest, is known as the *arrival curve* and it takes the form shown in Fig 3.5; this diagram also shows the shape of the graph of the current i_S at the sending end, the initial value of which is considerably greater than the steady value due to the sudden flow of charging current into the line capacitance.

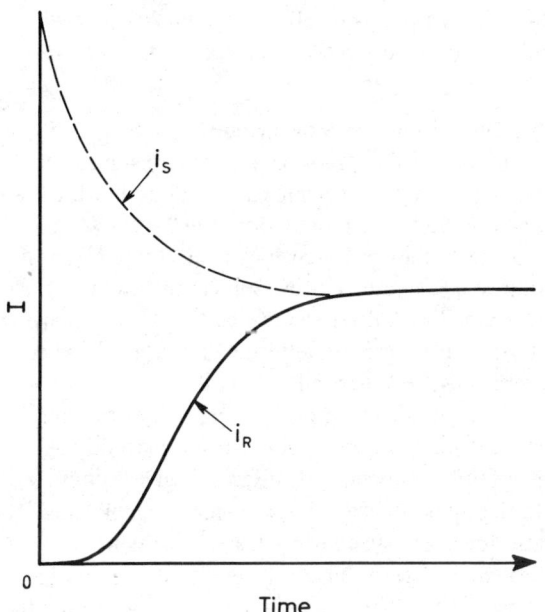

Fig 3.5 Arrival curve

From the classic formula associated with the name of Lord Kelvin, the instantaneous value of current, i amperes, at any time, t seconds, after application of a p.d., V volts, at the sending end is

$$i = \frac{V}{Rl} \left(1 - 2\,e^{-\pi^2 t/CRl^2} + 2\,e^{-4\pi^2 t/CRl^2} - 3 \ldots \right)$$

where C and R are capacitance in farads and resistance in ohms respectively, both *per mile*, and

l miles is the length of line. The product of total capacitance and resistance is CRl^2 and this factor governs the time required for current to reach any specified fraction of the steady-state value. For example, the current reaches half its final value when $t/CRl^2 = 0\cdot14$, or $t = 0\cdot14\ CRl^2$ and the steady state is virtually reached at time $t = 0\cdot55CRl^2$.

This dependence of the time delay upon CRl^2 is known as the 'CR law'.* Its particular significance is its regulation of the delay before a subsequent change of potential can be signalled, while ensuring adequate rise of current at the receiving termination. Although this law has little practical application to short landline systems operated at fixed modulation rates, an understanding of the basic principles is of value in studying d.c. telegraph-transmission problems.

3.4 Single-current transmission

This term is the traditional description for a method of transmission using unidirectional current. Signals are transmitted by making and breaking the circuit, sending pulses of current separated by periods of no-current according to the code being used. Two forms of single-current transmission are available depending upon whether the circuit is normally *closed* in the quiescent condition or whether normally *open*. Of the two, the closed-circuit method is the one usually employed. Because the transmitter and the (home) receiving electromagnet of a teleprinter are usually connected in series for single-current operation, the open-circuit arrangement would necessitate the use of a send–receive switch to enable signals from a distant station to be received. The reason for connecting transmitter and receiver in series is so that transmitted signals will also be recorded on the home receiver to provide the sender with a 'local record' of the outgoing message.

Fig 3.6 shows the basic circuit arrangement and also the signal waveform as generated by a single-current teleprinter transmitter in (*a*) a closed circuit and (*b*) an open circuit; the stop-signal condition is always that corresponding to the quiescent state of the line.

*Actually known as the KR law, K being the earlier symbol for capacitance.

Single-current transmission is rarely used in the United Kingdom but is widely used in telegraph networks of Continental Europe and elsewhere. Advantages of this method are: (1) the circuit arrangement

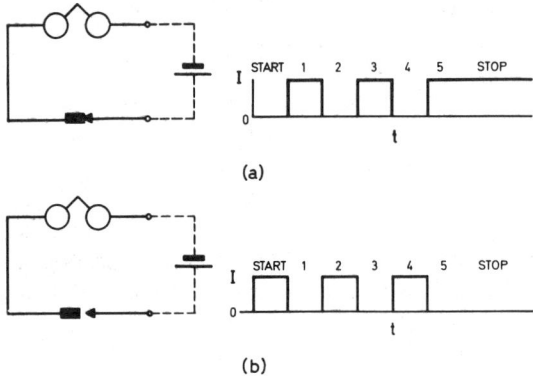

(a)

(b)

Fig 3.6 Single-current circuit: (a) closed circuit; (b) open circuit

is simple; (2) a simple make—break transmitter contact can be used; (3) a simplified transmitter comprising five separate sets of contacts joined in parallel can be used for 5-unit systems (the use of five sets of contacts is no advantage from a maintenance viewpoint); (4) the receiving device will operate whatever the direction of current; (5) a single battery only is required and this can conveniently be located at the central office; and (6) a local record of outgoing signals is obtained simply by connecting transmitter and receiver in series.

Disadvantages of the method are mainly related to the tendency to produce biased signals. The receiving device must be released when current ceases, either by mechanical force or else by local bias current in a second winding; ideally the bias control should be equivalent to that of one-half of the line current but, as the time-constant for the open-circuit condition ($RC/2$) is twice that for the closed-circuit condition ($RC/4$), this method of working leads to unbalanced signals: to counteract this it is necessary to increase the bias current beyond the half-value. With either method of release the restoring force may in service vary independently of line current which produces the operating force, and the difficulty of ensuring at all times the correct balance of forces produces a tendency to instability. When the transmitter contacts break, current falls to zero immediately, but when the

contacts make, current build-up is slow due to inductance of the receiver resulting in asymmetric signals being transmitted. High line capacitance causes considerable delay in decay of current on mark—space transitions. There is a different delay to the build-up of current due to the high-inductance electromagnet in series (unless semiconductor-input devices with resistive input are used): this causes inherent marking bias on long-loop circuits. It is often necessary to use a relay at the central office to provide two-way transmission; this constitutes an additional point at which signal bias may be introduced as well as being an additional maintenance liability. At modulation rates of 100 bauds upwards, single-current operation is unreliable, and the double current mode has to be used, because of difficulties with balance networks which are necessary in association with the 2-wire/4-wire conversion relays in the telegraph exchange.

3.5 Double-current transmission

In double-current transmission, binary signals are distinguished by reversing current direction for the two conditions of modulation. Fig 3.7 shows the essence of the method, which needs two batteries (unless a separate one for each circuit and a reversing transmitter were to be used). The receiving device must be sensitive to the direction of current, i.e. it must be polarized.

Though not so simple in execution as the single-current method, the double-current method has marked transmission advantages. It gives symmetry to the mark—space and space—mark signal transients; also the termination of each signal element is hastened by reversed current which discharges line capacitance more quickly than by disconnecting the circuit. Changes in line-insulation resistance would not produce signal bias such as occurs with single-current transmission. There are no variable biasing

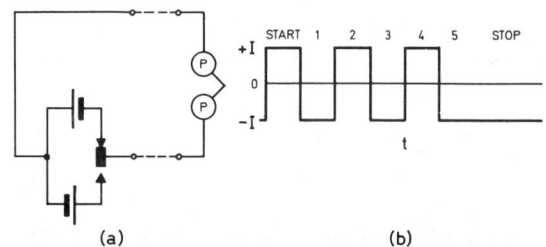

(a)   (b)

Fig 3.7 Double-current circuit

devices, operation and release of the receiving relay or electromagnet being brought about by identical, but opposite, current conditions. Receiving devices are given a 'neutral' adjustment which is their most stable condition. The tendency to residual magnetism in magnetic circuits is less marked than with single-current operation. As the receiver is always carrying current it is less liable to influences of leakage or inductive disturbances, compared with the single-current circuit when no current is flowing.

Double-current modulation does not readily allow signal transmission in both directions over a normal 'loop' circuit or pair of wires: two cable pairs would be required, one for each direction. For this reason, double-current operation (at 50 or 75 bauds) in the United Kingdom is provided over a separate, single wire for each direction of transmission, using the earth as the return path. The earth is a vast conductor of negligible resistance between any two points (apart from certain areas where the local resistivity of the earth may be high); the resistance of the earth-return path is virtually that of the connections themselves.

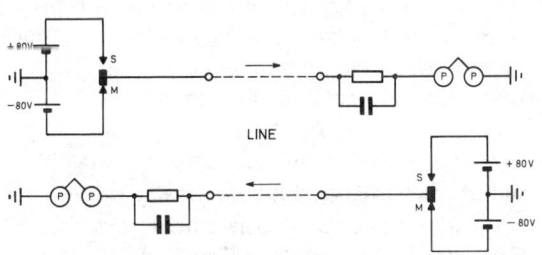

Fig 3.8 Two-way simplex teleprinter circuit (double-current)

At any central office it is economical, where possible, to use a single power installation of adequate capacity to serve all circuits. For double-current operation using earth-return circuits two batteries (or other source of power) are joined in series, the common centre-terminals being joined to earth (see Fig 3.8); alternatively, the arrangement may be regarded as a battery of twice the required voltage with a tapping at the centre-point connected to earth. The same arrangement is adopted at all teleprinter stations, irrespective of whether the power supply feeds one or more telegraph machines.

The voltage standardized by the British Post Office for telegraph operation is 80 V; elsewhere other voltages are used, for example 48 and 60 V in Continental European systems and 120 V in North America. The reason for the original choice of 80 V was that the applied voltage must be higher than that which is just adequate to give the steady-state current required by the receiving device, in order to overcome the delaying effects of inductance and capacitance. When teleprinters were first introduced to the network, all large central offices were equipped with a power installation which comprised a double-current 'universal' battery of −120 V + 120 V with tappings at ± 24, ±40, ±80 and ±120 V; these were used respectively for local circuits within the office, short-distance circuits, longer-distance circuits and the very long high-speed Wheatstone circuits. A 40-V supply was inadequate for teleprinter circuits while 120 V was regarded as excessive. Unlike a relay, the teleprinter electromagnet has to perform an appreciable amount of mechanical work, the energy for which is not adequately supplied at low voltage. With the teleprinter electromagnet connected directly to line, experiments showed that transmission at ±60 V appeared to give the optimum performance; consequently, ±80 V was selected for operating teleprinter circuits.

The basic telegraph circuit standardized for two-way operation over a pair of wires using the earth-return path is shown in Fig 3.8 and is symmetrical for transmission in two directions. At each transmitter, a battery (or mains-driven rectifier unit) supplies a voltage of 160 V, centre-tapped to give ±80 V with respect to earth potential. In the United Kingdom +80 V is used for the space/A/ start condition, −80 V for the mark/Z/stop condition.

Earth-return circuits are prone to provoke inductive interference into adjacent parallel circuits and are themselves sensitive to reception of interfering signals. With the precautions adopted, both senses of interference are controlled to harmless magnitudes.

Use of the abbreviation 'd.c.' to indicate double current is deprecated on account of confusion with the general use of d.c., meaning direct current; therefore, throughout this book any reference to d.c. is to direct current.

3.6 Simplex circuit

Over a simplex circuit between two stations information is transferred in one direction only at a time. If the transfer is permanently in one direction, the arrangement is called 'one-way simplex'. On the other hand, the direction of transmitting information may be reversed from time to time, messages being sent in each direction *alternately*: this two-way arrangement is called 'two-way simplex' (sometimes, *half-duplex*).

circuit shown may be adapted for different requirements: for example, message-relay systems use one-way simplex operation obtained by omitting the use of a send–receive switch, using only one channel – one direction of transmission only.

3.7 Duplex circuit

A duplex circuit is one which permits transmission of telegraph signals in both directions *simultaneously,*

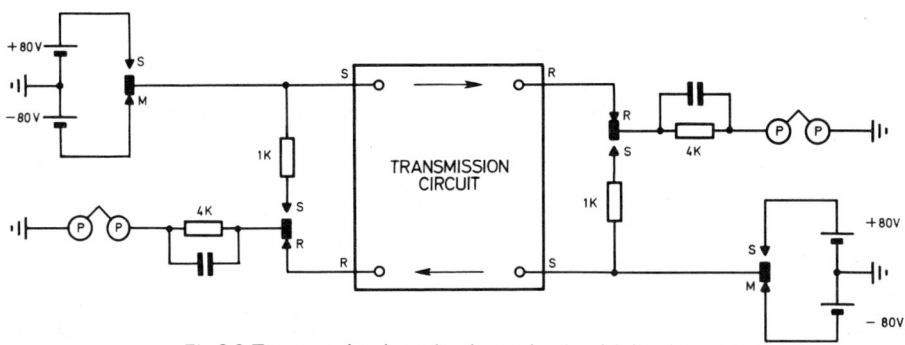

Fig 3.9 Two-way simplex teleprinter circuit with local record

Fig 3.9 shows basic principles of the two-way simplex teleprinter circuit. This is the standard arrangement provided in the United Kingdom for leased circuits, telex circuits and most circuits on the general telegraph service; it is symmetrical for each direction of transmission–*disymmetrical*. For present consideration the block marked 'transmission circuit' should be regarded as containing two metallic conductors for d.c. transmission, but the block could equally well represent a long-distance circuit provided by voice-frequency multiplex, HF radio, submarine cable or communication satellite. Signals can be transmitted from either station alternately; each time a key on the keyboard is depressed, the send–receive switch SR operates mechanically to shunt the local receiver across the circuit comprising transmitter and send wire to provide a local record of outgoing signals, at the same time disconnecting the receive wire. If both stations were to attempt to send at the same time both receive wires would be disconnected at intervals and, as transmissions from the two stations would usually be out of step, each receiver would print mutilated signals. As shown, the circuit provides for two-way simplex operation, the usual requirement. The basic

using the modulation of a continuous current. The duplex *circuit* is essentially a d.c. circuit, but the term duplex *operation* has come to mean operation of a telegraph circuit in both directions simultaneously, using any method of modulation – d.c. or a.c. (voice-frequency) systems.

The d.c. duplex circuit depends upon establishment of a balanced circuit arrangement and two basic methods of doing this have been used, the *differential* duplex and the *bridge* duplex systems. These circuits are now virtually obsolete.

Half-duplex,* a term originally introduced to describe a d.c. duplex circuit in which the terminal connections were arranged to inhibit both stations from receiving signals simultaneously, is now sometimes used to refer to the normal method of providing telegraph service, in which two disymmetrical simplex circuits are provided but transmission takes place in one direction only at a time, using one

---

* In the U.S.A., *simplex* means transmission in one direction only. *Half-duplex* means transmission in either direction (but only in one at a time). The term *full duplex* is used for duplex transmission – transmission in both directions simultaneously.

channel or the other according to the desired direction of transmission. This is in fact the two-way simplex circuit referred to earlier.

If a local record is required with duplex operation, two teleprinters are necessary at each station, one being used for reception only. All normal telegraph-type circuits in the United Kingdom networks are *capable* of duplex operation — it is only the arrangements at the terminal stations which decide the form of operation to be used. This is clarified in

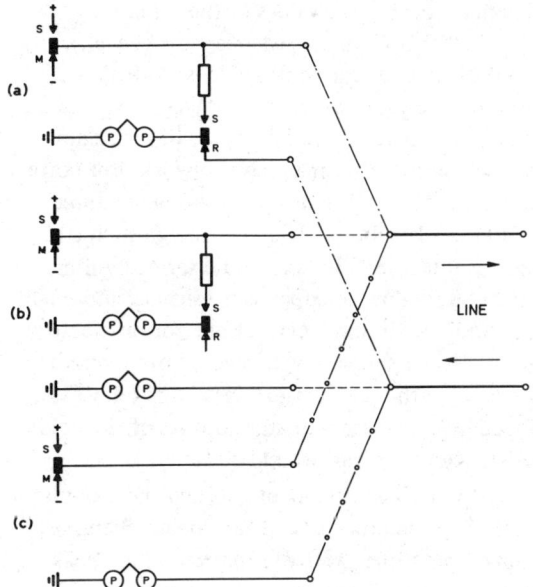

Fig 3.10 Teleprinter circuit terminations: (a) two-way simplex (half-duplex) with local record; (b) duplex with local record; (c) duplex without local record

Fig 3.10, which shows two-way simplex ('half-duplex') operation, and duplex operation, with and without local record, all capable of operating over a normal telegraph circuit.

3.8 Protective equipment
Voltages as high as ±80 V cannot be applied directly to lines at modulation rates of 50 bauds and upwards without including certain safety devices, not only for protection of telegraph apparatus against possibility of damage from excessive currents but to guard against propagating interference energy by line or radio. The essential ancillary devices are described in this section and

their proper use ensures satisfactory telegraph transmission with complete safety and immunity against causing interference with other circuits.

The Barretter. Even with a fuse in circuit it is undesirable to apply potentials of ±80 V to transmitter and relay contacts for modulation of line currents without including sufficient resistance to limit current to a safe value to prevent overheating or fire risk in the event of a faulty condition arising; faults which may occur in service are a 'full' earth contact on a line wire or a short-circuit across mark—tongue—space contacts. Inclusion of a high value of resistance in the transmitter circuit has to be avoided because this impairs transmission of telegraph signals.

In the barretter — also known as the *ballast resistor* or *resistor bulb* from the form in which it is used — a useful resistance characteristic is available for this purpose: at the normal low value of line current, approximately 20 mA, the cold resistance of the barretter is low, about 100 Ω but if, due to a fault, the full 80 V is applied across it, its resistance increases and current is limited to a safe value of about 100 mA. The barretter used in telegraph circuits consists of twin tungsten filaments in an atmosphere of nitrogen within a glass envelope about 2 in. high. The tungsten filament has a high positive temperature coefficient of resistivity; to some extent this increase in resistance, when the filament temperature is raised by the fault current, is influenced by pressure and density changes of the gas adjacent to the heated filament. Under fault conditions the filament glows sufficiently to indicate presence of the fault.

Ballast resistors are invariably included in the ±80 V feeds to relay and transmitter contacts. In complex switching-circuit diagrams these resistors (as well as fuses) have been frequently omitted in the interests of clarity, but their presence in the equipment should be assumed.

Low-pass Filter. Any square-wave signal can be analysed as a sinusoidal component at fundamental frequency together with other components at all *odd* harmonic or multiple frequencies. For example, at a modulation rate of 50 bauds, equivalent to a frequency of 25 Hz, the waveform is built up from components at 25, 75, 125, 175, 225, 275, ... Hz, the amplitude of these harmonic components falling as frequency rises.

Telegraph and telephone circuits are routed in the same local cables but, due to capacitive coupling between conductors in the cable (see Fig 3.11),

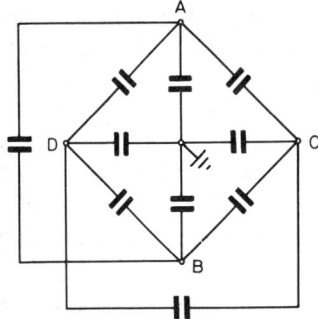

Fig 3.11 Capacitive coupling in a cable quad

energy from audible harmonic frequencies of telegraph signals would be induced into telephone circuits working on other conductors in the cable and cause interference or *crosstalk*. It is essential to prevent such interference and for this purpose a low-pass filter is included in series with every transmitting or modulating contact which feeds d.c. telegraph signals into a cable circuit. The low-pass filter attenuates higher-frequency components of the modulation so that crosstalk interference into adjacent pairs in the same cable is reduced to acceptable limits. The nominal cut-off frequency of the filter used (Post Office No. 4B) is 130 Hz; the attenuation/frequency graph for this is

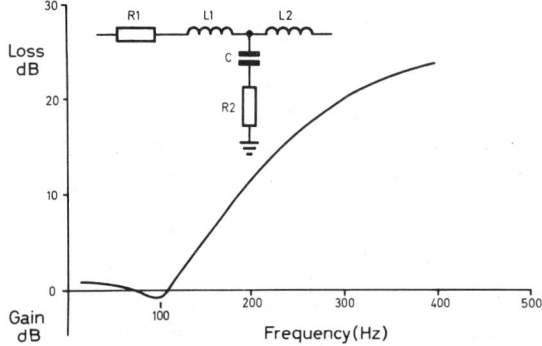

Fig 3.12 Attenuation/frequency characteristic of frequency filter No. 4B

reproduced in Fig 3.12, showing also its configuration. The filter is designed to produce insignificant characteristic telegraph distortion if connected in series with a similar filter: this condition arises on certain switched connections because not only must a filter be used with the transmitter at every teleprinter station, but a filter must also be inserted in all d.c. line circuits leaving the exchange, since these lines may be carrying service signals generated in the exchange, relayed signals, or signals from the receive relay of a voice-frequency channel. Component values of the filter are $L_1 = L_2 = 1 \cdot 3$ H (d.c. resistance = 50 Ω each), $C = 2$ μF. The design impedance is 1140 Ω – a compromise value which gives the best average impedance match to various types of line plant into which the filter may have to work. For correct matching, impedance of the modulation source should be also about 1100 Ω although such a high value is unacceptable as it would seriously limit current available for supervisory signals, and might also limit maximum length of line for satisfactory telegraph transmission which could otherwise be operated without resort to a repeating relay. The impedance of the modulator consists of the cold resistance of the barretter (100 Ω).

The circuit consisting of $L_1 C$ and the modulator forms an oscillatory circuit: an abrupt change of applied potential (±80 V) produces a high peak voltage followed by a steep dip or trough which, in extreme cases, may even temporarily reverse the polarity of the signal at this instant. These troughs occur at the midpoint of a signal element – the instant at which a receiver examines the polarity of an incoming signal. It is essential to damp this oscillatory 'ring' by reducing the Q-value of the circuit, which may be achieved by inserting suitable resistance at R_1 or R_2 in the $L_1 C$ circuit. There are practical problems associated with including R_2 (for which 510 Ω is found to be a suitable value) and the use of $R_1 = 200$ Ω has been adopted as a good compromise value.

Fig 3.13 shows typical waveforms relating to a start–stop 5-unit signal at 50 bauds transmitted over a d.c. circuit. At (*a*) are shown voltage variations from an ideal transmitter; (*b*) shows the waveform of transmitted current; the troughs which follow current build-up are due to the oscillatory nature of the filter circuit. At (*c*) is shown received current

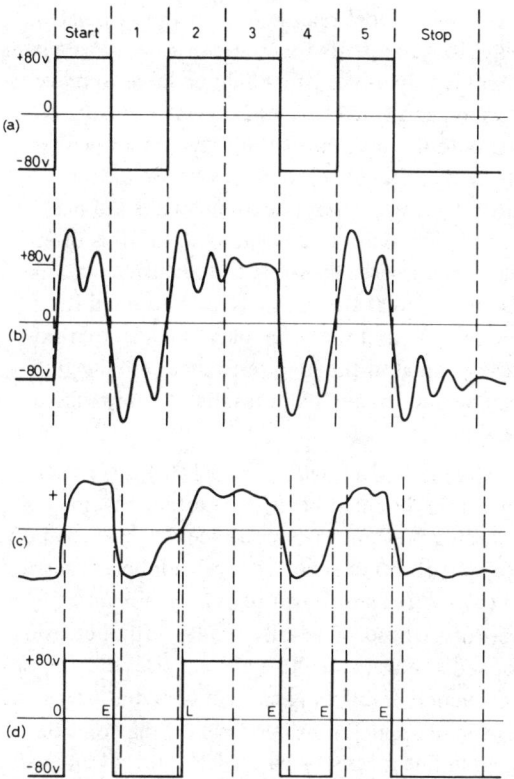

Fig 3.13 Waveforms in 50-baud teleprinter d.c. circuit:
(a) transmitter voltage; (b) sent current; (c) received
current; (d) restituted signal

(to a scale expanded in comparison with that used
for (b)) and at (d) is shown the restituted signal, the
significant instants of restitution being drawn by
reference to the zero crossing of received current
waveform. The start signal suffers early distortion;
other significant instants suffer early (E) or late
(L) distortion as indicated.

The line current received by a teleprinter electromagnet will depend upon line resistance. On circuits
composed of lightweight conductors, the static
value of this current can fall to about 8 mA under
adverse conditions. Even so, peak current (due to
the signal-shaping network) is of the order of 30–
40 mA, sufficient to operate the electromagnet
reliably. This low value of steady current has no
serious effect on reception unless it falls so low
that it is exceeded by induced fortuitous current
peaks such that intertransition misoperation of the
electromagnet occurs.

The spark Quench. When a current is interrupted
by opening a contact, the energy stored in an inductive device ($\frac{1}{2}LI^2$) or in any capacitor present,
including self-capacitance of the wiring ($\frac{1}{2}CV^2$), has
to be dissipated. When the interruption occurs, a
high voltage is induced which may be of sufficient
magnitude to break down the air dielectric, and a
spark or an arc is formed. Spark breakdown will not
occur unless voltage across the air gap exceeds
300 V. Conduction across the gap is by ionization
of air and vaporization of contact metal. The spark
is extinguished when the gap voltage is too low to
maintain the discharge.

An arc is a relatively large discharge current
across the air gap. For the duration of the arc, the
voltage across it is stabilized at about 15–20 V.
Conduction is largely due to vaporized contact
metal. The arcing may be a momentary continuation of the load current or a re-establishment of
current following spark breakdown of the air gap.
A molten bridge forms between the contacts and
this maintains current until either ruptured
mechanically or boiling of contact material takes
place. Extinction occurs when the energy transmitted via the contacts is insufficient to maintain
the discharge. For a given applied voltage, each
metal has a characteristic current value which, if
exceeded, will cause arcing when a resistive circuit
is opened.

The number of sparks and arcs at each contact
opening may be considerable, with high instantaneous values of current and voltage. High-frequency
oscillations may also be set up, capable of causing
interference with nearby radio receivers.

When contacts are closing, and the air gap becomes extremely short, the electric field may cause
a cold discharge of electrons to bridge the gap. The
bridge is usually ruptured by a relatively heavy discharge current from the reactance of the wiring.
The discharge takes the form of a damped oscillatory
current and inevitably causes contact damage as well
as being a source of radio interference. Contact
bounce and chatter naturally aggravate these ill-
effects.

Protection of contacts is important both as
regards prevention of contact failures and avoidance
of interference. The aim of a spark-quench circuit is
to limit the voltage across the opening contacts to
not more than 300 V. It is assumed that when con-

tacts open, the load current I is diverted instantaneously into the quench; voltage IR across the resistive element of the quench should not exceed 300 V, including any standing voltage (e.g. ±80 V).

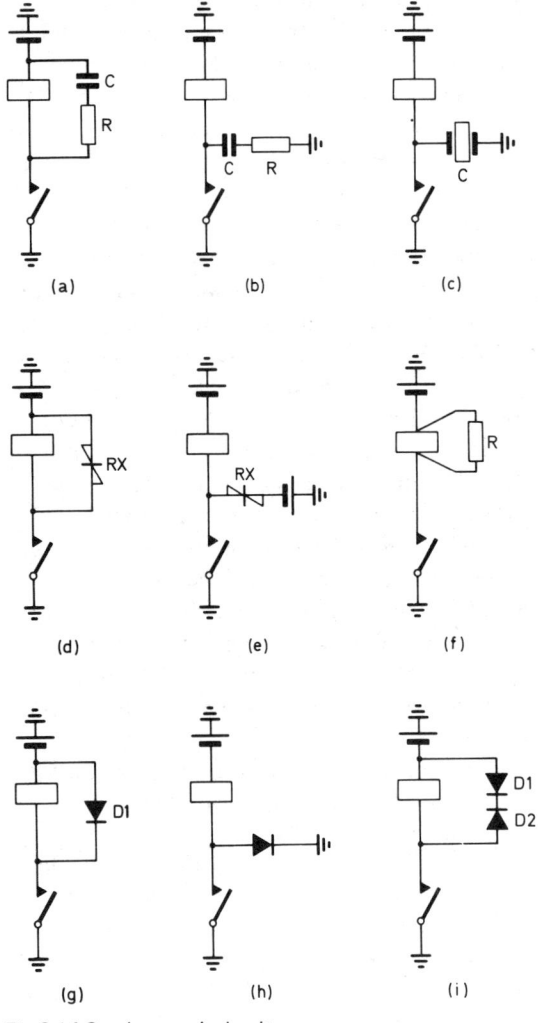

(a)   (b)   (c)

(d)   (e)   (f)

(g)   (h)   (i)

Fig 3.14 Spark-quench circuits

Various forms of spark quench are shown in Fig 3.14. The RC quenches are very effective in reducing contact erosion provided that values of R and C are correct. The quench circuit provides a bypass for continuation of circuit current, energy being dissipated by the resistance. The presence of resistance is essential to limit capacitor current on closure, otherwise contact welding could well occur; with the presence of coil inductance, the decaying current will be oscillatory. The quench

circuit may affect release time of the associated relay. Requirements for values of C and R are somewhat incompatible – C should be large to prevent sparking and small to avoid excessive closure current, R should be small enough for adequate quenching and large enough to prevent excessive closure current – for most applications values are not critical. The minimum value for C in μF is often taken as numerically equal to the current in mA disconnected at the contacts, and doubled for a safe margin, and the value of R is made approximately equal to the load resistance. Final values may be determined from results of comparative tests.

The method shown in Fig 3.14(a) is convenient for wiring, but that at (b) has certain advantages including prevention of noise feedback to the power supply; (c) shows a capacitor with inherent resistance. At (d) and (e) are shown non-linear silicon-carbide resistors, whose value falls rapidly with increasing applied voltage, and these are less effective than the RC quench in reducing contact wear but useful in light-load applications. Under working conditions the non-linear resistor passes leak current and cannot be connected across the contact; it also significantly delays release of an associated relay. The use of a non-inductive shunt as at (f) is simple and reliable, but disadvantages are additional energy loss, increased load and considerable increase in delayed release of the relay – a non-inductive shunt often being used for the latter object. A diode used as in (g) has a comparable effect to that of the non-inductive shunt resistor but without the waste of energy or increased load; the diode is also without the harmful charge–discharge currents which occur on closure with the RC quench. With contacts closed, the diode is in the high-resistance or reverse direction; when the contact opens, induced e.m.f. is in the forward, low-resistance direction. Armature release-time is notably increased. With the diode across the contact, as at (h), quenching takes place only while the coil voltage exceeds the avalanche breakdown of the diode; it is less effective than at (g) but the relay-release delay is less. Using back-to-back diodes as at (i), quenching takes place only when the avalanche breakdown of $D2$ is exceeded: it is less effective than at (g) but delay effect on relay release is reduced.

Two forms of standard spark-quench circuit

used in telegraph circuits are shown in Fig 3.15. The arrangement at (*a*) is used with polarized relays – 1000 Ω in series with 0·5 μF for each contact; at (*b*) is shown a single spark-quench circuit which serves both M and S contacts – the standard arrangement for teleprinter contacts working on an earth-return circuit. Fig 3.15 shows also the method of feeding ±80 V supplies through ballast resistors (fuses not shown), and inclusion of the low-pass filter 4B (with 200-Ω matching resistor) in the send-wire circuit.

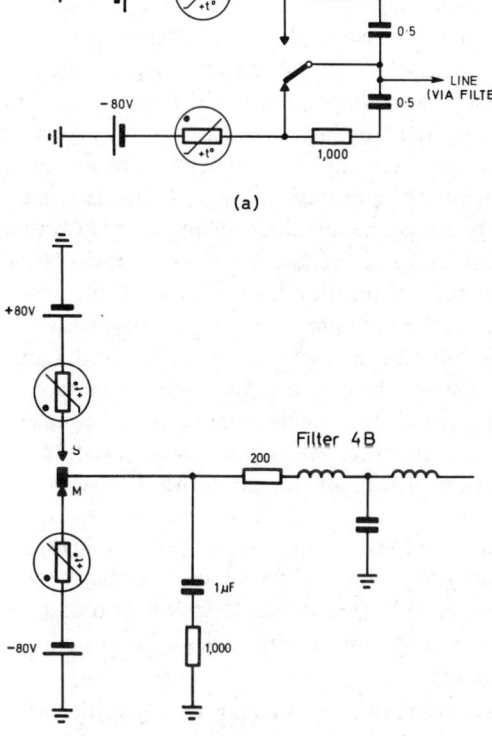

(a)

Filter 4B

(b)

Fig 3.15 Spark-quench, barretter and filter circuits: (*a*) polarized relay contacts; (*b*) teleprinter-transmitter contacts

Elimination of high-frequency discharges at contacts results in marked reduction in contact damage as well as in radio interference. Radio-interference-suppression (RIS) devices commonly used on contacts of teleprinters and polarized relays have proved very effective in extending the life of heavily-used contacts, such as those used in some teleprinter broadcast systems.

Radio-interference Suppression. The series of odd-harmonic frequency components of a square-wave signal extend into the spectrum of frequencies used for radio transmissions; although amplitude is small, these harmonics are a potential source of interference to radio reception. In addition to that from contacts of a telegraph transmitter or relay, interference can also result from rapid current changes which take place at commutator and governor contacts of the commutator-type electric motor which may be used to drive a teleprinter. The extent of interference from harmonic frequencies is influenced by the presence of resonant circuits due to inductance and capacitance present in the circuit.

Radio interference may be propagated in two ways: (1) by direct radiation from wiring in the vicinity of the source of disturbance; a short length of wire may act as an efficient radiator of radio-frequency energy. In a commutator motor, interference is radiated from armature and brushes and is accentuated if sparking is present. (2) With a motor driven from electric supply mains, interference is conducted along the supply wires as well as being radiated from them; interference may be picked up by receiving aerials at points near to the route of the power-supply cable.

It is preferable, since most effective, to suppress radio interference at source. Theoretically it is possible to prevent propagation of interference energy from any source by means of an electromagnetic shield and radio-frequency filters (Fig 3.16). The shield prevents direct radiation of energy from the source and the filter stops interference currents from spreading along power-supply wires. The ideal electromagnetic shield

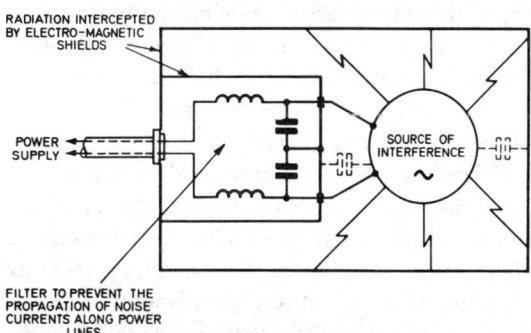

Fig 3.16 Principle of radio-interference suppression

would be a complete enclosure of continuous metal-sheet having zero impedance between any two points on its surface. Joints and removable sections of the shield are unavoidable and these discontinuities reduce attenuation afforded by the screen. In the split-metal casing of a teleprinter motor, maximum possible screening is ensured by a continuous low-resistance electrical contact over the whole of the joint. At radio frequencies, magnetic shunting of the shield is negligible; attenuation of the radio-frequency field is due entirely to currents induced in the shield material together with the cumulative effect of currents induced around the circumference of the shield (acting as a short-circuited transformer winding) and to eddy currents induced in the thickness of the shield material ('skin effect'). Such a shield is normally earth-connected in the interests of safety against electric shock, but earthing contributes nothing to attenuation of the radiated field.

It is seldom possible to achieve complete shielding of both the interfering source and associated wiring and the practice is generally to screen the interfering source and also to insert radio-frequency low-pass filters as close as possible to this source, bearing in mind that connecting leads are not only capable of acting as direct radiators but also as undesirable inductances in series with the capacitances.

Radio-frequency currents are propagated along wires in two ways: (1) *symmetrically,* i.e. out along one conductor, returning to the source by another, but at radio frequencies the two conductors are not usually sufficiently well balanced and terminated for interference currents in the two wires to become neutralized, and an asymmetrical component is developed; (2) *asymmetrically,* i.e. travelling in the same direction simultaneously along mains-supply conductors, returning to source via distributed capacitance to earth of the conductors and via capacitance of the earth frame-connection of the motor. These 'mains-borne' interfering currents are conveyed some distance along supply wires, and this occurs particularly at lower radio frequencies since higher-frequency currents suffer higher attenuation; on the other hand, the radiation efficiency of comparatively short lengths of conductor is high at higher frequencies and correspondingly low at lower frequencies. As a result, low-

frequency interference is primarily mains-borne while high-frequency interference is primarily directly radiated by wiring close to the source of interference.

A suitable filter would consist of low-impedance capacitors connected between supply conductors and the metal shield or frame of the source and also between the conductors themselves to shunt radio-frequency currents back to the source; alternatively, or combined with the capacitors, high-impedance inductors are connected in series with the conductors. The physical and electrical dimensions of inductors and capacitors used and their positions in the circuit depend upon the frequency concerned, mechanical construction of the interfering device and its associated wiring. Shunt capacitors are most effective when impedance of the source and of the supply circuit are high, whereas inductors are most effective where impedances are low. The maximum value of capacitor used may have to be restricted by safety considerations where connection between a.c. mains leads and exposed metal frames is concerned. Inductors are practicable only for small currents because they become very bulky and expensive when currents in excess of 5 A are involved. Capacitors should be placed so that large circulating radio-frequency currents in them are not coupled closely with other wiring and inductors should be placed so that input and output wires are well separated. The value of capacitance used varies from 1 μF (when safety permits such a value) to about 100 pF for suppression of frequencies between 150 kHz and 100 MHz. Values of inductance used for similar frequencies range from 2 mH to 10 μH.

Inclusion of capacitors is generally sufficient to suppress interference in low and medium-frequency radio-broadcast bands; for HF, VHF and UHF bands it is necessary to include inductors.

Standard arrangements for radio-interference suppression (RIS) applied to interfering sources of a teleprinter are shown in Figs 3.17 and 3.18. The source of radio-frequency (RF) currents in a typical teleprinter with a governed motor are (*a*) transmitter T contacts and associated send–receive switch SR, (*b*) commutator brushgear, (*c*) governor contacts; this latter interference is of considerable amplitude as comparatively large changes of current are involved. A typical form of suppression suitable

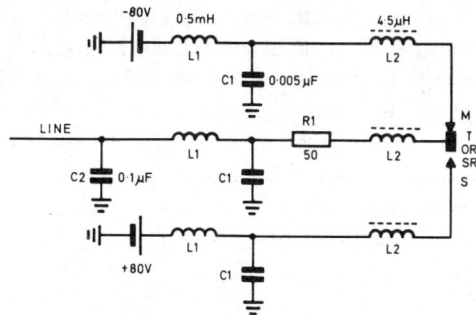

Fig 3.17 Radio-interference suppression — transmitter and send—receive switch contacts

for either transmitter contacts or send—receive switch is shown in Fig.3.17. Paper capacitors $C1$ and associated air-cored inductors $L1$ constitute low-pass filters which effectively attenuate RF currents below 30 MHz before they reach the line and the signalling power source. Resistor $R1$ limits radio-frequency current passing through the contacts, and capacitors $C1$ reduce contact erosion due to this current. The additional capacitor $C2$, which provides a low-impedance path to earth for RF currents, is necessary to reduce RF voltage applied to the line to acceptable limits, particularly in the range 200 kHz to 3MHz. For adequate suppression above 30 MHz, at which the $L1$ and $C1$ components are largely ineffective, inductors $L2$ are used

which are wound upon iron-dust cores and inserted as near as possible to the contacts; these present a high impedance to RF currents above 30 MHz.

Interference below 30 MHz, caused by sparking at commutator brushes, is reduced by capacitors $C3$ connected from each brush to the earth-connected motor frame (see Fig 3.18(a)). Above this frequency, interference is prevented by dust-cored inductors $L3$ in series with brush connections; connection of the armature between the two sections of series field-winding also assists. Capacitors $C3$ may be omitted if a mains filter is provided separately.

For governor-contact suppression, capacitor $C4$ and air-cored inductors $L4$ provide low-pass filtration effective up to 30 MHz; dust-cored inductors $L5$ attenuate RF currents above this frequency. These inductors are usually mounted for convenience outside the rotating governor, in series with the governor slip-ring brushes; although this is somewhat remote from the contacts, the inductors are sufficiently effective in this position. The purpose of resistors $R2$ is to limit RF current circulating through the contacts via capacitor $C4$.

In some equipment the foregoing arrangements are insufficient for the motor and governor to obtain adequate suppression as measured at the mains input to the teleprinter; as an additional precaution a mains filter is connected in series with the power cord. The components of this filter [Fig 3.18(b)] are enclosed in a metal case connected to the machine base and so to earth.

3.9 Transmission limits

When planning a telegraph network, whether for direct point-to-point circuits connecting two offices permanently or for station line circuits connecting teleprinter stations to the nearest exchange from whence they will be extended over other links to form through connections, it is necessary to specify the maximum length of d.c. line which can be permitted in various positions in the network to provide good transmission quality. The d.c. circuit is sometimes referred to as a *physical* or *metallic* circuit to distinguish it from a telegraph circuit provided by multiplex plant using a.c. (voice-frequency) transmission.

Telegraph circuits and networks are designed and planned so that signal distortion which results under

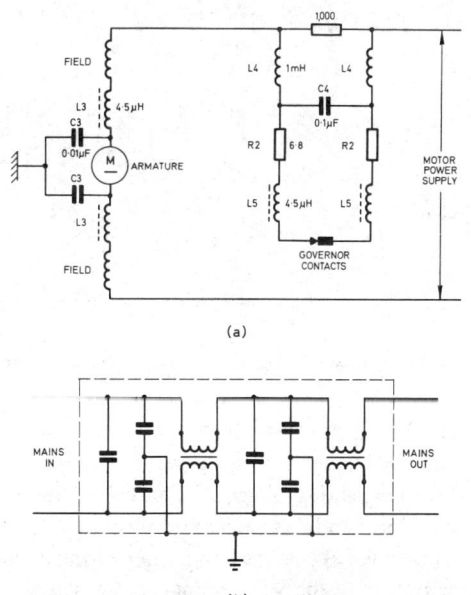

Fig 3.18 Radio-interference suppression: (a) motor commutator and governor contacts; (b) mains-cable cord

adverse conditions shall not exceed a specified limit. For any overall transmission path permitted distortion should not exceed 30% (P4),* in case it exceeds the minimum receive margin (35%) of the teleprinter, making due allowance for distortion in the transmitter. The transmission circuit should then make satisfactory reception always possible. Transmission limits which have been specified (see Table

Telegraph circuits, whether permanently-established direct circuits or switched through an exchange, may be operated from end to end entirely by direct currents if both terminal stations lie broadly within the same local area of a large town; long-distance circuits will not be operated by d.c. throughout their length.

The purposes for which d.c. line circuits are

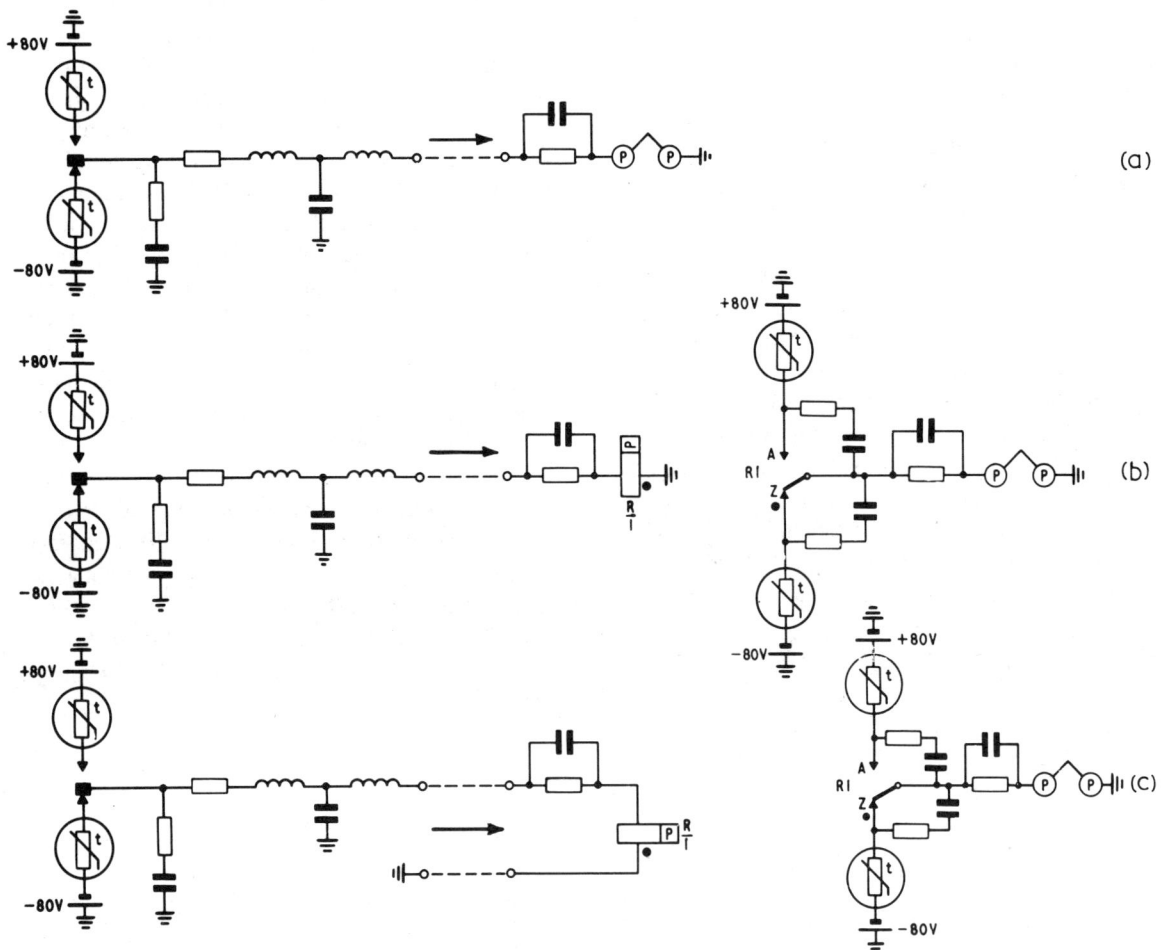

Fig 3.19 D.C. telegraph line circuits; (a) 2-wire unrelayed circuit; (b) 2-wire relayed circuit; (c) 2-loop relayed circuit

3.1) also take into account the need for adequate steady-state line current to operate any supervisory signalling relays present, as well as the teleprinter electromagnet and the static modulator in a voice-frequency system.

*P4 value: one transition in 10^4 is liable to suffer the quoted distortion.

used are: (1) most station lines connecting stations to their local exchange if in the same or an adjacent town; for long station lines, a VF circuit is used with d.c. 'local ends' at each extremity, as in (2) and (3) below; (2) the local ends connecting a teleprinter station to the VF terminal centre which serves it; (3) circuits which link an exchange with a VF terminal centre, if the two installations are in separate buildings in the same town; (4) circuits which may be required to link two VF terminal centres which are in separate buildings in the same large town.

　　With double-current transmission at modulation rates not exceeding 100–110 bauds, and within limiting line lengths which have been adopted, signal build-up times are symmetrical and signal distortion due to characteristics of the line itself is not significant. Using earth-return circuits, fortuitous telegraph-signal distortion is introduced by inductive interference from telegraph (or possibly telephone) circuits working on other conductors in the same cable – mainly between telegraph circuits working in the same cable quad or between adjacent cable pairs in unit-twin cables.

　　The conductors designated A, B, C and D of a cable quad (Fig 3.11) are inductively coupled to each other by mutual capacitances along the length of cable and also to earth; changes in signalling potential on any 'disturbing' conductor can result in high-amplitude potentials affecting waveform of signals being transmitted over a 'disturbed' circuit. The incidence of interference by capacitive induction occurs in a more or less random manner, and the resulting signal distortion is fortuitous; the effect is naturally greater under duplex operation than when signalling over one wire only of a pair.

　　Limiting lengths of line which give good quality transmission under all conditions have been ascertained and specified from a comprehensive series of measurements using artificial cables, with confirmatory measurements on actual cables. Statistical information on distortion distribution has been compiled from measurements by a distortion analyser on numerous circuits set up to present the most adverse conditions likely to be encountered in service.

　　Beyond the limiting length for which the resulting signal distortion is regarded as acceptable, it is necessary to insert a polarized telegraph relay at the receiving termination; the high sensitivity of the relay compared with the heavier armature and mechanical load of the teleprinter electromagnet enables the relay to respond earlier in the build-up period. The limiting line length against fortuitous distortion is approximately doubled by the use of a relay.

　　For the relatively few cases where line length

Table 3.1. Limiting lengths for d.c. transmission lines (simplex operation at 50 bauds)

Position in network	Unrelayed (2-wire)	Relayed (2-wire)	Relayed (2-loop)
1. Station line circuit	0–20*	20–40	>40
2. Circuit with 'negligible' distortion ($\leq 5\%$)	0–10	10–20	>20
3. Circuit with 'low' distortion (< 10%)	0–25	25–40	>40
4. Station line circuit – local end between station and VF terminal centre	as in (2) or (3) as necessary		
5. Station line circuit – local end between VF terminal centre and exchange	0–2†	2–15	>15
6. Trunk circuits – local ends between exchanges and VF terminal centres	0–5	5–15	>15
7. d.c. circuits connecting adjacent VF terminal centres	nil	nil	nil

* This limit is set by the possibility of connecting together two such station lines: this would give a maximum permitted unrelayed line of 40 miles.
† Independent of conductor weight. (This low limit is due to a brief unterminated condition at the exchange, which can occur during switching operations, the disconnection does not arise when a side-stable repeating relay is included in the circuit.)

Notes. 1. Lengths quoted are in route miles.
　　2. The above limits are based upon cable-conductor weights of 20 lb/mile. To convert to 20-lb equivalents, multiply length of 40-lb conductor by $\frac{3}{4}$; 10-lb conductor by 10/7; $6\frac{1}{2}$-lb conductor by 2; 4-lb conductor by 10/3.
　　3. For duplex operation the above lengths should be multiplied by 0·7.

for d.c. transmission must exceed the acceptable limit for a single-wire earth-return circuit, it is necessary to resort to loop working. With a well-balanced loop circuit, longitudinal interference voltages in the two conductors are equal and opposite, and are neutralized within the loop; such a circuit is virtually immune from interference. For two-directional transmission, whether simplex or duplex, it is then necessary to use two loops (four wires).

These three classes of d.c. line circuit (see Table 3.1), standardized in the United Kingdom for telegraph transmission over telegraph-type circuits, are illustrated in Fig 3.19. In practice, most circuits fall into the non-relayed category (*a*); a small proportion are of type (*b*) needing a receiving relay, while the 2-loop case (*c*) is rarely needed at 50 bauds.

In planning networks, application of limiting lengths involves a series of tables to cover different possible positions in the network at which d.c. circuits can occur. The essence of this information for integrating d.c. circuits into an overall VF-operated network is summarized in Table 3.1.

D.C. transmission at modulation rates exceeding 50 bauds

1. *D.C. transmission at 75 bauds.* The circuit components (including the low-pass filter No. 4B) used for 50-baud transmission can also be used for transmission at 75 bauds over earth-return circuits; interference experienced under 75-baud conditions is some 5% greater than at 50 bauds under corresponding conditions, shown by the following typical distortion measurements (*P4*) over a 40-mile (20 lb/mile cable) unrelayed earth-return circuit:

Transmission modulation rate (bauds)	Distortion due to Interference			
	50 bauds		75 bauds	
	% early	% late	% early	% late
50	22	25	25	32
75	23	22	30	32

Resulting from the increased degree of fortuitous distortion, the following limits have been recommended for d.c. transmission at a modulation rate of 75 bauds on earth-return circuits:

	2-wire unrelayed	2-wire relayed	2-loop relayed
'Negligible' distortion	0–10	10–20	>20
'Low' distortion	10–25	25–40	>40

Note. Lengths are in route miles of 20 lb/mile conductor weight. For other conductor weights the conversion factors of Note 2 to Table 3.1 apply.

2. *D.C. transmission at 110 bauds.* Again the same components used for 50-baud transmission (including the low-pass filter No. 4B) could be used for d.c. transmission at 110 bauds over earth-return circuits; the effects of interference and the resulting fortuitous distortion are so much more pronounced that it is always necessary to include a repeating relay in the line circuits. The following line-length limits are recommended:

	2-wire unrelayed	2-wire relayed	2-loop relayed
'Negligible' distortion	–	–	–
'Low' distortion	–	0 – 14	14 – 28
Overall point-to-point	–	0 – 20	20 – 50

Note. As for 75 bauds.

It is probable that if apparatus requiring 110-baud transmission comes into general use, voice-frequency transmission will be used in preference to d.c. methods, on grounds of greater economy and flexibility in planning, and with the advantages of standardization of methods.

3. *Transmission at 200 bauds.* At a modulation rate of 200 bauds, equivalent to 100 Hz, the lowest odd-harmonic frequency (300 Hz) is already in the audible-frequency spectrum and prevention of interference by filtration is out of the question. For this reason the d.c/VF interface for transmission at 200 bauds (possibly also at 110 bauds) and upwards has to be on the user's premises, the d.c. path being limited to a length of a few feet. With VF transmission, duplex operation becomes possible on a single cable pair using different carrier frequencies in the two directions.

References

1 Thorn, D. A. *et al.*, 'Radio Interference', *POEE Journal*, 50, p. 226 (1958); 51 pp. 41, 115 and 203 (1958); 52, p. 43 (1959).
2 Wheele, D. W. E. and Collier, E. G., 'Telegraph Distortion on Physical (D.C.) Lines and Telegraph Machines', *POEE Journal*, 52, p. 61 (1959).
3 Scott, W. L., 'Electrical Contacts in Telephone Exchanges: Contact Opening and Closing Phenomena and Quenching Techniques', ibid. 61, p. 263 (1969); 63, p. 179 (1970); 65, and p. 234 (1973).

4 A.C. Transmission

Telephone circuits are used as the 'bearer' circuits for multiplex telegraph systems and for phototelegraphy. The derivation and characteristics of these circuits are briefly described.

4.1 Noise

Electrical noise, inherent in any circuit, is a limiting factor in successful operation of a telecommunication circuit. Attenuation is not detrimental (apart from its variation with frequency) since losses can readily be made good by gain from an amplifier. When circuit attenuation is so high that signal level at the receiver or amplifier no longer exceeds noise by a sufficient margin, discrimination between signal and noise becomes very difficult. Audible in a telephone receiver as noise, this unwanted energy is also referred to as noise when present in telegraph circuits. Electrical noise has two main sources — that present in nature and that which is man-made.

In the first group is *Johnson noise*, also called *resistance noise, thermal noise* or *Brownian motion of electrons*, and is always present at temperatures greater than absolute zero. This noise is due to the ever-present cloud of free electrons in a conductor which are in rapid motion except when interrupted by collision with molecules or, more rarely, with other electrons. A charge in motion is an electric current so there is a large number of short pulses of current forming an aggregate which is observed as noise. Two outstanding properties of the Johnson noise are that it is Gaussian* and that its power spectrum is flat, i.e. contributions to its mean power are uniformly distributed along the frequency scale: it is known as *uniform-spectrum random (USR) noise*, formerly as *white noise*. There is no known

*Gaussian distribution: the instantaneous magnitude of the effect is distributed in accordance with the normal law of probability.

way of eliminating resistance noise at normal temperatures. Resistance noise power p is $p = kTB$ watts, where k is Boltzmann's constant, T the temperature in degrees Kelvin, and B the bandwidth in hertz over which the noise is measured.

In electronic valves, *shot noise* arises because under saturation conditions the current from cathode to anode is not continuous but is composed of discrete random arrivals of electronic charges.

A third form of natural electrical noise is *contact noise*, observed when current is transmitted across a boundary between two conductors and consequently present in current flow in semiconductors used in transistors. Fluctuations in density of current carriers cause conductivity to vary and lead to noise when current is passed through the semiconductor.

Radio transmission is affected in an irregular manner by atmospheric charges of static electricity.

In the second group is *crosstalk*, energy induced into a disturbed circuit from other disturbing circuits in the same multi-pair cable due to presence of mutual capacitance between conductors in the cable. In a coaxial cable, crosstalk is virtually absent because HF currents are confined almost entirely to the outer surface of the inner conductor and the inner surface of the outer conductor. Crosstalk between circuits is also caused by cross-modulation from other channels in a frequency-division multiplex system of telephony or telegraphy. In multichannel systems, it is important that common-system amplifiers are not overloaded otherwise higher-order modulation products rise rapidly and cause crosstalk in all channels. The station battery is a common impedance to all circuits which it serves, and adequate decoupling is essential. The limit for crosstalk ratio between telephone channels is about 62 dB; on bearer circuits to be used for VF telegraph systems the limit for crosstalk between transmit and receive channels is 40 dB.

The CCITT Recommendation for maximum

permissible noise, at zero level point, based on consideration of a specified hypothetical telephone circuit of length 2500 km, is 10 000 pW. Of this value, 2500 pW is allowed for inter-modulation noise in terminal frequency-translating equipment, the remainder, 7500 pW = 3 pW/km, being allocated to resistance noise in the HF line, together with crosstalk in repeaters.

A recommended value for very long submarine telephone cables is not yet published, but it has been suggested that noise on an intercontinental telephone connection should not exceed −43 dBm0p.* For a circuit of maximum terrestrial length, such as London−Sydney (26 000 km) routed in the Commonwealth transpacific cable (Compac) the above noise value is too great. The design figure which was met for this cable is 1 pW/km overall; at 26 000 km this is equivalent to 26 000 pW or −46 dBm0p. For satellite systems the maximum noise value is 10 000 pW0p (= −50 dBm).

For telegraph systems, noise objectives for bearer circuits on intercontinental cable routes are:

start−stop telegraphy at 50 bauds　　　　−40 dBm0
start−stop telegraphy regenerated
　　at 50 bauds　　　　　　　　　　　−35 dBm0
2-circuit synchronous telegraphs
　　at $82\frac{3}{7}$ bauds　　　　　　　　　−29 dBm0
3-circuit synchronous telegraphs
　　at $123\frac{3}{7}$ bauds　　　　　　　−44·5 dBm0
Facsimile telegraphy (FM)　　　　　　−41·5 dBm0

These figures are based on sending powers of −24 dBm0 per channel (alphabetic telegraphy), and −10 dBm0 (facsimile telegraphy) and refer to bearer circuits of nominal bandwidth 2800 Hz (in 3 kHz-spaced channels).

Other man-made noise sources include *mains hum modulation distortion* from electric power supplies, and *impulse noise*. The former is associated with use of a.c. power units, and interference arises from harmonics of 50 Hz (or 60 Hz) mains supply which become modulated along with wanted signals.

An excessive value of harmonic frequencies from the mains supply causes additional fortuitous distortion in a VF telegraph system due to modulation with the channel frequencies. Fig 4.1 shows spurious frequencies which result from presence of side

*See Appendix C.

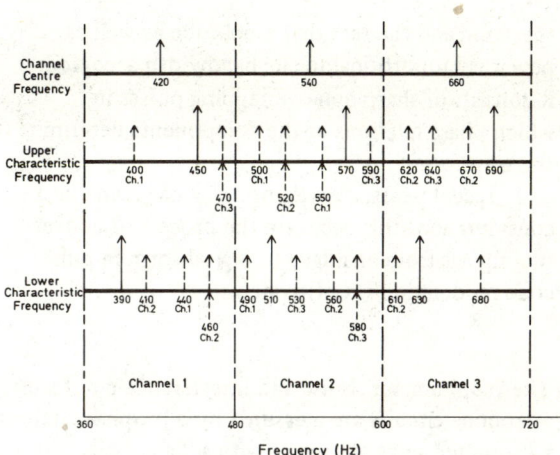

Fig 4.1 FM VF telegraph systems — presence of fundamental and second harmonic frequencies from 50-Hz power supply

frequencies which follow a harmonic series of the power-supply frequency 50 Hz. First and second harmonics of the power frequency modulate upper and lower characteristic frequencies of an FSM system by ±50 and ±100 Hz. Characteristic telegraph frequencies are shown by full lines and mains modulation components are shown by dotted lines. Spurious signal frequencies appear not only in a channel whose characteristic frequency is being modulated by mains harmonics, but channels either side are also affected by the product of telegraph modulation and power-frequency modulation, causing inter-channel interference.

Modulation by harmonic frequencies of the power supply is also detrimental to facsimile telegraphy, causing interference patterns to be present in a received picture.

Power-interference effects arise from inadequate smoothing of a.c. or d.c. supplies which feed carrier-generating equipment used in FDM telephone systems from which telegraph bearer circuits are derived. The level of power-supply harmonics must be at least −45 dB relative to carrier-signal level if interference is to be avoided.

Power-mains interference may also arise from parallelism with adjacent electric power lines through magnetic induction; this can be reduced to harmless values by providing balanced conditions on telecommunication circuits to neutralize power induction, and by other measures.

Impulse noise, which arises commonly from switching operations, is characterized by its wide spectrum and the fact that amplitude as well as power is proportional to the bandwidth accepted: it consists of sharp non-overlapping pulses in which phase relations of the components determine the waveform.

In recent years, *quantizing noise* has come into consideration: this occurs in the process of converting signals from analogue to digital form in pulse-code modulation (PCM) systems.

The Psophometer. Noise and interference e.m.f.s on telephone circuits are measured by a psophometer — a calibrated valve-voltmeter with substantially flat response/frequency characteristic. Of the sinusoidal components of a disturbance, those having the same amplitude but differing in frequency do not have equal disturbing effect upon the ear. For measurements on telephone circuits the psophometer is used in conjunction with a *weighting network* — an attenuator network tuned over the required bandwidth to simulate the joint response characteristics of telephone receiver and ear. The psophometer gives a reading integrated over a period of one minute for telephone circuits — a period related to the 3-minute average holding time of a telephone call. For telegraph circuits, a shorter integrating time, 5 ms, is used: this period is of significance in relation to 50-baud (20-ms) signal elements. These integrating times and characteristics of the weighting network are recommended by the CCITT. A complex voltage is measured in terms of an equivalent single-frequency signal which would produce the same subjective loudness when applied to a normal telephone line and receiver; measurements are usually made at a zero relative level point. The psophometric power of a noise is frequently referred to a power of 1 mW at the zero relative level point, the abbreviation dBm0p being used (p = psophometric).

For telegraph transmission, weighted psophometer readings are of no significance; measurements are made either using a flat, unweighted response, or alternatively a suitable allowance is made. The level of uniform-spectrum (flat) noise power in a 3·1-kHz band must be *reduced* by 2·5 dB to obtain the psophometric-weighted noise power — e.g. a power of −40 dBm0 (unweighted) would be taken as equivalent to −42·5 dBm0p (weighted).

From theoretical study, confirmed by measurements, it has been shown that distribution of telegraph distortion caused by resistance noise follows a normal error law, and the probability of a particular value of distortion being exceeded for a given signal/noise ratio is given by the complementary error function of a function of signal/noise ratio, distortion and build-up time of the signal envelope.

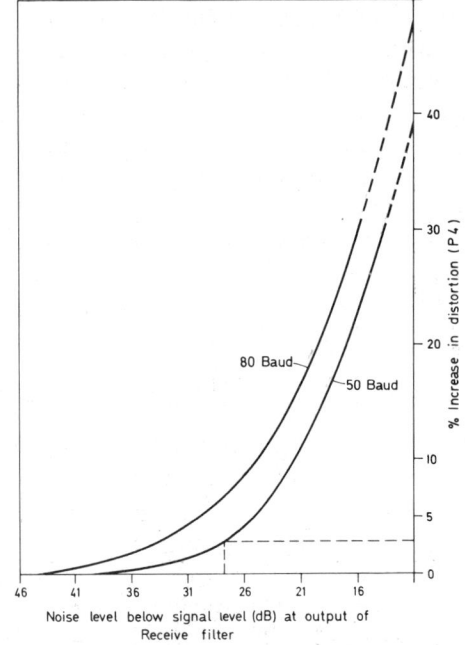

Fig 4.2 Effect of uniform spectrum random noise interference on VF telegraph channel

Fig 4.2 shows the percentage increase in telegraph distortion to be expected (at modulation rates of 50 and 80 bauds in 120 Hz-spaced channels) from progressive worsening of signal/noise ratio.* Suppose that an *increase* in distortion of about 3% could be tolerated due to presence of noise in the bearer circuit; from the graph this corresponds to a signal/noise ratio of 28 dB. If this is the signal/noise ratio measured in an effective channel bandwidth of 80 Hz (in a 120 Hz-spaced channel), this would be equivalent in a 3·1-kHz band to:

$$(28 - 10 \log_{10} 3100/80) \text{ dB} = (28 - 10 \log_{10} 38.75)$$
$$= 28 - 10 \times 1.588 = 28 - 16 = 12 \text{ dB}$$

*The ratio of signal S to noise N is

$$20 \log_{10} \cdot \frac{\text{rms signal voltage}}{\text{rms noise voltage}} \text{ dB}$$

The telegraph-signal level per channel at a relative zero point is −22·5 dBm0 for an FSM system. The noise level in a 3·1-kHz band at a zero relative point could be −22·5 − 12 = −34·5 dBm0. Allowing, say, 4 dB for variations in line level, the required performance should be obtained with a noise value not worse than about −38·5 dBm0 in a 3·1-kHz band.* If the telegraph power level used in transoceanic submarine-cable systems (−24 dBm0) were applied, this value would become −40 dBm0.

The Compandor. The word *compandor* means a compressor-expander, a device designed for the purpose of reducing the effect of noise (by about 25 dB) on long-distance telephone circuits. Though compandors are often present on long-distance circuits in the telephone network, their use is to be avoided on MCVF bearer circuits.

The compandor consists of two units — a volume-range compressor at the transmit terminal and an expander at the receive terminal; the gain of both units is a function of signal level. The inter-modulation distortion which results when two frequencies are applied simultaneously to a compandor increases significantly as the frequency difference falls below 200 Hz. If a compandor were present in the bearer circuit for an MCVF telegraph system with 120-Hz spacing it is possible for the crosstalk between adjacent channels to approach 25 dB and in this instance the presence of a compandor would be undesirable.

The presence of noise on an AM facsimile circuit produces a coarse-grain structure, accompanied by reduction of detail in dark areas of the received picture, an effect referred to as *noisy blacks*. These defects may be considerably reduced by including a compandor in circuit; on a noise-free circuit the presence of a compandor has no significant effect on a received picture.

4.2 Waveform

Use of steady-state theory of a transmission line implies that applied signals are periodic in character. Fourier's analysis shows that any periodic function can be represented as the sum of a series of simple

sinusoidal functions of given amplitude, phase and frequency. In particular, square waveforms common in telegraph signals consist of a fundamental sine-wave together with all the *odd* harmonic (multiple) frequencies:

$$e = 4E/\pi \left(\sin \omega t + \tfrac{1}{3} \sin 3 \omega t + \tfrac{1}{5} \sin 5 \omega t + \dots \right)$$

where e and E are instantaneous and peak values of applied e.m.f., and $\omega = 2\pi f$. The first term represents the fundamental frequency of a wave and has the same frequency as the square-wave reversals but $4/\pi$ times the amplitude. Amplitudes of harmonic components decrease as frequency increases.

The introduction of unwanted harmonic frequencies (e.g. by the non-linearity of an over-loaded amplifier) causes distortion in a signal wave-form. Fig 4.3 (*a*) shows at the dotted line the effect of adding third and fifth (odd) harmonics to a pure sinewave, tending to produce a square-wave. Fig 4.3 (*b*) shows the effect of adding second and fourth (even) harmonics to a pure sinewave, producing asymmetry. If the relative peak values, phase relationships or frequencies of one or more components present in a complex wave are subjected to change in any way the complex waveform will suffer distortion.

Equalizers: The response of any reactive network to an applied alternating e.m.f. varies with frequency. In propagation of complex waves along a transmission line, series and shunt reactive elements produce an attenuation loss which varies with frequency. Distortion which is due to a variation of loss (or gain) with frequency is known as *attenuation distortion*.

The time required to propagate energy over a given length of transmission line varies with frequency, lower frequencies being propagated at greater velocity while higher frequencies will lag at the receiving end. Distortion due to variation of propagation time of the system with frequency is known as *delay distortion*. In Fig 4.3 waveform components are shown in phase: the complex wave-form will be changed — distorted — if phase difference is introduced between components, or again if relative amplitudes of components are changed. In a complex wave, the time taken for fundamental and harmonics to pass from trans-

*CCITT Recommendation G153 quotes an unweighted noise-power level of −38·5 dB.

mitter to receiver must be equal if no distortion is to occur. In circuits used for speech, the ear pays no regard to relative phases of components of a complex tone and the effect of delay distortion is not harmful. In VF telegraphy, and particularly in facsimile transmission, effects of phase distortion can be serious.

in the line a dissipative equalizer — a network having a loss/frequency response, the converse of the line being equalized (Fig 4.5), which maintains the correct impedance. Equalization without qualification usually means attenuation equalization and is normal procedure in setting up a new long-distance circuit.

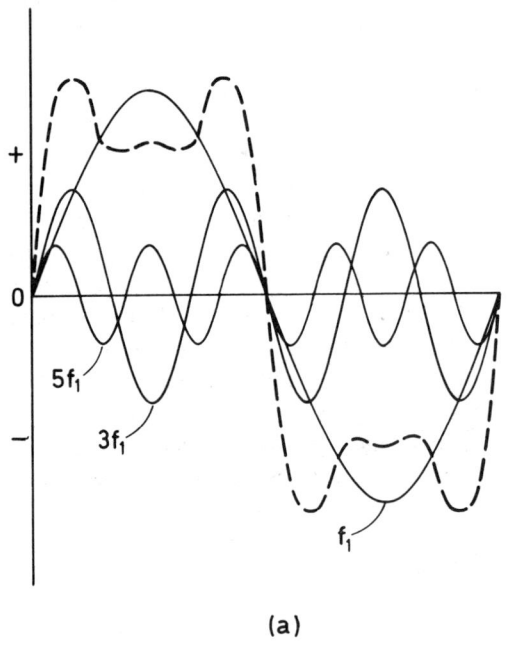

(a)

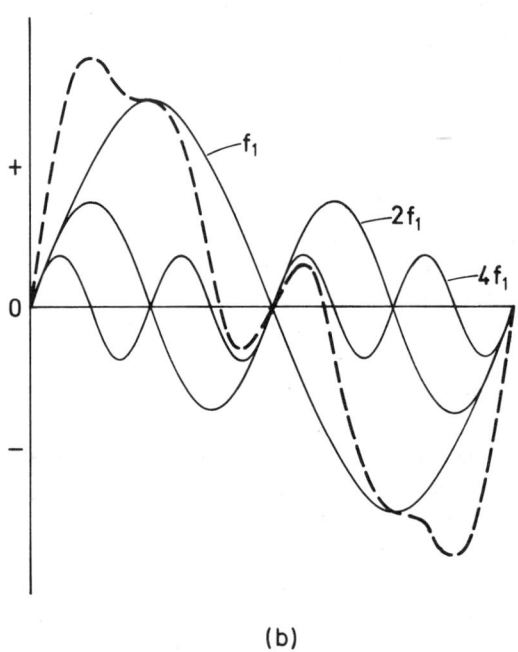

(b)

Fig 4.3 Waveform distortion by addition of (a) odd harmonics, (b) even harmonics

Attenuation Equalization. Before a long-distance circuit is put into service, end-to-end attenuation measurements in both directions are carried out at a number of spot frequencies from 300–3400 Hz (where applicable) to determine the attenuation/frequency of the circuit. All types of line plant have attenuation which varies with frequency, and if the attenuation/frequency characteristics do not meet the required specification it is usual to equalize the circuit by fitting a tuned network — an *attenuation equalizer* — which has inverse attenuation/frequency characteristics of those of the line circuit. Equalization is normally applied at the final repeater station of the receive channel. There are two general methods of attenuation equalization: (1) by arranging for the gain/frequency response of the amplifier connected at the receiving end of a section of circuit to be the mirror-image of the insertion-loss/frequency response (Fig 4.4); or (2) by inserting

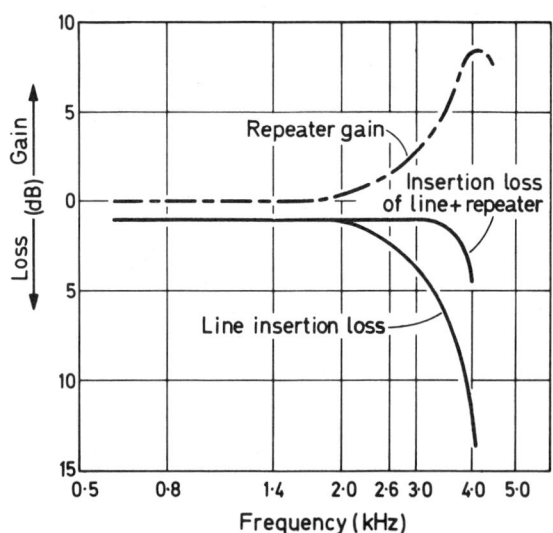

Fig 4.4 Attenuation equalization using tuned amplifier equalizer

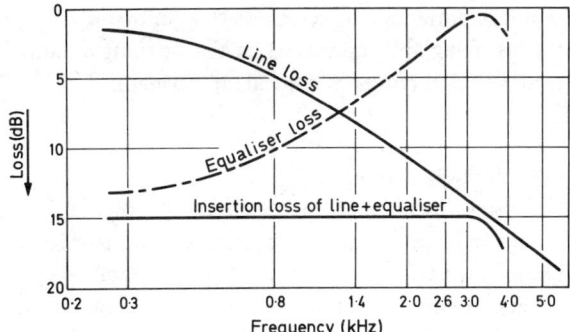

Fig 4.5 Attenuation equalization using equalizing network

Phase Equalization. Though not required for circuits used for telephony, this is often essential on those to be used for facsimile telegraphy and data transmission and also for circuits used for VF telegraph systems carrying TDM systems if delay distortion exceeds 2 ms. Compared with the relatively straight-forward tuned circuit of an attenuation equalizer, design of a phase equalizer is a complex and arduous task, nowadays left to a computer.

Non-linearity of the phase/frequency characteristic is assessed by measuring the group-delay/frequency distortion of the circuit. Delay results from a finite propagation velocity, the two being in inverse proportion. *Phase delay*, T_p, applies to a steady sinusoidal signal only: it is the ratio of total phase-shift β radian divided by angular frequency, ω radn/s: $T_p = \beta/\omega$ s

To transmit information, a change in the sinusoidal signal must be made; group delay T_g is a measure of transmission time of this change through a system: $T_g = d\beta/d\omega$ s.

Because group delay affects a modulated waveform envelope it is sometimes referred to as *envelope delay*.

In an ideal communication channel, phase-shift would be linearly related to frequency and zero at $f = 0$; all frequencies in a complex waveform would be delayed by the same propagation time and group delay T_g have a constant value equal to T_p. In a practical communication channel the phase/frequency characteristic always possesses a degree of non-linearity and propagation time varies with frequency, T_g being greater than T_p.

Pulse distortion occurs when group delay is not constant within the frequency band of that pulse, so that group-delay/frequency distortion exists

through the transmission medium. The effect of this depends on the modulation system used, but whenever a sinusoidal carrier frequency is modulated by more than one sinusoidal frequency, group delay/frequency causes distortion of the signal.

Group-delay/frequency distortion arises from non-linear phase/frequency characteristics which in turn result from filters in the transmission path. The low-pass filter effect of a loaded audio circuit causes group delay to increase with frequency within the audio bandwidth. In HF cable systems, band-pass filters in the translation equipment contribute increasing group delay towards both upper and lower frequencies of the audio band. It is particularly high in channels which occupy the 'edges' of the group-frequency spectrum where delay effects of through-group filters or stop filters may also be encountered. If two or more such circuits are connected in tandem, the delay distortion is cumulative. As far as possible, use of edge circuits is avoided for telegraph systems. Delay distortion, not noticeable in a speech circuit, constitutes the limiting impairment for facsimile telegraph systems. Fig 4.6 shows the group-delay/frequency characteristic for typical telephone-channel translation equipment. The important factor is group-delay/frequency distortion – differential group delay over the band considered, relative to the minimum value in that band. Group delay is directly additive: the delay distortion introduced by several links or items of equipment can be found by adding

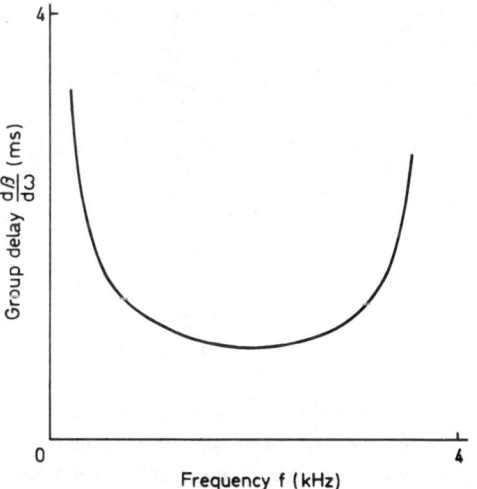

Fig 4.6 Group-delay/frequency characteristic for a typical HF channel

the distortions at spot frequencies through the audio band of each individual item and relating the results to the minimum value.

The positioning of attenuation/frequency equalizers in a line is influenced by signal/noise considerations which make it necessary to limit the level to which a signal is allowed to fall. Group-delay equalizers can be placed at any convenient point along a line.

Excessive delay distortion causes marked degradation in facsimile telegraph reception: at a sharp black/white transition to a picture, delay distortion results in some signal energy arriving late, with the result that a blurred transition takes the place of a sharp boundary of tonal change. The practical effect is not unlike that caused by echo effects due to mismatched impedances. For photo-telegraph transmission using FM, envelope-delay distortion should not exceed 0·3 ms over the band 1300–2500 Hz – the characteristic shown in Fig 4.7. The delay equalizer must not introduce atten-uation distortion. For high-speed facsimile news-paper transmission, differential delay present can be critical: it should not exceed ±25 μs (group circuits) or ±5 μs (supergroup circuits).

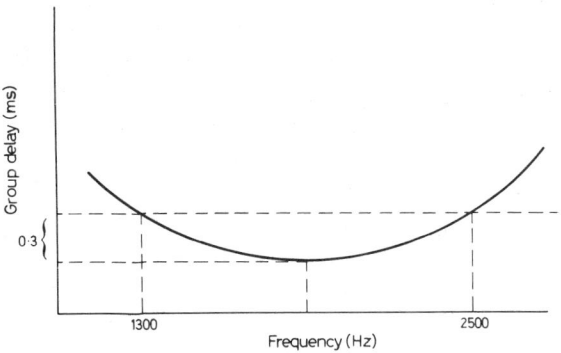

Fig 4.7 Limits of envelope delay for phototelegraphy

Differential-delay distortion can cause increased telegraph distortion – mostly bias distortion – in VF telegraph systems, FSM systems being more sensitive to this effect than are AM systems; rise and fall times of the demodulated signal are increased by differential delay. At 50 bauds, differential delays up to 18 ms for AM systems and 10 ms for FM systems would cause little additional distortion – perhaps 2%. Where appreciable delay distortion is

met it is preferable to complete the 'lining-up' of the VF telegraph equipment on the working circuit rather than relying on 'in-local' adjustment.

4.3 The 4-wire circuit

At a subscriber's or renter's station a telephone circuit requires one pair of conductors to provide a circuit for two-way communication. On long tele-phone circuits, amplifiers are required at intervals to compensate for attenuation losses. An amplifier is a one-way device, and a means has to be devised for two-way amplification. This is achieved by the *hybrid* termination, a device which divides the circuit into separate uni-directional transmit ('go') and receive ('return') channels at terminal repeater stations.

The 2-wire/4-wire termination comprises a pair of matched transformers and a network to balance the characteristics of the 2-wire line. Their purpose is to prevent as far as possible any coupling between the two amplified transmission channels. Due to energy dissipation in the balance network and transformers, signals passing in either direction through the hybrid termination suffer an attenuation loss of approxi-mately 4 dB.

The standard method of operating long-distance telephone circuits is to maintain separate transmit and receive paths over the whole distance between terminal repeater stations, usually associated with switching points. Although use of a 4-wire circuit doubles the plant requirement, it results in very stable long-distance circuits and it is the form used for all transmission media – land or submarine cable, radio, microwave or communication satellite. The general form of a 4-wire repeatered circuit is shown in Fig 4.8. Due to their stability, these can be lined-up (i.e. adjusted from end-to-end) for a nominal zero overall loss, power at the 2-wire point at reception being nominally exactly equal to that at the 2-wire point on transmission, all attenuation losses being compen-sated by amplifier gains. Standard power levels in dBm0 required at various points to achieve a zero-loss circuit are shown in the diagram, the two channels being disymmetrical. Figures in brackets are amplifier gains and line losses, the latter assumed to be 30 dB each. Amplifier gain is usually fixed, the required output level being obtained during the lining-up process by adjustment of an attenuator.

The distance separating repeater stations along a line depends upon bandwidth requirement and cable characteristics. The lowest level to which a signal may be permitted to fall before amplification becomes necessary depends upon minimum acceptable signal/noise ratio. Typical repeater spacings are 50 miles for audio plant, 3 or 6 miles for coaxial

The *low-pass* filter has a transmission band extending from zero frequency up to cut-off frequency – the frequency at which attenuation rises from approximately zero to a very high value and the filter ceases to transmit.

The *high-pass* filter has a transmission band extending from cut-off frequency up to infinite

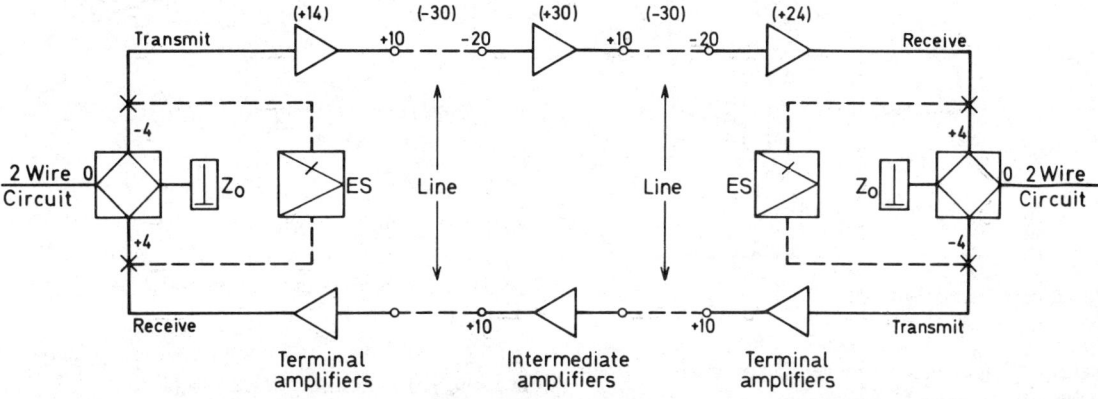

Fig 4.8 4-wire circuit

land cables and 6–25 nautical miles for submerged repeaters on submarine cables.

Circuits are lined up by applying an 800-Hz test tone with a power of 1 mW (O dBm) at the 2-wire point which is the *zero-level reference point* (O dBm0).

The 2-wire/4-wire terminations are removed from circuits used for MCVF telegraphy; and also for facsimile telegraph leased circuits if 4-wire ends are made available to the user, leaving two separate associated channels, one for communication in either direction. Although facsimile telegraphy requires only one direction of picture transmission, the terminations are removed to reduce the presence of reflected energy which would have a serious effect on a received picture by repeating or echoing tonal transitions a short time after the true transition, so causing a blurred effect on the picture. For MCVF bearer circuits, telephone-signalling terminations and echo suppressors (if present) are also removed.

4.4 Filters
Electric-wave or frequency filters, used extensively in all telecommunication equipment, exist in four basic types.

frequency. (The terms *high* and *low* apply in a relative sense.)

The *band-pass* filter has a single transmission band with two cut-off frequencies, neither of these being zero or infinity.

The *band-stop* filter, of more limited application, has a single *attenuation* band, neither of the cut-off frequencies being zero or infinity.

In the ideal case, filter elements are non-dissipative – inductances have no resistance and capacitances have no losses; no power can be absorbed in the filter itself. If the frequency of a generator applied to the input terminals of the filter is varied, input impedance changes from being resistive to reactive at a specific frequency. When input impedance is resistive, the generator delivers power to the filter and, since this power cannot be absorbed, it is transmitted to the load. The range of frequencies for which this occurs is the *pass band* or *transmission band* of the filter. When input impedance is reactive, the filter takes no power from the generator and none is delivered to the load: the range of frequencies for which this occurs is the *stop band* or *attenuation band* of the filter.

The description of the action of a filter is more easily explained by using the half-section, rather than a *T*- or other section. Using the symbols of Fig 4.9 (the value $2Z_2$ is shown because it would be shunted by a similar $2Z_2$ in the related half-section),

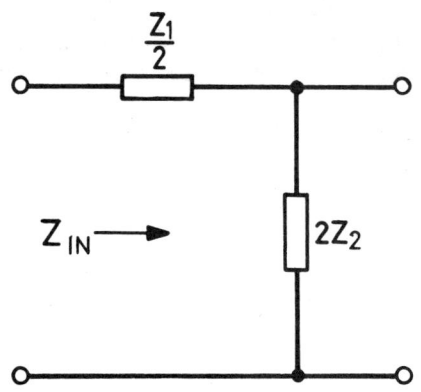

Fig 4.9 Basic frequency filter

the input impedance is the geometric mean of open and short-circuited terminations:

$$Z_{IN} = \sqrt{\frac{Z_1}{2} \cdot \frac{Z_1}{2} + 2Z_2} = \sqrt{\frac{Z_1^2}{4} + Z_1 Z_2}$$

$$= \sqrt{Z_1 Z_2} \cdot \sqrt{1 + \frac{Z_1}{4Z_2}}$$

For a high-pass or low-pass filter the series and shunt arms are conjugate reactances — one is an inductance and the other is a capacitance. If $Z_1 = j\omega L$ and $Z_2 = 1/j\omega C$ (a low-pass filter):

$$Z_{IN} = \sqrt{j\omega L \cdot \frac{1}{j\omega C}} \cdot \sqrt{1 + \frac{j\omega L \cdot j\omega C}{4}}$$

$$= \sqrt{\frac{L}{C}} \sqrt{1 - \frac{\omega^2 LC}{4}}$$

Over the frequency range:

if $\frac{\omega^2 LC}{4} < 1$, Z_{IN} is real or resistive: this represents the pass range (for the low-pass filter being considered). If $\frac{\omega^2 LC}{4} = 1$, Z_{IN} is zero, i.e. a short-circuit to the generator. No power reaches the load; the frequency at which this occurs is the cut-off frequency.

If $\frac{\omega^2 LC}{4} > 1$, Z_{IN} is unreal, i.e. reactive. No energy is supplied to the load; this represents the attenuation band.

For a high-pass filter, $Z_1 = 1/j\omega C$ and $Z_2 = j\omega L$ and:

$$Z_{IN} = \sqrt{j\omega L \cdot \frac{1}{j\omega C}} \sqrt{1 + \frac{1}{j\omega C \times 4 j\omega L}}$$

$$= \sqrt{\frac{L}{C}} \cdot \sqrt{1 - \frac{1}{4\omega^2 LC}}$$

The pass band occurs when $4\omega^2 LC > 1$, and Z_{IN} is real and positive.

The cut-off occurs when $4\omega^2 LC = 1$, and $Z_{IN} = 0$.

The stop band occurs when $4\omega^2 LC < 1$, and Z_{IN} is unreal and reactive.

Fig 4.10 shows for a low-pass filter, (*a*) half-section, (*b*) basic *T*-section configuration, (*c*) attenuation/frequency characteristic, (*d*) Z_{IN}/frequency characteristic, and (*e*) pass (*P*) and

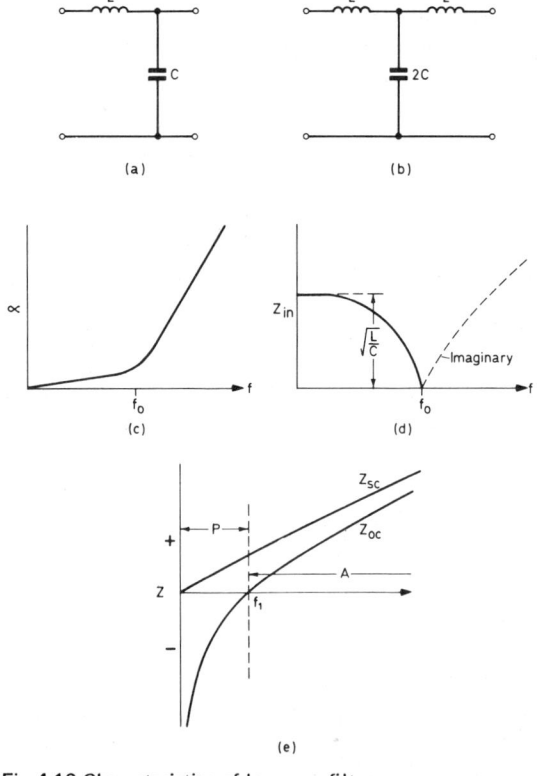

Fig 4.10 Characteristics of low-pass filter

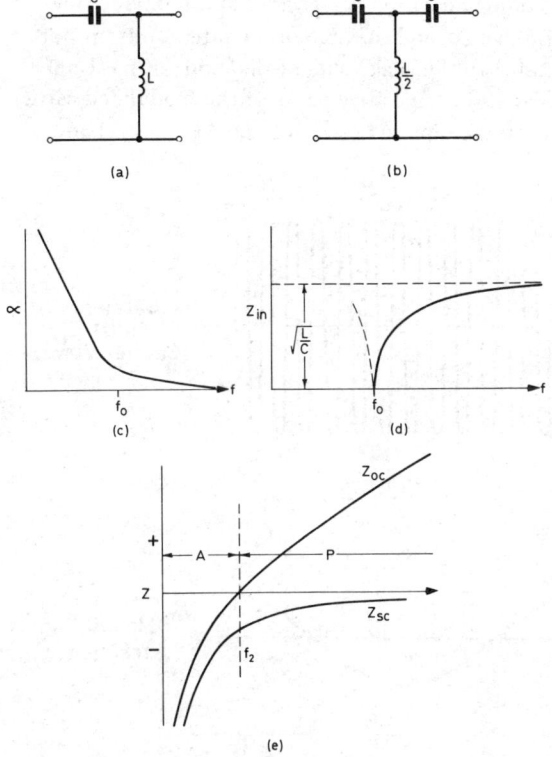

Fig 4.11 Characteristics of high-pass filter

resonant frequency f_4. The frequency band between resonant points is designed according to the pass range required of the filter. Impedance/frequency graphs are shown at (e). The attenuation/frequency and Z_{IN}/frequency graphs are shown at (c) and (d) respectively.

In practical filters components have resistance and dielectric losses, consequently attenuation and transmission characteristics are somewhat modified from those described.

4.5 Modulation

In telecommunication, information is conveyed by introducing a change in electrical condition at the transmitter in a form which may be recognized at the receiver. The process by which the electric state is changed in accordance with characteristics of the signal to be conveyed is termed *modulation*. All modulation systems depend upon changing the electric state with time; information carried by a wave — the *carrier wave* — is determined by the

attenuation (A) bands and their relation to the open-circuited and short-circuited terminations. For the short-circuited termination of the half-section low-pass filter, only inductance L is present and Z_{IN} is ωL (the graph Z_{SC}): the open-circuit termination is the series-tuned circuit LC (the graph Z_{OC}).

Fig 4.11 shows the comparable characteristics of the high-pass filter. In each case the resonant frequency is that of the series *a.c.* circuit wherein:

$$\omega L = 1/\omega C, \ \omega^2 LC = 1, \ \omega^2 = 1/LC, \ \omega = 1/\sqrt{LC}$$

and $f_0 = \dfrac{1}{2\pi\sqrt{LC}}$

In Fig 4.12 (a) and (b) show the half-section and full-section band-pass filter. In (a), the short-circuited condition is the series-tuned circuit $L_1 C_1$ with resonant frequency f_2; the open-circuited condition includes the series-tuned circuit $L_1 C_1$ in series with a parallel-tuned circuit $L_2 C_2$; the component values are so chosen that as frequency increases from zero, the filter firstly has a series-resonant frequency f_1 then a parallel-resonant frequency f_3 and another series-

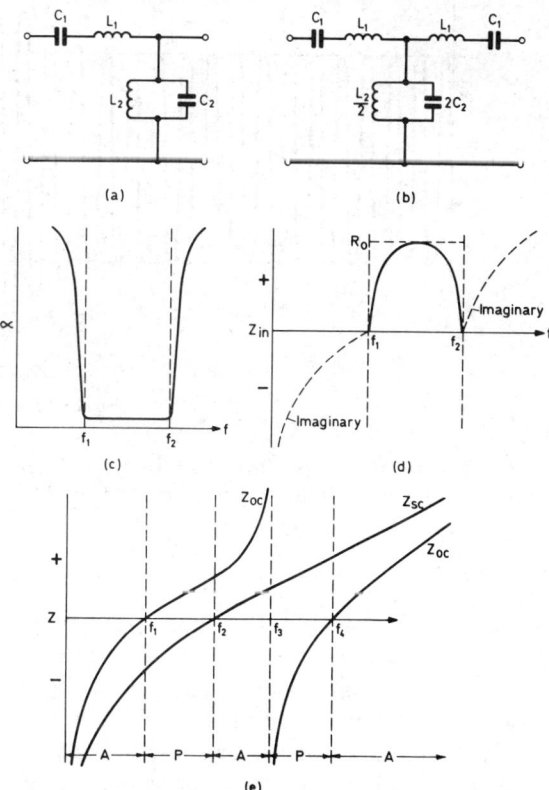

Fig 4.12 Characteristics of band-pass filter

manner in which the instantaneous value of the wave varies with time. Variation may be introduced into the amplitude, frequency or phase of the wave. The particular case of d.c. modulation for binary telegraph signals is discussed in Chapters 2 and 3.

modulated carrier wave varies at the modulating frequency, and the amount of variation is proportional to the peak value of the modulating signal. Variation of positive peaks of the modulated carrier wave takes place between limits $(V_C + V_M)$ and

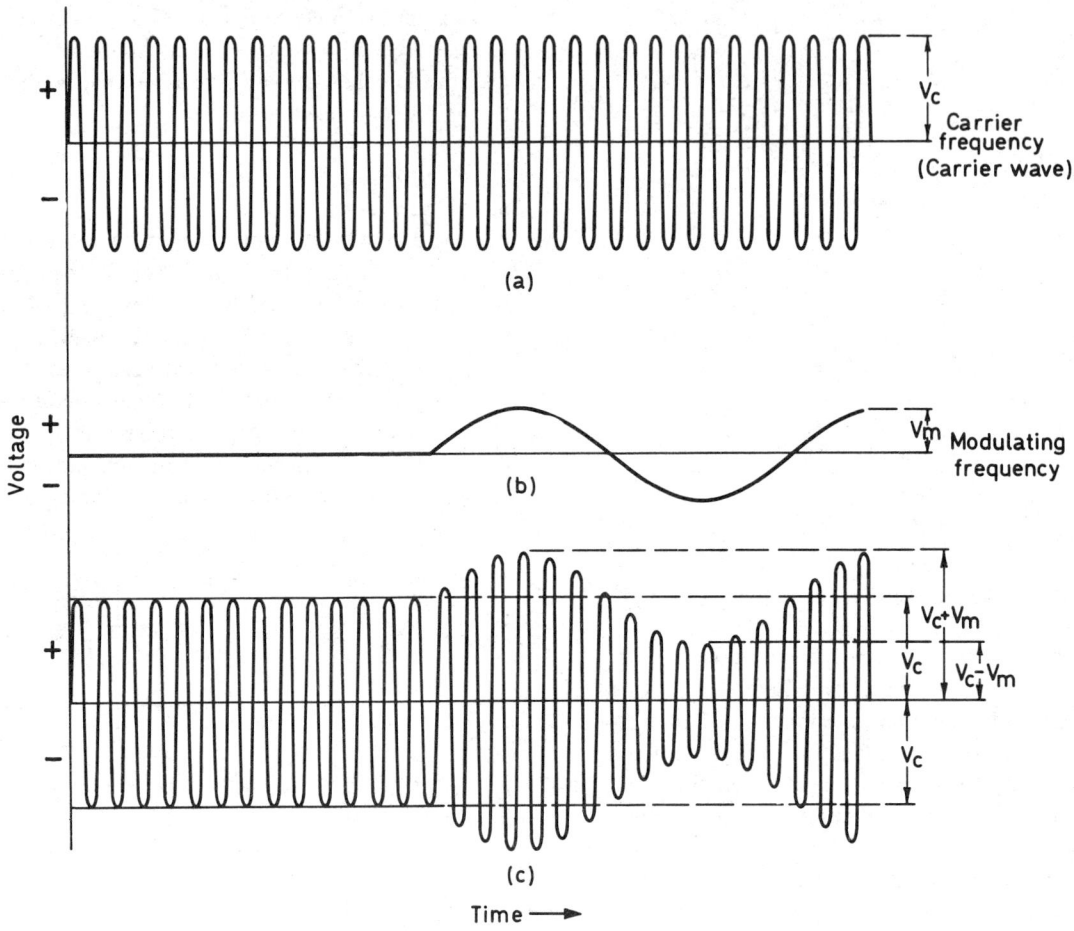

Fig 4.13 Amplitude modulation

Modulation of VF signals for alphabetic and facsimile telegraphy follows the principles used for modulation in telephony.

Amplitude Modulation (AM). In this process, amplitude of the carrier wave is modulated by the signal which is to be transmitted. In Fig 4.13, (a) represents a carrier wave having peak amplitude V_C volts; (b) represents the information signal, of peak value V_M volts, with which the carrier is to be modulated. The modulated wave at (c) shows that the peak value of

$(V_C - V_M)$, and of negative peaks between $(-V_C + V_M)$ and $(-V_C - V_M)$ volts. The area between these limits is termed the *modulation envelope* and it represents the sum of peak values of the carrier and the instantaneous values of the modulating frequency.

Using the suffix letters M for modulating signal of frequency $p/2\pi$, and C for carrier wave of frequency $\omega/2\pi$, the instantaneous value of carrier-wave voltage is $v_C = V_C \sin \omega t$, and that of the modulating wave is $v_M = V_M \sin pt$. When the peak value of carrier wave is made to vary in accordance with the modulating wave its value is $(V_C + V_M \sin pt)$ and the instantaneous value of the modulated carrier-wave voltage V_{MC} is:

$$v_{MC} = (V_C + V_M \sin pt) \sin \omega t$$
$$= V_C \sin \omega t + V_M \sin \omega t . \sin pt \text{ volts}$$

Using the form of expansion of the trigonometrical identity

$$2 \sin A . \sin B = \cos (A - B) - \cos (A + B)$$

$$v_{MC} = V_C \sin \omega t + \frac{V_M}{2} \left\{ \cos (\omega - p)t - \right.$$

$$\left. \cos (\omega + p)t \right\}$$

$$= V_C \sin \omega t + \frac{V_M}{2} \sin \left\{ (\omega - p)t + \frac{\pi}{2} \right\}$$

$$+ \frac{V_M}{2} \sin \left\{ (\omega + p)t - \frac{\pi}{2} \right\}$$

This shows that the modulated wave may be considered as the sum of three waves, each of

constant peak value, having frequencies $\omega/2\pi$, $(\omega + p)/2\pi$ and $(\omega - p)/2\pi$ Hz. The first of these components has the same frequency and peak value as the original carrier wave. The other two waves have frequencies equal to the sum and difference of carrier and modulating frequencies; they are termed *side frequencies* and known as upper side frequency and lower side frequency respectively.

The peak value of each side frequency is $V_M/2$, i.e. half that of the original modulating signal. The form of a carrier wave, upper- and lower-side frequencies and the resultant amplitude-modulated carrier wave are shown in Fig 4.14.

The waveform of the modulating signal is usually complex, containing many component sinewaves, each of which produces an upper and a lower side frequency when modulated. The frequency band in which these component frequencies lie appears as an *upper sideband* and a *lower sideband* following the modulation process. The carrier frequency is usually higher than any of the modulating frequencies and upper and lower sidebands are symmetrically situated with respect to frequency above and below

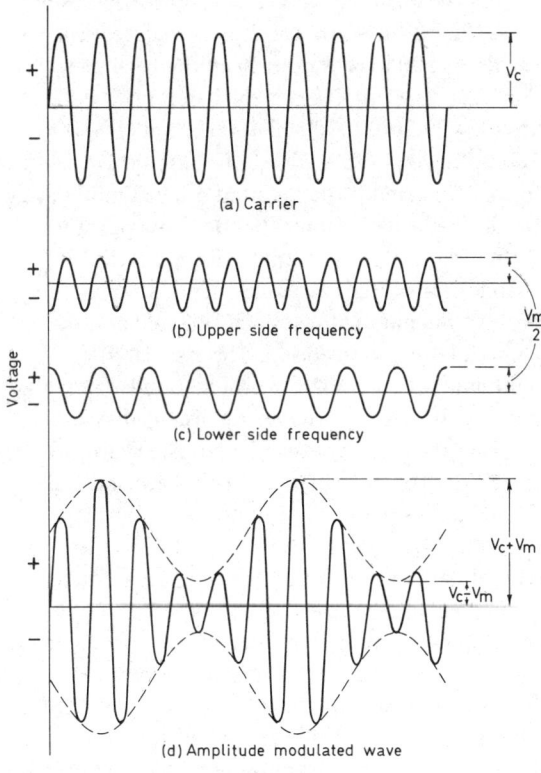

Fig 4.14 Amplitude-modulated carrier wave, showing side frequencies

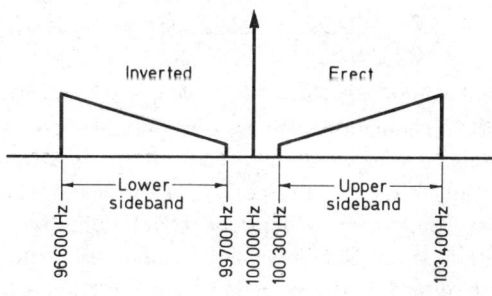

Fig 4.15 Graphical representation of carrier and sidebands

the carrier frequency. Taking a practical example, if a carrier wave of 100 kHz is modulated by a telephone speech-band signal of 300–3400 Hz, the products are: (1) the original carrier frequency, 100 kHz; (2) an upper sideband of 100 kHz + (300 to 3400 Hz) = (100 300–103 400 Hz); and (3) a lower sideband of 100 kHz − (300 to 3400 Hz) = (99 700–96 600 Hz). In the lower sideband, the *lowest* modulating frequency (300 Hz) becomes the *highest* modulated frequency (99 700 Hz), while the highest modulating frequency (3400 Hz) becomes the lowest modulated frequency (96 600 Hz) — the lower sideband is said to be *inverted*, while the upper sideband remains *erect*. This condition is shown graphically in Fig 4.15. For purposes of identity in diagrams, it is conventional to show a lower amplitude for the lower frequency of the original modulating frequency, and a higher amplitude for the higher frequency; this is helpful when considering systems wherein more than one stage of modulation is employed and the inverted sideband is not necessarily the lower one.

When the carrier has a higher frequency than any of the modulating frequencies — the usual case — the bandwidth required for the modulated wave, with its sidebands, is twice that of the modulating frequency. This does not hold if the modulating frequency is greater than the carrier frequency.

The relationship between peak amplitudes of carrier and modulating wave is expressed by the modulation factor (m), defined as *the ratio between the difference and sum of the numerical values of the largest and smallest amplitude in one cycle of modulation,* which is usually expressed as a percentage. Using the symbols of Fig 4.13:

$$m = \frac{(V_C + V_M) - (V_C - V_M)}{(V_C + V_M) + (V_C - V_M)} \times 100\% = \frac{V_M}{V_C} \times 100\%$$

For example, if $V_M = \frac{1}{2}V_C$, $m = \frac{1}{2} \times 100\% = 50\%$. If 100% modulation were used, amplitude of the modulated wave would fall to zero at points corresponding to the negative peak value of the modulating signal. The greater the depth of modulation, the greater is the peak value of side frequencies; if the depth of modulation exceeds 100%, distortion will result.

At 100% modulation, sidebands have each one-quarter of the energy of the carrier wave, i.e. the information power carried in the two sidebands is only one-third of the radiated power. Since each sideband carries the same full information from the original modulating signal, and the carrier component contains no information other than its frequency, *singe-sideband* (SSB) transmission systems are frequently used in which the carrier component is eliminated (by using a balanced modulator) and one sideband is suppressed (by filtering); this not only conserves approximately 50% of bandwidth, it also conserves power. In *double-sideband* trans-

mission (DSB), the carrier component may be transmitted at a much reduced power, sufficient for its frequency to be used at the receiver for accurate *demodulation*.

Single-sideband-suppressed carrier systems are used extensively on landline and radio systems. The original modulating signal may be recovered at the receiving station in a *demodulator*, using the received sideband to *modulate* a locally-produced carrier frequency. This process is the same as the modulating process, apart from the input signal

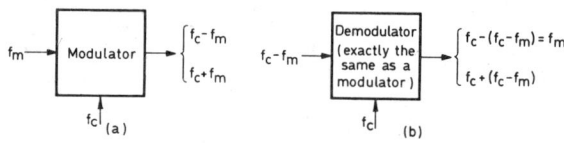

Fig 4.16 Frequencies in (*a*) modulator; (*b*) demodulator

which is fed into the modulator (see Fig 4.16). Taking the example used earlier, if the lower sideband (96 600–99 700 Hz) is demodulated using a carrier of 100 kHz, output from the balanced demodulator contains two sidebands (100 000– 96 600–99 700) = 300–3 400 Hz and (100 000 + 100 300–103 400) = 200 300–203 400 Hz. The former sideband is the required original modulating signal, the latter is eliminated by a low-pass filter.

In circuits which comprise disymmetrical channels, the same carrier-frequency supply is fed both to the modulator for transmission and the demodulator for reception. The associated modulator—demodulator combination is known as a *modem*. If there is a frequency discrepancy δ between the carrier frequency f at the transmitter and the carrier frequency ($f + \delta$) at the receiver, the demodulated frequency will not be f_M but ($f_M + \delta$). Any small frequency error such as this in the demodulated signal, while not noticeable for telephony will cause distortion in frequency-modulated telegraph systems, the frequency error producing a bias distortion. In high-grade cable circuits frequency-translation error will not exceed ± 2 Hz. The relative phase of the re-applied carrier at the receiver is of no consequence provided that only one sideband has been transmitted.

The information-carrying characteristic of an AM

signal is its amplitude. For this reason, AM signals are vulnerable to changes in transmission level and also to impulse noise which may equal or exceed the information amplitude and duration. In AM VF telegraph systems, use of automatic gain control is essential to neutralize effects of unwanted changes in amplitude brought about by level changes in the transmission system.

In *vestigial (asymmetric) sideband modulation* systems, the lower sideband and carrier are transmitted, together with a vestige of upper sideband; compared with double-sideband transmission this gives economy in bandwidth. A disadvantage of such a method is that the carrier frequency is located towards one end of the available frequency band where it is likely that attenuation and delay distortions will be more severe than at the centre of the band. A high degree of delay equalization is essential for successful application of vestigial sideband technique. It is used in systems for which very low modulating frequencies are present (e.g. facsimile telegraphy); in these cases, filtering to retain one sideband and reject the other is not practicable. The general form of the vestigial sideband is shown in Fig 4.17.

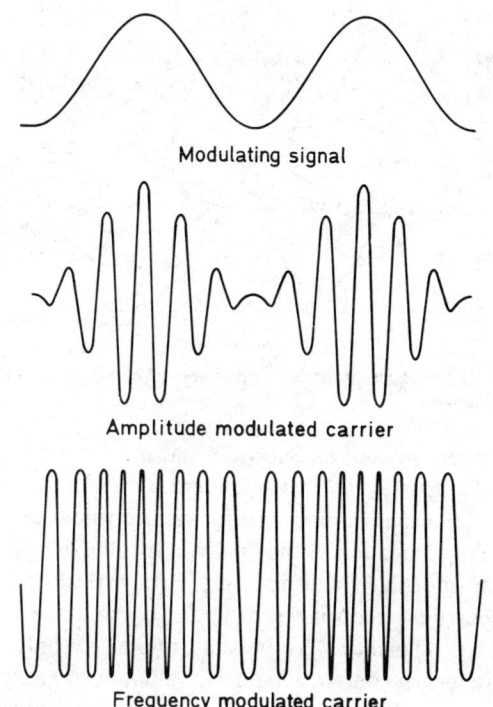

Fig 4.18 Comparison of AM and FM waves

Modulating signal

Amplitude modulated carrier

Frequency modulated carrier

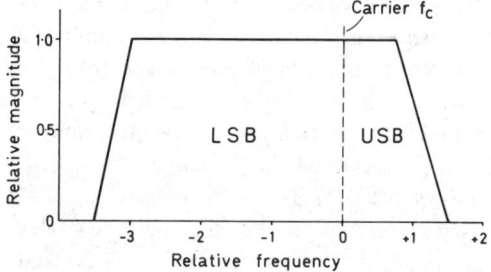

Fig 4.17 Vestigial sideband

Frequency Modulation. Frequency modulation (FM) is *angle modulation of a sinewave carrier in which the instantaneous frequency of the modulated signal differs from the carrier frequency by an amount proportional to the instantaneous value of the modulating wave.* The effect of the modulating signal is to vary the frequency of the carrier above and below the mean value f_C by an amount depending upon the amplitude of the modulating signal. Fig 4.18 shows a comparison between the modulated signals produced by AM and FM from the same modulating signal.

If Δf is the frequency deviation above or below the mean carrier frequency and f_M is a modulating frequency, the instantaneous voltage v of the modulated signal is:

$$v = V_C \sin\left\{ 2\pi f_M t + \frac{\Delta f}{f_M} \sin 2\pi f_M t \right\}$$

The mathematical expansion of this expression shows that not only is the first pair of side frequencies $(f_C \pm f_M)$ generated, but other pairs appear at frequencies $(f_C \pm 2f_M)$, $(f_C \pm 3f_M)$, . . . , $(f_C \pm nf_M)$ up to infinity. The magnitudes of these side-frequency components are determined by the ratio of frequency deviation (Δf) to maximum modulating frequency (f_{Max}), termed the *modulation index* m: hence $m = \Delta f / f_{Max}$. Fig 4.19 is derived from the mathematical expansion of the above equation in terms of Bessel functions and shows how, for values of f_C greater than either f_{Max} or Δf, amplitudes of carrier and side-frequency components are dependent upon the value m of the modulation index. For example, at values of approximately $m = 2.4$ and 5.5 the carrier component vanishes, and so do various pairs of side-frequency components

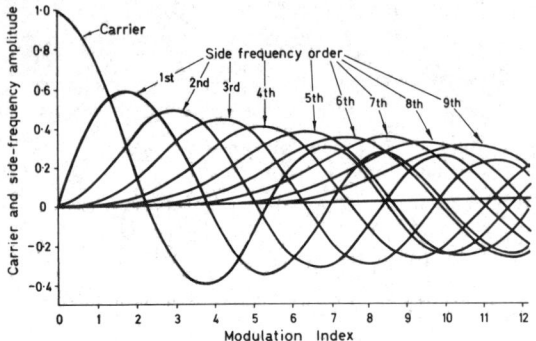

Fig **4.19** Frequency modulation: levels of carrier and side frequencies

at other values of m. Shown in another form, Fig 4.20 illustrates the energy content of carrier and side-frequency components for various values of modulation index. Although theoretically an infinite number of side-frequency pairs, symmetrically disposed by frequency separation f_M about the central frequency f_C, is possible, in practice only those which contain a significant amount of energy, relative to the unmodulated carrier, are important: the bulk of these are contained within the frequency range $f_C \pm (f_M + \Delta f)$, a bandwidth of $2(f_M + \Delta f)$, indicated by the arrowed lines in the diagram. When the modulating signal has a complex waveform, side frequencies exist at intervals in the frequency spectrum corresponding to each component of the modulating frequency.

In MCVF systems, where it is desired to restrict the bandwidth of the telegraph channels to obtain the maximum number on one bearer channel, it is possible without undue increase in telegraph distortion to use a bandwidth not much greater than that

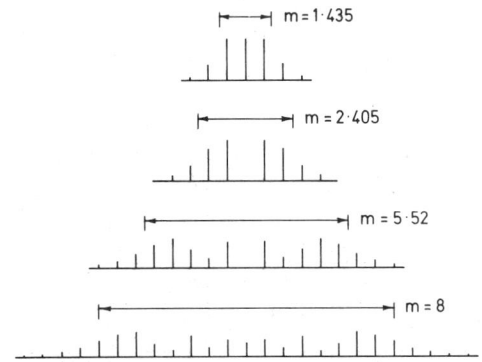

Fig **4.20** Frequency modulation: energy content in carrier and side frequencies (f_M constant, Δf varied)

necessary to cover the frequency deviation, provided that the maximum modulating frequency f_M does not exceed the frequency deviation Δf, i.e. the modulation index is not less than one.

A great advantage of an FM system is its relative insensitivity to noise. An FM signal is transmitted at constant amplitude and during transmission this amplitude may be varied by various forms of interference, but at the FM receiver the signal is first amplified and its amplitude is then cut to a uniform value by a limiter: though this may not completely remove the effect of the noise picked up, a marked improvement in signal/noise ratio (about 6 dB compared with an AM system) is obtained.

4.6 HF cable telephone systems

By a process of modulation, the telephone-channel bandwidth (300–3400 Hz) can, by appropriate choice of carrier frequency, be translated to any desired position in the frequency spectrum for the purpose of transmission, the normal audio-frequency range being recovered at the receiving station by a complementary process of demodulation.

The use of filters enables telephone-circuit bandwidths, which have been translated to occupy different frequency bands, to take up a series of adjacent positions in the frequency spectrum. In this way, very many telephone channels can be assembled for operation over a single transmission path — a process known as *frequency-division multiplexing* (FDM). This is the basis of the standard method of providing long-distance telephone circuits on carrier and coaxial cables (land and submarine), and also on radio microwave and satellite systems.

Carrier telephone systems operate on pairs of conductors in conventional multi-quad cables. The characteristics of a coaxial-cable pair render it suitable for carrying upwards of 1000 telephone channels at high frequencies over a wide transmission band. The method of assembling this number of channels side-by-side in the frequency spectrum is to use successive modulation processes, each step handling a wider band at a higher frequency until the final single HF wideband is achieved. Three or four stages of modulation (and demodulation) suffice, and in this way the number of separate carrier frequencies to be generated is restricted. They are all derived from a basic crystal oscillator of high stability to which all carrier frequencies are locked. Single-sideband transmission with suppressed carrier is used to make economical use of frequency band and reduce loading on common wideband amplifiers.

Frequency allocations for a 3-Mhz system are shown in Fig 4.21. Telephone channels are first assembled in 12-channel groups (60–108 kHz) using carrier frequencies 64, 68, . . . , 104, 108 kHz. Five such groups are then group-modulated, using carrier frequencies of 420, 468, 516, 564 and 612 kHz (i.e. 48-kHz spacing) to form a *supergroup* in which 60 channels now occupy the frequency band 312– 552 kHz, using the group lower sidebands (erect). Finally, a number of supergroups, according to the system capacity, are assembled to form the aggregate

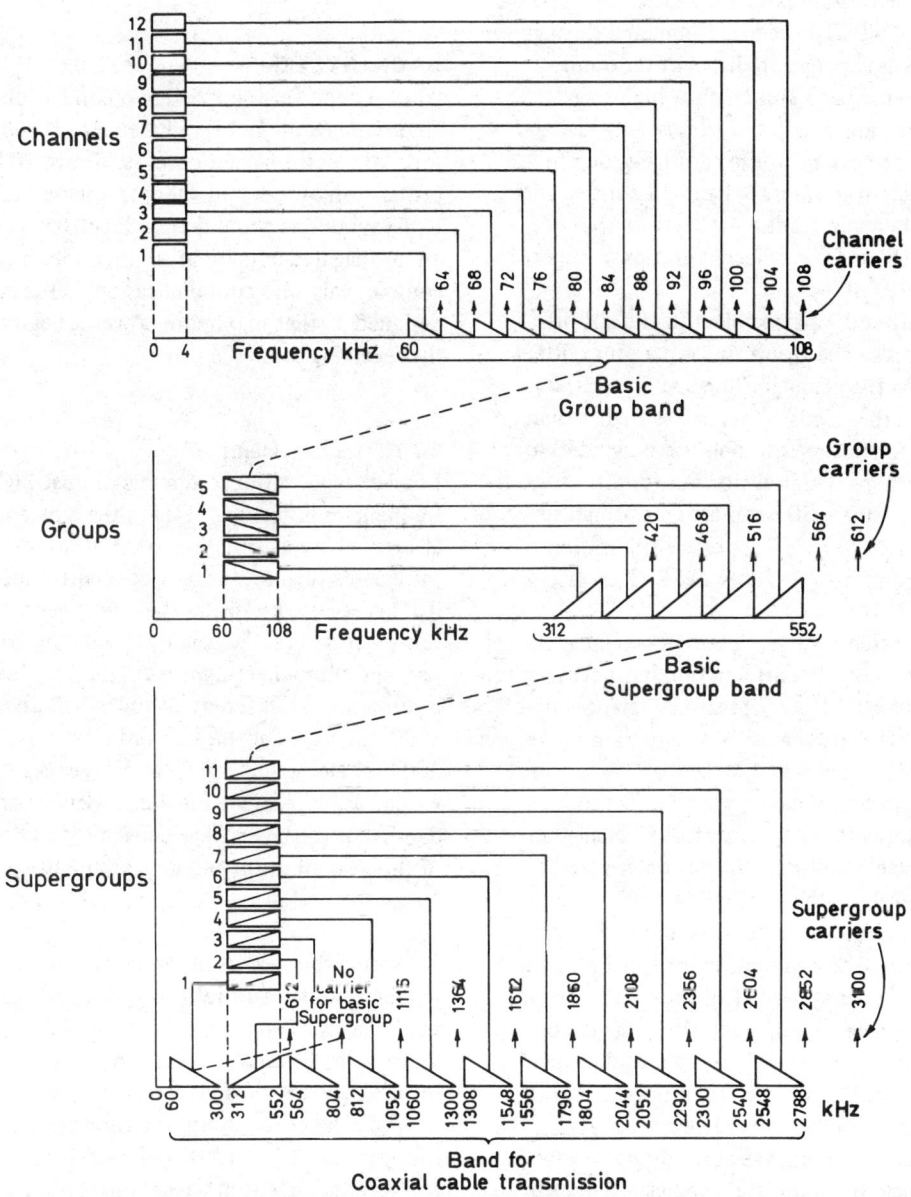

Fig 4.21 Assembly of 4-kHz channels in wideband transmission system

baseband, using a further group-modulation process with carrier frequencies of 612, 1116, 1364, 1612, 1860 . . . (spaced mainly at 248 kHz); the lower sidebands (inverted) are selected by filters. Exceptionally, the second supergroup (the only erect sideband) is inserted, without group modulation at this stage, in the range 312—552 kHz: this is the original frequency range of a supergroup, and it falls between first and third supergroups modulated with carriers of 612 and 1116 kHz respectively.

For routing links in tandem, groups (48 kHz) or supergroups (240 kHz) can be connected through at the group or supergroup distribution frame; they can also be made available for high-speed facsimile telegraphy or data systems. In wideband systems it is important to use through-group or through-supergroup filters — band-pass filters with steep cut-off characteristics — to ensure that no interfering signals from adjacent groups or supergroups are present.

Channels 1 and 12 are subject to the cut-off characteristics of the group and supergroup filters in addition to the channel filters. As a result, delay distortion on these 'edge' channels is higher than normal and for this reason their use is avoided for telegraph systems. Delay distortion for a through-group filter is 200—250 μs and for a through supergroup filter 15—20 μs; after phase equalization these values are reduced to 6—20 μs and 2—7 μs respectively.

A buried cable is subject to an appreciable annual temperature cycle. The attenuation has a temperature coefficient of +0·2% per °C; seasonal changes up to ±10°C result in a 4% change in attenuation which can mean 33 dB/100 miles at 4 MHz. Compensation for changes in cable attenuation with temperature is effected automatically by a controlled change in gain of the line amplifier, which is determined from the power level of a pilot signal. In earlier systems the overall gain of the system was controlled by automatic insertion and removal of fixed equalizing networks from the system. This provoked abrupt changes of perhaps 0·5 dB each time equalization changed; though not of serious magnitude when taken alone, the cumulative effect if a number of such circuits are linked in tandem could cause transmission level to change abruptly by several dB.

Pilot signals at various frequencies are transmitted from end to end of HF systems for control purposes.

A group-reference pilot is injected in the formation of all 12-channel groups as an integral part of the group until final demodulation. Its purpose is to keep the transmission level within close limits and is also useful in fault locating. For a group-reference pilot, the ideal position is in the middle of the group-frequency spectrum — 84 kHz. As this value corresponds with the carrier frequency for channel 7, the pilot is made 84·080 kHz — the '84-plus-delta' frequency.

RADIO SYSTEMS
When a.c. power is applied to a conductor a certain amount leaves it and is radiated into space in the form of electromagnetic waves. Above 10 kHz the proportion of radiated energy increases, especially with favourable aerial design. Electromagnetic waves are propagated equally in all directions but, for commercial radio communication, aerial systems are designed so that maximum power is beamed towards the receiving aerial.

4.7 HF radio systems
Long-distance radio communication at high frequencies is dependent upon the existence of electrified layers situation 100 km and more above the earth's surface, forming concentric shells with it. This region, known as the *ionosphere*, consists of low-pressure gases which are ionized by cosmic rays and ultra-violet light from the sun. Ionized layers occur at different altitudes — D and E layers at 90—150 km above earth and F layers at 200—300 km. During daylight, the F layer separates into F1 and F2 layers (see Fig 4.22). Heights and depths of the layers vary constantly and the pattern of their distribution and ionization changes daily, seasonally and in sympathy with the 11-year sunspot cycle.

Absorption occurs at the D layer, mainly during daylight hours. Auroral zones, centred near the magnetic poles, introduce additional absorption in radio paths traversing them. Absorption bears an inverse ratio to transmission frequency.

The E layer functions as a good conductor at lower frequencies and with the earth forms a spherical waveguide within which long LF waves are largely confined. The shorter waves (HF) penetrate the E layer and are returned to earth by reflection at, or bending due to refraction within, the F layers. The refractive index of the ionosphere is a function of radio frequency.

Due to dependence upon sunlight, the extent of ionization varies with altitude and also throughout the 24 hours. Over long terrestrial distances, daylight and ionization conditions vary continuously: the degree of ionization near points of transmission and reception is of greater importance than at intermediate points, the controlling factor being the condition nearest the darker end of the transmission path.

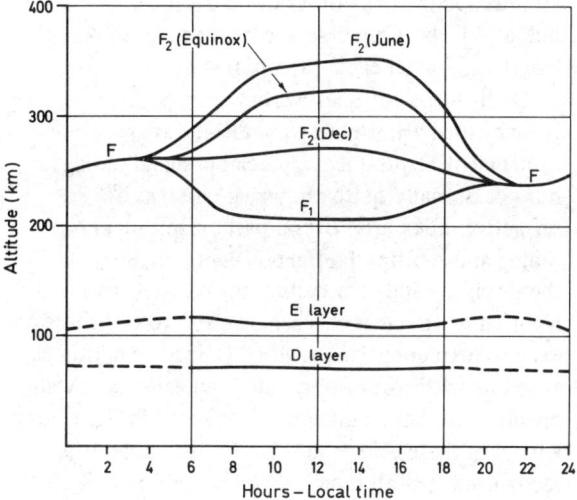

Fig 4.22 Typical daily variations in ionospheric layers (mid-latitudes in northern hemisphere)

Depending upon frequency, angle of incidence and state of ionization, electromagnetic waves propagated upward from the earth's surface may pass through the ionosphere and be lost, or be refracted by varying amounts before being reflected back to the earth's surface at some considerable distance from the source and with a large proportion of their original energy retained; attenuation suffered within the ionosphere depends upon frequency, altitude and time of day. Over great distances, reception may be due to multiple reflections involving two or more 'hops' between earth and ionosphere; attenuation occurs at the earth's surface, where scattering too may occur. The waves may follow a number of separate paths and return to earth at various distances from the source.

Attenuation of the direct or ground wave increases with distance from the source and a limiting point is reached when little or no radiated energy can be detected; beyond this limit exists a zone termed the *skip distance* which extends to the point where radiated energy begins to arrive at the earth's surface after reflection from the ionosphere.

From international systematic measurements by ionospheric sounding it is possible to predict ionospheric conditions and to recommend the *maximum usable frequency* (MUF) — the highest frequency which in a particular set of conditions can be used to propagate radio waves over a given route. There is a relatively small range of useful frequencies lying between those limited by reflection and those limited by absorption. For reliable HF communication over distances beyond range of the direct wave, it is necessary to select the transmission frequency which gives optimum reflection and least absorption. To maintain communication it is necessary to change transmission frequency during a 24-hour period and also with the seasons.

Because of the irregular and unstable structure of the ionosphere, HF signals propagated by this medium tend to take a number of separate simultaneous paths and to arrive at the receiver as a group of independent signal components with random phase relationships. The resulting wave interference causes wide fluctuations (about 30 dB or more) in signal amplitude, known as *fading*. During periods of fades, noise level rises, aggravated by automatic gain-control usually present in a receiver. Noise results from multiple transmission from regions of high thunderstorm activity and from e.m.f.s generated by electrified layers moving in the earth's magnetic field; the noise is subject to the same variation as other types of signal.

Slow fading (day-to-day variations in received signal strength) is due to absorption and skip-zone variation. Rapid fades (those occurring in periods measured in seconds) are due to wave interference and changes in polarization. The amplitude distribution of rapid fading is regarded as following a Rayleigh distribution.* Apart from normal daily

*$P(v) = \epsilon^{-0.693}\left(\dfrac{v^2}{v_M}\right)$, where $P(v)$ is the proportion of time for which the amplitude v volts is exceeded and v_M is the median value. The distribution is uniquely determined by the choice of the value for v_M.

variations in absorption, sudden ionospheric disturbances, known as *Dellinger fades*, occur due to solar flares during periods of maximum sunspot activity.

Signals may arrive at a receiver after undergoing different numbers of reflections, travelling different distances over different paths with different transmission times; this *multipath* propagation effect tends to repeat or prolong signals. The time-spread between first and last significant arrivals becomes a limiting factor in telegraph reception. Multipath delays are least at frequencies close to the MUF value and lie typically between 0·5 and 5 ms; values of 2 ms are common. These delays are readily observed on a facsimile telegraph test chart in which straight lines normal to the line of scan bear a superimposed ripple, the width of which is proportional to the difference in propagation time of the shortest and longest paths that predominate from instant to instant.

With the restricted number of transmission frequencies available to be shared between all users the HF band is severely congested and some interference is to be expected between transmissions. Fortunately, after a period of heavily increasing traffic approaching saturation over the HF band, some relief is now being afforded by the availability, for certain routes, of transoceanic submarine cables and communication satellites, both of which provide a more stable service compared with HF radio transmissions. For HF radio-teleprinter systems, the use of some form of error-detection equipment is essential.

4.8 Microwave radio-relay systems

At very high frequencies radio waves follow quasi-optical paths. With relation to the curved surface of the earth, the effective horizon for these radio waves is slightly beyond the normal optical horizon, by a factor of about 1·33. This results from refraction in the lower regions of the earth's atmosphere due to the relative permittivity which is somewhat greater than unity, due to the presence of water vapour; with increased height and decreasing atmospheric pressure and temperature, the relative permittivity and refractive index of the air decrease. As a result, the radio waves are subjected to a refraction which bends the waves slightly away from the earth's surface and extends the radio horizon.

This method of propagation is now employed extensively for normal provision of long-distance telephone circuits over *radio-relay* or *line-of-sight* systems.

Radio-relay systems operate in the VHF, UHF and SHF bands. Systems working at frequencies beyond 1000 MHz are referred to as *microwave* systems in which wide frequency bands are available which can be used for FDM transmissions providing up to 1800 telephone channels of standard 300–3400 Hz bandwidth from a single broadband radio channel. The radio path has an attenuation that is independent of frequency over the bandwidths required for multiplex telephone working. The path attenuation is of the order of 60 dB for a 25-mile link at 4 GHz — much lower than an equivalent length of coaxial cable (about 160 dB at 3 MHz).

On line-of-sight microwave paths, departures from normal atmospheric conditions are generally insufficient to produce appreciable signal fading, but occasionally quite abnormal patterns of refractive index arise over a path, resulting in both fading and multipath effects. Due to changes in the refractive index, a defocusing effect between the highly directional aerials can give rise to transmission losses. At frequencies above 10 GHz attenuation can arise due to absorption by rain, hail, snow and water droplets suspended in cloud or fog, but in the United Kingdom these effects are almost negligible. It is recognized that all radio-relay links will be affected to some extent by fading.

With such high concentration of traffic from several broadband channels carried on one microwave system, the question of reliability assumes great importance. It is usual to provide a number of standby 'protection' channels, one of which is automatically switched into service to replace a faulty working channel; the switching is controlled by pilot signals. The resultant break in transmission occupies only 1 ms or less, but the operate time of the fault-detection equipment is up to 20 ms.

4.9 Satellite systems

A communication satellite orbiting the earth may be either *passive*, simply reflecting or scattering signals beamed to it from earth, or *active*, in which signals received from earth on a given carrier frequency are amplified from a source of power in the satellite before being retransmitted back to earth on a different frequency. Line-of-sight propagation paths are used between earth stations and satellites.

The time for a satellite to make one revolution around the earth increases progressively with the orbit height. At 22 300 statute miles (35 800 km) the true period of orbit is 24 h, the same as the period of the earth's rotation; a satellite in orbit at this height, and travelling in the same direction as the earth, appears to be stationary with respect to an observer on earth and is known as a *geo-stationary* or *synchronous* satellite.

The (active) synchronous satellite system is now one of the main media for international tele-communication: it is complementary to and fully integrated with submarine-cable and HF radio systems. A circuit provided by a synchronous satellite system gives a very stable performance and has the normal characteristics of any telephone circuit, except for possibly a slightly higher noise level and, more important, a long propagation time. In the 'hop' from earth-station to satellite to earth-station the distance is 44 600 miles, and at 186 000 miles/s the time for this is 240 ms. For telephone speech, the delay between a talker speaking and receiving an answer from the correspondent is over half a second, a period which can cause confusion in conversation, so special echo suppressors were designed for these long-delay circuits. Serious difficulty would arise on a circuit over two such satellite hops in tandem. For telegraph systems this long propagation time is no disadvantage, apart from the slightly longer delay (which can also arise on HF radio circuits) in receiving answer-back and telex supervisory signals.

4.10 Bandwidth economy

Information theory indicates that it is always theoretically possible to economize in bandwidth at the expense of signal/noise ratio, though the principle is usually exploited in the opposite sense.

The need for economy arises where bandwidth is scarce or its cost is high — conditions usually co-incidental and which arise in long submarine-cable systems. For a telephone system the problem can be attacked by development of voice-coding methods for speech analysis and synthesis (e.g. the vocoder). For other requirements, including telegraph systems,

the solution has been to reduce bandwidth at the expense of quality, at the same time using more efficient filtering to increase the efficiency of using available bandwidth.

With standard telephone-channel spacing at 4 kHz the nominal useful bandwidth is 3100 Hz (300–3400 Hz). The loss of bandwidth, 900 Hz in 4000 Hz, amounts to $22\frac{1}{2}$%, though these frequency gaps have been used for various purposes. Loss of bandwidth in these gaps can be reduced to about 5% by special filter techniques which have been adopted for long submarine cables; for this purpose 3-kHz spacing in 16-channel telephone groups is normal. The nominal bandwidth available is 2850 Hz (200–3050 Hz) — only 250 Hz less than the 3100 Hz available in the 4 kHz-spaced channels.

4.11 Circuit characteristics

For telephony, important circuit parameters are attenuation/frequency response, degree of intelligible crosstalk and psophometric noise. Compared with speech transmission, telegraph transmission demands a much greater precision in circuit performance, particularly in delay distortion, impulse noise and abrupt level changes, to all of which the ear is more tolerant.

Recommendations have been published for the performance to be expected from bearer circuits for telegraph transmission and the more important of these are outlined below. In CCITT recommendations, Category A circuits are less than 2500 km in length; Category B circuits are those exceeding 2500 km.

Standard International Telephone Circuit (CCITT M58 – formerly M61). This class of circuit is provided for international telephone circuits in the general telephone service, but it could be used for a phototelegraph transmission over the normal telephone network; also it could be nominated as a VF telegraph bearer *reserve* circuit.

The attenuation/frequency distortion should not fall outside the hypsogram limits shown in Fig 4.23 for 3 kHz- and 4 kHz-spaced channels. The return loss must be at least 20 dB over the transmission range. The near-end signal/crosstalk ratio between the two directions of transmission must not exceed 43 dB. Permissible noise values are −50 dBm0p

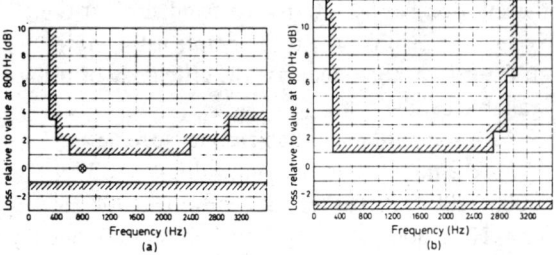

Fig 4.23 Attenuation/frequency hypsogram for international telephone circuit (CCITT M58): (a) 4 kHz-spaced channels; (b) 3 kHz-spaced channels

(category A at 4 pW/km), or −44 dBm0p (Category B = 40 000 pW). Group-delay distortion is not specified.

400, 600, 800 (single-frequency test), 1000 (permitted), 1400, 2000, 2400, 3000, 3200 and 3400 Hz (the two latter being excluded from the 3 kHz-spaced channels).

Relative levels of normal and reserve lines, at change-over point, should preferably be the same, and attenuation/frequency characteristics of the two lines at this point should not differ at any frequency by more than 2 dB.

Use of the edge channels of a coaxial group (channels 1, 12 or 16) is to be avoided since the presence of group-filter characteristics may introduce greater distortion than on other channels in the group.

The near-end signal/crosstalk ratio between the two directions of transmission should be at least

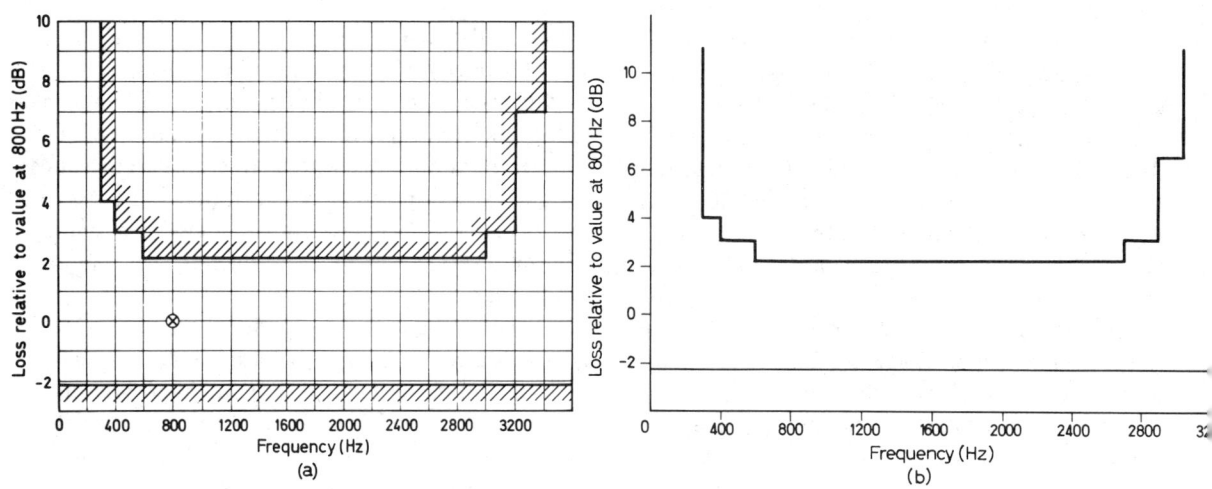

Fig 4.24 Attenuation/frequency hypsogram for international 24-circuit VF telegraph bearer circuit (CCITT M81): (a) 4 kHz-spaced channels; (b) 3 kHz-spaced channels

International Bearer Circuits for MCVF Telegraphy at 50, 100 or 200 bauds (CCITT M81). This is a 4-wire circuit constituted entirely on coaxial or carrier plant, either Type 1 (4 kHz-spaced channels) or Type 2 (3 kHz-spaced channels). Values for Type 3 (audio circuits) are specified, but not quoted here. The circuit, normally set up on carrier or coaxial plant as a channel of a direct group (i.e. a single link) is free of terminating units, signalling equipment and echo suppressors.

The limits of attenuation/frequency distortion are as shown in Fig 4.24. Test frequencies are 300,

43 dB. The relative power level at the input of the VF telegraph link should not exceed −3·5 dBr* and should be at least +3·5 dBr at the output of the link. The permissible noise value is −41 dBm0p. For VF telegraph bearer circuits it is desirable to specify the unweighted noise values. If noise is uniformly distributed over the band, and there is no significant noise outside this band, the unweighted noise is 2·5 dB higher than the weighted value (e.g. the value −38·5 dBm0 is equivalent to −41 dBm0p). For bearer cir-

*See Appendix C.

cuits carrying synchronous VF telegraph systems, higher noise values can be tolerated, e.g. −30 dBm0p (for a particular system). Limiting values for group-delay distortion are unnecessary for VF telegraph bearer circuits, and are not specified. The level of any interfering harmonic frequency due to mains supplies should not exceed −45 dB. The difference between sent and received audio frequencies for each direction of transmission would not exceed ±2 Hz.

Leased International Circuits for Facsimile Telegraphy (CCITT M88).

Permanent facsimile telegraph circuits are set up and lined up as 4-wire circuits between renters' offices.

A part-time leased circuit is a nominated circuit normally in use for telephony; it is extended to the phototelegraph station on a 4-wire basis, with terminating equipment, echo suppressors, etc. disconnected. The overall attenuation levels of the extended circuit would be in accordance with the values shown in Fig 4.23 for the international telephone circuit.

For a phototelegraph transmission using AM, the attenuation/frequency distortion between photo-telegraph stations should not exceed 8·7 dB over the transmitted frequency band.

To limit phase-frequency distortion, differences in group-delay time of the circuit, over the photo-telegraph transmission range, should not exceed $\Delta t \leqslant 1/2f_P$, where f_P is the maximum modulating frequency for the definition and scanning speed concerned.

Differences between group delays at various frequencies and limitation of the transmission band give rise to transient phenomena which limit photo-telegraph transmission speed. At a given speed, circuit quality depends upon the number of 12-channel HF links which may be used in tandem. For a world-wide chain of 12 circuits in tandem, limits of difference between group delays over the transmitted frequency band should not exceed the values in Table 4.1. The use of channels 1 and 12 (or 16) of an HF system are avoided for phototelegraph trans-

missions, due to the extra phase distortion introduced by filters. The above details apply to Category A circuits.

Table 4.1 Limits of differential group-delay distortion

Circuit	At lower frequency limit (ms)	At upper frequency limit (ms)
International chain Each national	30	15
4-wire extension	15	7·5
Whole 4-wire chain	60	30

Wideband (Group) Circuits for Facsimile Telegraphy (CCITT H14).

Attenuation/frequency distortion over the band 64·2−103·8 kHz should not exceed 3 dB. Group-delay frequency distortion over the range 64·2−102·5 kHz should not exceed 25 μs, rising to 45 μs at 103·8 kHz.

References

1 Cobbe, D. W. R., 'Noise Interference from External Sources', *POEE Journal*, **48**, p. 39 (1955).
2 Bordiss, H. J. K. and Davies, A. P. 'A 12 Mhz Coaxial Line Equipment (CEL 8A)', *ibid.* **54**, p. 73 (1961).
3 Halsey, R. J., 'The Commonwealth Transpacific Telephone Cable (Compac)', *IEE Journal (E & P)*, **10**, p. 379 (1964).
4 Myerson, R. J. W., 'St. Margaret's Bay − La Panne 120 circuit Submarine Cable System with Transistor-type Submerged Repeaters', *POEE Journal*, **60**, p. 1 (1967).
5 East, F. R., 'The Properties of the Ionosphere which Affect HF Transmission', *Point-to-Point Telecommunication*, **9**, p. 5 (1965).
6 Jones, D. G. and Edwards, P. J., 'The Post Office Network of Radio Relay Stations', *POEE Journal*, **57**, pp. 147 and 238 (1964).
7 Baker, A. E. and Brice, P. J., 'Radio Wave Propagation as a Factor in the Design of Microwave Relay Links', *ibid.* **62**, p. 239 (1970).
8 Halsey, R. J., 'The Economic Usage of Broadband Transmission Systems', *ibid.* **51**, p. 212 (1958).
9 Back, R. E. G. and Withers, D. J., 'The Development of the Intelsat Global Satellite Communication System', *ibid.* **62**, p. 207 (1970); **63**, p. 1 (1970).
10 Taylor, F. J. D., 'Intelsat − The International Telecommunications Satellite Consortium', *IEE Journal (E & P)*, **17**, p. 8 (1971).
11 Simpson, W. G. *et al.*, 'The 60 MHz FDM Transmission System', *POEE Journal*, **66**, p. 132 (1973).

5 Relays

This chapter is restricted to relays not described elsewhere: the regenerative repeater — a distortion-correcting relay — is included.

5.1 Polarized relays

This type of relay is used when it is required to repeat telegraph signals in d.c. circuits; in the United Kingdom these signals are in the double-current

after its designer). The sensitivity of these relays is due to the differential characteristic of the magnetic circuits, by which the flux due to the operating current has only to overcome the unbalance flux present while the armature tongue is resting against either fixed contact; the armature is accelerated to the opposite contact under the influence of the combined fluxes due to the operating current and a permanent magnet.

Fig 5.1 Polarized relay No. 4199, showing cover and jack (*Courtesy of STC*)

mode. Although the directional response of the relay is important, design is basically aimed at producing a highly-sensitive relay with a balanced moving system which is very stable both magnetically and mechanically, and whose performance is independent of the direction of its armature movement so that negligible signal distortion is introduced. With switching relays, operation depends on magnetomotive force, but release depends upon mechanical force.

The two main designs of polarized relay in use are known as the 'bridge-type' (from the form of magnetic circuit) and the Carpenter type* (named

*Detailed descriptions of bridge-type and Carpenter relays will be found in *Introduction to Telephony and Telegraphy* (E. H. Jolley), Pitman.

The STC Relay No. 4199. With the increased tendency towards using semiconductors and other small components, the need arose for a polarized relay which is small yet shows high sensitivity, good stability, low signal distortion and freedom from interference by adjacent components. These requirements are met in the STC relay No. 4199 (Figs 5.1 and 5.2) which measures little more than $2 \times 1\frac{1}{2} \times 1$ in. (50 x 40 x 30 mm), excluding terminal tags, and weighs 4·6 oz (132 g).

The relay is enclosed in a moulded cover of polypropolene secured to the relay by a captive screw. A wire handle, normally flat against the cover, is used to withdraw the relay from the jack if maintenance is required. The performance of the relay is not influenced by the presence of the cover.

The material of both fixed and moving contacts is copper–palladium. The fixed contacts are dome-shaped and riveted to cantilever springs which are insulated from the relay frame by moulded bushes. The spring is tensioned against an adjusting screw which locates in a groove on a moulded insulating stud near the free end of the spring: this provides for adjustment of the contact gap.

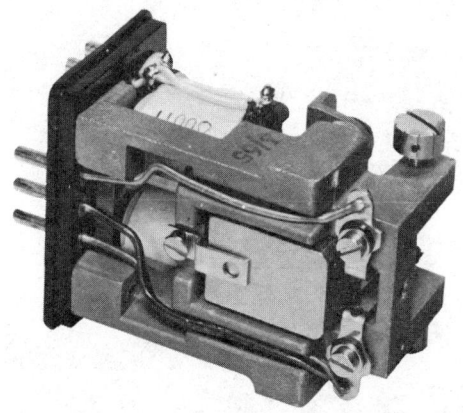

Fig 5.2 Polarized relay No. 4199 — rear view (*Courtesy of STC*)

The permanent magnet, of sintered alnico, gives high values of BH_{max} while not suffering from the drawbacks of the otherwise more efficient aniso-tropic alloys. The single L-shaped magnet is held in place by a clip.

The armature, which carries undirectional flux only, is of annealed iron, and is the balanced type, hinged at its gravitational centre by a thin torsional spring of stainless steel or beryllium–copper secured between the laminations of the armature. This assembly is mounted on a yoke which in turn is positioned on the relay frame in such a way that the mechanical and magnetic neutral axes of the relay coincide. The magnetic strength, together with proportioning the gaps in the magnetic circuit, ensure that the polar pull on the armature swamps any mechanical stiffness in the armature suspension. This form of construction results in a high degree of side stability with a consequent reduction in the tendency for contact bounce. Bias variation over wide ranges of frequency and excitation is also reduced to small amounts by the method of armature balancing described. At its front end the armature carries a pair of anti-chatter springs

to which are riveted the disc contacts; the contour of these springs and the frictional tension between them contributes to the general high stability and performance of the relay. The armature carries two thin beryllium–copper springs to which are riveted the disc-shaped contacts. The distance travelled by the armature between fixed contacts is set at 0·004 in. by use of the fixed contact-spring adjusting screws; rotation of the calibrated screw-head by 1/10th turn corresponds to a contact movement of 0·001 in. The adjusting screws are self-locking, due to a nylon insert in the frame.

The poles are of Permalloy C of large cross-section since they carry both alternating and unidirectional flux. Provision for adjustment of pole gaps after manufacture is unnecessary.

The core on which the windings are assembled — the only part of the magnetic circuit to carry alternating flux — is of Permalloy C laminations, a material of high permeability at low excitation, low hysteresis and small remanence.

The winding is disposed on two moulded spools which are a close sliding fit on the tape-wound core. Each spool can accommodate two independent coils, which allows considerable flexibility in winding arrangements to meet such requirements as parallel, balanced, concentric or bifilar windings, low capacitance or high insulation; the coils are wound with the exact number of turns with ±15% tolerance on resistance. Particular attention is given to chemical purity of all materials to avoid corrosion which might otherwise occur due to minute leakage currents at high (anode) voltages, and to permit operation in tropical climates. The winding is designed to dissipate 2 W.

The magnetic circuit (Fig 5.3(*a*)) is of the bridge type, shown by analogy in Fig 5.3(*b*). The flux in the 'detector' arm of the bridge is sensibly constant but the flux generated by the coil fluctuates. This magnetic circuit has the advantage that the direct and alternating flux circuits are well decoupled. The core of the coil is not saturated by permanent flux and it carries only a small portion of the polarizing flux; the core is thus of small cross-section. The operating flux from the coil passes direct to the pole gaps by pole pieces of large cross-section, with but little flux leakage via the back gaps. The permanent flux is conveyed to the pole pieces through the back gaps BG and thence through the pole gaps PG and armature back to the opposite pole of the magnet through the armature pickup shoes and the front gap FG. The magnet gaps BG create a degree of decoupling between the polarizing and exciting circuits.

In operation, the armature, being a series element in the polarizing circuit, has an induced magnetic polarity. As the pole gap is in series with the magnetic circuit, the pole pieces exhibit a magnetic

for bounce-free operation is 10 AT at modulation rates up to 200 bauds. Negligible signal distortion is introduced by the relay.

5.2 Testers for polarized relays

After the contacts of a polarized relay have been cleaned and adjusted it is necessary to check that the sensitivity, neutrality and transit time are

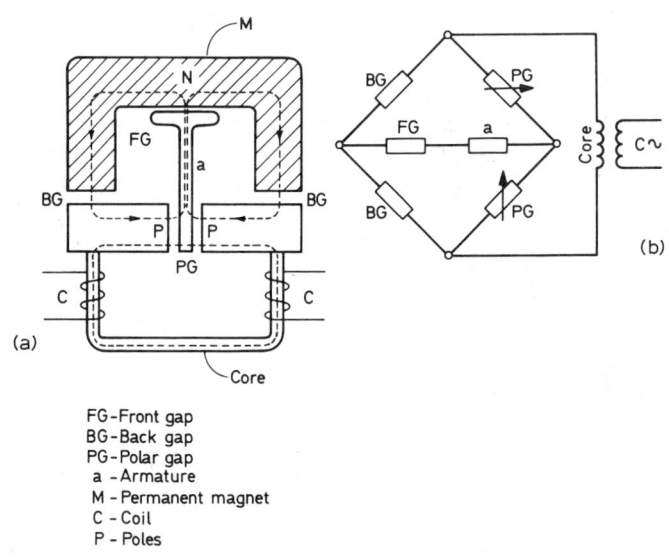

FG—Front gap
BG—Back gap
PG—Polar gap
a —Armature
M — Permanent magnet
C — Coil
P — Poles

Fig 5.3 Magnetic circuit of No. 4199 relay

polarity opposite to that of the armature tail lying in the gap; both pole pieces are of the same polarity, being connected to the same magnet pole. If the armature is held centrally in the pole gap it is under equal attraction from each pole. Consequently, when the coils are energized the flux generated adds to the strength of one pole and subtracts from the strength of the other; the armature is more attracted to the strengthened pole than to the weakened one. Once balance has been destroyed, the permanent flux is sufficiently large to ensure that the armature remains on that contact until reversal of the coil flux unbalances the pole strength in the opposite direction: the armature then changes over to the other contact.

The relay is available in plug-in form (12-way) for transmission equipment and 3000-type relay plates; alternatively, as a wired-in relay. Adjacent apparatus has very little influence on operation of the relay. The minimum recommended excitation

within limits. A relay-test set comprises jacks suited to the relays to be tested, keys to set up the testing conditions and a centre-zero ammeter.

Sensitivity Test (Fig 5.4(a)). This test also checks the continuity of the windings. By throwing appropriate keys, all windings can be either connected in series or tested separately, resistors being switched into circuit to limit the current to the correct minimum operating value. The tongue, in series with the meter, is connected to the centre-point of a potentiometer across the ±80 V supply. After the d.c. key KDC is operated, use of the MARK/SPACE key KM/KS reverses the direction of current in the windings, movement of the tongue to either contact being indicated by a reading of 20 mA on the centre-zero meter. The side-stable characteristics can be checked by restoring the MARK/SPACE key to normal: there is now no operating current in the

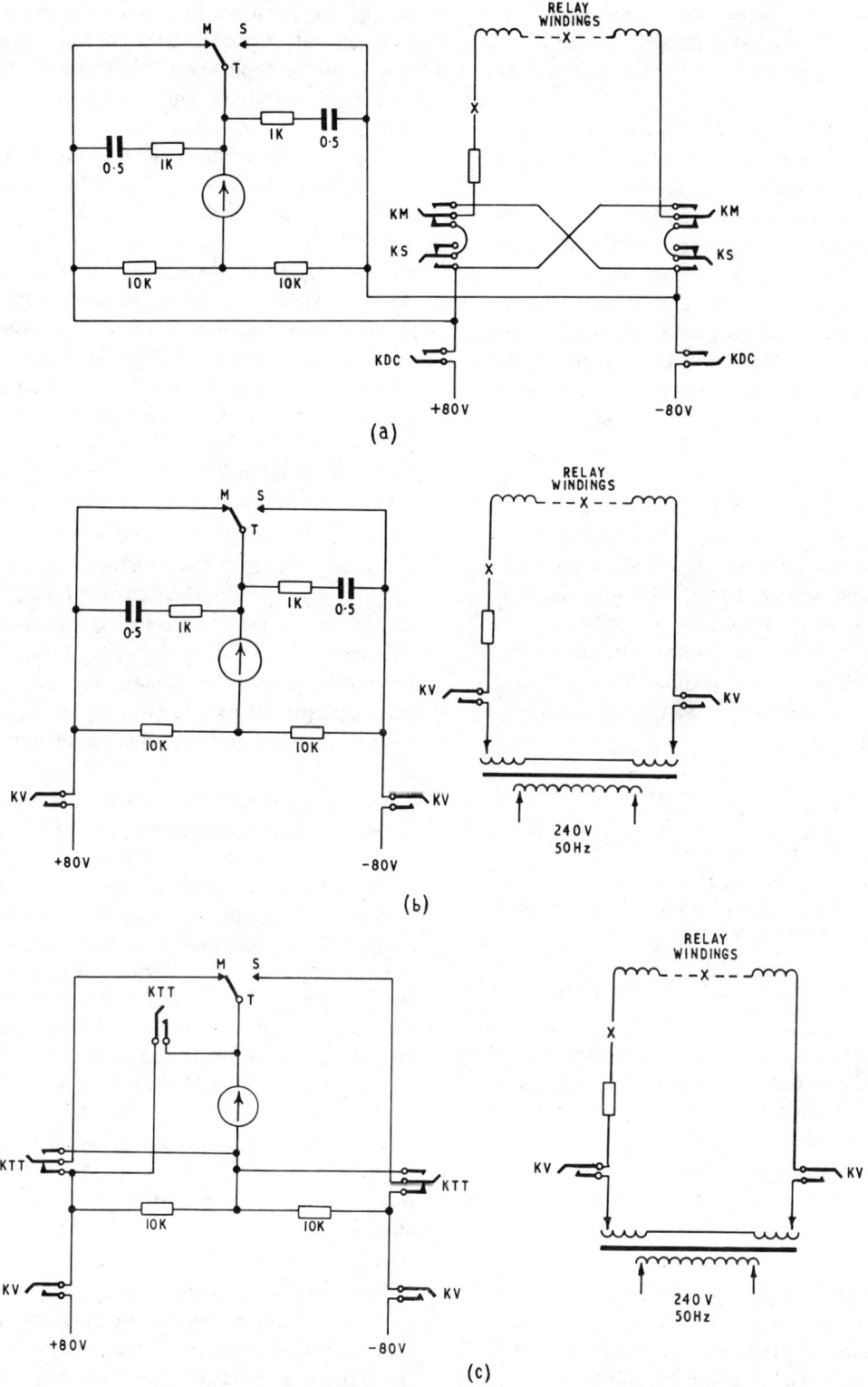

Fig 5.4 Testing polarized relays: (*a*) sensitivity; (*b*) neutrality; (*c*) transit time

windings and the tongue should remain on the contact to which it was last deflected.

Neutrality Test (Fig 5.4(b)). Operation of the VIBRATE key KV applies an adjusted value of a.c. at 50 Hz to the windings (as selected) from a transformer connected to a.c. mains supply. If the relay is neutrally adjusted, the tongue will spend equal times on mark and space contacts and the meter will vibrate on the zero position. If bias is present, the mean position of the needle will give a reading b mA to left or right of zero. The percentage bias δ is $\delta = b/D \times 100\%$, where D is the steady meter reading (20 mA) shown by the sensitivity test.

Transit-time Test (Fig 5.4(c)). In this test, the VIBRATE key KV is again operated to apply a.c. to the relay windings; the TRANSIT TIME key KTT is also operated. The meter is now short-circuited while the relay tongue is resting on either the mark or space contact, current flowing in the meter only while the tongue is in transit between contacts. The transit time, T s, can be calculated from:

$$T = \frac{1}{2f} \times \frac{(D-d)}{D} \text{ s}$$

where f is the supply frequency (50 Hz), d mA is the measured deflection and D mA is the steady meter deflection (20 mA). For example, if the meter shows a deflection $d = 18$ mA:

$$T = \frac{1}{2 \times 50} \times \frac{20-18}{20} = 0.001 \text{ s}$$

The protective resistor bulbs will glow if any contact bunching occurs. A chart accompanying the tester shows bias and transit times corresponding to various meter deflections. If a relay has auxiliary windings, continuity and sensitivity tests are provided for checking performance of these.

5.3 Contact materials

Metal suitable for relay contacts is selected from a small number of materials: only the noble metals – gold, platinum, palladium and rhodium – can be regarded as corrosion-resistant. Silver is cheaper,

has high electrical and thermal conductivity and is the general-purpose material for low-current lightly-worked applications, but its low melting and boiling points and low hardness render it unsuited to high current and voltage. Platinum is hard with high melting and boiling points; compared with silver it has better minimum arc-current limits and is suitable for heavy loads and higher rates of operation. Palladium is superior to silver in the properties mentioned, though not quite to the same extent as platinum which is, however, the more costly. Copper–palladium has the ability to resist the transfer of metal between contacts during high-speed operation. Tungsten has high melting and boiling points, is very hard, and has higher minimum arc-current and voltage limits than the foregoing materials. These properties make it suitable for heavier loads at high rates of operation but it is prone to the formation of high-resistivity oxide films.

Contact resistance for given materials is a function of contact pressure, the current flowing and surface conditions. The presence of current reduces resistance because the heating effect softens the materials at the point of contact, the pressure then causing plastic deformation to increase the area of contact. A wiping action is essential to all types of contact.

Surface contamination by poorly-conducting material is the main factor in any problem concerning use of contacts at low voltage. Such contamination may consist of dust particles, grease or films produced by adsorption or corrosion on relay contacts due to the reaction between contact metal and environment – including insulating material used for relay coils; phenolic materials and paints not free from solvents are particularly undesirable. The effect of films may be aggravated by the occurrence of arcing and sparking which facilitate the formation of low-conductivity deposits; these films and deposits have to be broken down or dislodged by application of sufficient potential. Atmospheric pollution, particularly by sulphur dioxide in industrial areas, and high relative humidity contribute to film formation.

Films due to tarnishing usually consist of semi-conducting oxides or sulphides commonly encountered with silver, copper and their alloys. The contact resistance associated with these films decreases rapidly with temperature and voltage rise. Cases exist in which an otherwise unsatisfactory contact is made serviceable by superposing a steady current of a few mA drawn from a supply having a potential of several tens of volts. This technique is known as *contact wetting*; a wetting current of 5 mA from a 50-V exchange battery is often used.

In telegraph and telex exchanges all switching is performed at the ±80-V d.c. points of connection and not at low-voltage VF a.c. potentials.

Purely mechanical wear of contacts is negligible. Erosion of contact material occurs from arcing and sparking, metal being transferred from one contact to the partner contact resulting in formation of corresponding pip and crater with consequent reduction in contact pressure and conductivity.

5.4 The mercury-wetted reed relay

In this relay the contacts are maintained continually wet with mercury. Fitted with a change-over contact unit, this relay can be used to replace the conventional polarized telegraph relay, with the advantage that attention to the contacts is neither necessary nor possible. The contacts are highly reliable in service and when used within the ample current rating they have an extremely long life estimated to be 10^9 or more operations. The relay is small enough to be mounted directly on printed-circuit boards.

Magnetic action follows the principles for traditional polarized relays. The nickel–iron reed (Fig 5.5), tapered in width, is welded to a metal tube sealed

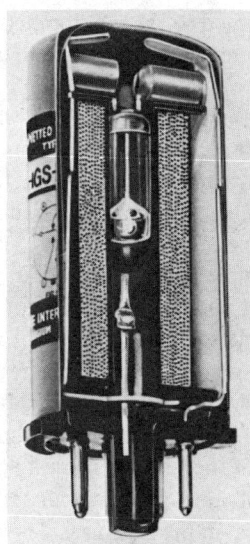

Fig 5.6 Section of mercury-wetted reed relay (*Courtesy of C. P. Clare International*)

into one end of a glass tube. Fixed pole-pieces of a magnetic alloy, carrying platinum ball contacts, are sealed into the opposite end of the glass tube so that the tip of the reed lies between them and slightly overlaps their ends; platinum is very slow to oxidize and readily wets with mercury. The platinum ball contacts are located close to a nodal point for a principal mode of vibration of the reed: this reduces contact bounce sufficiently to enable the mercury surfaces to sustain contact. The reed is cut with a series of fine parallel longitudinal grooves to act as capillary channels; the surface is given a granular finish and plated with a metal such as nickel which is capable of being wetted by mercury. The metal tube supporting the armature is used to introduce the mercury and to evacuate and pressurize the capsule with hydrogen at 150 lb/in.^{2} or more, prior to sealing the glass tube. The hydrogen provides an environment that removes oxide films and also cools the contact region. A sectional view of a mercury-wetted relay is shown in Fig 5.6.

In use the relay must be mounted within 30° of the vertical plane, though some have been designed for mounting at 45° or even horizontally. Mercury is fed to the entire surface of the reed by capillary action of the grooves in the way a lamp wick is fed with oil. When the reed touches either platinum contact, a mercury bridge is formed between the reed contact and the pole-piece contact. Wetting in

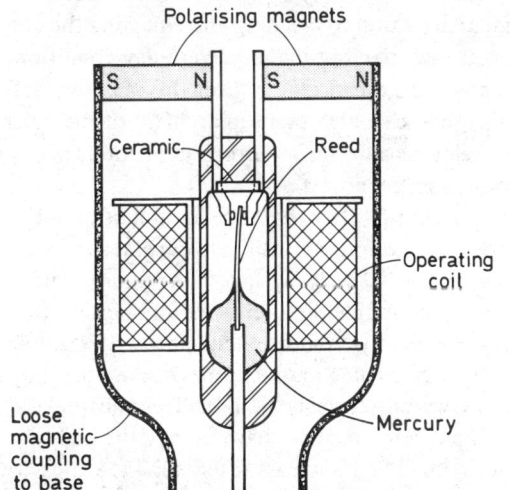

Fig 5.5 Principle of mercury-wetted reed relay

this area is confined to the platinum contact by oxidizing the pole-piece prior to sealing it in the glass and by fitting a ceramic wetting-resistant detail between the pole-pieces. The contacting area is larger than the usual small point contact because a fillet of mercury surrounds the contacting surfaces. When the two surfaces separate, the mercury stretches into a thin filament and then breaks, giving a clean snap-action completely free from bounce. It is characteristic of mercury to break, not in the middle, but at two points so that a thin rod of it is isolated, quickly changing into a sphere by surface tension. This minute ball of mercury (about 0·001 in. diameter) falls from the contacts to the mercury pool at the bottom of the tube. If sparking occurs on breaking, a small amount of mercury is vaporized and it condenses on the glass capsule to return eventually to the reservoir. The loss of mercury disturbs the equilibrium of the capillary system and more of it is fed up the reed: in effect a new mercury-contact surface is provided for every operation. If sparking is excessive, a deposit may be formed on the metallic surface, wetting the contact area may cease and contact deterioration ensues: use of a spark-quench circuit is very important.

Small permanent magnets are soldered to the pole-piece projections and extended by Permalloy plates to complete a loose magnetic coupling path to the lower end of the reed; this high-reluctance path is necessary to prevent magnetic saturation. The reed moves to one or other of the pole-pieces and so partially completes the magnetic circuit on that side. Energizing the operating coil with current in the appropriate direction will change the direction of reed magnetism and so cause the reed to move with very short transit time to the other pole-piece where it will remain until a current of opposite direction is passed through the coil. This bistable operation is achieved with carefully-balanced strengths of the permanent magnets.

By careful control of contact gaps and the amount of mercury in this area, bridging of all three contacts can be avoided. Some protection must be provided against bridging due to the possible presence of a drop of mercury between contacts when the relay is first used after being transported or inverted. The contact loading is limited only by the ability of the mercury to remain liquid and not be vaporized by the power dissipated

— a current up to 5 A with a limiting load of 100 VA is possible, subject to adequate external protection for the contacts; operation and release times are of the order of 1 ms. The mercury-contact reed relay is less sensitive than the traditional form of electromagnetic polarized relay, energization of 15 AT — or more if the relay is one-side-stable — being required. The relay can be operated at frequencies almost up to the reed resonant frequency of 100 Hz.

A valuable property of this relay is that the contact resistance falls from the open-circuit value (in the range 15—50 mΩ immediately contact is made between two clean mercury surfaces, and remains constant within 2 mΩ throughout the life of the relay. With the dry-reed relay the contact resistance is relatively high on first light-closure, falling comparatively slowly while contact pressure builds up. In telegraph equipment the mercury-wetted relay is used in 62-type MCVF circuit cards and in broadcast units.

5.5 Electronic relays

The polarized relay has always been a vital component in telegraph equipment. On busy circuits these relays are heavily worked, often at high modulation rates, their contacts controlling highly reactive loads. On synchronous (ARQ) radio-teleprinter systems such relays operate continuously at modulation rates up to 192 bauds. Despite the use of protective devices, contact erosion occurs leading to signal distortion and eventually to relay failure. Therefore regular attention to cleaning and adjusting the contacts is essential, but under present-day conditions the costs and effort required for this maintenance have become so high that replacement of the polarized relay by a device requiring no maintenance becomes an economic necessity.

Very shortly after the introduction of MCVF systems it was found possible to supersede the polarized SEND relay by a *static relay* or modulator, using copper-oxide rectifiers; the function of this relay was simply to short-circuit the carrier supply to effect amplitude modulation. On the other hand, the MCVF output relay — and indeed most telegraph polarized relays — have usually the highly reactive load of a low-pass filter No. 4B, a capacitive cable circuit, a pulsing relay, or the electromagnet of a selector or teleprinter.

Availability of new materials and devices, particularly the semiconductor, has opened new possibilities towards development of a telegraph relay virtually free from wear. The situation is somewhat fluid and development has proceeded along two paths — firstly a replaceable device such as the mercury-wetted relay which has a long but limited life requiring no maintenance. The relative insensitivity of this device prevents its use as a direct replacement for every type of polarized relay. Secondly, a semiconductor device of indefinitely long life, also needing no maintenance, has been developed. The use of semiconductors to control potentials of ±80 V standardized for vast existing networks presents some difficulties, particularly as the device must safely withstand the effect of a full-earth fault on the line. Electrical isolation between input and output is also a frequent requirement.

A number of designs for an electronic switch operating at ±80 V have been produced. Because of the lack of a transistor capable of withstanding the high voltage, initial designs employed either a number of transistors in cascade to share the voltage, or a pair of thyristors. The former approach led to complex designs using a large number of components, particularly if isolation were required, and the cost of electronic switches to operate at ±80 V tended to be high. The electronic relay may also need more power to operate it compared with a sensitive electromagnetic polarized relay.

Thyristors have been successfully used but not without some disadvantages. It is not difficult to turn a thyristor ON by a pulse and gate, but a separate control is required to turn it OFF. If either control fails for any reason, then either two thyristors will be left ON, short-circuiting the telegraph supplies and effectively earthing the output of the device, or both thyristors will be turned OFF, leaving the device in an unswitched state. Due to complexity of design and high cost of thyristors, this type of electronic switch tends to be expensive although isolation can be readily achieved.

Using transistors, one technique for controlling output by input and providing isolation between them is to use the input signal to switch one of a pair of oscillators whose outputs are alternatively coupled via transformers to produce the control voltage.

With the availability of high-voltage transistors a number of successful designs of electronic switch have been produced. Since a transistor is fully controlled through its base electrode, the output of the device can at all times be made dependent upon the input condition in such a way that when one side is in the ON state — 'bottomed' — the other side is held OFF as a consequence, and the possibility of both output 'sides' becoming switched ON at the same time is eliminated. Figure 5.7 shows an electronic switch which accepts signals at ±6 V from ARQ synchronous equipment and repeats them at ±80 V for onward transmission over a cable network or for teleprinter operation.

With no bias applied and the potential of the input wire negative relative to earth, TR1 is cut off and TR2 conducts. As a result, current flows into the base of TR5 and it conducts. The emitter current of TR5 is the base current of TR7; transistor TR7 is bottomed and in this condition provides a low-resistance connection between the output wire and the −80-V supply. A negative potential is fed back via R9 to the input circuit to maintain the device firmly in the −80-V switched condition until the input wire becomes positive.

The feedback circuit also serves to trigger the relay into the required state. On a transition from positive to negative potential at the input, as soon as the potential has become negative by only a small amount changes occur in the conducting states of TR2, TR5 and TR7, causing a negative potential to be fed back from the output. Regenerative action round the loop then quickly puts the device fully into the −80-V conducting state. Because of this very rapid changeover it is possible to use transistors that are rated for only low-power dissipations.

The reason for including TR5 in the circuit is to obtain the gain needed to hold TR7 in the bottomed condition, even under fault conditions. If TR7 departed from the bottomed state, the voltage across it would rise and the power dissipated in it would exceed the safe limit. Transistors TR5 and TR7 are both n-p-n types capable of withstanding a backward voltage of 200 V, a value in excess of the voltage they are required to withstand in the present circuit.

When a positive potential is applied to the input, TR1 conducts and TR2 is cut off. With TR2 in a state of non-conduction, TR5 and TR7 also cease to conduct. The collector current of TR1 produces a voltage across R3 so causing base current to flow in

TR3. The voltage across R7 due to the collector current of TR3 causes base current to flow in TR4. The collector current of TR4 is also the base current of TR6 which now conducts and connects the output wire to the +80-V supply. A positive potential is fed back via R9 to the input circuit to maintain the device firmly in the +80-V switched condition.

of TR4. However, the collector of TR3 and the emitter of TR4 are each at earth potential and the desired voltage limitation is achieved.

Diodes D1—2—3—4 are included in the circuit to prevent damage to the output circuit if the device is connected to its load via a low-pass filter 4B. The input voltage to such a filter is likely to assume

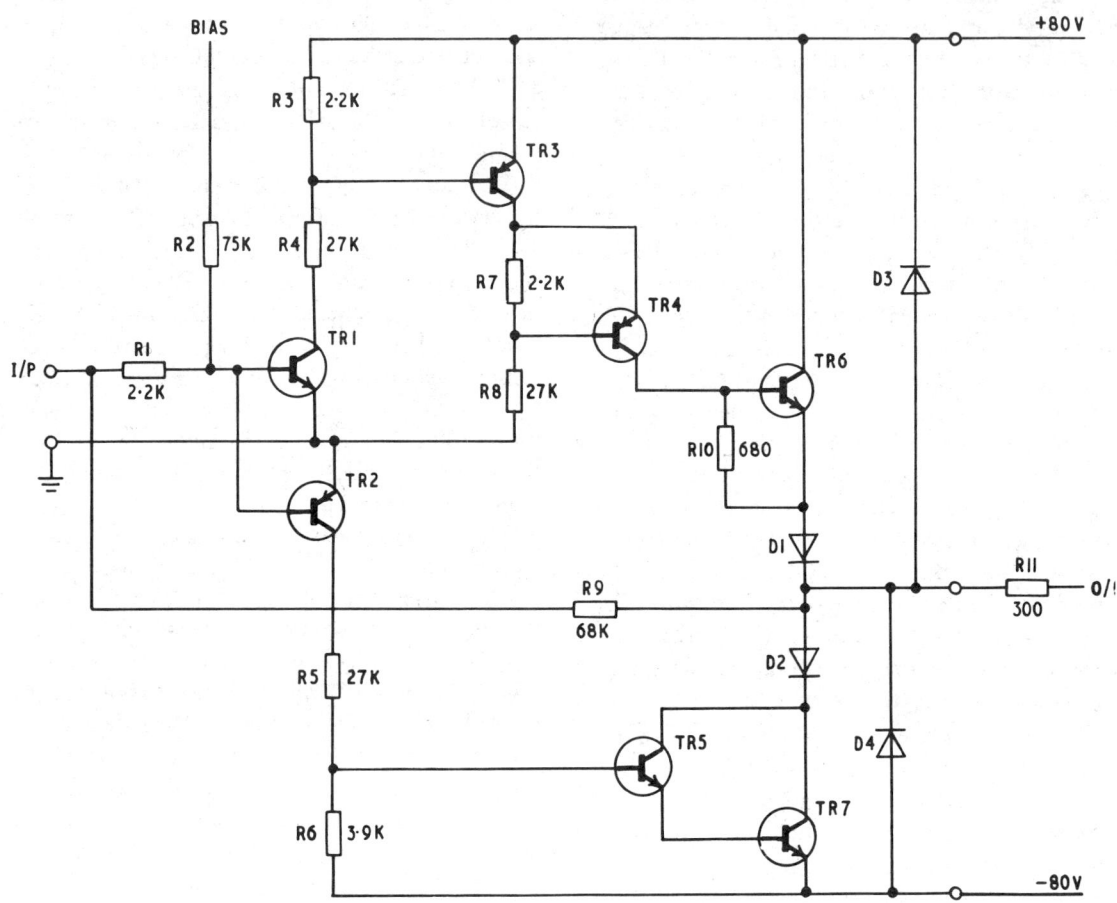

Fig 5.7 Electronic relay

Transistors TR3 and TR4 are necessary to obtain the current gain needed to hold TR6 in the bottomed condition, even under fault conditions. Fundamentally, one transistor could effect this, but inexpensive p-n-p transistors that can withstand backward voltages of 160 V are not at present obtainable. The transistors used will withstand a 125 V backward potential and the circuit design ensures that this value is not exceeded. When the output is switched from +80 to −80 V a potential of 160 V exists between the emitter of TR3 and the collector

instantaneous values exceeding 80 V and diodes D1—2 prevent reverse potentials being applied to the output transistors. Diodes D3—4 act as voltage clamps and provide a low-impedance path for the reverse current, the energy being dissipated in the 300-Ω resistor.

A biasing arrangement, consisting of resistor R2, may be connected to the input circuit if required. Values may be chosen suitable for R2 such that, by applying the appropriate potential, the output may be switched as desired to either the positive or negative condition if the input circuit is left open. By varying the value of R1, which limits the current into the base of TR1, it is possible to operate satisfactorily over a wide range of nominal input voltages.

In this electronic switch, silicon transistors are used: the relay is virtually distortionless at modulation rates up to 200 bauds. Isolation between input and output is neither required nor provided by this design.

On account of the higher cost compared with mercury-wetted relays, use of the electronic switch tends to be restricted to applications requiring modulation rates in excess of 50 bauds. For newly-designed equipment the electronic relay no longer needs to be in plug-in form to replace an existing electromagnetic relay and the output relay tends to lose its separate identity and becomes integrated with the output stage in the overall equipment — a factor which tends to lower cost.

5.6 Regenerative repeaters

A well-designed and adjusted telegraph relay will repeat signals without introducing additional distortion; it follows that any distortion present in the input signal will not be corrected by the conventional relay but will be repeated into the forward channel or equipment. Telegraph networks are designed so that aggregate distortion from transmitter and transmission channel is well within the margin of the receiving device. In a minority of cases this cannot be achieved except by regenerating the signals at some point and repeating them free from distortion into the succeeding channel or equipment.

Basically, two types of regenerative repeater are required, depending on whether a channel operates in the start—stop mode or in the synchronous mode. A regenerative repeater is a combined receiver—transmitter: on the receive side it is designed to accept distorted signals with a receive margin approaching ±50%; on the transmitting side the signals are retransmitted virtually free from distortion.

The conventional mechanical teleprinter has a receive margin slightly greater than ±40%. Because of the inconsiderable difference of their margins, there is little to be gained from using a regenerative repeater at the receiving end of a teleprinter circuit

to repeat signals into the teleprinter. The position in a line circuit where a regenerator can prove advantageous is near the midpoint of a connection between transmitter and receiver, over which signal distortion otherwise exceeds teleprinter receive margin; the regenerator splits the channel into two sections, neither of which should now introduce distortion in excess of the respective receive margins of regenerator and teleprinter. If a connection between two teleprinters is required to pass over more than six VFT channels in tandem it would be necessary to insert a regenerative repeater midway in the chain of channels: signals would then pass into the second section of the transmission path with zero distortion. Again, if a telex station were remotely located with respect to the transmission network so that the specified transmission limits could not otherwise be met, it would be necessary to include regenerative repeaters in the station line at the local telex exchange to cover both directions of transmission.

If two radio-telegraph links are connected in tandem, considerable advantage can arise from inclusion of a regenerative repeater at the junction point.

At radio-receiving stations, distortion on signals arriving over the radio path may be so high that regeneration is needed before feeding the signals over a VFT channel to a telegraph installation.

At a radio-transmitting station, if signals from the telegraph station have arrived over a landline channel which introduces a significant amount of distortion, it is well to regenerate the signals so that they are emitted from the aerial free from distortion.

The start—stop Regenerative Repeater* If signals are being received over a radio link and passed into a start—stop regenerative repeater, a code element will occasionally suffer excessive distortion (up to ±50%), or its significant condition may be changed due to noise or fading. Without error-correction equipment it will be impossible to determine the correct character and one error will result. Only if this error happens to result in the loss of a shift signal or a carriage-return signal will succeeding printing

*Recommendations R60—61.

errors result from the single propagation error. However, if the faulty element is the stop signal, the essential rest condition of start—stop working is absent and a further, false, character cycle will immediately follow. If, during this false cycle, a genuine character cycle commences to arrive, the start element may well be recorded as though it were a code element and a false character will result. If further characters are received from an automatic transmitter at maximum character rate it is possible for the receiving device to remain out of step with received signals for a considerable period, during which many false characters will be recorded.

This type of mis-operation is illustrated in Fig 5.8 for a regenerative repeater receiving the characters THE with a mutilated stop element at X; symbols A and Z indicate the positions of start and stop signals respectively. In the circumstances shown, this mutilation results in reception of the characters T CARRIAGE RETURN E.

If, while a channel is in the idle (mark) condition, a line interruption causes a brief transition to the space condition, a regenerative repeater may record this as a genuine start signal. If, during the false cycle which follows, genuine signals arrive, they will be examined out of phase and incorrect reception will result. For the particular example shown in Fig 5.9, the transmitted characters THE are incorrectly recorded as combination 32, CARRIAGE RETURN E. The diagrams assume that phase is restored by the time the third character arrives; in practice, out-of-phase conditions could exist during reception and regeneration of several characters.

If the incidence of the above types of fault is high, the error rate could in fact be higher *with* the regenerative repeater in circuit than without it, since

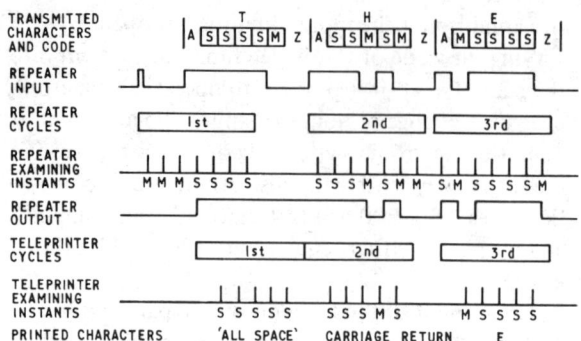

Fig 5.9 Effect of false short-space signal

each false cycle could result in a number of subsequent errors due to incorrect phasing of the start—stop instruments. For this reason, especially when used on radio channels, the design of a regenerative repeater may include two supplementary features: (1) *short-start rejection,* a device in which all start signals are examined, and if their duration is less than a specified amount (e.g. 10 ms) the start signal is rejected as being false and the repeater awaits a genuine one before commencing a character cycle; (2) *automatic stop insertion,* by which a stop signal of not less than 20 ms (at 50 bauds) is automatically inserted in each regenerated character at the appropriate period.

On radio channels using automatic transmission, the short-start rejection feature can actually be a disadvantage since it can throw the repeater out of phase on receipt of a genuine, but excessively shortened, start signal. If signals emanate from a teleprinter keyboard there will usually be sufficient time between transmitted characters for the regenerative repeater to complete a false cycle and come to rest before the arrival of the next genuine character. With signals at full rate from an automatic transmitter, if a very shortened, but otherwise genuine, start signal is rejected, the regenerative repeater will commence its cycle of operations on the first *code* element of space condition (provided that it too is not excessively shortened); the regenerator will then remain out of start—stop synchronism until a character is received which has no space elements in the latter part of the code. In general, it may be undesirable to use short-start signal rejection on circuits where automatic transmitters are normally used. For other cases the maximum duration for the start-signal-rejection

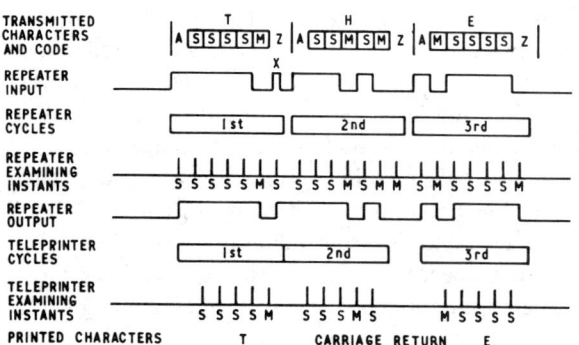

Fig 5.8 Character with mutilated stop signal

decision may well be adjusted according to characteristics of the individual radio path. On the contrary, it may be desirable to arrange the regenerator for automatic *start* insertion — the acceptance of even the shortest start signal as genuine and its regeneration and retransmission with full time value. A teleprinter, by virtue of the inertia of the receiving mechanism, is unresponsive to very short signals and has some degree of immunity from the effect of spurious short start signals.

On switched networks it is necessary to transmit the clear signal — a space condition of at least 300 ms. If regenerative repeaters are used on such networks, it is arranged that automatic insertion of the stop signal is suppressed provided that no signal transitions have occurred during the cycle up to the instant $t = 130$ ms, at which examination for the stop signal is made. This arrangement allows for transmission of the ALL-SPACE combination 32 if used.

The inclusion of a regenerative repeater in a channel restricts the use of that channel to the type of signals for which the regenerative repeater is designed, e.g. 50-baud 5-unit start—stop signals; the channel is no longer 'open' for transmission of signals having other characteristics. On circuits used in switching systems which require transmission of dial pulses, it is arranged that the start—stop regenerator is automatically switched into circuit only when the call becomes effective, i.e. the mark condition is present on both channels of the connection. This treatment implies that dial pulses (which are at a lower modulation rate than teleprinter signals) rarely require regeneration or, alternatively, that they can be regenerated within the exchange switching equipment.

A regenerative repeater functions by examining the significant condition of the incoming signal at its theoretical midpoint, e.g. $t = 10$ ms for 50-baud signals. It follows that retransmission of signals through a regenerative repeater is delayed by 10 ms (at 50 bauds): this is usually of little practical significance.

The value of the minimum character period (in effect the duration of the stop signal) is of considerable importance.

The functions performed by a start—stop regenerator suitable for 5-unit 50-baud signals are:

(1) Its cycle of operations commences on receipt of the start condition.

(2) The duration of the start signal is checked and if it is less than the selected period it is rejected as spurious, and the repeater reverts to the rest condition and awaits a new one. The rejection feature is optional: on some models of regenerative repeater the period for short-start rejection is adjustable from 1—10 ms in steps of 2·5 ms.

(3) The significant condition of the incoming signal is sampled at $t = 30, 50, 70, 90$ and 110 ms, counting from $t = 0$ at the beginning of the received start signal; the polarity for the retransmitted code elements is correspondingly determined.

(4) At $t = 130$ ms the output is forced to mark condition for the stop signal unless all five code elements have been read as space. If this occurs, a stop signal is retransmitted only if one is present in the received signal at $t = 130$ ms, otherwise an uninterrupted space condition is retransmitted until the received signal returns to mark.

(5) At any time later than $t = 140$ ms the repeater is ready to receive, and check, a new start signal. The earliest time at which the new start signal can have been checked is therefore $t = 150$ ms, and this is also the earliest at which the repeater can commence to retransmit a new start signal. Since the stop signal commenced at $t = 130$ ms it follows that the minimum length of stop signal is 20 ms.

The logic circuit for a typical regenerative repeater is shown in Fig 5.10. The multivibrator is the basic timing element of the repeater. In conjunction with the incoming signal it controls the timing for operations of bistable triggers (BST), a monostable trigger (MST), an OR* gate (GO), five AND* gates (GA) and amplifiers. A polarized output relay is used to repeat the signals at ±80 V.

Character Timing. The waveforms and timing are shown in Fig 5.11. The 50-baud start—stop signal applied to the input is first limited; the limiter output is shown at A. After passing through the pulse-shaping network PS, the edges are improved and the signal is inverted. The input amplifier, IA, inverts the signal and the leading edge of the start pulse causes BST3 to change over. The output at D switches on the multivibrator which produces a 100 Hz output, shown at E, which is fed into a 4-stage binary

*See Appendix B.

counter BST4—7. Since only the positive-going edge of the multivibrator output is used, the mark—space ratio is of no consequence. From the 4-stage binary counter the wave shapes shown at H and J are available. When these are applied to the AND gate GA5, together with the wave at E, the resulting output

goes positive, MST is triggered and at the output a negative-going 10-μS pulse at 20-ms intervals appears, as shown at L. These pulses, applied to the AND gates GA1 and GA4, are sampling pulses which test the polarity of the input signal at the midpoint of each signal element. It is on this process that the

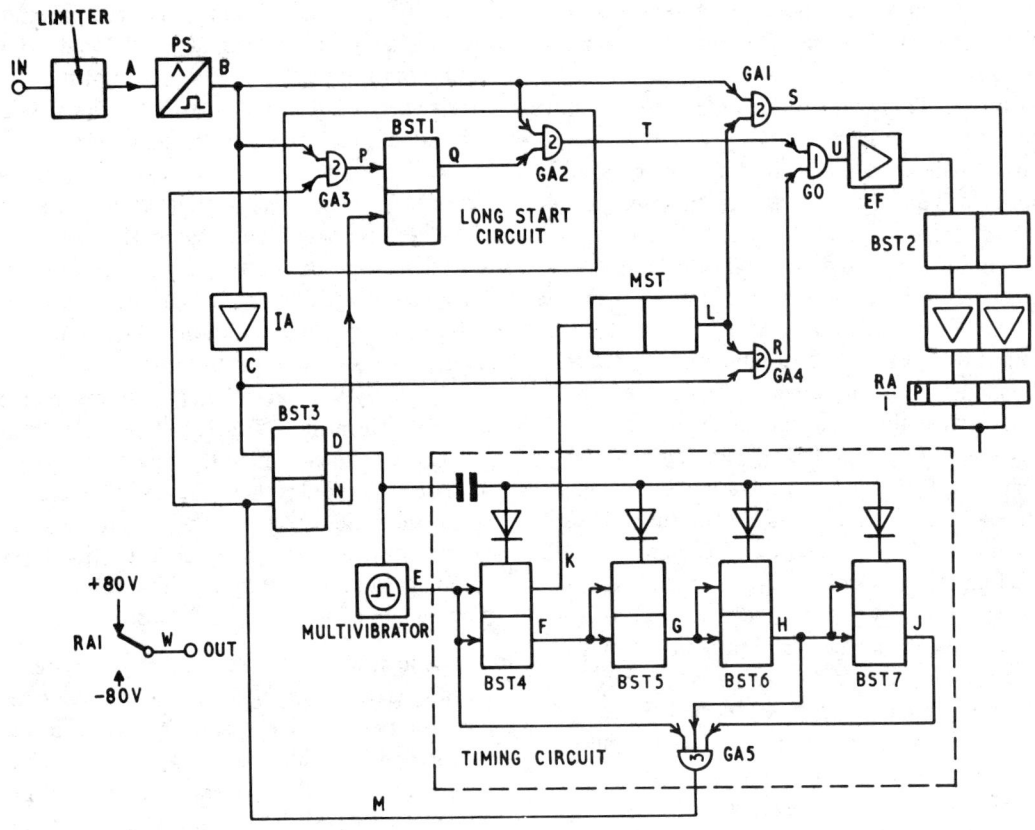

Fig 5.10 Start—stop regenerative repeater — logic circuit

wave is as shown at M. The positive-going edge of this wave switches over BST3 and the wave shape at D is generated. The sum of E (5 ms), H (40 ms) and J (80 ms) is 125 ms. The next positive E pulse (5 ms) stops the multivibrator at $t = 130$ ms and the positive-going edge resets the 4-stage binary counter. One character interval has been timed and the circuit is ready to receive the start signal of the next character.

Character Regeneration. When the start signal is received, the signal shown at K is generated for the duration of the character. Each time this waveform

regenerator depends, for by examining the input elements at their midpoints by the use of a very short examining pulse, considerable distortion on the element transitions can be tolerated.

When the input signal is positive, the wire at B is negative and a negative 10-μs pulse emerges from the output wire of the AND gate GA1 (see waveform S). When the input signal is negative, the wave at C is also negative and a negative 10-μs pulse emerges from the output wire of the AND gate GA4 (see waveform R). Waveform S is fed directly to one side of BST2 and the R wave is fed to the other side of BST2 via the OR gate GO1. The trigger circuit BST2 operates and drives the d.c.

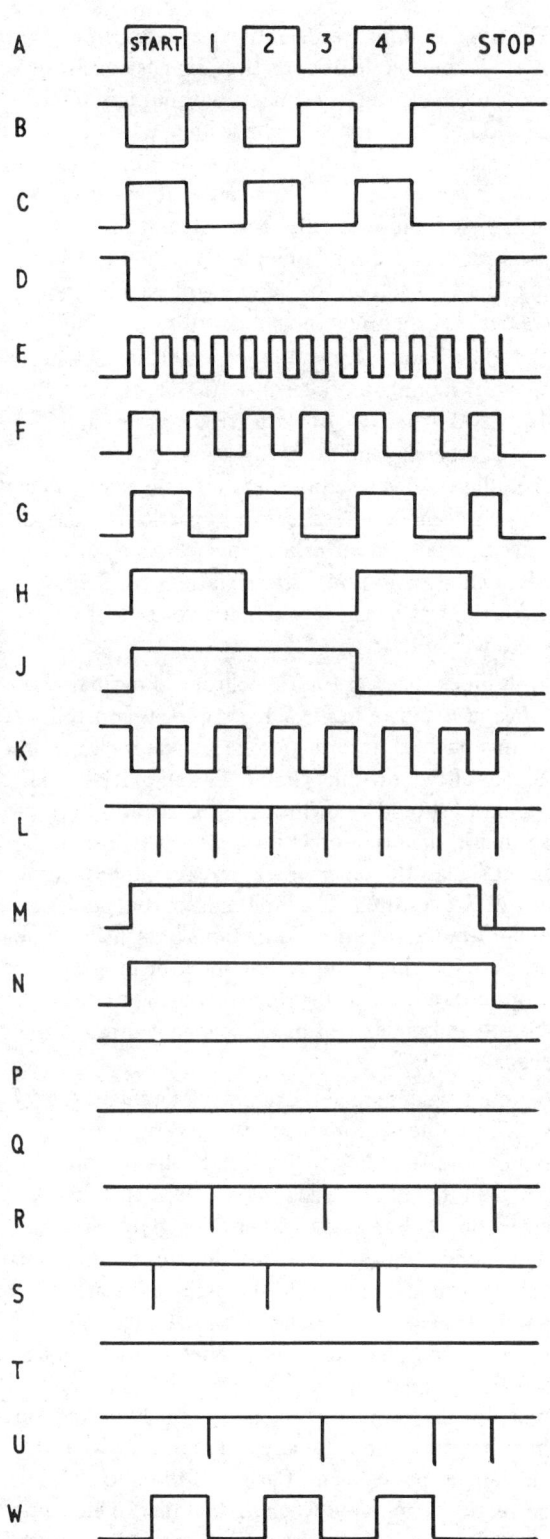

amplifiers and polarized relay so that a regenerated version of the input signal (delayed by 10 ms) is obtained from the relay tongue. The output wave shape is shown at W.

On circuits used for switched networks a disengaged VFT channel may remain for long periods in the space condition. With this design of regenerative repeater, a long-space circuit is incorporated for priming it so that the series of inspection pulses will be applied when the channel reverts to the mark condition; this long-space circuit is also instrumental in permitting a clear signal to pass through the regenerator. Under normal conditions, when a character is received at the input no pulse emerges from the AND gate GA3: the wave shapes at B and M are such that no pulses can emerge. The trigger BST1 is therefore not changed over and as a consequence the AND gate GA2 remains closed. The long-start circuit plays no part when a normal character is applied to the input.

Following transmission of a final start—stop signal and reversion to the long-space condition, the input wave goes to start polarity and remains in that condition for a period in excess of one character before returning to stop polarity. The diagram of Fig 5.12 shows the waveforms and timing for the long space (about 180 ms shown for convenience of this diagram) followed by reversion to the space condition.

The timing circuit operates as described for a normal signal and seven inspection pulses are generated — see wave shape L. Both input (A) and output (W) waves continue to be at start polarity. Because the input wave has this polarity, the wave at B is negative and it primes the gate GA3 so that when the M wave is also applied to the gate, an output is obtained as shown at P; when the M wave momentarily rises, the P wave does likewise. This positive-going pulse switches the trigger BST1 over and the Q wave therefore changes its polarity. The B and Q waves now being negative cause the T wave to change over. The change passes through the OR gate GO1 to give a change in the U wave to the emitter follower EF; this negative-going change has no effect on BST2. When the A wave returns to stop polarity, the B wave goes positive again and as a consequence the P, T and U waves go positive. The trigger BST2 is switched by the positive-going change so that the output relay is switched to stop polarity.

Fig 5.11 Start—stop regenerative repeater — timing chart

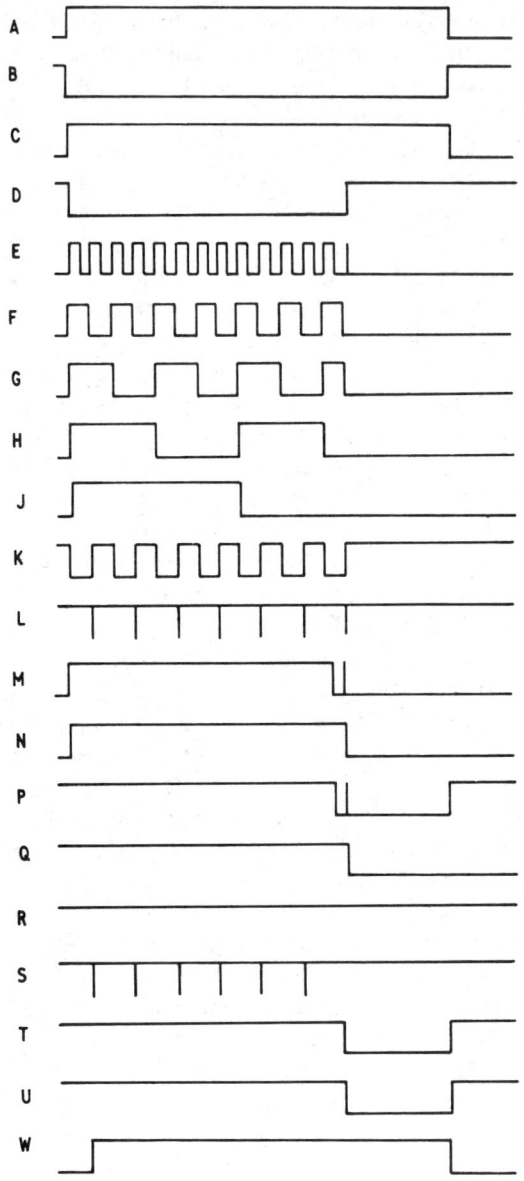

Fig 5.12 Timing chart for long-space condition

The Synchronous Regenerative Repeater. In the synchronous mode of operation, signals consist of a continuous stream of elements without any reference point comparable to the leading edge of the start signal in start—stop systems. As with synchronous receivers, the regenerative repeater must phase itself automatically from integrating the phase displacement of a number of signal elements in order to maintain the sampling of incoming elements at their nominal midpoint.

Synchronous regenerative repeaters usually need to cover modulation rates 80—100 bauds or 160—200 bauds for use with ARQ systems at 96 and 192 bauds (and other slightly different rates). At radio-receiving stations it is sometimes convenient to incorporate the regenerator as part of the output stage of a radio receiver. Regenerators may also be required at radio-transmitting stations for feeding distortionless signals into the drive circuit of a radio transmitter.

A block diagram for a synchronous regenerative repeater is shown in Fig 5.13. Timing is controlled from a phase-shift oscillator via a squarer circuit and a scale-of-two dividing circuit. The stability of the oscillator should be of the order of 10^{-6} or better so that in the event of an intermittent failure of the input signals the repeater will remain in phase until signals are restored. The input and output toggles and output gate perform functions of signal examination comparable to those described for the start—stop repeater. Resulting from this signal examination there is a delay of half an element through the repeater.

Input signals are also examined at a phase gate which produces, from the instants of restitution, pulses corresponding to any out-of-phase condition between the input signals and the oscillator. These correction pulses trigger either the early or late monostable trigger, the output deriving an integrated voltage which is applied to the oscillator-control amplifier. The precise value of this voltage will depend upon phase relationship between incoming signals and repeater and it is used to control the phase of the oscillator. The rate of phase correction in either direction is usually of the order of 1% of an element per element. Change of direction of correction is initiated when the number of elements distorted in one direction exceeds those distorted in the opposite direction by about six. The time con-

The repeater is now in a condition to receive the next character and produce a regenerated version at the output. All waveforms except one are returned to the polarity which should exist in the stop condition. The waveform which is not returned is the Q wave which is returned to its normal condition at the start of the next character. When the multivibrator is switched on by BST3, a positive change N is transmitted to BST1 which changes it over and restores the Q wave to its normal condition.

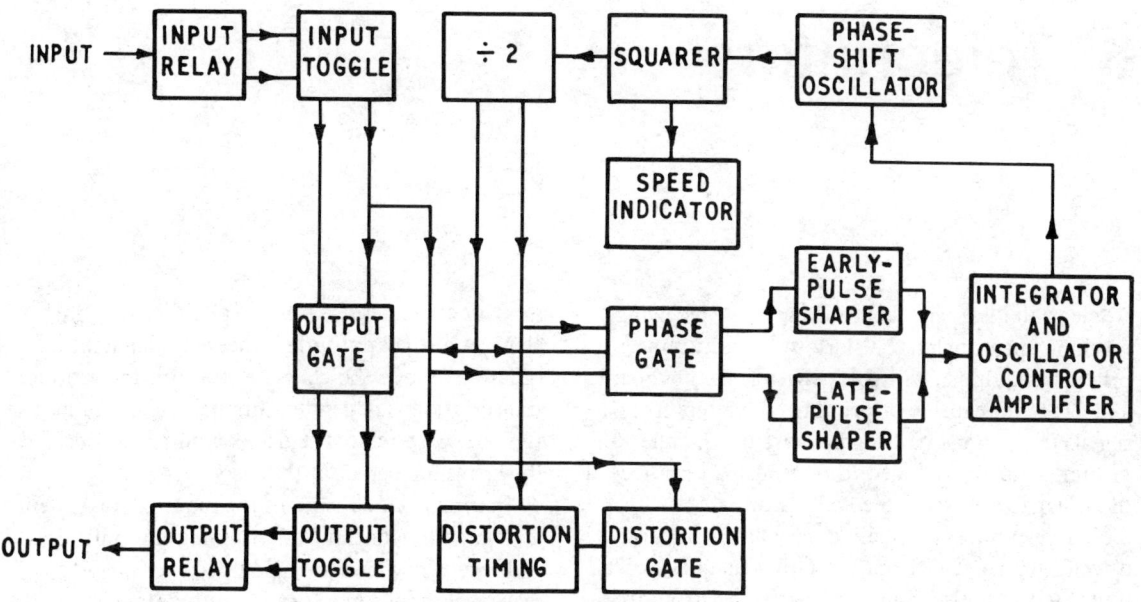

Fig 5.13 Synchronous regenerative repeater

stant of the phase-controlling circuit is somewhat critical: if correction is applied too quickly any short-term variations in phase of the incoming signal may be passed on to the outgoing channel, while a long time-constant will prevent the regenerator from giving maximum performance immediately after a break in reception should synchronism be lost. A time-constant of about 1 s is usually found to be an optimum value.

References

1 Scott, W. L. 'Electrical Contacts in Telephone Exchanges: Contact Opening and Closing Phenomena and Quenching Techniques', *POEE Journal*, **61**, p. 263 (1969).
2 Fairweather, A. and Frost, E. J. 'The Design and Testing of Semipermanent Metallic Contacts for Use at Low Voltages', *ibid.* **53**, p. 26 (1960)
3 Rudeforth, S. 'Contact Resistance and its Variation with Current', *ibid.* **42**, p. 65 (1949).
4 Sallis, R. T. G. 'A Device to Replace Conventional Polarized Relays', *ibid.* **59**, p. 181 (1969).
5 Bigg, R. W. 'The Quest for a Telegraph Relay of Greater Reliability', *ibid.* **61**, p. 196 (1968).

6 Teleprinters

A teleprinter is a telegraph transmitting–receiving machine operating on the start–stop (arhythmic) principle, equipped with a typewriter-like keyboard and a receiver capable of printing characters from signals received over a telegraph circuit; the teleprinter is usually arranged to print also a local record of all signals sent from the keyboard.

The teleprinter is usually a mechanical apparatus driven from an electric motor. Other main electrical units are transmitter contacts, controlled from the keyboard, and receiving electromagnet, controlled by signals received from the line or the local transmitter. With expanded facilities of the newer alphabet No. 5 and the tendency towards higher speeds, designs of semi-electronic teleprinter are becoming available.

As an adjunct to the teleprinter, paper tape is used in which character signals are stored as holes, perforated according to the telegraph alphabet. A tape may be perforated by a keyboard perforator (an off-line mechanical instrument) or by means of a reperforator, controlled from electrical signals. In the *printing reperforator* the message is printed on the perforated tape so that the start, address and end of a message can be quickly identified. Signals stored in a perforated tape are transmitted by an automatic transmitter or a tape-reader. The reperforator and automatic transmitter, which are start–stop instruments, may be designed as integral, optional units of the teleprinter.

Most teleprinters in use throughout the world are of the 5-unit type, operating at 50 bauds (400 characters per minute) to alphabet No. 2.

6.1 The start–stop principle

In alphabetic-printing telegraphy, each character signal is coded according to a telegraph alphabet and transmitted as a unique group of signal elements; the receiver examines the binary state of each signal element at the appropriate theoretical instant in order to decode the character for printing. Correct transcription is dependent upon coincidence in time between related signal-examining periods and the arrival of signals at the receiver.

In synchronous systems the required coincidence is ensured by continuously-running transmitters and receivers which are held in correct phase relationship by some form of automatic phase correction which is both complex and costly. For fixed point-to-point systems carrying heavy traffic between installations where skilled maintenance staff are available, such systems are well justified. For large numbers of scattered stations working on switched networks or private circuits, the flexibility and simplicity of the start–stop system provide a satisfactory solution.

In the start–stop system, though the driving motors may be running, the sending and receiving devices are normally held at rest in a zero-phase position. They are started in phase for each individual character which is sent and then stopped in the zero-phase position. For each character sent, the transmitter automatically prefixes the group of code elements with a *start* signal which serves to initiate a character cycle in the distant receiver mechanism in phase with the transmitter mechanism. The transmitter automatically terminates the group of code elements with a *stop* signal, which restores the receiver mechanism to the zero-phase rest position in readiness for receipt of a subsequent character signal. Transmitter and receiver are brought to rest on completion of every character cycle.

Equality in duration of transmitter and receiver character cycles is determined by the speed of their driving motors (or of a time-base); any discrepancy between them is restricted to a single character cycle and it cannot become cumulative over successive cycles. The speed tolerance on motor speed recommended by the CCITT is ±0·75%;* the British Post Office adheres to a maximum of ±0·5%.

The start and stop signals correspond to the binary modulation states. The stop signal must be the idle, quiescent state of the line – the mark (Z) condition; the start signal corresponds to the space (A) condition.

The duration of the start signal is equal to one unit element of the modulation. The minimum duration of the stop signal is a very important feature of the receiver to ensure satisfactory start–stop operation under adverse conditions when receiving from an automatic transmitter: for 50-baud 5-unit systems, it has been fixed at a minimum of 1·4 units, preferably 1·5 units.* The complete character cycle comprises 7·5 units (a 1-unit start signal, five 1-unit code elements, and a 1·5-unit stop signal); at 50 bauds (20 ms per element) the character cycle occupies 150 ms. The stop signal persists until the start signal of a subsequent character is originated at the transmitter; a minimum duration of stop signal is of consequence only when using an automatic transmitter in which one character can be followed immediately by another at the maximum rate (60/0·150 = 400 characters per minute).

Since transmitter and receiver always start from the zero-phase position of rest, and the sequence of subsequent timing operations for any character takes place at instants determined by the commencement of the start signal at transmitter and receiver, the propagation time of signals over the line is of no importance, provided that it remains constant throughout the character cycle.

The optimum performance of the receiver is secured by designing it so that the examining period for each code element takes place during a very short time about the ideal centre. Provided that a signal transition has occurred not later than the beginning of the examining period for that element, it will be correctly recorded, provided also that the transition of the following element is not so early that it occurs before the end of the examining period appropriate to the previous element. This capacity of the receiver to accept distorted signals and yet produce perfect selection of the required character is termed the *receiving margin* of the machine; its theoretical maximum

* Recommendation S3 (50 bauds); S3bis (75 bauds); S3ter (100 bauds).

value is ±50%, but a maximum value for a mechanical receiver would lie in the range ±40–±45%. The CCITT limit* for margin is not less than ±35% (50 and 75 bauds) or ±30% (100 bauds).

Duration of Stop Signal. The lengthened stop signal was introduced to permit use of a regenerative repeater (a back-to-back receiver–transmitter) at an intermediate point in the line. Phasing errors on the two links may be additive and to guard against this a stop signal greater than 1·0 unit must be maintained between two successive cycles. If the regenerative repeater has a *receive* cycle of 6·5 units and a receiving margin of ±50%, a distortion of ±0·5 unit (10 ms at 50 bauds) may be tolerated on each link.

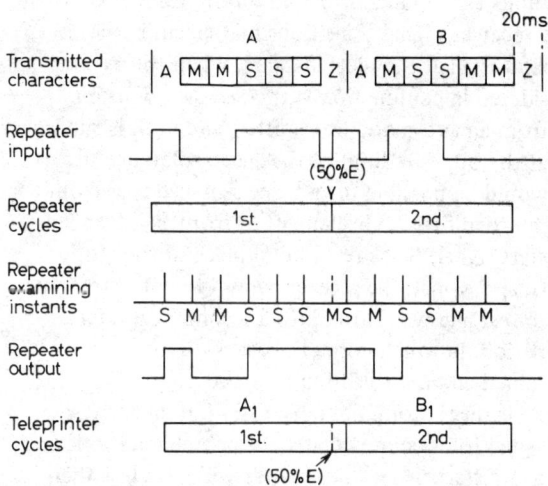

Fig 6.1 Transmission through start–stop repeater

Fig 6.1 shows the timing of two consecutive 50-baud 7·0-unit start–stop characters A and B transmitted through a start–stop regenerative repeater under limiting conditions; transmission time does not affect the conclusions. Owing to early distortion in the first link, the start of the second character B occurs 10 ms early (50% early distortion) at the repeater. The receiver of the regenerative repeater will therefore also be released 10 ms early at Y, immediately following the receive cycle for the first character A. As the receive margin is ±50%, correct reception results.

The output cycle of the repeater will be similar to the input cycle, but delayed by 0·5 unit or 10 ms (this is characteristic of a start–stop regenerative repeater): the output signal will be as shown for A_1. If the start of the second character B_1 when retransmitted is again subjected to 50% early distortion over the second link, the teleprinter receiver is released to start the second character before coming to rest after the first, and the machine will be out of phase. Though the likelihood of the same transient being affected by such distortion in transmission over both links is small, it is statistically significant and if not taken into account would result in an increase in transmission errors. For this reason a minimum stop signal of 1·4 units is specified.

Duration of Receiver Cycle. Telegraph signals are subject to timing distortion during transmission: at a receiver, signal transitions may occur earlier or later than at the ideal instants. When this is considered in conjunction with a receiver working from an automatic transmitter, and each is running at the opposite limit of the speed tolerance, it would be possible for a space signal (either from a retarded fifth code element or from an advanced start signal) to be received in place of the stop (mark) signal; the receiver would then fail to come to rest and would run into a second character period, in which correct phase is no longer maintained and mis-selection may occur.

Under automatic transmission, if an early start signal follows immediately after a character having a late start signal, the receiver will have less than the normal time to complete its cycle for the first character before it is again released for reception of the second.

To prevent a transmitter from over-running a receiver under adverse conditions, the nominal duration of the receive cycle must be made *less* than that of the transmitter cycle. This difference must allow for (1) maximum speed difference of transmitter and receiver, (2) maximum permissible signal distortion caused by line and equipment, and also (3) operating delay of the receiver clutch, or similar device, which puts the receiver cycle into effect after detecting the start signal.

If a transmitter running 0·75% fast is working to a receiver which is running 0·75% slow the result will be that signal transients produced by the transmitter will be successively 0·75% early, while selecting examining instants in the receiver will occur 0·75% late, compared with the theoretical timing. A cumulative timing discrepancy between transmitter and receiver will occur throughout the character cycle, attaining a maximum value of 2 x 0·75% x 150 = 2·25 ms at the end. If, in addition, the start of the next character is subject to 35% early distortion, the stop element will be further shortened by 20 x 35/100 = 7 ms, giving an effective total shortening of 9·25 ms – almost half a unit. Provided that the receiving cycle is terminated by the stop signal, the maximum discrepancy will be confined within one character period.

To preserve the start–stop feature under such conditions, and prevent two connected machines from running out of phase, the receiver is designed to complete its character cycle in a period at least half an element shorter than that of the transmitter, i.e. 7·0 units for the receiver against 7·5 units for the transmitter. However, at the time when it was decided that this safeguard should apply, it was possible when working over international circuits to encounter older-pattern machines which had a stop signal of only 1·0 unit (a total character cycle equal to 7·0 units). To receive 7·0-unit transmissions with the prescribed 0·5-unit safeguard, the receiver must be brought to rest in the period corresponding to 6·5 units. Although older transmitters have by now largely been withdrawn from use, the prescribed values remain for 5-unit 50-baud machines – namely a 7·5-unit transmitting cycle and a 6·5-unit receiving cycle.

A 7·5-unit transmitter cycle with a 6·5-unit receive cycle permits a time displacement of one unit (20 ms at 50 bauds) between successive start signals without endangering correct termination of the receiver cycle by receipt of the stop signal.

The self-synchronizing feature of the start–stop system has made for extreme simplicity in operation and it enables any two compatible start–stop instruments to be connected together at will, a fundamental requirement for any switched service such as telex. The main disadvantage of the start–stop system is the redundancy of start and stop signals, which appreciably reduces information rate or, alternatively, increases modulation rate and bandwidth requirement.

Figures quoted above relate to the 5-unit 50-baud system; the same principles would apply to comparable start–stop systems. Machines designed to use 7-unit alphabet No. 5 usually have a 2-unit stop signal, which together with a 1-unit parity element and 1-unit start signal results in an 11-unit character signal.

6.2 The No. 7 teleprinter

Though obsolescent, this machine is still extensively used over a wide field. Designed primarily for page-printing from a roll of paper, the paper-carriage unit is detachable and can be replaced (as on printer-grams positions) by a paper-tape carriage. Alphabet No. 2 is used, with modulation rate of 50 bauds and a 1·5 unit stop signal, giving 6·6 char/s. Both 3-row and 4-row keyboards are used, the latter with numerals on a separate row of keys.

MODEL 7A, long obsolete, received 7·5-unit signals only, with a receive cycle of 140 ms.

MODEL 7B,* the basic version (with a receive cycle of 6·5 units), is used on leased circuits requiring no special features.

MODEL 7D included modifications needed for telex service, e.g. motor-on-speed contacts, a new answer-back unit with precision timing and off-normal contacts, etc.

MODEL 7F has an overlap cam unit fitted to the receiver to print a final character which, in the earlier models, remained stored in the combination head. This unit also results in a better receive margin – at least ±40%, compared with ±35% for earlier models in which the receiving camsleeve controls selection, translation and printing functions. These used a ratchet and pawl clutch on account of the heavy load imposed by the last two functions. This clutch has 60 teeth and, as the shaft makes one revolution in 130 ms, pickup time varies between 0–2·2 ms, depending on the relative positions of the pawls to the rotating ratchet teeth at the moment of release to the start signal. This is equivalent to a loss of margin of 2·2 x 100/20 = 11%, or ±5·5%. Again, because of insufficient time to carry out all functions during one revolution of the cam, selection and translation are carried out during one revolution and printing

* For principles, see *Introduction to Telephony and Telegraphy* (E. H. Jolley), Pitman; for details see *Telegraphy* (J. W. Freebody), Pitman.

during the next (while the next character is selected and translated).

The overlap cam unit overcomes both disadvantages by dividing the function between three cams, each being released in turn, following a pilot cam. The load imposed by selection is small, and a friction clutch is used for the selector cam, eliminating the variable pickup time and allowing increased margin. The two other cams are driven through ratchet clutches, but for these functions accurate timing is not required. Although the selector cam still rotates in 130 ms, the total time available for rotation of the cams is 250 ms, so that there is time for printing to take place before the last cam comes to rest, and the last character is not stored. When receiving at maximum speed, printing will still take place while the next character is being selected. The cam unit is provided with an orientation device which enables the early and late margins to be balanced: this provides an approximate means of measuring margin.

The overlap cam unit comprises four clutch-driven cams, each with a detent-release mechanism: (1) orientation pilot cam; (2) selector-cam assembly; (3) comb-setting camsleeve; and (4) printing camsleeve. Each cam is released by the preceding one in the series, except for the pilot cam which is released by the electromagnet armature, and completes its revolution in 130 ms.

The orientation pilot cam determines the instant at which the selector-cam assembly starts to rotate and the instants at which incoming signal elements are read. The pilot cam is driven through a friction clutch with a constant pickup time of a fraction of a millisecond.

The selector-cam assembly, also driven through a friction clutch with a small and constant pickup time, reads and interprets the five incoming code signals; during the fourth element it releases the comb-setting camsleeve.

The comb-setting camsleeve is actuated by a ratchet and pawl clutch whose pickup time (0–2·17 ms) is variable but unimportant. It raises the bellcranks of the combination head, allowing the typehead to rotate; it then raises the comb-setting fingers already set by incoming signals to position the receiving combs. Bellcranks and fingers are then lowered, in that order, and the printing camsleeve is released.

The printing camsleeve, driven by a ratchet and pawl clutch, causes the carriage to feed, and after a suitable interval of time (to allow the typehead to rotate and latch on the selected bellcrank) it operates the typehammer to print the selected character.

Timing of the four cam cycles is arranged so that the machine will function correctly even when receiving from an automatic transmitter sending 7·0-unit signals. When receiving undistorted signals, each cam is at rest for at least 10 ms, even at maximum speed of operation, since the time of rotation is 130 ms and the character duration with 7·0-unit signals is 140 ms. With distorted signals the rest period is reduced, but adequate.

6.3 The No. 11 teleprinter* (Creed Model 47)

This is a tape-printing machine used on the TAS inland telegram service which uses alphabet No. 2 at a modulation rate of 50 bauds. The transmitter produces 7·5-unit signals; the receiver will accept either 7·5- or 7·0-unit signals with a stop signal of 1·5 units, hence character rate is 6·6 char/s.

The keyboard is designed on the sawtooth principle in which combination bars with projections like sawteeth are moved directly by depression of the keys, in contrast to the No. 7 model in which restoration of combination bars is effected by motor power. The sawtooth keyboard is claimed to produce a freer touch due to absence of the locking bar; it adds considerable complexity to the mechanism, not only in the transmitter unit but also in the answer-back unit. A separate WRU key is provided which can be operated only after pressing the FIGURES SHIFT key, and the D-key cannot then be operated. To send the identification signal a HERE IS key is mechanically connected to the answer-back unit. A RUN OUT key gives continuous transmission, while depressed, of the signal combination corresponding to the last key operated.

Though a tape machine, CARRIAGE RETURN and LINE FEED keys are provided for working to page-printing machines, e.g. on telex. A visual warning is given to the sending telegraphist when the page attachment on the distant teleprinter is approaching the end of a line so that the

* Described in detail in *Telegraphy* (J. W. Freebody), Pitman.

CARRIAGE RETURN and LINE FEED keys may be operated. An end-of-line counting device records the number of key depressions after each operation of the CARRIAGE RETURN key. When 55 impressions have been recorded (14 before the end of a line of 69 characters), the device closes a pair of contacts to light a lamp.

A tape-fail alarm operates should the tape break, become jammed or slip to such an extent that characters are overprinted. The correct feed for each character is 0·125 in.; the alarm device measures the feed as it occurs, operating the alarm contacts if movement of the tape falls below 0·06 in.

Printing and tape tear-off positions are on the left-hand for operating convenience; paper-feed and ink-ribbon mechanisms are placed to enable the roll of paper tape and the ink-ribbon to be changed without need to remove the main dust cover. The tape-roll holder is fitted with a handle to facilitate removal and replacement and is supported on a bracket at the rear of the main base, locked in position by a simple catch; a brake prevents the roll of tape from spinning and causing loose turns; a message tray fits on top of the cover and a dialling-code bulletin on the front.

The Keyboard Unit. When a key is depressed, the combination bars are moved to represent the corresponding code and are mechanically held in position until transmission of the combination is completed; this lock is imposed indirectly by a locking frame which acts on the selecting levers in the transmitter unit. With sawtooth combination bars it is unnecessary to provide special arrangements to prevent operation of a second key during transmission because, while combination bars are set and held in position for one character, their projections prevent full operation of keys other than that for the character concerned.

Keybars are shaped as in Fig 6.2. They are supported by a common pivot bar at the rear of the unit and held in their normal position against the tops of slots in the front guide-plate by helical springs. Each keybar has a knife-edge formed on its underside so that when they are operated the knife-edges engage the sawteeth of the combination and functional bars.

The five combination bars are provided with saw-tooth projections according to the alphabet. The projections each have one vertical and one sloping face; the sloping face may be either on the left or right of the vertical face, according to the movement required. When a keybar is depressed the knife-edge on its underside exerts a downward

slots in the undersides of the bars (Fig 6.2). The bars are not restored between characters: after a key has been depressed, some will be in the left-hand position (space), others in the right-hand position (mark).

Fig 6.3 represents a part of the combination bars. Depression of the U key would cause bars 4

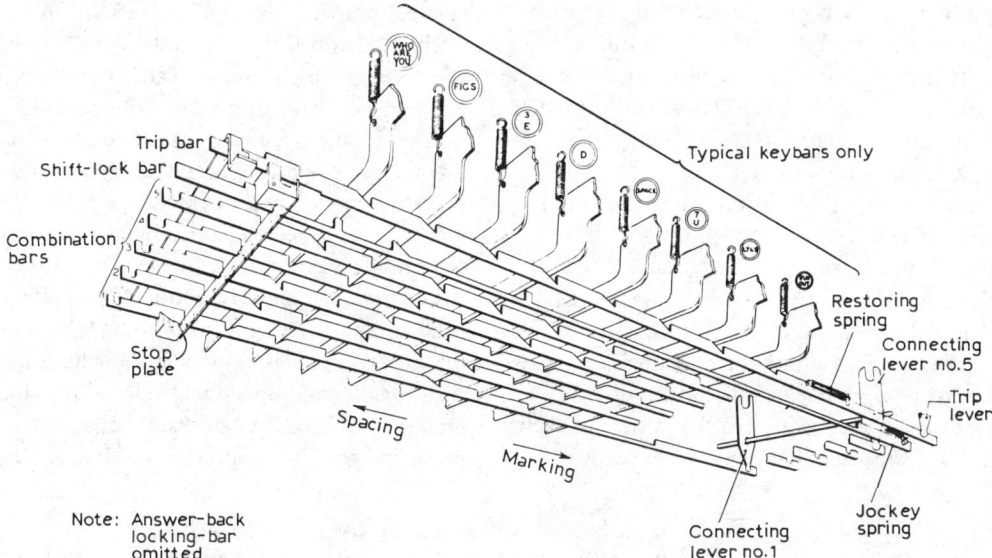

Fig 6.2 Teleprinter No. 11: keyboard combination bars

pressure on the sloping side of a tooth on certain bars (depending upon the combination) and shifts them as required. Combination bars are supported in guide blocks and are free to move a small distance from left to right or vice versa, the movement being limited by a stop plate which fits into

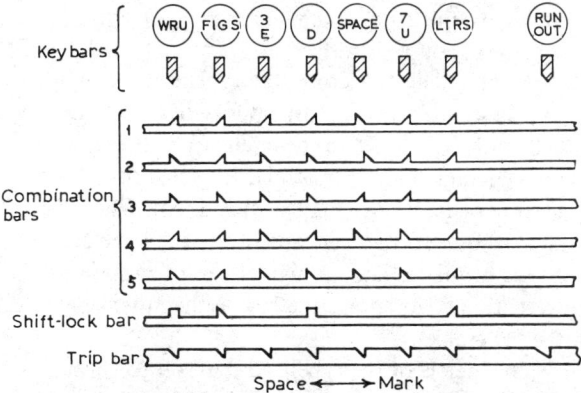

Fig 6.3 Teleprinter No. 11: keyboard combination bars — schematic

and 5 to move to the left for combination MMMSS; subsequent depression of the LETTERS SHIFT key (MMMMM) would move these bars again into their right-hand positions. The right-hand end of each combination bar is coupled to a connecting lever (Fig 6.2).

The shift-lock bar is mounted in front of the five combination bars and has two sawtooth and two rectangular projections. The FIGURES and LETTERS SHIFT keys each act on the sloping face of one of the sawtooth projections so that depression of the FIGURES SHIFT key causes the bar to take up its left-hand position, and the LETTERS SHIFT key moves it to its right-hand position. One of the rectangular projections obstructs the D key to prevent its operation while the shift-lock bar is in the left-hand (figures) position; the other obstructs the WRU keybar while the shift-lock bar is in the right-hand (letters) position.

The transmitter is of the striker type, similar to that on the Model 7.

Orientation. Due to the orientation facility, instants at which selections for the five code elements take place can be varied with relation to commencement of the start signal. This enables early and late margins of the receiver mechanism to be balanced for best reception.

Reception of a start signal releases a pilot cam which in turn controls delayed release of the receiving camsleeve. The delay introduced may be varied by adjustment from $11·7-31·7$ ms. The timing performed by the cam, after commencement of the start signal, may be varied from 10 ms early to 10 ms late in relation to corresponding timing of a receive cam without orientation.

By adjustment of the relationship between the pilot cam and levers which control release of the receiving camsleeve, instants of selection of the code elements may be varied by ±10 ms (50% of a code element) in relation to the start-signal commencement. This allows the selecting mechanism to be correctly phased so that selection is effected during a short period in the middle of each code element, ensuring that margins for early and

holds the assembly together. Next to the typehead are the thrust washer and thrust collar, the latter keyed by diametrically-opposite lugs to the boss on the latch arm. The stop arm is assembled between the latch arm and the driving plate. Between the latch arm and the stop arm is a friction washer; a second one is placed between the stop arm and the driving plate. The friction-damping spring assembled between thrust collar and latch arm, thrusts the latch arm, stop arm and two damping washers towards the driving plate, so that any movements of latch arm and stop arm are heavily damped.

The driving plate has a projection – the centralizing pin. Shock-absorbing springs of the latch arm and stop arm apply a clockwise torque to the stop arm and an anti-clockwise torque to the latch arm so that edges A1 and A2 are normally thrust against the centralizing pin. The action of stopping the typehead and positioning it for printing is seen in Fig 6.5. Rotation of the clutch and typehead before it is stopped by a dropped bellcrank is in a clockwise direction and is driven by the typehead clutchdrum and the clutch spring expanded and in con-

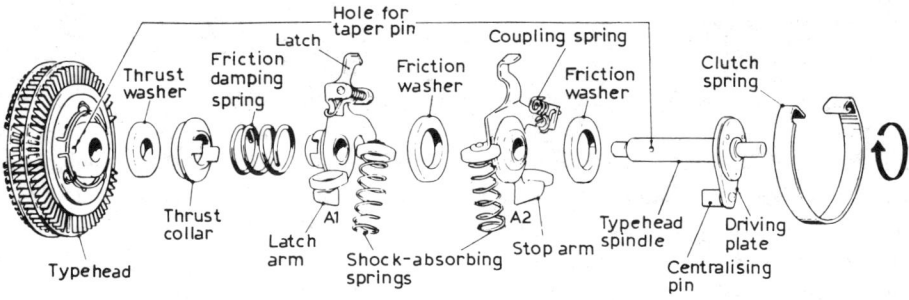

Fig 6.4 Teleprinter No. 11: typehead and clutch

late signals are symmetrical. The orientation facility can be used to check the receiver mechanism. The overall margin can be measured by feeding undistorted signals into the electromagnet and progressively moving the setting of the orientation adjustment, first in one direction and then in the other, until correct reception fails; the overall margin is found by subtracting one reading from the other.

The Typehead Clutch (Figs 6·4—5). This is a special shock-absorbing type designed for long life. The boss on the typehead is pinned to the spindle and

tact with the drum lining which engages with the typehead coupling spring.

After a bellcrank drops, the typehead continues to rotate until the stop arm approaches the fallen bellcrank. The latch on the latch arm first engages the bellcrank and is pressed back against the tension of its spring (Fig 6.5(*a*) and (*b*)). When the stop arm meets the bellcrank at (*c*), the latch arm passes beyond it and, jumping forward again under the action of its spring, latches on the other side of the bellcrank.

The inertia of the typehead causes it to continue rotating and the centralizing pin, rigid with the typehead, continues to move in a clockwise direction, thrusting against the bottom of the latch arm and compressing both shock-absorbing springs. During this motion some energy of the typehead is dissipated in the two damping washers, the rest being stored in the shock-absorbing springs which then expand, the centralizing pin moving in an anti-

6.4 The No. 12 and No. 75 teleprinters

The No. 12 machine is a special adaptation of the Creed Model 75 teleprinter, selected for use in the telex cordless switchboard on account of its compact design.

This model uses a keyboard detached from the

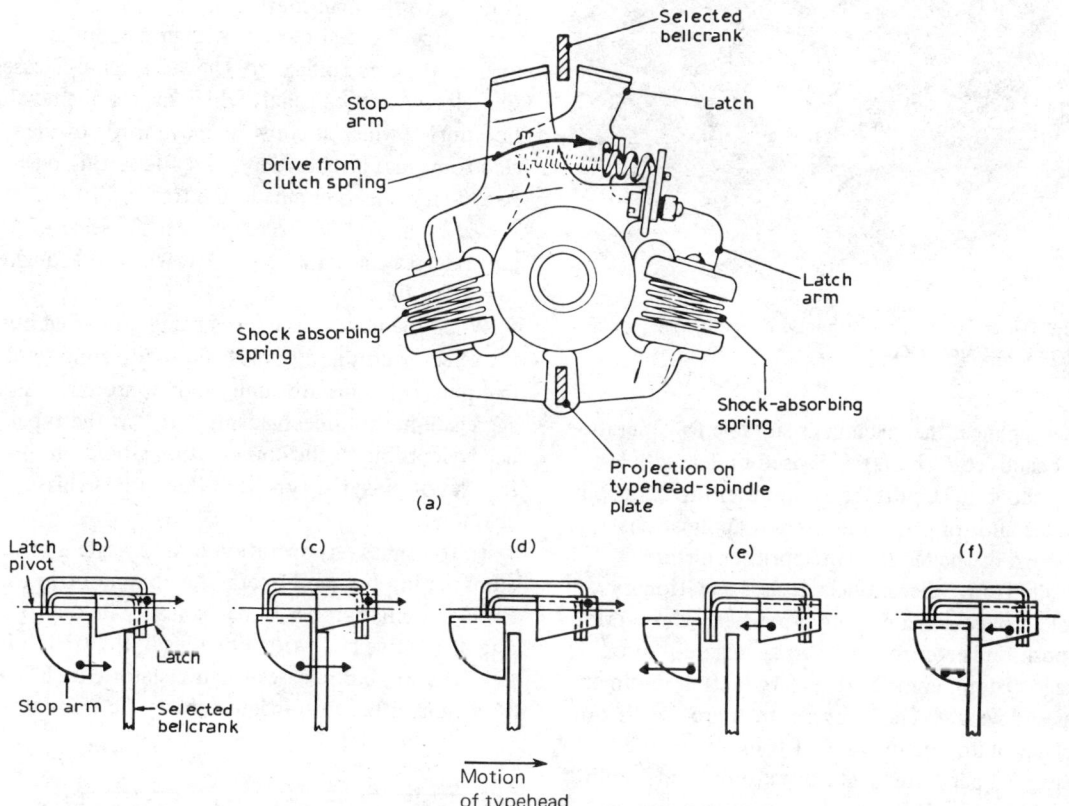

Fig 6.5 Teleprinter No. 11: clutch operation

clockwise direction until it makes contact with the bottom of the stop arm. The stop arm is moved and again causes the shock-absorbing springs to be compressed, though not to such an extent as previously, due to the latch, because more of the energy of the typehead has been dissipated in the damping washer.

As shown at (d), (e) and (f), the typehead executes a rapidly-damped oscillatory movement, coming to rest with the selected type opposite the typehammer; its final position is determined by the facts that the stop arm and the latch arm are at rest against each side of the bellcrank, and the centralizing pin, firmly held between the two arms, is rigid with the typehead.

receiver so that it can be fitted at the front of the switchboard keyshelf convenient to the operator. The receiving unit is behind a perspex panel at the rear of the switchboard position and fitted with a special paper take-off roller and winder to display several lines of the printed signals to the operator. The keyboard provides a simultaneous 5-wire output (ON for mark) together with a start pulse. The 5-unit signals are formed electronically in external equipment which supplies a 'stop' signal to an electromagnet on the keyboard for resetting the combination bars.

The Model 75 (Fig 6.6) was designed primarily for operation at 75 bauds; using the 5-unit alphabet No. 2 this gives a speed of 10 char/s. By a simple

Fig 6.6 Teleprinter No. 75 with tape transmitter
(*Courtesy of Creed & Co. Ltd*)

gearing change, the machine is suitable for operation at 50 bauds (6·6 char/s). It is used on leased telegraph and data circuits, as a computer input/output machine and for other automation applications.

Design of the Model 75 teleprinter differs radically from other models. It uses a stationary paper carriage and a moving typewheel whose printing position is controlled by an aggregate-motion linkage system, contributing to reduction in dimensions and weight. The local record is provided from mechanical interconnection of transmitter and receiver; an accurate local copy in this event is not necessarily proof of accurate transmitted signals. Mechanical local record also precludes duplex operation. The machine is available with optional, integral, reperforator attachment, automatic transmitter or tape-reader, or as a printing reperforator and caters for sequential or simultaneous input and output signals. Either 3- or 4-row keyboard can be provided; RUN OUT, HERE IS and WRU keys are provided. Additional features include interior illumination for the printed page, end-of-line lamp, two-colour printing, sprocket-feed platen and automatic carriage return and line feed.

Outline of Operation. The main units are interconnected as shown in Fig 6.7. The driving motor may be a.c.- or d.c.-governed (4200 rev/min), or a synchronous motor (3000 rev/min for 50 Hz). An automatic switch which stops the motor after a period of 90 s without line signals, is switched on either by the first incoming start signal or by depression of any key on the keyboard. The machine has two camshafts — the selector shaft and the translator shaft — both linked through friction clutches to the mainshaft.

Considering first reception, signals from line operate the electromagnet. The start signal releases the selector-unit camshaft which makes one revolution during which it converts movements of the electromagnet armature into a static setting on five binary selecting pins in the translator unit. Towards the end of this revolution the selector unit releases the translator shaft, which is brought to rest by the stop signal.

While the next character is being processed by the selector unit and transferred to a second set of five pins, the translator unit, acting through the aggregate-motion mechanism, positions the typehead according to the combination stored on the first set of pins; the typehead then prints this character.

In transmission, operation of a key sets up five combination bars and operates a common trip bar. Each combination bar is mechanically linked to one of the five code-selecting pins in the translator unit; the trip bar releases the translator clutch. The transmitter unit, located above the translator

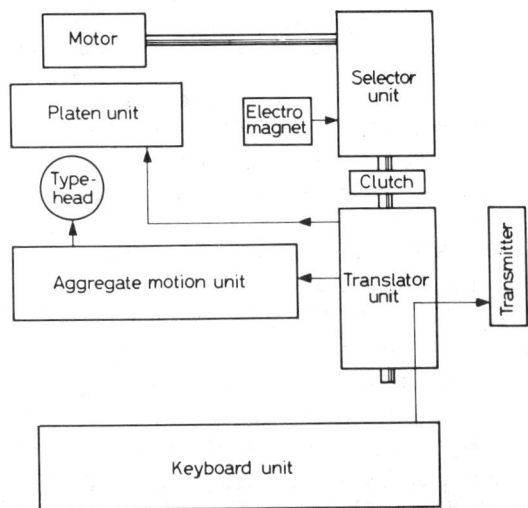

Fig 6.7 Teleprinter No. 75: block diagram of mechanism

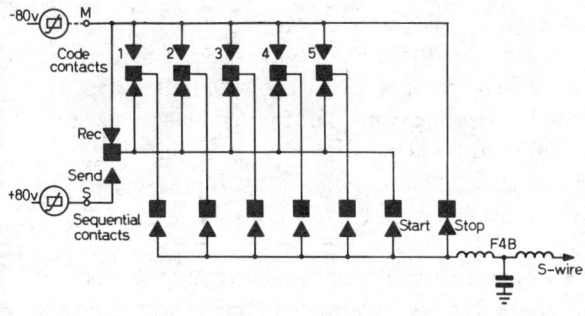

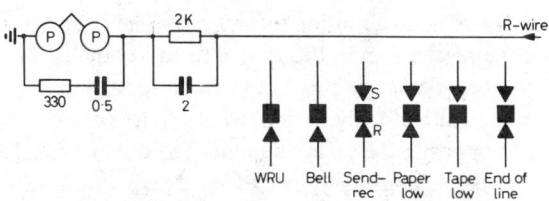

Fig 6.8 Teleprinter No. 75: signalling connections

unit, consists of two banks of electrical contacts, mechanically independent but electrically interconnected (Fig 6.8). One bank consists of five change-over contacts which reflect the combination on the selecting pins and which operate simultaneously. The other set comprises seven make contacts which operate sequentially; five of these are linked to the associated change-over contacts and the remaining two produce the start and stop signals. Once the code for the depressed key has been established on the selecting pins and the translator clutch released, the change-over contact bank takes over the combination setting. The seven sequential contacts, driven by cams on the translator shaft, then scan this setting at the significant instants of modulation to transmit the start—stop 5-unit signal.

A consequence of using the translator camshaft for controlling keyboard transmission is that the transmitter signal is basically that of the receiver, i.e. 6·5 units. The transmitted stop signal is extended, by means of a 'lag-weight' from 0·5 to 1·5 units.

A depressed key is released at about one-third of the duration of the transmitter cycle. This permits a subsequent key to be depressed early and gives the operator a measure of freedom to type irregularly about the maximum speed of the keyboard.

The Selector Unit. The main function of the selector unit is to convert the incoming signal combination

into a setting on one or other of two sets of five pins in the translator unit. It consists primarily of a striking mechanism along the line of pins; both mechanisms are driven by cams on the selector camshaft. An interposer, controlled by the electromagnet, can block the striking action to prevent a particular pin from being set.

If the first code element is mark the interposer permits the striker to push the first pin inwards, where it is retained by a spring. If the signal is space the interposer will block the forward movement of the striker and so prevent the pin from being set. While the striker is moving across to take up position in front of the next pin, the next code element is being received and examined. After the fifth code element has been registered, a trip linkage operated from the selector-shaft retention lever releases the translator camshaft.

The pin box which holds the two sets of pins is arranged to move endwise, forwards and backwards, to produce the required changes of control. This mode of operation permits the teleprinter to print each character immediately after selection.

The selector unit includes an orientation device by means of which the selection instant can be varied. The receive margin is ±40% at 50 bauds and ±35% at 75 bauds. For dual-speed machines the selector unit can also incorporate a manually-operated 2-speed gearbox.

Translator Unit. The main function of this unit is to examine the code setting established at the pin box by the traversing pecker controlled from the selector unit; it then conveys the result of this examination to the aggregate-motion unit which decides from this information which type in the typewheel shall print. Simultaneously with this examination of the pinbox setting, another examination of the pin box is taking place to establish whether the particular code setting corresponds to any of the non-printing functions such as CARRIAGE RETURN. If this shows the code to be a functional one, printing and feeding actions are inhibited and the appropriate function is carried out.

For non-printing functions a blank type on the typehead faces the paper.

Other mechanisms which form part of the translator unit are linkages to operate non-printing functions (carriage return etc.); the transmitter contact bank, including the send—receive switch; the case-shift control mechanism; the cam which resets the keyboard combination bars; the spring-controlled lag-weight which introduces an additional 20 ms for the stop signal, and also for the send—receive switch in the send position.

The Typewheel. The lightweight typewheel is mounted vertically and has four layers of type, each containing 16 types. In the rest position, to which it returns after each character has been printed, the top of the typewheel is below the level of the printing point, giving good visibility. To enable it to select and print a character, the typewheel has four movements.

mechanism, the action of which is to lift the typewheel so that the layer of types selected by the combined action of the two control rods is opposite the printing point. The four combinations of conditions for element No. 2 and the shift mechanism, corresponding to the four layers of types, are shown in Fig 6.9.

3. *Rotary movement* under control of elements 1, 3, 4 and 5 of the code combination. These elements determine the positions of four more control rods on the aggregate-motion mechanism. The action of the mechanism in this case is to rotate the typewheel clockwise or anti-clockwise, through not more than half a revolution, to bring the correct type opposite the printing point. The direction of rotation is determined by element No. 3, and the angle of rotation by a combination of elements Nos 1, 4 and 5.

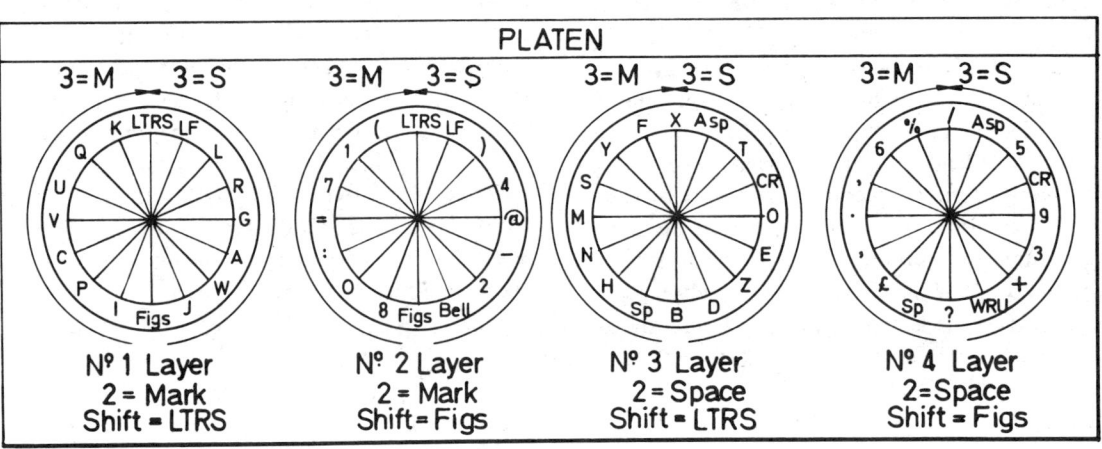

Fig 6.9 Teleprinter No. 75: typewheel allocations

1. *Lateral movement* under control of a rack and pawl mechanism. The typewheel is caused to traverse the length of the platen a character at a time, except when a functional combination is received. If this combination is the CARRIAGE RETURN signal the typewheel returns sharply to the beginning of the line where it is smoothly brought to rest by a small piston-type dashpot.

2. *Vertical movement* under dual control of the letters/figures-shift mechanism and code element No. 2. Each of these determines the position of a control rod of a link-type aggregate-motion

4. *Forward movement* under control of the printing mechanism. When the selected character is opposite the printing point, the frame supporting the typewheel is swung forward. This raises the ribbon into line with the selected type and causes the typewheel to strike forward at the platen and print the character (a separate typehammer is unnecessary). At the end of the printing operation the typewheel is restored to its rest position and the ribbon lowered.

The Aggregate-Motion Mechanism (Fig 6.10). This mechanism consists of two similar but independent arrangements of links and rods which act together to control lift and rotation of the typehead and so bring the selected type round to face the paper.

Components concerned with vertical positioning of the typewheel include rack A, idler gear B, gear N are at their maximum lift, and rods L and M are consequently at the limit of movement to the left. The position which floating pivot Q now takes up corresponds to the top layer of types being in line with the printing point. When shaft P is released at

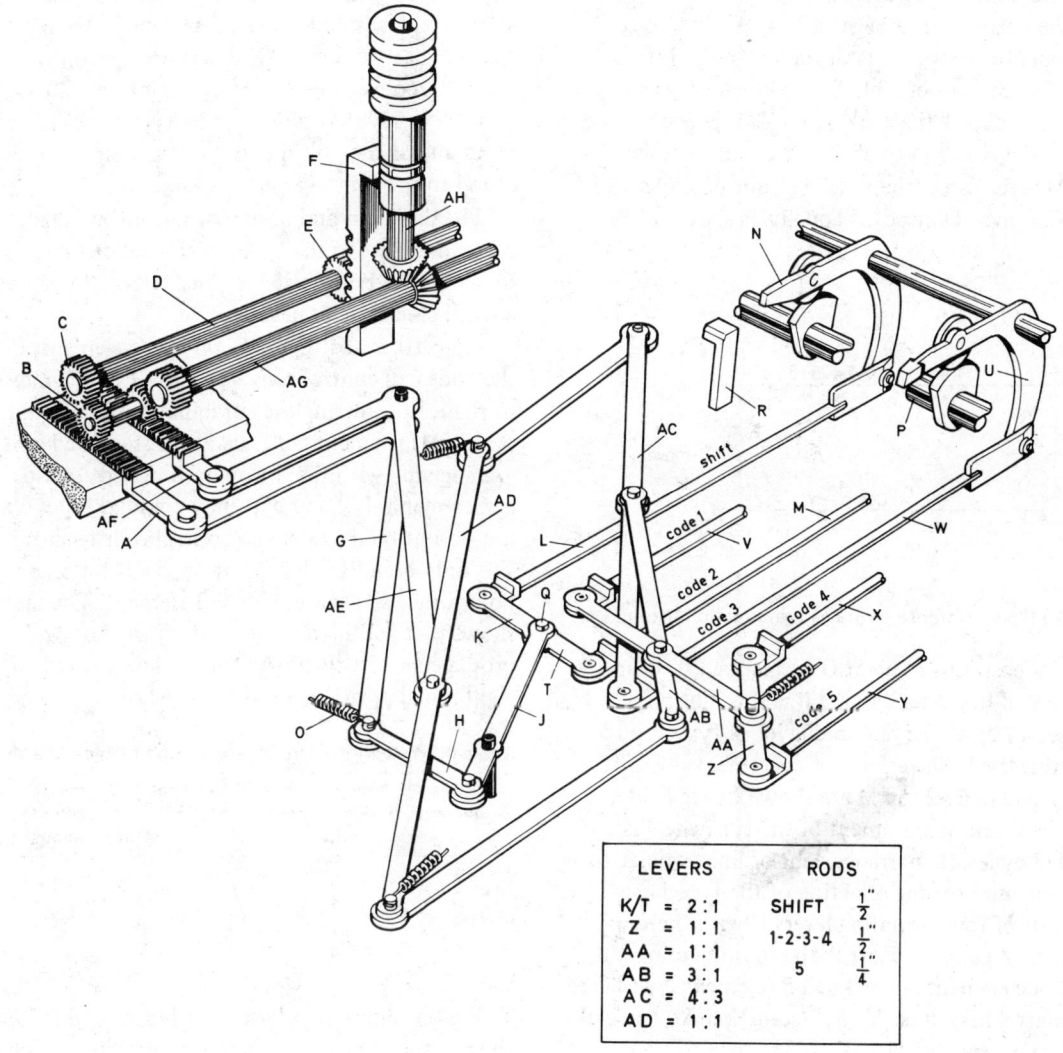

LEVERS	RODS
K/T = 2:1	SHIFT $\frac{1}{2}$"
Z = 1:1	1-2-3-4 $\frac{1}{2}$"
AA = 1:1	5 $\frac{1}{4}$"
AB = 3:1	
AC = 4:3	
AD = 1:1	

Fig 6.10 Teleprinter No. 75: aggregate-motion mechanism

C fixed to splined shaft D, gear E and rack F, links G, H, J and KT, and rods L and M. Rods L and M are controlled by individual operating levers of which lever N is typical. Levers N are in turn operated by cams carried on the translator shaft P. Spring O, acting through links H and J, urges rods L and M to the right, so holding levers N on their cams.

In the rest position of shaft P, operating levers the end of the selection cycle, levers N attempt to turn under the action of spring O. Whether they are able to do so depends on the position of case-shift latch R and of the selecting pin (not illustrated) associated with element No. 2 of the code. If latch R is in figures position (as illustrated) and selecting pin No. 2 is at space, both levers N will rotate and rods L and M will both move to the right. If latch

R is in letters position (blocking the path of lever N) and pin No. 2 is at space, only rod M will move and link KT will turn anti-clockwise around what has now become a fixed pivot at the end of rod L. The reverse of this action takes place if rod M is blocked and rod L free to move.

The purpose of linkages KT, J, H and G is to combine the movements of rods L and M into a single movement of rack A. Since pivot Q is so sited on link KT that link K/T = 2/1, and since the range of movement of shift rod L and code No. 2 rod M is the same, there will be four possible positions of pivot Q disposed equally as shown in Fig 6.11.

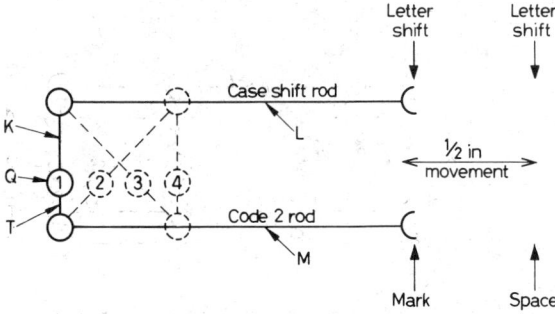

Fig 6.11 Teleprinter No. 75: aggregate-motion lift linkages

The positions of pivot Q correspond to four degrees of lift at rack F and it is this motion which is used to bring the required layer of types up to printing level. The effects of movements of shift rod L and code 2 rod M are shown in Fig 6.11.

The rotating movement of the typewheel is controlled by another arrangement of links similar to the one used for the selection of the type layers. It consists of four operating levers U with their associated cams on shaft P; the four selecting pins for code elements, 1, 3, 4 and 5 (element 2 was used to control lift); links V, W, X and Y; links AA, AB and AC connected via AD to AE; rack AF, and splined shaft AG linked by bevel gears to the vertical splined shaft AH which controls rotation of the typewheel. The bevel gear associated with splined shaft AG, and spur gear E associated with splined shaft D are free to slide along their shafts as the typewheel traverses the printing line.

The cams on shaft P, which control rods associated with code elements 1, 3 and 4, each produce a half-inch left-to-right movement when a space setting is read off selecting pins. The cam associated with rod

Y differs from the others in that it gives only a quarter-inch movement when space is registered on pin 5.

Since each of these four rods associated with typewheel rotation can have mark (blocked) or space (free) actions, there are 16 possible positions all of which combine at lever AE. Rack AF and splined shafts AG and AH convey the setting on level AE, converting it in so doing into clockwise or anti-clockwise rotation of the typewheel to bring one of the 16 types on a particular layer round to the printing point.

The direction and degree of movement that each rod can impart to the typewheel is set out in Table 6.1: a clockwise rotation is considered as positive, an anti-clockwise as negative.

Fig 6.10 shows that only on space elements of the code can control rods V, M, W, X and Y move to right. Considering the combination for letter B (MSSMM), element No. 2 can be disregarded as it is associated with the lift operation, and so also can elements 1, 4 and 5 as they are marks and do not permit their associated control rods to move. The only effective element from the rotational point of view is space on No. 3 element. The table shows that half-inch movement on this rod will produce an anti-clockwise rotation equivalent to eight radial positions on the typewheel.

Table 6.1 Teleprinter No. 75 — selection of type in a layer

Code element	Movement on space element (in.)	Typewheel rotation	Degree of movement (characters)
1	$\frac{1}{2}$	+	+4
3	$\frac{1}{2}$	−	−8
4	$\frac{1}{2}$	+	+2
5	$\frac{1}{4}$	+	+1

Fig 6.9 shows that letter B is located 180° from the printing point. The −8 movement brought about by the half-inch left-to-right movement of rod M will bring letter B round to face the paper.

When more than one space occurs among elements 1, 3, 4 and 5 of a combination, subsequent movement of the typewheel can be determined by taking the algebraic sum of individual effects of space-control rods. Considering the code for letter Z (MSSSM), element No. 2 and all marks can be disregarded. Elements Nos 3 and 4 are space and cause −8 and +2 movements respectively; the algebraic sum of these two movements is −6. Reference to the typewheel layout chart shows that anti-clockwise rotation of six type-positions magnitude will bring letter Z to the printing point.

Although rotational and vertical movements of the typewheel have been described separately, they are produced by the same camshaft P and occur together. The combined effects of lift and rotation mechanisms produce any required selection from alphabet No. 2.

Keyboard Transmission. The keyboard has no separate transmitter nor camshaft associated with it, but employs the receive camshaft to drive the multi-contact transmitter; adjustment and cleanliness of so many contacts is a maintenance hazard. The keyboard contains conventional motorized combination bars which move to right for mark elements and, via bellcranks, set up a combination direct on to the translator pin box. A trip-bar releases the translator shaft clutch and the lag-weight. The combination set up on the pin box is then examined by levers which operate the transmitter contact assembly.

After the translator camshaft starts to rotate the five moving blades of the code-contact bank (Fig 6.8) move to mark and either return to space or remain, depending upon the code-combination set up on the selecting pins; this action takes place while the start signal is being transmitted. Code contacts remain in positions to which they have been set for the remainder of the cycle, during which other cams on the translator shaft operate the sequential contacts. Mark contacts are finally returned to rest (space) 2 ms after the translator shaft comes to rest.

Since contacts of the transmitter are driven from cams carried on the main translator camshaft and they read off the code-combination direct from the code-selecting pin box, these contacts are operating while the machine is *receiving*, as well as transmitting. It is necessary to use the send—receive switch to inhibit the transmitter electrically during reception to prevent retransmission of received signals (see Fig 6.8).

Reperforator Attachment. The teleprinter No. 75 can be supplied with an optional reperforator

attachment fitted to the side of it to record, as 5-unit perforations in a paper tape, all incoming and outgoing messages simultaneously with the page-printed copy. The local record facility enables the machine to be used as a keyboard perforator to prepare a perforated tape, off-line.

The holes are perforated by punches beneath a punch block. The punches are selected by bars according to the combination set up on the translator pin box, and are operated together by a punching arm. The return stroke of the punching arm withdraws the punches from the tape and operates a ratchet which drives the tape-feed mechanism.

The roll of paper tape is housed in a drawer in the teleprinter base; perforated tape feeds towards the operator and 'chads' are collected in a cuttings box. The reperforator is controlled by a mechanical ON/OFF switch. A BACKSPACE key, for off-line operation only, enables perforating errors to be erased by overpunching a faulty character with the letters-shift combination MMMMM.

The Printing Reperforator. This machine supplies a perforated tape from either the incoming signals or from the keyboard in local; the message is printed on the tape between the centre feed holes. Due to positioning of printing and perforating points, the printed character is displaced 8½ feed holes behind the corresponding perforated character. A RUN OUT key (for combination 29 or 32) is provided for feeding out tape to a tear-off point which will be many character spaces away from the printing point.

The machine suppresses perforation of the WRU signal because an automatic transmitter reading the tape would not wait for a reply after transmitting this signal from the tape, and the next 20 characters in the tape would be mutilated by signals from the answer-back unit of the distant teleprinter. Suppression is effected by disabling the tape feed for one character so that the WRU combination overpunches the preceding FIGURES SHIFT combination (without changing it). This can be done because the tape when punched is not fed out of the punch block until the next character cycle; this feed is then suppressed to permit the overpunching. If an answer-back unit is fitted to the printing reperforator, this will be operated and the 20-character answer-back code will be punched in local on the tape.

Special slim-faced types are used so that legibility is not impaired when the feed holes are later perforated along the printing line. Recommended symbols of Table 1.4 can be fitted to the typewheel to give a printed record of functional combinations.

Simultaneous Input or Output Signals. This teleprinter lends itself to use of 5-wire simultaneous input or output signals as an alternative to sequential signals. By a simple re-arrangement of connections (Fig 6.8) to eliminate the sequential contacts, a simultaneous output can be read from code contacts, single or double-current; duration of the output pulse is 60 ms. A 5-wire input signal can be used if the teleprinter is fitted with five code solenoids and a trip magnet operated from a sixth wire to release the translator shaft; the print-out speed is 11·5 char/s. Modifications are necessary to the selector translator and other units. Each solenoid controls one of the

to the solenoids which when energized lift a shoe on the associated selector frame and set a pin in the pin box to register a mark. Further rotation of the translator camshaft causes the pin box to present its code setting to the aggregate-motion control mechanism which positions the typewheel.

The transmitter includes a single-make master-contact operated by the function bar in addition to the contacts previously described; once a combination has been set up in the pin box, code elements are read off by levers associated with the five contacts.

6.5 The No. 15 teleprinter* (Creed Model 444)

This teleprinter was introduced to provide improved performance and facilities at telex subscribers' stations and for private circuits. It uses 5-unit alphabet No. 2 at 50 bauds, and can be modified by a simple change of gear ratio to operate at 75 bauds. At both rates, the machine is designed to accept 7·0- or 7·5-unit signals.

The design differs radically from previous Creed

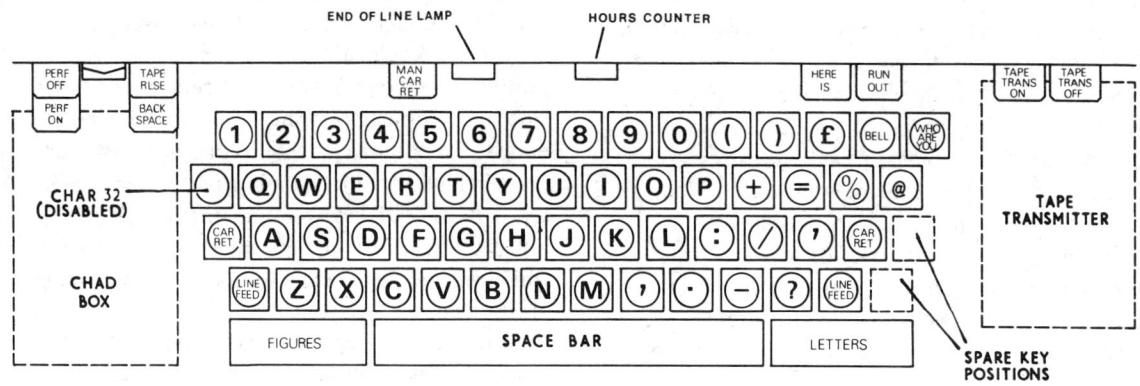

Fig 6.12 Teleprinter No. 15: keyboard layout (*Courtesy of Creed & Co. Ltd*)

five selector bars which are used to carry the keyboard code setting up to the translator-unit pin box.

In reception, a positive trip pulse is applied to the electromagnet which has its armature biased mechanically to mark. The first mark/space transition of the armature lifts the selector-clutch detent and allows the clutch to engage the drive; the translator release cam lifts the translator-clutch detent 15 ms after the selector shaft has started to turn.

Simultaneously with the trip signal, code pulses from the controlling equipment are fed, in parallel,

teleprinters. A traversing typewriter-style type basket is used for printing on to the stationary platen unit which bears the paper roll. The type basket holds up to 29 type bars, each of which carries two types with a maximum capacity of 56 printed characters; a character is printed on receipt without waiting for a following character.

A 4-row keyboard is used for telex since this

---

* Described in detail in *The International Telex Service* (R. N. Renton), Pitman.

layout most nearly resembles the standard type-writer keyboard, but a 3-row keyboard can be supplied. Shift pads are at either end of the space bar; separate HERE IS, WRU and RUN OUT keys are fitted. The 4-row keyboard (Fig 6.12) has 57 keys plus three pads, with a separate key for each character and mechanical case-shift lock: with this, all secondary (or primary) keys are locked following depression of the LETTERS SHIFT (or FIGURES SHIFT) key; the 3-row keyboard has 34 keys plus one pad. A total of 63 key slots is available, the 63rd position being reserved for a key to control a possible third case shift.

Two-colour printing is standard — red for transmission and black for reception. Paper to the international standard width (8·3 in. (210 mm)) is used. Externally-mounted paper packs may be fed through a slot in the rear of the cover. The machine can be supplied with a friction-feed platen or with a dual-purpose friction/sprocket-feed platen with retractable pins which can be set to accept either plain paper from a 5-in. maximum diameter roll within the cover or margin-perforated printed stationery; printing force is adjustable for up to five carbon copies. There is an external manual platen knob, and also a manual carriage-return key. Three degrees of line-feed spacing can be selected by the operator. The line length is adjustable from the standard 69 to 76 characters at 0·01-in. pitch. The operator views the printed page through a moulded acrylic window designed to avoid reflections from room lighting reaching the operator's eyes; the top of the window incorporates a paper-knife. On the front of the cover is a transparent plastic message lectern which has a horizontal spring-loaded cursor which serves to retain message forms and to act as a line guide. The lectern can be slid to the right to give an unobstructed view of the printed line.

An amber lamp, mounted above the keyboard in line with the function control keys, gives an end-of-line warning by glowing at the 55th character in a line; if the keyboard is operated thereafter the lamp flashes until the carriage is returned.

An automatic motor ON—OFF switch is included; this can be readily disabled on telex and other installations not requiring the facility.

A range of optional features is available to enable the machine to be used for data processing and automation applications.

The teleprinter is arranged to provide integral (optional) reperforator and tape-transmitter facilities. These attachments, which can readily be added on site, do not increase the overall dimensions as they are contained within the cover.

The reperforator, fitted on the left, is controlled from four organ-type keys grouped about the tape-exit point; keys are for reperforator ON and OFF, BACKSPACE and TAPE RELEASE. Perforated tape feeds out towards the operator over a removable cuttings box. A knife-edge gives an open V-shaped tear-off edge which distinguishes the end of a tape message, the leading edge being in the form of a closed V. The 8 in. diameter tape roll is housed within the cover. A tape-reel-low alarm is provided.

Provision is made for automatic suppression of punching on four selected combinations, in either or both shifts. Two of these combinations can be employed for mechanically controlling the REPERFORATOR ON and REPERFORATOR OFF keys; there is also provision for remote control of these two keys from two electromagnets, one for each key.

The automatic tape transmitter, fitted at the right, is controlled by two ON—OFF organ-type keys. Perforated tape is fed in from right to left, but an angled guide deflects the tape by 90° after being read so that it feeds out towards the operator. The tape transmitter has its own striker transmitter, which for most purposes is wired in series with the keyboard transmitter, though the two transmitters can be used independently. The two transmitters share the automatic *send—receive* switch, but this connection can be disabled if independent operation is needed. When the tape transmitter is in use, the keyboard is mechanically inhibited — the key-bars are free but ineffective. Electrical indication of the TRANSMITTER ON condition is provided by a change-over contact operated by the ON key. Another change-over contact indicates when the electromagnetic break-in mechanism is used, or when TAPE OUT is sensed. Both sets of contacts restore when the TRANSMITTER OFF key is operated. A TAPE OUT sensing mechanism senses the tape once during each cycle at a point just behind the reading line — the attachment will read the last combination in the tape and then stop.

Machine design permits the use of a 6-unit code (total of 8·5 units), with case shift, throughout the mechanism, including tape attachments; it also allows for simultaneous input or output signals in either 5- or 6-unit code.

Automatic carriage return and line feed can be included. Other special features available are horizontal (by repeated character space) and vertical tabulation.

A character-recognition unit (sometimes referred to as a *stunt box*) enables up to 44 characters, in either or both shifts, to be recognized mechanically and operate individual electrical contacts which can be used to control switching operations – to switch the attachments on or off, to change printing colour, etc. Except for BELL and WRU, which are change-over contacts and provided as standard, these contacts are fleeting make contacts, operating for 55 ms at 50 bauds (37 ms at 75 bauds) in each receive cycle. Contacts of this unit can be used to convert incoming sequential signals into simultaneous signals.

Radio-interference suppression components are fitted to all transmitter and other contacts where radio-interfering signals would otherwise be radiated.

The drive motor may be an a.c.- or a d.c.-governed series motor (3750 rev/min if a.c. or 3000 rev/min if d.c.); or a synchronous motor (3000 or 3600 rev/min for 50 and 60 Hz respectively). A.c. motors governed to run at a speed corresponding to the power frequency show a tendency to remain in synchronism for a while and then to fall abruptly out of step. A speed of 3750 rev/min is sufficiently spaced from synchronous speeds for 50- and 60-Hz supplies.

In the United Kingdom the national electric-power supply from an integrated frequency-controlled grid network can be relied upon to remain at constant frequency within limits required for teleprinter motor speed; use of synchronous motors for teleprinters has now been widely adopted. This brings considerable advantages such as elimination of speed-checking and governor adjustments; absence of brush-gear, carbon dust and commutator wear; reduction in generation of radio-frequency interference due to the absence of governor contacts and commutator; and reduction in fortuitous distortion caused by governor operation.

Keyboard Unit. Operation of the keyboard sets up the mark–space combination of the selected key on five groups of comb bars, and simultaneously activates a trip bar common to all keys which releases the transmitter camsleeve to make one revolution. During this time the transmitter mechanism inspects the combination sequentially and converts it into a train of mark–space signals which make up the 5-unit combination for the key depressed. The transmitting mechanism, which electrically consists of a single side-stable change-over contact, automatically inserts the start and stop signals at the appropriate time; it also switches the send–receive switch to SEND before beginning transmission and returns it to RECEIVE when transmission of a character is concluded.

The synchronous start–stop distortion on transmitted signals is ±3% at 50 bauds (±5% at 75 bauds).

When a key is pressed it produces a trip action to activate the transmitter and also simultaneously sets five comb bars which are read by the transmitter to form the 5-unit signal. For clarity, only one of five pairs of comb bars is shown, at A and B in Fig 6.13. Keybars C and D are held in a comb (not shown) and are pivoted at the rear end; at rest they are held above the comb bars by leaf springs E.

For code setting, operation of keybar C will press it down on the top edge of each of five pairs of comb bars A and B, which are linked together at each end by T-shaped levers G. The depressed keybar engages a tooth on comb bar A, so depressing the bar and at the same time raising the other comb bar B, which has a cut-out at this point. This parallelogram motion of comb bars gives high rigidity in the operating direction and low friction. Levers G turn anti-clockwise, so sliding linkbar H to the right. The operator cannot depress two keys simultaneously because of the rocking action of pairs of comb bars and the interference of their teeth. Depending on the position of teeth and cut-outs on the other four pairs of comb bars their associated links H will slide to left to set up a mark condition and to right for space. The underside of the depressed keybar C also operates trip bar F which lies beneath all keys, sliding trip link I to the left. This movement releases the transmitter-unit camsleeve.

Shiftbar J can be positioned to block either letters keys or the figures and symbols keys. It is moved to the left (to block letters keys) when the figures-shift

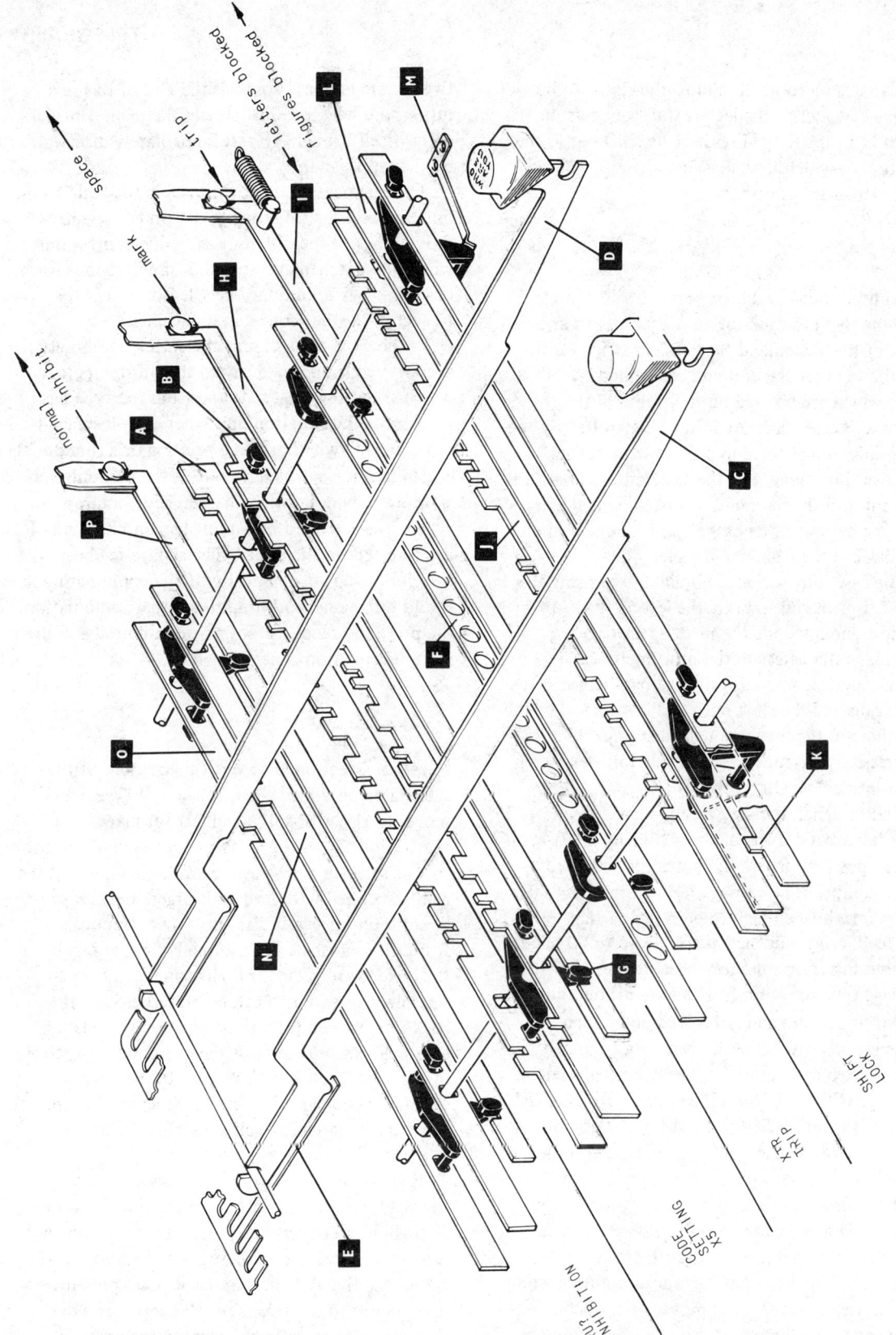

space

trip

letters blocked

figures blocked

mark

inhibit

normal

P

A

B

H

I

L

M

O

D

C

J

F

N

G

K

E

WRU?
INHIBITION

CODING
SETTING
X5

X-TR
TRIP

SHIFT
LOCK

Fig 6.13 Teleprinter No. 15: keyboard mechanism (*Courtesy of Creed & Co. Ltd*)

keybar acts on tooth K, and to the right (to block figures keys) when the letters shift key acts on tooth L. Leaf spring M retains the shift bar in the position to which it was last set.

A striker transmitter is used.

Reception. Individual type bars are selected by allowing them to drop into a slot across an arrangement of five castellated comb bars carried on the type-basket carriage unit and controlled by five vanes which are pivoted on the frame of the machine. These vanes are in turn controlled by the combination registered in a selector unit. This downward movement of the selected type bar brings it into the path of a print bail which lifts the type bar out of its basket, so impressing the type against the ribbon and paper.

The case-shift action is obtained by raising the type-bar pivot rod to bring the letters to the printing line, and lowering the pivot again to bring secondary characters to the printing line.

The type basket runs on three parallel rails (two guide rails and the shift rail), and is returned to the beginning of the line by a helical spring which is progressively extended as the basket moves along the printing line. On receipt of a CARRIAGE-RETURN signal, the feed mechanism is disengaged and the basket drawn back rapidly to the left-hand margin position, where the impact is absorbed by an air dashpot. The carriage-return time (less than 200 ms) is short enough to permit a carriage-return, line-feed, print sequence at speeds up to 50 bauds without inserting a dummy character to allow the carriage time to return from a long-printed line.

Printing action and letter-feed action can be made inoperative, either individually or jointly, on up to ten combinations. These normally include CARRIAGE RETURN, LINE FEED, BELL, WRU, LETTERS and FIGURES SHIFTS and combination 32. The remaining three combinations are available for accents or diacritical signs. Character-feed and dashpot mechanisms are reversible to provide right-to-left traverse for non-Roman scripts.

The start signal releases the selector-unit camsleeve which makes a half-revolution, during which it converts the subsequent movements of the electromagnet armature into a static code-setting on five 2-state latches. Towards the end of this revolution, the selector unit releases the main camshaft and is itself brought to rest immediately afterwards by the stop signal.

The main camshaft now transfers the code-combination established sequentially on the selector-unit latches to the link unit and code-control unit where it is latched and stored in simultaneous form. Once this has been done the selector unit is free to accept another code-combination.

In the code-control unit the stored combination is conveyed simultaneously to the print selectors which select the required type from the type basket, and to the function unit where it is inspected to determine whether it is a printing or a functional combination, such as line feed. If the combination is found to be a printing character, the appropriate type bar is activated by a print bail and its symbol is impressed on the paper, the carriage-feed operation following automatically. If the combination is found to be a non-printing functional combination, the print and feed operations are inhibited and the required function is performed.

Receiver Selection. The vertical armature of the side-stable polarized electromagnet A (Fig 6.14) moves to right for space and left for mark. The illustration shows the armature in the mark position.

The selector camsleeve (cams C, Q, G, R and the cam associated with lever I) is urged in a clockwise direction by two felt discs driven continuously from the mainshaft, but normally held at rest by the lower arm of detent B engaging stop cam C. The pivot of detent B can be moved around the edge of stop cam C by rotating knob D. This action advances or retards the examining operations of the selector unit with respect to the beginning of the start signal and so provides a means of estimating the receive margin, which is ±42% at 50 bauds (±38% at 75 bauds).

Receipt of the start-signal moves armature E to the right, so withdrawing support for trip lever F. This allows detent B to move away from stop cam C and the selector camsleeve starts to revolve. As it does so, five flats on cam pack G are presented in sequence to the five sequential levers H, which then pivot inwards. In the same time sequence,

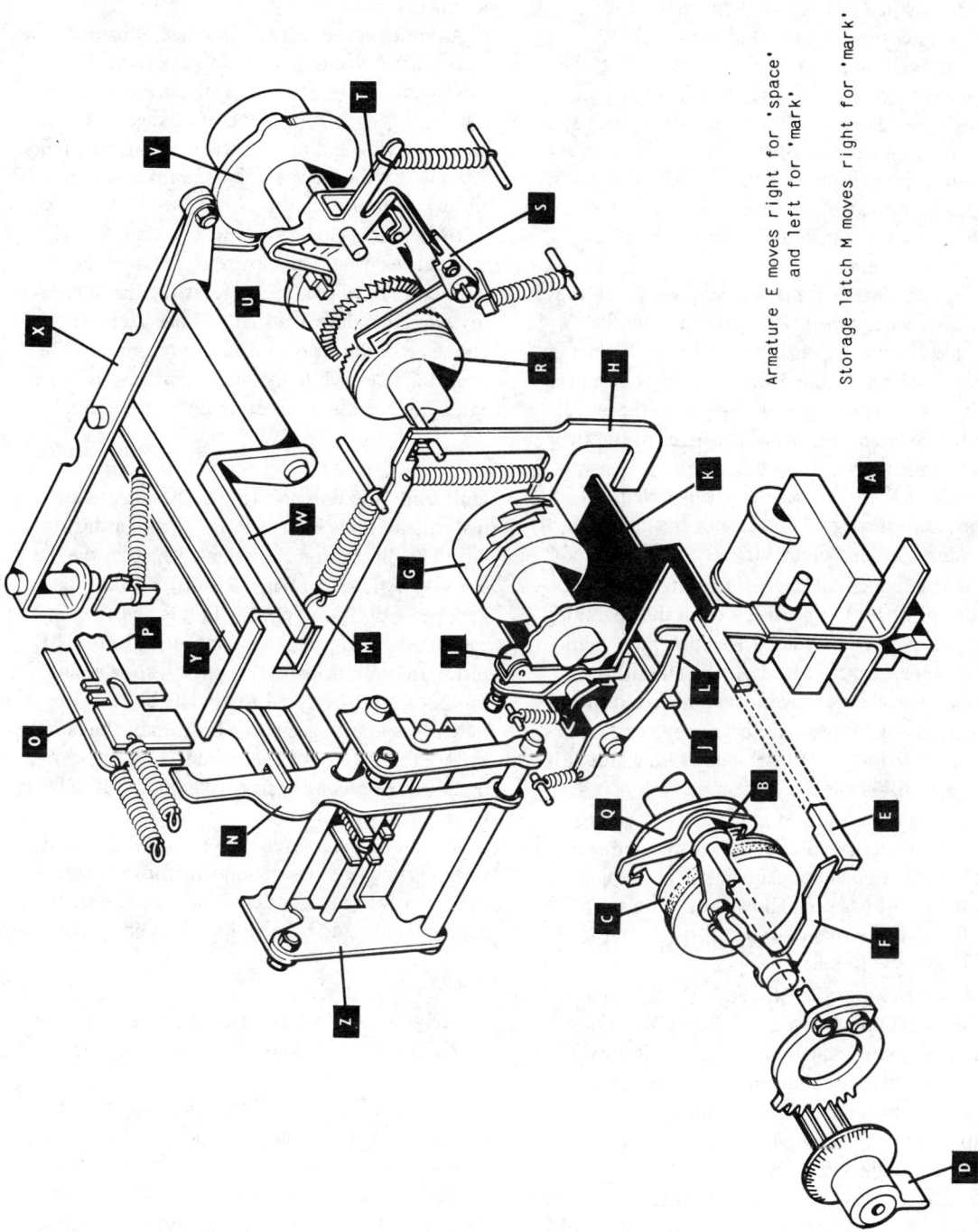

Armature E moves right for 'space' and left for 'mark'

Storage latch M moves right for 'mark'

Fig 6.14 Teleprinter No. 15: selector unit (*Courtesy of Creed & Co. Ltd*)

recesses in the read cam allow lever I, in conjunction with bail-lifting lever J, to lower read lever K if it is free. Whether K is free to move or not depends upon the position of armature E.

If the armature is at space, i.e. to the right, the downward movement of read lever K is blocked, although lever J can still lower space-lock lever L down the left-hand side of the armature to prevent it returning prematurely to mark. In the space position, read lever K will not affect sequential levers H, which just pivot in and out beneath it in response to their cams.

If the armature is at mark, read lever K can move down, so engaging the lower arm of sequential lever H as it pivots inwards beneath lever K. The subsequent downward movement of lever H moves the projection at its top end away from the associated storage latch lever M, which slides to the right to lock lever H down. The downward movement of read lever K also causes its sensing extension to pass down the right-hand side of the armature, so locking it against premature return to space.

The five sequential levers H operate at 30, 50, 70, 90 and 110 ms respectively from the beginning of the start signal; they are driven down for mark and remain at rest for space. At the end of the fifth element, the fleeting blocked or free responses of the sequential levers have been converted to a static code-setting on storage latches M, whose left-hand ends enfold the five code-transfer levers N.

When a storage latch M is at space, i.e. to the left, the upper cranked end of its associated transfer lever N is positioned opposite link O in the link unit. If latch M has been allowed to move to the right for a mark signal, the top of its transfer lever will be opposite link P.

After the end of the fifth code element, the stop signal will move the armature back to mark. Reset cam Q rotates detent B to bring its lower arm into the path of stop cam C to arrest the camsleeve, and at the same time rotates trip-reset lever F anti-clockwise. The angled end of lever F displaces armature E slightly as it rises, but not sufficiently to move it to space. The armature then moves back beneath lever F to maintain detent B in the stop position.

At the end of this selection cycle, cam R turns trip lever S clockwise, so withdrawing the step from beneath the horizontal arm of detent T; this lifts

to release ratchet-clutch coupling U and to connect power to the main camsleeve which carries cam V, among others.

A hump on the rear vertical face of cam V now presses lever X and linkage Y to the right. This movement causes frame Z, which is compliantly coupled to the left-hand end of linkage Y, to turn clockwise and press levers N against the links P and O. By this means, the combination established on transfer levers N is passed to the link unit where it is latched and stored. Cam V then rotates lever W anti-clockwise to press all the storage latches M to the left, so resetting the latches of any sequential levers H that had been set to mark. The selector unit is now ready to respond to the next character, while the main camsleeve is translating the combination set up on the link unit.

Link Unit. The link unit (Fig 6.15) consists of five pairs of links such as R and S, and a shift link. The illustration shows link S held in the mark position by latch A. When a transfer lever P takes up a space position opposite link R and then moves forward, subsequent inward movement of link R turns latch A clockwise (via pin C), so releasing link S, which is restored to the left by its spring. Latch A is urged against the bottom of the latching surface (at the end of leader line S) by the spring on link R. A second space strike on link R will be ineffective.

When a transfer lever P is set to mark to bring it opposite link S, its subsequent forward movement sets S to mark, where it latches as shown. A subsequent mark strike on link S will be ineffective.

Printing Translation. The mechanism has to discriminate between printing and non-printing functions.

The right-hand end of each link S (Fig 6.15) is connected to a bellcrank D which simultaneously controls (in conjunction with bellcrank F) the *vertical* movement of code slats W which select the required type, and the *horizontal* movement of comb bars H which select non-printing functions such as carriage return,

Code slats W have a U-section channel along their top edges in which run the rollers of the five

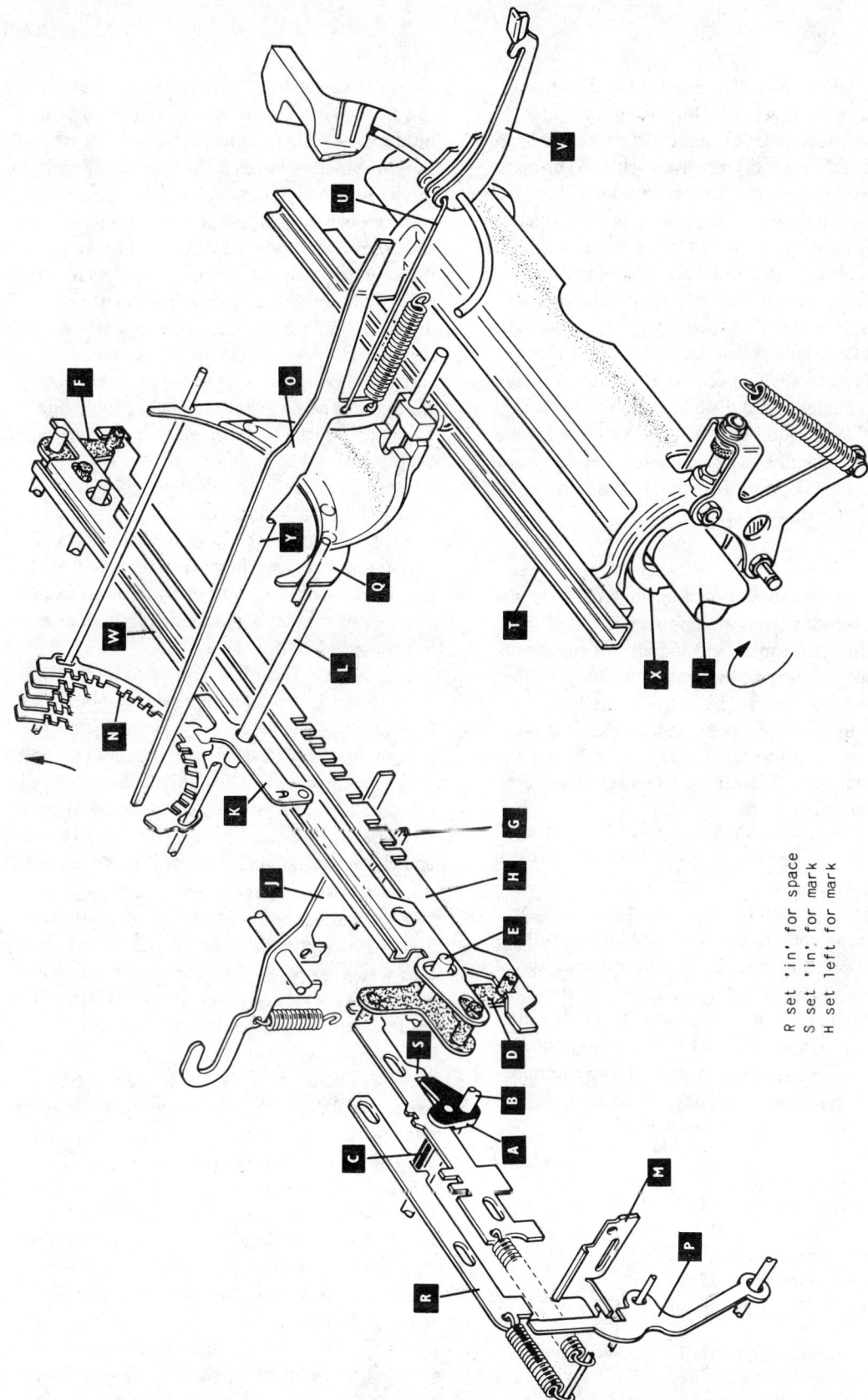

R set 'in' for space
S set 'in' for mark
H set left for mark

Fig 6.15 Teleprinter No. 15: link unit and typebar selection (*Courtesy of Creed & Co. Ltd*)

selector bellcranks K that control the type-basket sectors N, swinging the sectors to right for mark and to left for space. Arranged across sectors N are the code-seeker bars O, one for each two (primary and secondary) printing characters. These bars are spring-loaded down on to printing bail Q, which holds them just clear of the sector teeth.

All received combinations are passed to sectors N, but only when it is a printing combination will print bail T move to the rear, carrying with it type-bar bail Q. All the seeker bars are lowered on to the code sectors N, but only one will find a path through the teeth and drop low enough for its notch Y to catch on type bail Q. Further rearward movement of bail Q will swing the required type-bar V, via link U, out of the type basket and up on to the ribbon and paper.

Function Translation. Function control levers, such as J, lie beneath the five comb bars H which are moved to left for mark and to right for space by bellcrank D. When a combination is set up on the teeth below comb bar H, control levers J are allowed to turn anti-clockwise by a bail that normally rests on lower arm G of lever J. The upper horizontal arms of levers J then sense the pattern of the comb-bar teeth.

If the setting on comb bars H is for a printing combination there will be no path across the teeth and the function-control levers will be lowered again. Carriage feed takes place automatically after the printing operation. If the combination is that of a non-printing function, a path will open across the comb-bar teeth and the selected function-control lever will move up into it. This extra anti-clockwise motion of lever J has three consequences: (1) its lower vertical arm triggers the appropriate machine function; (2) carriage feed is suppressed; and (3) printing action is suppressed.

6.6. Teletype machines* T32, 33, 34, 35
This series of teleprinters from the Teletype Corporation includes Models 32, 33, 34 and 35.

The even numbers – 32 and 34 – operate to alphabet No. 2; the odd numbers – 33 and 35 –

* Described more fully in *Data Telecommunication* (R. N. Renton), Pitman.

are for alphabet No. 5. The lower numbers (32 and 33) are described as low-cost light-duty models; the higher numbers (34 and 35) are heavy-duty machines. Models 32 and 33 are basically of similar design, as are the 34 and 35, the change of alphabet being accomplished by the addition of code bars and associated items, and by changing the keyboard and typehead. For studies in economics, 'light-duty' has been defined as not more than two hours' daily use, which corresponds to about 2500 h in a 5-year period. Assuming 50% activity, this corresponds to about 30 million character cycles, a point at which trouble due to worn parts might be expected to begin and where overhaul would be recommended. If the teleprinter is used more heavily the overhaul point will naturally be reached sooner.

Model 32 is available with reperforator and tape-transmitter attachments to form the ASR set (automatic send–receive), or without attachments as the KSR set (keyboard send–receive), or as a receive-only (RO) model.

The Models 32 and 34 (5-unit code) transmit a 7·5-unit signal at modulation rates of 50 or 75 bauds (and at other rates used in the United States).

The Models 33 and 35 (7-unit code) transmit an 11·0-unit signal (which includes an even parity bit (M) and a 2·0-unit stop signal) at modulation rates of 74 bauds (6·6 char/s) and 110 bauds (10 char/s), and also lower rates. (In the Model 33 the parity bit is not used as such, and is always a mark.)

Keyboards for alphabet No. 2 are available in 3- or 4-row versions, the latter with partial shift-lock which is operative on the numerals row and on the related row of primary keys. HERE IS (for station-identification code) and REPEAT (RUN OUT) keys are provided.

A keyboard layout for the Model 33, basically using alphabet No. 5, is shown in Fig 6.16. In this alphabet the numerals and common punctuation signs are not secondary characters and no shift-key operation is necessary between letters and figures. Other, secondary, symbols are transmitted with the shift key held down *simultaneously* with the character key. The keys for purely functional combinations are operated simultaneously with the CONTROL key.

Many of the usual facilities, such as two-colour printing, sprocket-feed platen, low-paper alarm

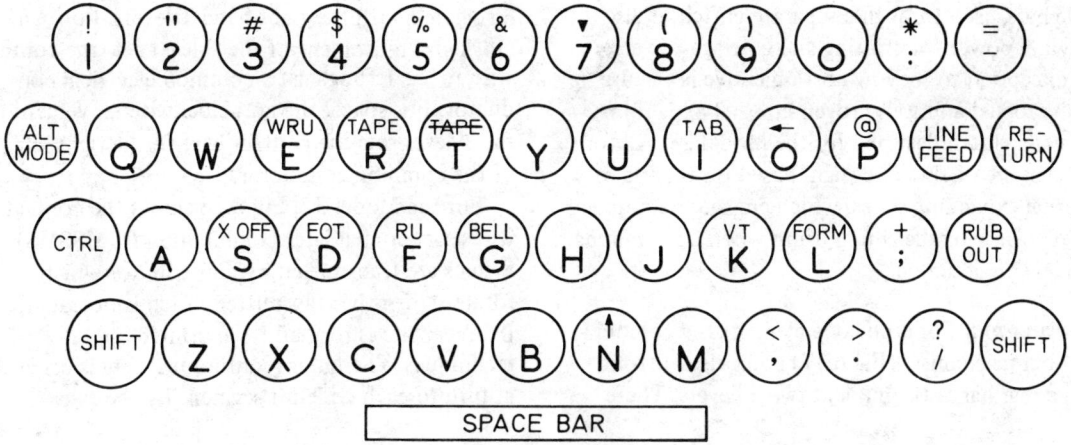

Fig 6.16 Teletype T33: keyboard layout

and automatic motor switch, are available, some as options. A *stunt box* (a set of electrical contacts operated by the receipt of specific function signals) is available for the Models 34 and 35. An answer-back unit (20 characters) is an optional device. The light-duty machines will take one carbon copy, the heavy-duty machines up to eight.

Tape attachments use $\frac{11}{16}$-in. (17·5-mm) tape for 5-unit, and 1-in. (25·4-mm) tape for 8-unit code. Control of the reperforator is by ON/OFF keys with BACK SPACE for deleting errors and RELEASE for easy tape insertion. The tape transmitter is controlled by a START and STOP switch with a FREEWHEEL position for ease of tape insertion. A TAPE OUT device also stops the transmitter tape.

Teletype Models 32 and 33.

Power Drive. The motor drives a distribution shaft, via intermediate gears and a toothed rubber belt, and the shaft drives the keyboard transmitter, through a clutch. The distributor shaft also drives the main shaft, through gears. The main shaft drives the receiver selector unit, function unit and

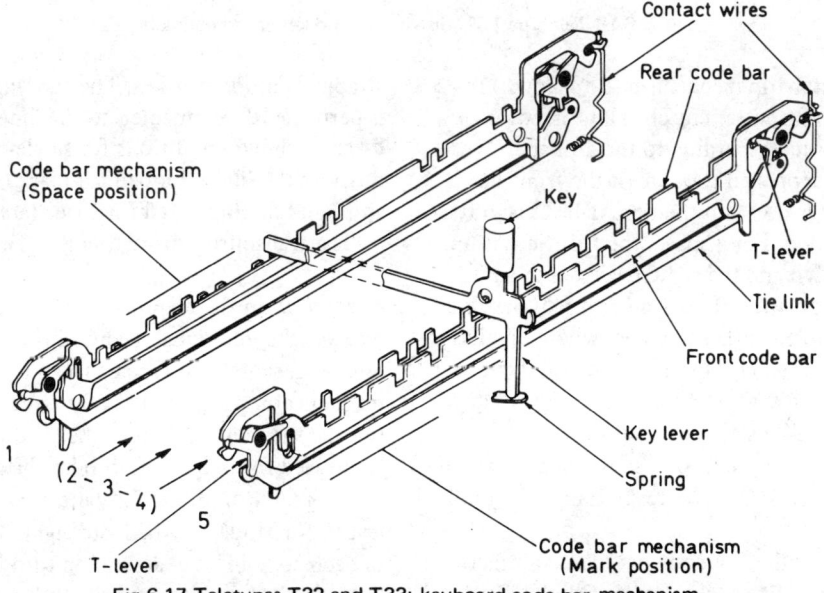

Fig 6.17 Teletypes T32 and T33: keyboard code-bar mechanism

code bars, each through a separate clutch; it also provides power for printing and carriage traverse. An exception to the low-cost objective is the use of the robust and well-proven internal-expanding all-metal clutch designed for the Model 28. All four clutches are of similar design, based on the use of internal-expanding shoes, which engage the notched inner surface of the clutch drum when a restraining trip lever is actuated.

Transmission. For each code element the combination-bar mechanism (Fig 6.17) comprises a front bar, a rear bar, a tie link and two T-levers. There

associated with the right-hand T-levers. For mark, clockwise movement of the T-lever sets the contact wire to the left against a common electrical conductor. For space, the anti-clockwise movement of the T-lever sets the contact wire to the right, clear of the common conductor.

For the Model 32, Fig 6.18 shows the contact wires set for character combination D (MSSMS). Signals are transmitted by a distributor which rotates when the transmitter clutch is operated; the faceplate is formed from printed-circuit techniques. The inner (continuous) ring is bridged in turn to each code-bit segment by the two

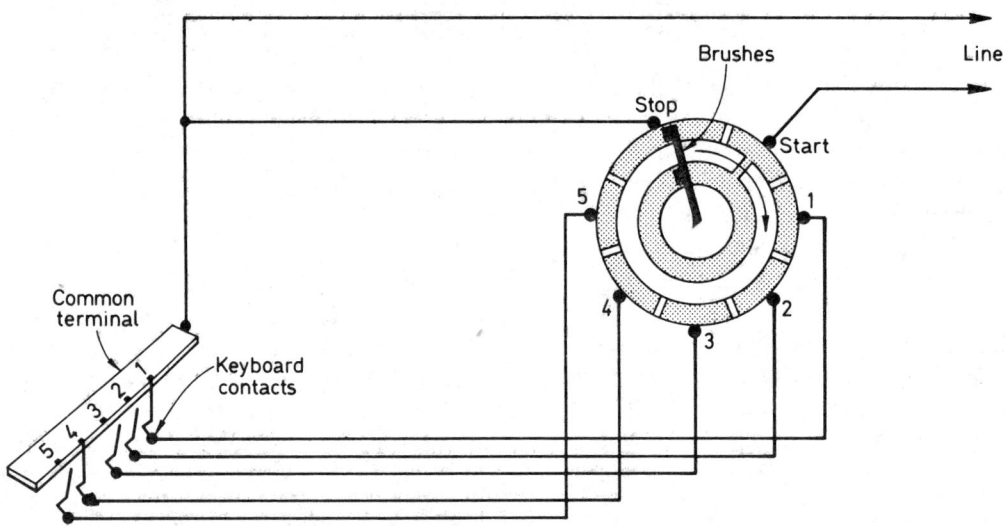

Fig 6.18 Teletype T32: distributor and keyboard contacts

is also a character-trip mechanism (not shown) for operating the transmitter clutch. The top edges of the bars are slotted according to the alphabet. When a key is depressed for a mark the front bar falls, the rear bar rises and the right T-lever is in its clockwise position. For a space, the front bar is up, the rear bar down and the right T-lever is in an anti-clockwise position. The front and rear bars are coded in a complementary manner: where one has a slot the other is solid. This arrangement serves as an interlock to prevent depression of a second key. The combination-bars are not reset between characters, and are moved for a subsequent character only if the new code bit requires a change of condition.

Movements of the five code-bar mechanism control the setting of five contact wires (Model 32)

strapped carbon brushes. The start-signal segment is permanently connected to the inner ring. The diagram shows conditions for single-current transmission; the line is looped for mark (1, 4 and stop) and disconnected for space (start 2, 3 and 5).

The transmitter distortion is ±5%.

Reception. In the receiver, an electronic unit receives the line signals (neutral or polar, i.e. double-current) and repeats them to the selector magnet (non-polarized) as 500-mA single-current signals; the telemargin is ±40%. The selector unit is released for one revolution by the arrival of a start signal. For Model 32 there are five selector levers; for Model 33 there are eight. The polarity of each code bit is examined in turn by the cam-offered selector lever, and the state found is stored

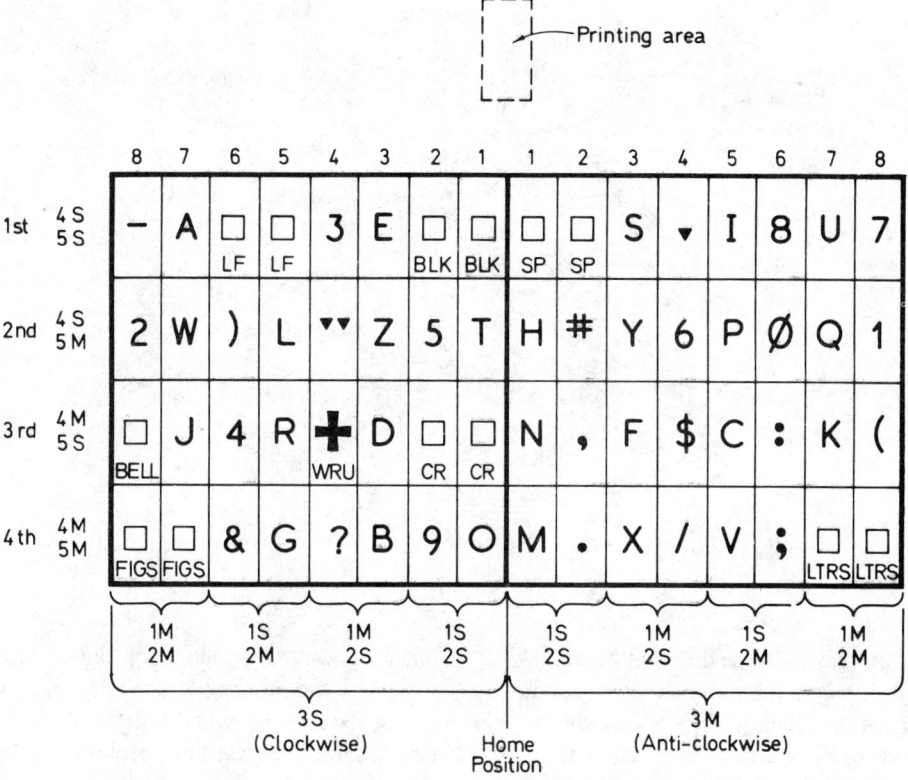

Fig 6.19 Teletypes T32 and T34: arrangement of symbols in typewheel and typeblock

in a set of two-state push levers. The push levers in turn control a set of two-state blocking levers, whose function is to control the setting of the receiver code bars. Towards the end of the selector cycle, the code-bar clutch is tripped to enable code bars to sense the positions of blocking bars and so reconstitute the combination. They are all reset towards the end of the combination-bar cycle. This cycle continues after the character start—stop cycle and enables printing to take place without waiting for a further character cycle.

Printing. The operation of setting the typewheel is similar for both models, 32 and 33, though different numbers of combination-bars are involved. In the following description, the second figure, in parentheses, refers to the Model 33, the first figure to the Model 32.

The typewheel is cylindrical; in Fig 6.19 it is shown in developed form, in the rest position, for the Model 32 machine; the arrangement shown

is for alphabet No. 2 (except for the V and Z secondaries). Characters are arranged in 16 vertical rows, each of four characters (including blanks). To move a selected typeface to the printing area, separate, but simultaneous, vertical and rotary (clockwise or anti-clockwise) motions are imparted to the typewheel. For both motions, movement must be controlled to select the required type. To print letter D the typewheel must be rotated 2½ rows clockwise and raised three rows (plus the distance below the printing area). A print hammer (with a soft head to avoid damage to the type) then drives the typewheel forward to print letter D in the printing area. The lower end of the typewheel shaft is pivoted to permit this action.

The *direction of rotation* is decided from No. 3(4) code element: clockwise for space and anti-clockwise for mark (Fig 6.20). For a mark the No. 3(4) code bar moves up and to the left. A following slide, through a linkage, moves a rotary drive arm down so that it engages the

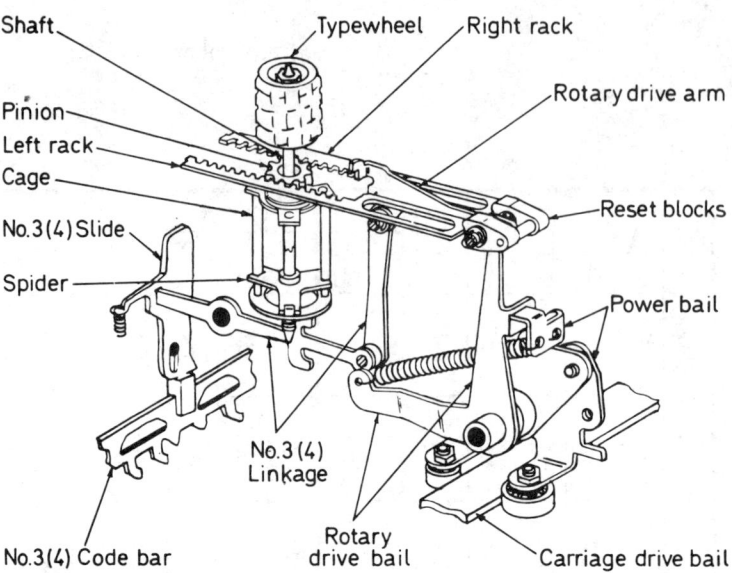

Fig 6.20 Teletype T32: rotary-positioning mechanism (1)

left-hand rotary rack, but not the right-hand rack. If the No 3(4) bit were space the alternate condition would exist, in which the No. 3(4) code bar was down and the drive arm was up, where it would engage the right-hand rotary rack but not the left.

A rotary-drive bail is held against the power bail by a spring. As these bails rock clockwise (viewed from the left) during the first part of the function cycle, the drive arm, which is attached to the rotary-positioning bail, moves towards the front. If the arm is down (No. 3(4) a mark), it pulls the left rack with it, and the rack rotates a rotary pinion, a cage, a spider, a shaft and the attached typewheel anti-clockwise (viewed from the top). On the other hand, if the link is up (No. 3(4) is space) it pulls the right-hand rack towards the front and rotates the typewheel clockwise. As the power bail and rotary-drive bail rock back to their stop position during the latter part of the function cycle, two reset blocks on the drive bail return racks and typewheel to their stop position.

The extent of *rotary movement,* in either direction, is determined by the No. 1(2) and No. 2(3) code bits and the No. 0(1) code bar. (In the Model 32, the No. 0 code bar is the shift code bar.) Referring to Fig 6.21 when the No. 1(2) and No. 2(3) code bits are both space, corresponding code bars and their following rotary stop slides

remain down. A common stop slide, which is moved by the other two, also remains down. As one of the racks is pulled frontwards, the opposite rack is driven rearward by the pinion and strikes the common stop (the one nearest the front). This permits the pinion to rotate the typewheel enough for the No. 1(2) or No. 2(3) vertical row (depending on the position of the No. 0(1) code bar) of either the clockwise or anti-clockwise field to be aligned with the printing area at the time of printing. When a rack is stopped, the rotary bail

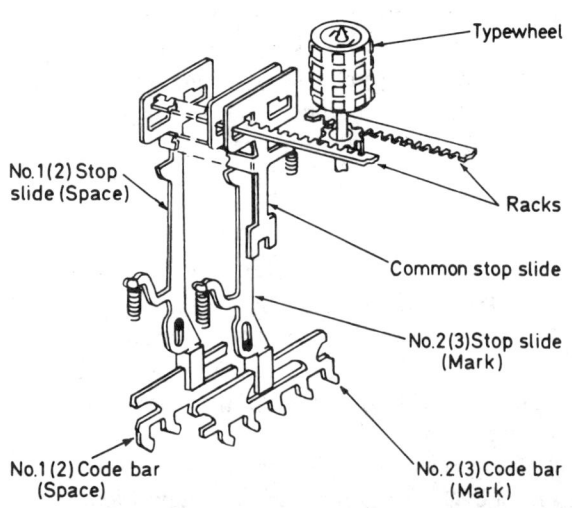

Fig 6.21 Teletype T32: rotary-positioning mechanism (2)

stops and its spring extends as the power bail continues its travel.

If the No. 1(2) bit is mark, and the No. 2(3) is space, the No. 1(2) code bar and slide move up, and the No. 1(2) slide lifts the common slide. In this position, holes in the common slide permit whichever rack is moving towards the rear to pass through and strike the No. 2(3) slide (second

even rows are selected; when it is down the odd rows are selected. In Model 33 machines, the No. 2 code element controls the position of this bar.

Vertical positioning of the typewheel is determined by element Nos 4(5) and 5(7). Referring to Fig 6.22, a vertical-drive bail is held against the power bail by a spring. When these bails rock clockwise (viewed from the left) during the first part of

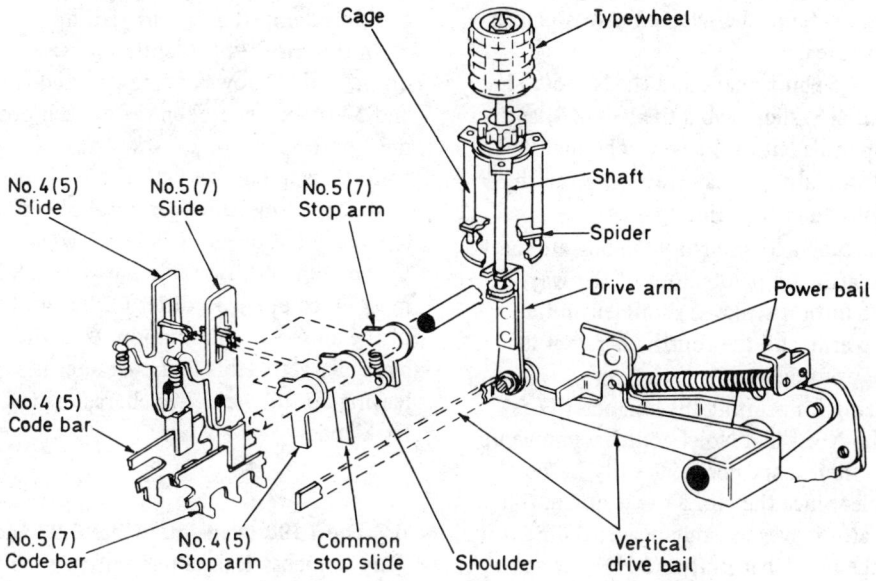

Fig 6.22 Teletype T32: vertical-positioning mechanism

from the front), which is down. The No. 3 or No. 4 row is aligned with the printing area.

If the No. 2(3) code bit is mark and the No. 1(2) is space, the No. 2(3) slide is moved up and lifts the common stop. The rack that is moving rearward passes through holes in these slides and strikes the No. 1(2) slide (third from the front), which is down. The No. 5 or No. 6 row is aligned with the printing area.

If both the No. 1(2) and the No. 2(3) code bits are mark, the No. 1(2) and No. 2(3) slides and the common slide are moved up and permit the rack to pass through holes and strike the shift slide (farthest from front). The No. 7 or No. 8 row is aligned with the printing area.

Whether an odd or even row is selected depends on the position of the No. 0(1) code bar. In Model 32 machines this bar is controlled by a letters/figures shift mechanism. When this bar is up the

the function cycle, the vertical drive bail, through a drive arm, lifts the spider, typewheel shaft and typewheel. How far the typewheel is raised is determined by three stop arms which are positioned in response to the No. 4(5) and No. 5(7) code bits. When the drive bail encounters an arm, it is stopped and its spring extends as the power bail continues to pivot. The spider moves up and down the bars of the cage and so permits rotary motion to be transferred to the typewheel regardless of its vertical position. During the latter half of the function cycle, the parts are returned to their stop position.

When a code-combination is received in which the Nos 4(5) and 5(7) code bits are space, the Nos 4(5) and 5(7) code bars and their following vertical slides remain in their space (down) position and no motion is transferred to the stop arms. As the rear extension of the drive bail rises it strikes the common stop arm, which is the longest. This permits the typewheel to be raised to the point where the No. 1 character in the selected row is in the printing area at the time of printing.

If the No. 4(5) bit is space and the No. 5(7) is mark, the No. 5(7) code bar moves its following slide up early in the code-bar cycle. The slide pivots the common and No. 5(7) stop arms rearward out of the way of the drive bail. The bail strikes the No. 4(5) arm (the second longest) and the second character in the selected row is placed in the printing area.

If the No. 4(5) bit is mark and the No. 5(7) is space, the No. 4(5) slide pivots the No. 4(5) and common stop arms out of the way. The bail strikes the No. 5(7) stop arm (the shortest) and the third character is placed in the printing area.

If both the No. 4(5) and No. 5(7) bits are mark, all three stop arms are pivoted out of the way. The bail moves up until it strikes a shoulder on the common stop arm, and the fourth character is placed in the printing area.

Summarizing for printing the character D (MSSMS): the No. 1 and No. 4 code bars move up to the left to their mark positions early in the code-bar cycle. Since the No. 3 bar is down, the rotary-drive arm moves up and engages the right rack. Since the No. 1 bar is up and No. 2 is down, the No. 1 and common stop slides are moved up. Since the printer is in the letters condition, the 0 code bar keeps the shift slide down. Since the No. 4 code bar is up and the No. 5 down, the No. 4 and common stop arms are pivoted out of the way of the vertical-drive bail. As the power bail rocks clockwise (as viewed from the left) during the first part of the function cycle, the following operations occur. The rotary drive arm pulls the right rack frontward and rotates the type-wheel clockwise. The left rack, driven by the type-wheel pinion, moves rearward, passes through a hole in the common slide, and strikes the No. 2 stop slide. The stop slides and their guides are moved rearward, and the wider outline of the shift slide strikes the front stop surfaces of the stop plate. The No. 3 vertical row is aligned with the printing area. Concurrently with the rotary positioning, the vertical-drive bail moves up until it strikes the No. 5 stop arm, and the third character in the No. 3 row is placed in the printing area.

The upward component of motion of the code

bars is used to control positioning of the typewheel; the lateral movement is used to determine which function, if any, will be used. The underside of each code bar is coded by notches and projections for function control.

The carriage unit is carried on a toothed rubber belt which for each printed character (and character space) moves one space against the tension of a powerful carriage-return spring. The carriage-return action is damped by an air piston.

In the answer-back unit, a universal plastic coding drum, with 21 rows of tines, is used for Models 32 and 33. A tine is broken off in each position which must correspond to a mark. A set of wire-spring contacts is in parallel with those of the keyboard, a contact being held open by the presence of a tine (space) but allowed to close for an absent tine (mark). In the home position, the answer-back contacts must all be open: In order to prevent transmission of the all-space combination 32 as the first answer-back character, this first character is suppressed, leaving 20 fully-selective characters for the answer-back code.

6.7 The T100 teleprinter (Figs 6.23 and 6.24)

This Siemens–Halske teleprinter is a 5-unit machine for alphabet No. 2 at 50 bauds although adaptation to a modulation rate of 75 bauds is possible, and a version operating at 100 bauds is available. This machine provides all normal features, including two-colour printing, automatic motor switch, paper-

Fig 6.23 Teleprinter T100 (*Courtesy of Siemens & Halske AG*)

Fig 6.24 Teleprinter T100 — cover removed (*Courtesy of Siemens & Halske AG*)

fail alarm, sprocket-feed platen, tabulation and a counter for hours of service. The machine is designed to incorporate tape transmitter and reperforator attachments.

The receiver uses a traversing typebasket with stationary paper platen. In common with Continental teleprinters, lower-case printing is normal. The machine is designed for single-current (neutral) operation, though double-current (polar) transmitter and polarized electromagnet can be supplied.

In the standard 4-row keyboard a separate key is allotted to each character; in the condensed 4-row keyboard, numerals keys occupy the rear row while keys in the two front rows have both a primary and a secondary function. In both keyboards, full or partial mechanical case-shift lock is provided to prevent operation of any key unless the appropriate shift key has first been operated. HERE IS and RUN OUT keys are provided. A character-recognition unit can be added to give electrical signals on receipt of any combination.

Motor. The machine is driven by an a.c.- or d.c.-governed motor (3750 rev/min) or, alternatively, by a synchronous motor (3000 or 3600 rev/min for 50- or 60-Hz supplies). Motor-starting time is 500 ms; the a.c. motors are fitted with a thermal-protection device.

Transmission. In the transmitter a three-pawl clutch is used, pawls being displaced by one-third of a tooth pitch within an internally-toothed ratchet wheel to reduce pickup delay. The transmitter uses a single-make (single-current) contact or a single change-over contact and send—receive switch (double-current). Transmitter distortion is less than ±5%. Radio-interference suppression is provided, the motor and the transmitter being in screened containers to assist in suppressing interference at frequencies up to 300 MHz.

Reception. The receiver camshaft is driven via a sealed friction clutch. Just before the camshaft is arrested by the stop signal, it releases the typing unit for printing the received character without awaiting further signals.

The single-current (non-polarized) electromagnet has twin armatures which have completely independent functions. The first serves solely for starting and stopping the receiver camsleeve. As it has to perform work only once per character it is designed with a large travel and is slowly restored by the motor during the receiving process. The second armature is offered to the magnet under mechanical power and is required only to release or lock the selector lever according to the signal state during the examining periods of the five code elements. The short armature travel here required is of great advantage, because the signal-examining process occurs in a very short interval of time. This mechanical arrangement gives a good receive margin (better than ±42% at 50 bauds and better than ±40% at 75 bauds). An orientation device or *range-finder* is included.

In the single-current model, each armature has a restoring spring and in the idle condition current flows in the magnet and both armatures are attracted. Fig 6.25 shows operations performed by the start—stop armature. The receiver shaft is driven by friction clutch FC. In the rest condition, receiver camshaft CS is arrested by locking-lever LL and release lever RL. When the start signal arrives (no current) both armatures release; the start—stop armature SSA drops on to the crescent-shaped extension of release lever RL which is caused to disengage from locking lever LL. Cam C is now able to pivot locking lever LL and the receiver camshaft is free to turn for one revolution. As the

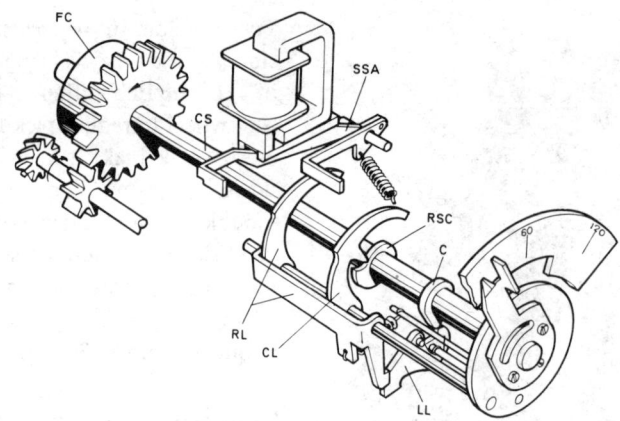

Fig 6.25 Teleprinter T100: starting and stopping the receiver shaft

receiver shaft revolves, the crescent-shaped lever CL, raised by resetting cam RSC, applies the start–stop armature SSA to the magnet poles. During the stop signal, the armature is held attracted by the magnet and lever CL returns to its rest position. At the end of the revolution of the receiver shaft the nose of cam C runs up against locking lever LL and is unable to push it aside and the receiver shaft CS is arrested.

For examining the five code elements (Fig 6.26) selector cam SC offers the sensing armature A for each code element to magnet M. If the magnet is energized at this instant (mark current), armature A is held attracted while shaft CS continues to turn. Selector lever SL pivots in a clockwise sense as it rides down the lobe on cam CA and is able to follow the contour of it. While edge SLB of selector lever SL swings upwards, code lever CL is turned anti-clockwise by cam CB. The operating edge of selector lever SL tips the sword S anti-clockwise. Selector lever SL and code lever CL are returned so that the sword is fixed in its position by guide G. Should the magnet be de-energized (no-current space) while the armature is being offered, armature A will ride down the lobe on cam SC. When selector lever SL is being turned by cam CA, its extension SLA abuts against the tip of armature A. As code lever CL is now turned anti-clockwise, sword S will be held by a spring at the right-hand side of the guide. After the five code elements have been examined, the code levers are released by their cams. The swords move upwards under the action of the code-lever springs to set the five transfer bars.

For double-current operation, the modification when using a polarized electromagnet (with single armature) is illustrated in Fig 6.27. On receipt of the start signal both levers are unlatched and allowed to drop. The start–stop lever trips the receiver in the manner performed by the start–stop armature in the non-polarized magnet. The sampling lever, mechanically offered to the armature for each code element, is either latched or released by the armature, depending on the polarity of the signal elements. Its position is examined by the selector lever, as for the single-current receiver. Retaining the 'edge-decision' principle of the neutral version necessitates an asymmetric position of the polarized armature between the pole-pieces, together with a delay winding (shunted by a resistor and diode) to compensate for the resultant mechanical bias. The receiving margin is but slightly impaired.

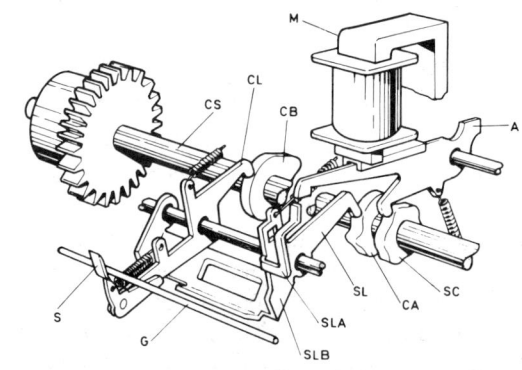

Fig 6.26 Teleprinter T100: sampling the 5-unit combination

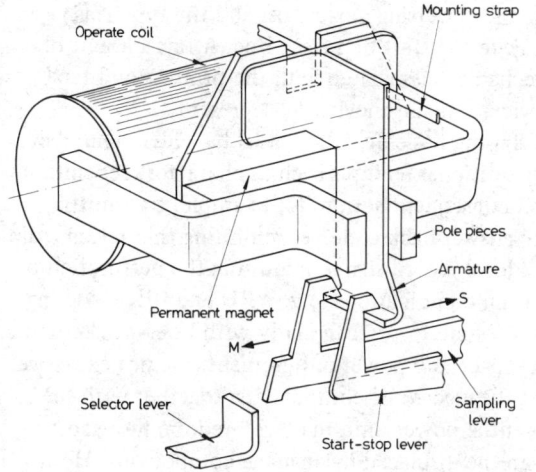

Fig 6.27 Teleprinter T100: polarized receiver electromagnet

The answer-back unit does not give full freedom of choice for the 20 characters* of the answer-back code. Only 18 wards are fitted to the drum; the 1st and 20th characters of the code are fixed as LETTERS-SHIFT (MMMMM) because the answer-back drum is in the rest position and the keyboard comb bars are not opposite any ward in the drum.

Storage Transmitter. The model T100S has a storage transmitter capable of storing up to six characters. The purpose of this is to insert automatically the LETTERS-SHIFT or FIGURES-SHIFT signal when any key requiring a change of case is pressed, particularly useful for national alphabets containing more than 26 letters. The effect of the storage transmitter conforms closely to typewriter operation and gives more operating freedom since it permits a temporary higher keying speed. Each character is transferred into a mechanical store before being picked up by the transmitter shaft which continues to revolve so long as any character is in store. Speed bursts up to 12 char/s can be accepted for transmission at 6·6 char/s, with proportionate values for the 75-baud model.

Fig 6.28 illustrates the storage operation. Code bars CB (and the setting members they engage), displaced in the usual manner by the operation of a key, actuate the contact control levers CL, not directly but through the medium of a storage drum

*Recommendation S6

D and the sensing mechanism SM. The storage media are radially-displaceable pins which are latched in their end positions and arranged axially in rows of five each. The sending mechanism is interlocked with the storage drum and follows its rotation while signals are being stored. Having left its rest position it releases transmitter shaft TS which causes it to jump over to the next row of pins, against the sense of rotation of the storage drum. It actuates the contact-spring sets SS by means of the contact control levers in accordance with the code-combination set up on the drum. One contact spring-set is provided for each code element; a sixth contact supplies start and stop signals.

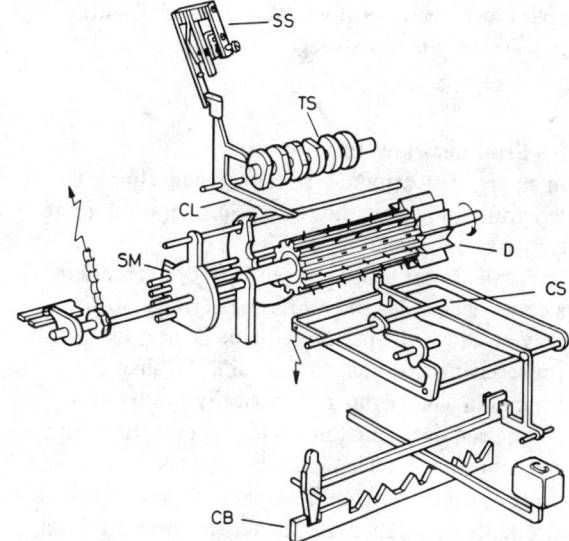

Fig 6.28 Teleprinter T100: principle of storage transmitter

Rotation of the storage drum is initiated by depression of key levers, while motion in the opposite direction of the sensing mechanism is controlled independently by the revolving transmitter shaft. This revolves without stopping as long as the sensing mechanism is outside its rest position. The storage process is controlled by camshaft CS which, tripped by the keyboard, performs one revolution in 1/12 s. Only the second half of each revolution is used: the first half is made available for storing case-shift signals. Operation of a key necessitating case shift causes a sixth code bar to be displaced, and as a result the storage drum is advanced one step during the first portion of the revolution of camshaft CS either without a pin being pressed in (LETTERS SHIFT) or with the third pin being pressed in (FIGURES SHIFT). The next key can be operated again at a spacing of 1/12 s, in spite of insertion of the case-shift signal. The pins pressed in are released before being returned to the storage position. When case-shift signals are being inserted, or when keys are operated rapidly, the storage accepts more signals than can be cleared by the transmitter. The sensing mechanism moves away from its rest position more and more, and approaches it again as the intervals between key operations increase. In the rare case when storage capacity determined by the crescent-shaped sensing members is exhausted, one more key can be depressed. Following this, a keyboard lock becomes effective temporarily, until the next step of the storage.

6.8 Semi-electronic teleprinters

In an era of electronic technology, a mechanical teleprinter might, without due consideration, be regarded as an anachronism.

A teleprinter consists essentially of a keyboard, a coder, a decoder and a printing carriage; to these may be added the perforated-tape devices used for receiving and transmitting, and the answer-back unit. The keyboard is basically mechanical; coding and decoding can readily be performed by electronic means. The major problem concerns the printing device and, until some cheap and reliable alternative medium arises, percussive printing must be performed by a mechanically-driven type-hammer. Xerographic, magnetic, electro-chemical and other printing methods such as matrix character formation are available, but their high cost restricts application to such fields as high-speed computer-output printers. Methods which require the use of special sensitive paper are costly to use and those which radiate energy at radio frequencies are unacceptable. Several carbon copies of a message are usually required and, although these could be provided by a separate copying machine, the cost is again higher. The present solution for a teleprinter seems to be to retain the conventional printing unit and to apply electronics as far as possible to the coding, decoding and function controls. The printing/paper carriage must jointly be given a two-co-ordinate feed movement, with a means of bring-

ing the traversing device (usually the typehead) back to the start of a fresh line. A fair amount of mechanism associated with the printing and tape devices remains inevitable at present.

To the basic decoding must be added a number of additional features such as character recognition and consequent action; for example, transmitting the answer-back code yet inhibiting this action from the local record signal; inhibiting the perforation of certain combinations (e.g. WRU and BELL). Using semiconductors, particularly with integrated circuits, the space and power requirements are not excessive, yet the electronic components, together with the essential power unit, may well require more space than the displaced mechanical components. However, the choice of location for printed-circuit cards is not conditioned, as with mechanical components, by the need to place them adjacent to related components.

Use of electronic decoding dispenses with the need for clutches and eliminates those mechanical parts which are of most critical adjustment and subject to most wear, resulting in a simplification in maintenance and increase in stability and life of the machine. It is generally accepted that a modulation rate of 200 bauds – corresponding to 5-ms elements, with character periods of about 50 ms and speeds up to 20 char/s represents the upper limit at which mechanical teleprinters or teleprinters having a mechanical typing unit can operate while still retaining the necessary reliability and machine life; it is also more difficult to achieve the required degree of receive margin with such short element durations. The performance of current mechanical teleprinters (typically ±3% emission distortion and ±40% receiving margin) has reached a high order and could be improved only marginally by electronic means.

At moderate speeds, up to about 75 or 100 bauds, reliable and relatively inexpensive mechanical solutions with good performance are possible; the field for electronic solutions is probably better suited to applications at higher speeds with extended codes and alphabets. Ultimately it is a question of economics.

Some semi-electronic teleprinters in commercial use are described in the following pages. One essential requirement is the isolation, for reasons of human safety, of the telecommunication line from the mains-drive electronic circuits: the solution for this is coupled with the provision of correct interface conditions between the electronic system and the line, typically driven at 6 and 80 V respectively.

6.9 Creed semi-electronic teleprinter

The ENVOY dataprinter operates to alphabet No. 5 at a speed of 10 char/s and is suitable either for line communication or as a computer input/output terminal set. An 11-unit signal is generated (start, seven

effect if used together with these keys. Other keys have only a single meaning which is the same whatever the prevailing SHIFT or CONTROL condition.

If a non-parity combination is recognized, an open-diamond symbol is printed: receipt of a DELETE signal is recorded by printing an open-square symbol. A full range of facilities is available (some optional) such as tape transmitter, tape punch, controlled two-colour printing, vertical and horizontal tabulation, and form-feeding. The form-

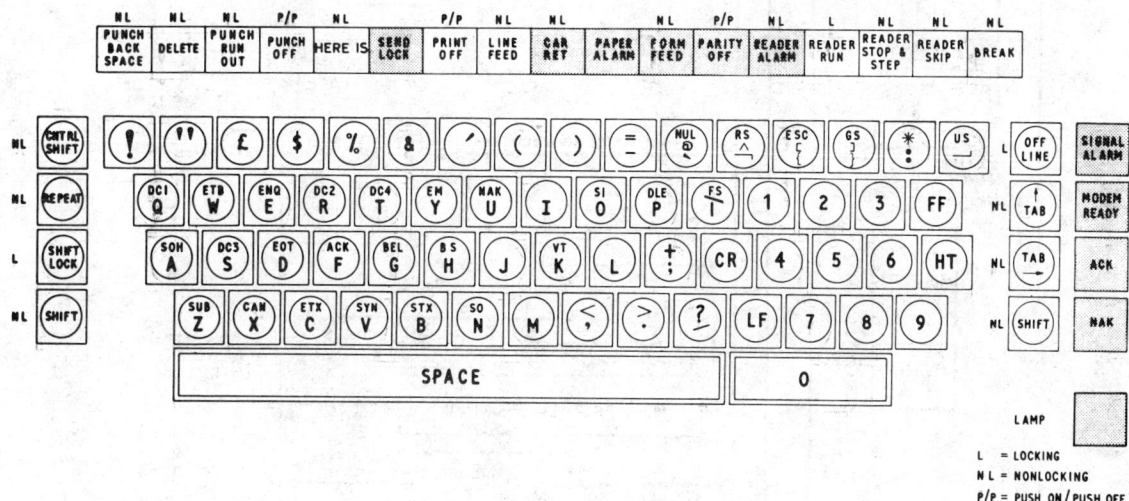

Fig 6.29 Creed Envoy dataprinter — keyboard layout

code element, one even-parity M bit, and a two-unit stop signal), resulting in a modulation rate of 110 bauds. The machine can also be provided for modulation rates of 50 and 75 bauds, with operating speeds of 4·5 and 6·8 char/s respectively. The four-row motorless keyboard generates all 128 combinations of alphabet No. 5. The typewheel, similar to that used in the Creed Model 75 teleprinter, but with six type levels each containing 16 types, prints all 96 graphic characters (including upper- and lower-case letters) of the alphabet. The keyboard layout (Fig 6.29) includes 62 keys; keytops are colour coded to assist operation. The specific use of SHIFT and CONTROL keys is the same as described for the Teletype Model 33. Some keys have three meanings, depending upon whether they are operated (1) alone, (2) with the SHIFT key, or (3) with the CONTROL key; a few have two meanings, depending upon whether they are used with or without the SHIFT key. The CONTROL key has no

feeding action is stopped by a sensing hole in pre-printed stationery. Provision is made for external paper rolls and sprocket-feed fanfold stationery. The platen has retractable pins to make it suitable for friction or sprocket feed and a maximum of five printed copies is possible. A non-printing character should always follow the carriage-return signal to allow time for the printing head to return to the start of the line.

A set of control keys, some with inset lamps mounted at a raised level behind the keyboard, provides full control facilities for such features as off-line tape preparation, duplication, interpretation and editing. The functions of these keys and lamps are self-explanatory from their labelling; the keys provide for versatility in use of the printer, punch and tape transmitter. The four left-hand and four right-hand keys control respectively the adjacent punch and tape transmitter attachments. The SEND LOCK key enables the keyboard to be locked by

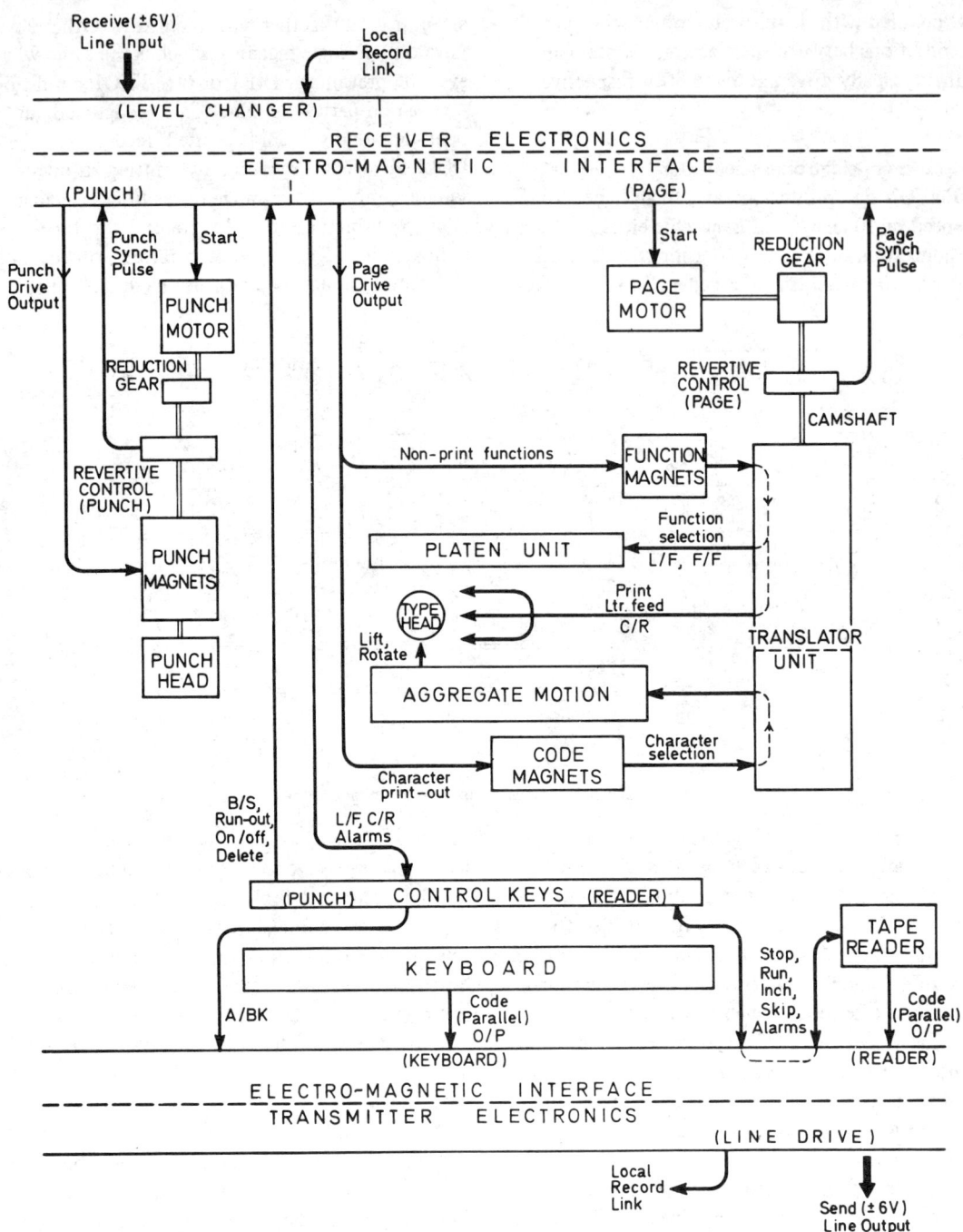

Fig 6.30 Creed Envoy dataprinter — electromechanical interface

an external signal. The LINE FEED key, for local machine control, causes continuous line-feed action as long as the non-locking key is held down. The CARRIAGE RETURN key is also for local control only; the CARRIAGE RETURN lamp glows

for the final quarter of a 72-character line of print. Operation of the READER SKIP key causes the tape transmitter to step one character without transmitting signals. The READER ALARM lamp indicates TIGHT TAPE or TAPE OUT or parity error;

in the latter case the tape transmitter is automatically stopped. Depressing the READER ALARM key cancels the parity-error condition and restarts tape transmission. Operation of the BREAK key sends a clear (space) signal. The BELL signal produces a tone in a miniature loudspeaker. Receipt of a tabulation signal causes the printer mechanism to feed to the next tab stop. If the local TABULATION pushbutton is operated, the printer mechanism advances to the next tab stop, and the appropriate number of character-space or line-feed signals is sent to line.

Transmitter and receiver can be used with signals in either the sequential or simultaneous mode.

The gross start–stop distortion of the transmitter is less than 2·5% and the receive margins not less than ±47·5%, both at 110 bauds.

Mechanical units such as clutches, the transmitter unit and the selector unit have been eliminated and replaced by integrated-circuit electronic units. The mechanical units which remain (Fig 6.30) perform relatively simple functions such as printing and paper transport. As a result the sound level is low compared with a mechanical teleprinter and the number of mechanical adjustments needed is small. The rotation and lifting adjustments of the typewheel can be made while the machine is printing. A bank of seven electromagnets is used to convert the received signals into equivalent mechanical displacements for lifting and rotating the typewheel, using the aggregate-motion principle. Code signals are applied to the magnets while the armatures are offered to the pole faces by mechanical power; the armatures of electromagnets which receive space elements are then released. Mechanical operation of the typewheel is the same as for the Model 75 machine, except that in the present case there are six levels of types. Printing is inhibited on receipt of any of the 32 control combinations.

The electronic units perform all the timing, code-conversion, storage and control functions of the teleprinter. Two identical, but independent, induction generators, each consisting of a rotating disc carrying a permanent magnet past two stationary induction coils, provide revertive control signals. One generator is associated with the page printer and rotates with the printer camshaft; the other is associated with the tape punch and rotates with the tape-punch drive shaft. Each generates a continuous train of pulses to control and time the

transfer of information from the electronic units to the clutchless printer and tape punch.

A block diagram of the electronic units is given in Fig 6.31.

Transmission. Signals originate from the tape transmitter, from the answer-back unit, or by depression of a key on the keyboard. The SHIFT or CONTROL key is operated simultaneously with another key when required. The keyboard generates code-combination signals by means of transducers, the keyboard being locked until the signal combination is transferred to the electronic circuits. All generated signals are in 7-unit simultaneous form. They are then applied to the parallel-series converter which adds the start, parity and stop elements when converting the signals to sequential form by pulses from the time-base unit. The line driver converts the low-voltage electronic signals to double-current signals suitable for line transmission.

Reception. Incoming signals enter the logic-level converter where they are converted to low-voltage signals suitable for the electronic circuits. Synchronization with the incoming signal is maintained without a clutch by cycling the printer marginally faster than is necessary to print at 10 char/s, and introducing the required number of idle cycles.

The start signal is first checked for duration by the short-start-signal detector which rejects spurious start signals. The time-base is switched on, and succeeding bits are gated to the series-parallel converter in order that they may later be presented simultaneously to the seven code magnets. When the last code bit has been received in the converter, the simultaneous signals are transferred to the code store, provided that code parity is confirmed by the parity-check unit. Transfer of the code-combination from the store to the data-printer takes place when this unit is ready to receive the new combination. The electromagnets hold (M) or release (S) to reconstitute the signal combination. Their pattern preselects the character to be printed out later in the cycle.

Functional combinations in the store are recognized by the functional decoder, which controls a group of five separate electromagnets to select the appropriate functional actions required to accompany or substitute for a printing cycle.

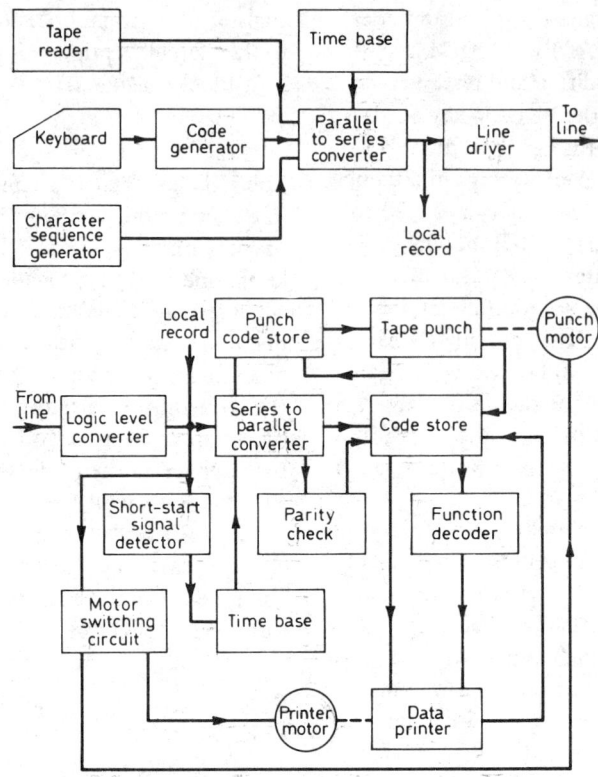

Fig 6.31 Creed Envoy dataprinter — electronic units

If the tape punch is in circuit, each combination transferred to the printer-code store is received also by the punch-code store; signals are fed at the appropriate instant to the set of ten electromagnets. Eight of these select the code and parity punches in each punching cycle; one is used for punching the feed hole and for feeding the tape through a one-hole pitch. The tenth magnet operates in place of the feed magnet when it is required to backspace the tape. The tape-punch attachment is driven by a separate electric motor.

The printer and punch motors are controlled by an automatic motor ON/OFF switch. The motors reach correct operating speed in 1·5 s.

The tape transmitter is driven electomagnetically by a feed-rake mechanism. Tape holes are sensed by peckers which operate code contacts. TIGHT TAPE and TAPE OUT contacts are provided.

The answer-back unit, which has no moving parts, will provide up to 30 characters, electronically generated. The unit can be adapted to transmit two separate codes which could be used, for example, as the head and end of message codes in a tape-relay system.

A character-recognition unit detects certain signal combinations to provide function controls, to light lamps, or to provide 5-V control signals.

6.10 The Sagem SPE semi-electronic teleprinter (Figs 6.32, 6.33)

This teleprinter uses electronic circuits for coding and decoding, the keyboard and translator printer being of conventional mechanical form. The electro-mechanical assembly comprises motor, keyboard and translator printer. Printing is performed by a stationary typebasket with a traversing paper carriage. Electronic units comprise time-base, sending and receiving circuits.

The teleprinter is for alphabet No. 2 at modulation rates of 50 and 75 bauds, with a 4-row or 3-row keyboard; characters are printed (69 to the line) on a 6-in. wide paper roll. Transmitter distortion is less than ±2% and receive margin exceeds ±47%, both at 50 or 75 bauds. Normal facilities, such as two-colour printing, automatic motor ON/

Fig 6.32 Sagem SPE teleprinter (*Courtesy of Sagem SA*)

OFF switch, answer-back unit, paper-fail alarm and tabulator are provided. The electronic equipment, with sub-assemblies on a modular concept, is housed in the base of the machine.

To provide isolation and voltage interface, send and receive lines terminate on polarized relays. Electronic circuits determine the significant instants in send and receive cycles, sequential timing of signals set up on the keyboard, tape-transmitter or answer-back unit, recognition of incoming signals, control of mechanical-printing components, tape-stepping and recognition of WRU and BELL combinations

The electronic equipment comprises a time-base oscillator (100 Hz for 50 bauds or 150 Hz for 75 bauds) with dividing stages, together with a number of bi-stable triggers for keyboard and tape transmitter, send–receive switch, transmitter, receiver, storage and a number of diode-gate matrices. A schematic diagram of the electronic equipment is shown in Fig 6.34 where mark and space polarities accord with CCITT convention, i.e. positive for mark.

Reception. A start signal operates receive trigger BdR, which starts the oscillator via gate PO. The time-base makes one cycle, resetting storage triggers E1–E5 to mark and successively opening receive gates PR1–PR5 at the theoretical mid-element examining periods. Space elements switch storage triggers, and the code-combination is reconstituted on stores E1–E5. The send–receive trigger BER being in the receive condition, the time-base is

directed to transfer stored code elements simultaneously to control amplifiers for magnets C1–C5 which position mechanical code bars. The transfer operation also biases the amplifier which controls the printing magnet, and the character is printed as a follow-on operation. The clutch magnet EMB is operated from the time-base via the control for the five combination-bar magnets. The time-base is reset by the stop signal.

Transmission. When any key on the keyboard is operated, send–receive trigger BER is operated to SEND, via anti-bounce trigger BaR, common trigger BU and gate PE. The send–receive trigger starts the oscillator via gate PO. A similar train of operations results from pressing the ON button to operate the keyboard/tape trigger BCL.

With the send–receive trigger at SEND, space gate PS is opened after 10 ms for the start signal to operate the send relay via send trigger BdE. At this stage storage triggers E1–E5 are reset to mark. Transfer gates PT1–PT5 are then opened according to the code-combination for simultaneous transfer of signals, either from keyboard or tape transmitter, to storage triggers E1–E5. Gates PER1–PER5 (mark), or PET1–PET5 (space) are successively opened, as determined by the five storage triggers E1–E5, at instants 30, 50, 70, 90 and 110 ms pulsed from the time-base. The send relay trigger

Fig 6.33 Sagem SPE teleprinter — cover removed (*Courtesy of Sagem SA*)

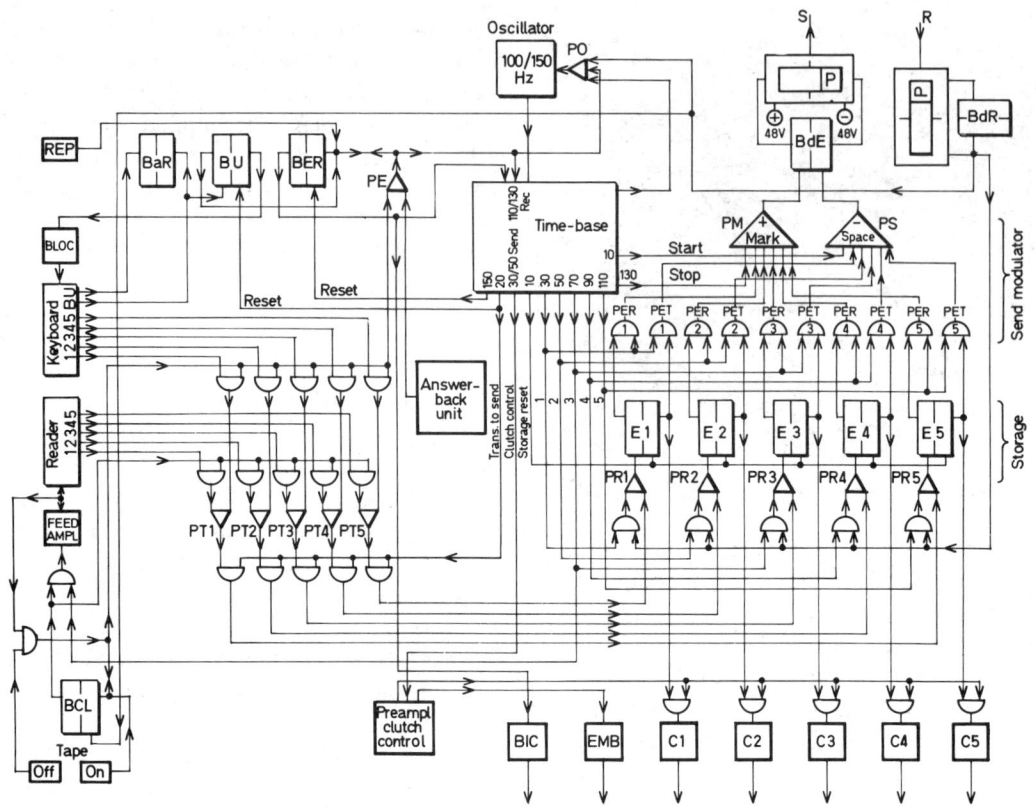

Fig 6.34 Sagem SPE teleprinter — schematic diagram

BdE is operated accordingly via mark and space gates PM and PS and signals are transmitted to line via the polarized send relay.

While the send—receive trigger is at SEND, storage triggers E1—E5 also bias the amplifiers controlling code-selection magnets C1—C5, and this provides the local record. The stop signal is provided by the time-base to switch send trigger BdE to the stop condition. The send—receive trigger is returned to the RECEIVE condition and the oscillator stops.

While the send—receive trigger is in the SEND condition, the electromagnet BIC shifts the ink-ribbon control to print in red: in the RECEIVE condition this magnet is not energized.

Receipt of the FIGURES SHIFT combination opens a contact in the (mechanical) translator assembly. This contact, together with subsequent storage of the WRU combination (the send—receive trigger being at RECEIVE), opens a gate to trigger the answer-back unit. The BELL signal is recognized in a similar manner.

Mechanical Assembly. The mechanisms are controlled by four cams on a camshaft driven by a motor via a toothed belt and two pulleys. The camshaft is released for the receive cycle (130 ms at 50 bauds, 84 ms at 75 bauds) by a magnet-controlled clutch.

The translator—printer device includes five code magnets and five receive code bars with latching devices. A rack lowers the set of type-selector levers, one of which will enter the aligned notches in the code bars; the rack restores later, and code bars are reset to normal. A printing bail engages the end of the type-selector lever which is in the aligned notch and the selected character is printed in a follow-on cycle.

A shift mechanism responds to the LETTERS and FIGURES SHIFT signals to lower or raise the type basket for printing from the selected case.

On the keyboard, five code bars operate electrical contacts to set up the required combination; the code bars remain in position until a subsequent key requires a change of code condition. A trip bar operates a contact shortly after the code contacts have closed, and it starts the electronic send cycle. The keyboard is disabled by a transistor-operated electromagnet which blocks the trip bar against re-operation until the code-combination has been read into the five storage triggers. It is similarly disabled when the carriage reaches the 68th character position: a magnet operates a blocking bar which prevents operation of any key other than LETTERS SHIFT, FIGURES SHIFT, CARRIAGE RETURN or LINE FEED. The keyboard is fitted with a shift lock to prevent numerals keys from being operated while in letters shift, and vice versa. A REPETITION (RUN OUT) key is provided.

The answer-back unit is tripped by an electromagnet and driven from the translator—printer assembly. Five contacts, operated by the tines on the answer-back drum, transfer code signals direct to the storage triggers. A trip lever ensures a single rotation of the drum, and off-normal contacts are operated by a cam. The answer-back unit electromagnet is tripped either on receipt of the WRU signal or by the HERE IS key.

The tape transmitter is controlled by ON/OFF buttons and a TAPE OUT contact. The tape is driven step-by-step by an electromagnet energized from a transistor amplifier controlled by the time-base circuit. Five hole-sensing levers operate simultaneous contacts to transfer the electrical signals direct to storage triggers E1—E5.

The reperforator is controlled by ON/OFF keys and a BACKSPACE key; alternatively, it may be controlled remotely by two electromagnets. Punches are operated by links directly connected

to the code bars and the tape is stepped by mechanical drive.

Conversion from 50- to 75-baud operation is made simply by transposing the position of a pair of pulleys, which transmit mechanical power, and removing a strap in the oscillator capacitance circuit.

6.11 Perforated tape

The use of perforated tape fulfils two functions — it is a convenient means of storage for messages in code, and it permits signal transmission at the maximum rate for which a circuit is designed — faster than can be achieved by manual operation. As a storage medium, it avoids the need for manual re-transmission.

Paper tape is used for 5-, 6-, 7- and 8-unit codes. In equal-length codes it is perforated for each code-combination transversely to the direction of feed. The holes represent mark elements; it is unnecessary to punch start and stop signals because, being invariable, they can be automatically inserted by the tape transmitter. A line of small central feed holes at character pitch serves to step the tape while undergoing transmission. Standardized tape widths are $\frac{11}{16}$ in. (17·5 mm) for 5-unit, $\frac{7}{8}$ in. (22·2 mm) for 6- and 7- unit, and 1 in. (25·4 mm) for 8-unit signals. In standard tape the centre hole is in line with the code perforations, although a few machines use an advanced feedhole and some tape transmitters are adaptable to accept this. Standard dimensions and form for perforated tape are shown in Fig 6.35 for 5-unit tape. Adjacent holes are $\frac{1}{10}$ in. apart in hori-

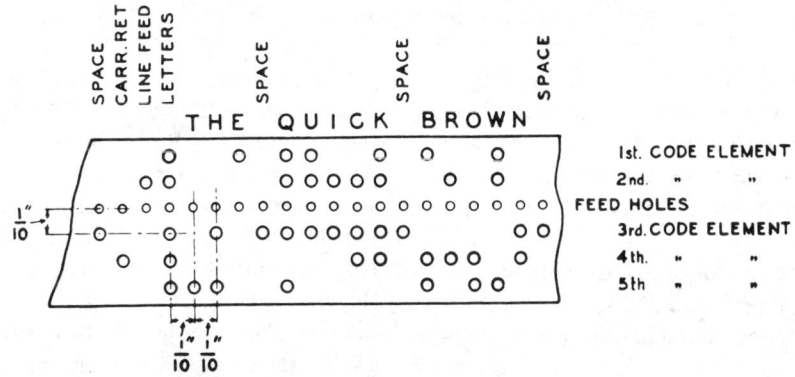

Fig 6.35 Characteristics of 5-unit perforated tape

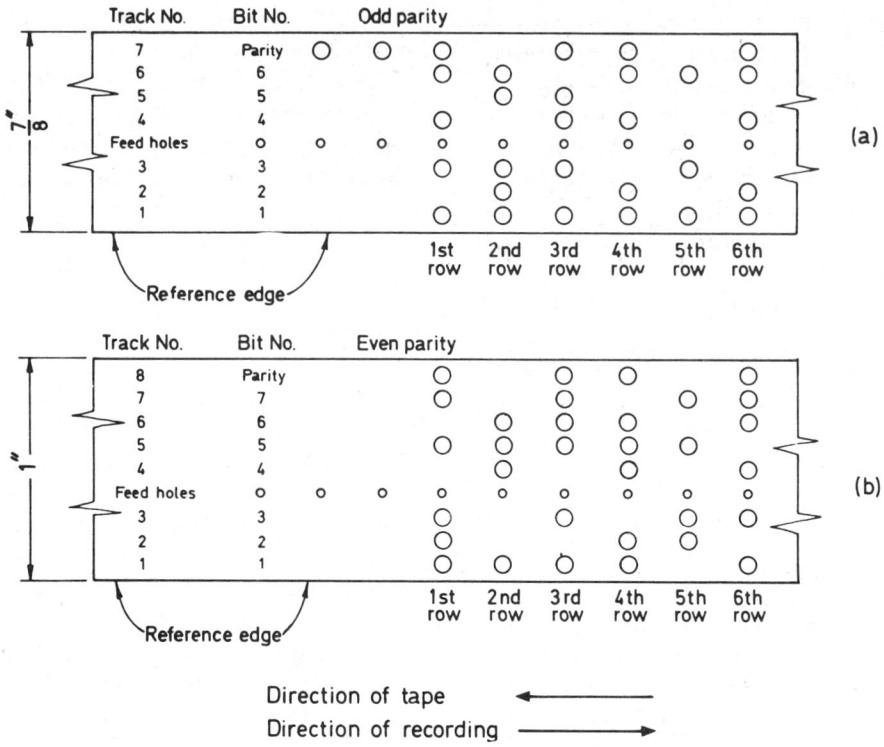

Fig 6.36 Characteristics of 6-unit and 7-unit perforated tape: (a) 6-unit code with odd mark parity; (b) 7-unit code with even mark parity

zontal and vertical directions. The track layout* for 6- and 7-unit tapes (with an odd and even mark parity element respectively) is shown in Fig 6.36: the reference face of the tape is the one illustrated. The paper-tape tear-off edge of most perforating devices

identified without need for the operator to memorize the telegraph alphabet. Printing may be either on the edge of an extra-wide tape (e.g. $\frac{7}{8}$ in. for 5-unit code) or more usually between the feed holes, using a slim typeface. In either case, printing takes place

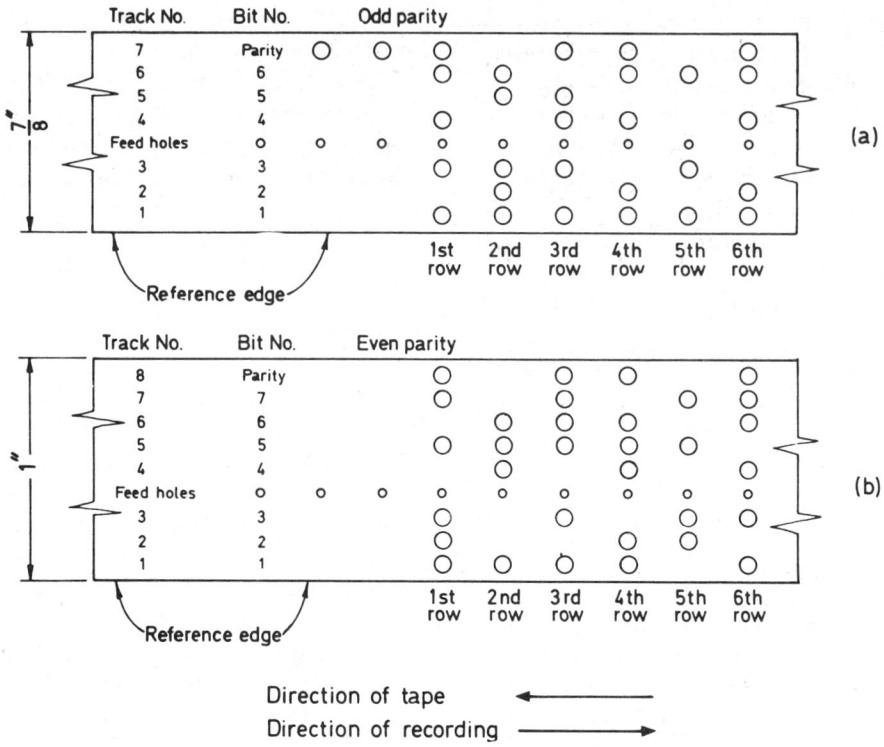

Fig 6.37 Chadless printed—perforated tape: 5-unit

is designed in a V-shape so that the tape leading edge has a closed-V and the trailing edge an open-V, the complete message tape having the form of an arrow showing the correct direction for transmission. The tape is made from parchment paper to give adequate strength.

A printing reperforator prints the message on the tape in addition to perforating it, so that the address and other information on the heading can be quickly

necessarily several characters behind the related perforation position owing to the minimum physical separation needed for the print and punch heads. Some reperforators produce *chadless* tape — the small paper circles or *chads* are not completely punched out and the chad remains hinged to the tape by a narrow neck of paper forming a paper lid (Fig 6.37). The advantage of this is that printing can take place over the chads without the need for a wider tape for edge printing. Disadvantages are that chads become caught up when the tape is

* British Standard: BS 3880 and 3967: 1966.

wound, giving paper-feed troubles; also it cannot be used with photo-electric reading systems. Chadless tape has largely fallen into disuse.

At large offices where considerable quantities of tape have to be punched, keyboard perforators are used. More commonly, a teleprinter with a reperforator attachment is utilised (off-line) for this purpose, the printed page providing a check of correct punching. Tape verifiers and tape comparators are also available for checking error-free punching.

at rest until sufficient tape is fed to permit the control arm to drop and restart the motor. At the normal speed for 50-baud machines, control contacts operate between 90 and 180 times a minute. As a precaution against tape breakage, due to over-running at high speed if the tape supply is suddenly stopped, the tape-control arm is arranged to cause the motor to be reversed by S_2 and S_3 before the arm reaches the limit of its travel. Tape is then unwound until the arm drops sufficiently to operate the S_1 contacts,

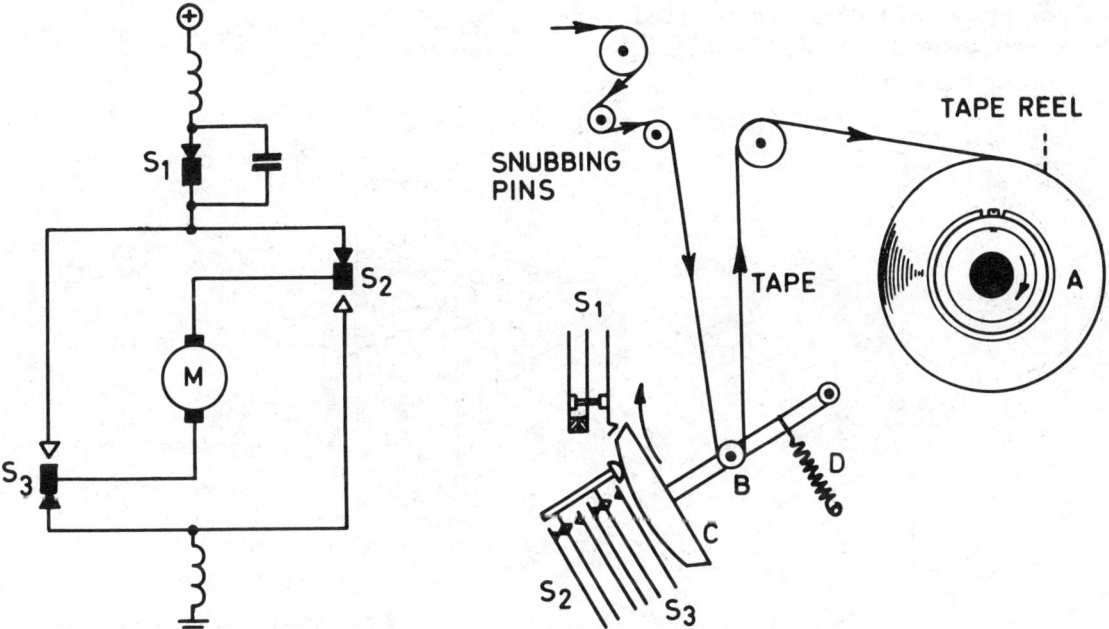

Fig 6.38 Tape winder No. 3C — functional details

Tape winder. Where large amounts of perforated tape are being produced, motor-driven tape winders are used to reel the tape as it comes off the punching machine. The winder is driven through reduction gearing from a commutator motor designed for a.c. mains or 80- or 160-V d.c. supplies.

To keep the winding tension within fairly close limits, the winder (Fig 6.38) is fitted with control arm B restrained by spring D. An insulated cam block C on the end of the control arm mechanically operates three sets of contacts S_1-S_3. The roll holder winds the tape at a speed much higher than it will be fed from the receiving machine, and as the loop is taken up the tape-control arm is raised and the motor is stopped. The roll holder remains

which then stop the motor. When tape feed resumes, the tape-control arm drops, restores first the S_2-S_3 micro-switches and then the S_1 contact. The motor then runs in the forward direction for normal operation.

Occasionally it may be necessary to unwind the tape from the roll holder by hand while it is still on the machine and to do this a free-wheeling facility is provided.

6.12 Tape transmitters and tape readers
A tape transmitter produces a sequential output signal and rarely operates at a speed greater than 10 char/s; tape readers provide simultaneous (multi-

wire) signals and operate at various speeds from 25 char/s upwards*. Most tape readers will take 5-, 6-, 7- or 8-unit tape.

Automatic Transmitter No. 2D (Creed Model 6S/5). Nowadays, a tape transmitter is available as an attachment to the teleprinter. For other purposes this automatic transmitter is commonly used for 7·5-unit transmission at 6·6 char/s. The machine (Fig 6.39) is driven by a 160-V d.c. series-wound motor at a constant speed of 1500 rev/min controlled by a centrifugal governor; alternatively, a universal

Fig 6.39 Automatic transmitter No. 2D — cover removed (*Courtesy of The Post Office*)

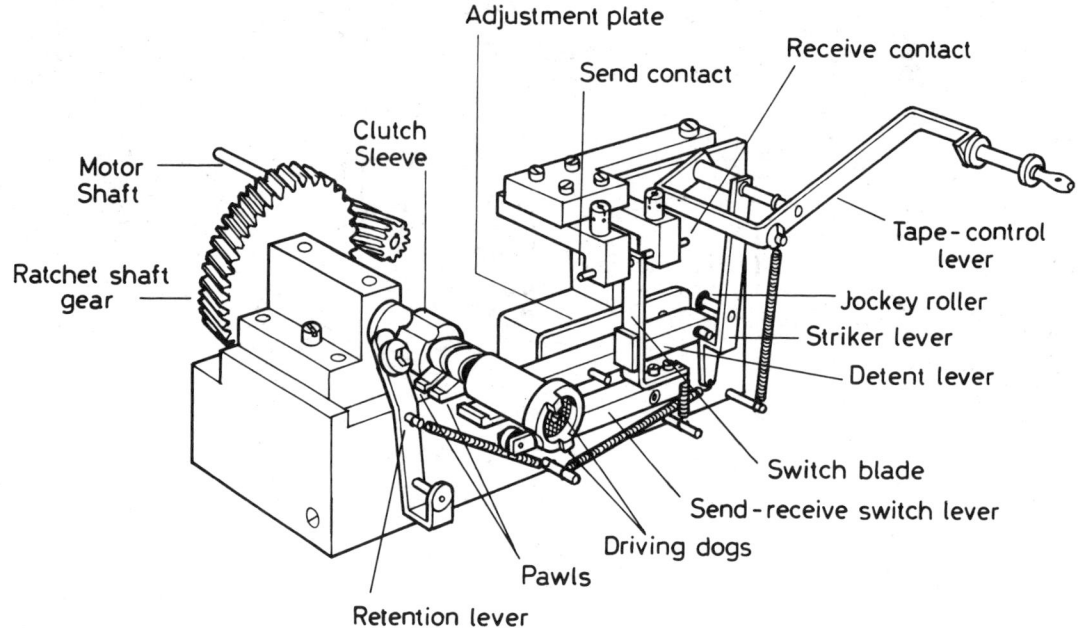

Fig 6.40 Automatic transmitter No. 2D — tape-control unit

commutator motor suitable for either a.c. or d.c. supplies is available.

The motor drives the transmitter head through a ratchet and pawl clutch which, in the idle position, is held out of engagement by a lever which can be operated either by hand or controlled by the tight and slack paper tape being fed to the transmitter head. The tape is fed forward step-by-step during

transmission of the stop signal; it is stationary during transmission of the code signals. The tape-control unit (Fig 6.40) includes the transmitting-shaft clutch, the device for controlling the pawl abutment by the tension of the paper tape, and the send–receive switch; the clutch mechanism is of the ratchet and pawl type. A pinion on the motor shaft meshes with a fibre gearwheel on the ratchet shaft which revolves continuously while the motor is running; surrounding the ratchet shaft is the clutch sleeve which carries the pawls. The pawl abutment holds the pawls out of engagement with the ratchet; a

* Descriptions of medium- and high-speed tape readers and punches will be found in *Data Telecommunication* (R. N. Renton), Pitman.

retention lever rests in a notch in the clutch sleeve when it is in its normal position.

When operated, the pawl abutment is withdrawn from the pawls which are allowed to engage with the teeth on the ratchet; the clutch sleeve and the transmitting head rotate with the driving spindle. The pawl abutment, or detent lever, is pivoted at its centre and has its movement limited by a pin which projects into a hole in the lever. At the end remote from the abutment face, the lever is provided with a knife-edge having three inclined surfaces. Movements of the pawl abutment are controlled by the tape-control lever which consists of an inclined lever, pivoted at its left-hand end, and a vertical lever which carries a jockey roller. One spring tends to pull the inclined lever down against a stop, while another acts on the vertical lever so that the jockey roller is held in engagement with the knife-edge of the detent lever. The transmitter head is declutched from the motor drive by lifting the inclined lever into its uppermost position. The jockey roller rides on the top inclined surface of the detent lever, the angle of which is such that it supports the lever against tension of the springs. In this position, the pawl abutment is turned clockwise and the pawls are withdrawn from engagement with the ratchet when they encounter the abutment face.

To begin transmission the tape-control lever is lowered by hand, causing the jockey roller to bear on the lowest inclined surface of the detent lever, which is turned anti-clockwise leaving the pawls free to engage with the ratchet. While transmission is in progress the tape is pulled through the reading head and it is necessary that, if the flow of tape to the head is checked for any reason, the head is declutched to prevent damage to the tape. Before passing through the transmitter head, the tape is fed over a fixed guide and then under the roller on the tape-control lever. If the tape is checked, the loop in which rests the roller on the control lever is drawn tight, the roller is raised sufficiently to bring the jockey roller on to the middle inclined surface of the pawl-abutment lever and the transmitter head is declutched. Should the tape become slack, the inclined lever falls under the influence of its spring and the transmitter head is again operated.

A send—receive switch is fitted so that the line is connected to a receiver when the transmitter is at rest. In this state a roller on the end of the send—receive switch lever rests in an indentation in the clutch sleeve and the switch blade rests against the receive contact; immediately the transmitter head is coupled to the motor for transmission, the clutch sleeve rotates and the roller rides out of the indentation causing the switch blade to change over to the send contact.

The tape passes over the pecker guide plate and is held in contact with it by the tape-retaining plate. It is drawn through the transmitter head one feed-hole at a time by means of a starwheel, which is advanced by one tooth during transmission of each stop signal by means of a ratchet and pawl operated from a cam which is secured to the rear of the transmitting spindle.

The transmitting camsleeve (Fig 6.41) has six flats cut in its surface and these are arranged so that the pecker levers, if otherwise free to do so, may rise successively under tension of their associated springs at the correct instants as the camsleeve rotates; a rack is provided to give lateral stability for the pecker levers. The camsleeve has a timing cam secured on its rear end with seven indentations in its periphery and secured to the camsleeve so that these are in correct phase relationship with respect to the operating portion of each signalling cam.

The five pecker levers are mounted side-by side on a spindle. In its normal position, each lever is horizontal and extends under the transmitting camsleeve, the upper edge of the lever being kept in contact with the surface of the camsleeve by spring tension. A pecker is attached near to the right-hand end of each of the five levers and the extreme right-hand end of each lever projects under an extension of the common lever. The movements of each pecker lever control the position of the common lever which in turn positions the striker with respect to the tongue knife-edge. A sixth, the start—stop lever, is mounted on the same spindle as the pecker levers and situated between levers Nos 2 and 3, vertically beneath the train of feed holes in the tape. This lever is similar in design to the pecker levers, except that it has no pecker associated with it and is free to ride on to the flat on its associated cam track on every revolution of the camsleeve. The position of the camsleeve ensures that when the pawls have been declutched and the sleeve is at rest the start—stop lever is in a raised position.

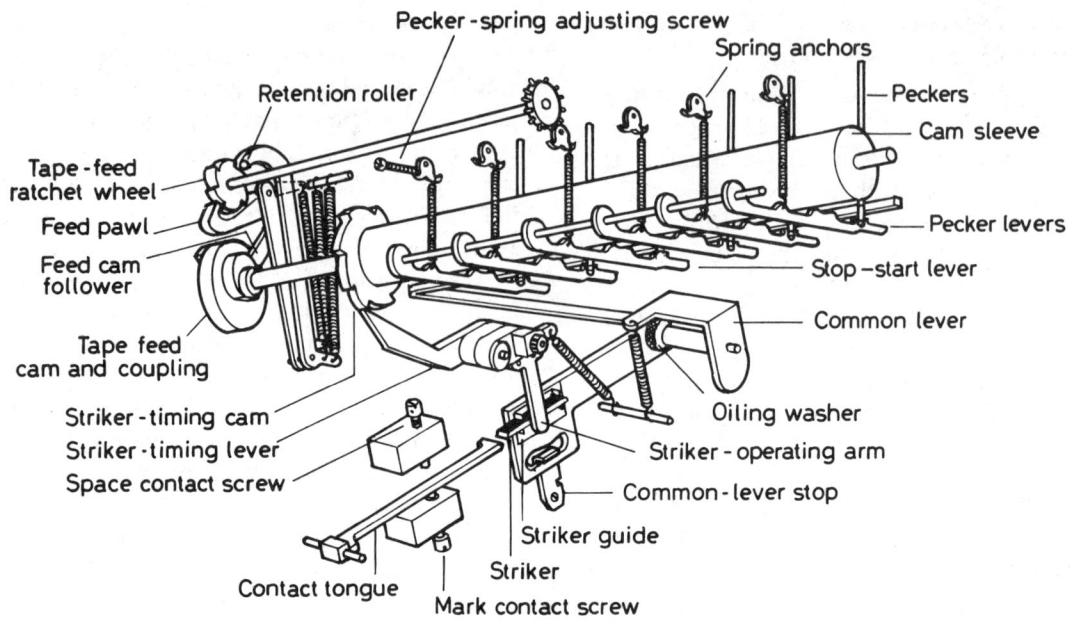

Fig 6.41 Automatic transmitter — striker transmitter

The five peckers are thin steel rods, each attached at its lower end to one of the pecker levers; the upper end of each rod passes into its respective hole in the pecker guide plate. The function of the peckers is to sense the tape for perforations, this being carried out by each pecker successively as its associated lever rides on to the flat of the camsleeve. When the pecker levers are in their lowered position, the top ends of the peckers are located just below the upper surface of the guide plate and do not touch the tape.

The common lever is a bellcrank lever pivoted at its angle with an extension on its upper arm projecting towards the front of the machine and this is situated just above the right-hand ends of the pecker levers; the lower arm carries the striker in its insulated guide. Movements of the common lever are controlled by movement of the pecker levers which in turn move according to perforations in the tape; the common lever positions the striker over one side or the other of the transmitter-tongue knife-edge.

The striker-timing lever consists of two parts — a cam follower and an operating arm — mounted on an insulating block through which the pivot spindle passes. The cam follower is maintained in contact with the timing cam by spring tension and the operating arm engages with a slot in the striker.

When the cam rotates, the timing-lever cam follower follows the contour of the timing cam, operating once for each signal element, causing the striker knife-edge to chop the tongue over from mark to space or vice versa. The striker is a small dart of rectangular cross-section with a knife-edge at one end and is carried in an insulated guide situated on the lower arm of the common lever. The transmitter tongue has two knife-edges at the free end, one for the striker operation and the other for the jockey roller. The jockey-roller pressure assembly consists of a frame, one end of which is pivoted on an adjustable eccentric pivot while the other end carries the jockey roller on a spindle. Spring tension holds the roller in contact with the front knife-edge to ensure rapid transit and no bounce.

While the transmitter-head mechanism is in the rest position, the striker-timing lever cam follower is depressed by a lobe of the timing cam to hold the striker clear of the tongue. When the tape-control lever is lowered the clutch pawls are released, engage with the ratchet-shaft teeth, and the transmitting cam commences to rotate. As it does the start—stop pecker lever is depressed by the cam, causing the common lever to rotate anti-clockwise about its pivot under the action of its spring, and move the striker to a position over the lower right-hand face of the contact-tongue knife-edge.

The timing cam is rotating with the transmitter cam and, shortly after the common lever has come to rest against its stop, the cam follower of the striker-timing lever enters the first indentation of the timing cam. It then rotates about its pivot, under the action of its tension spring, and sharply drives the striker downwards against the right-hand face of the tongue knife-edge. This blow forces the tongue from the mark (right-hand) to the space contact to commence transmission of the start signal. The jockey roller, acting upon a second knife-edge, holds the tongue firmly in whichever position it is set by the striker.

The camsleeve continues to rotate and causes the second lobe of the timing cam to depress the follower of the striker-timing lever. The operating arm of the timing lever lifts the striker clear of the contact tongue, which remains held in position against the space contact by the jockey roller. The first pecker lever rides up on to its cam flat and its associated pecker rises to sense the tape. If the first code element is mark, the pecker is able to rise through the hole in the tape. The pecker lever comes into contact with the arm of the common lever, partially rotating it in a clockwise direction, to position the striker over the upper left-hand face of the tongue knife-edge. The follower of the striker-timing lever enters the second indentation of the timing cam, allowing the striker-timing lever to rotate about its pivot under the action of its tension spring and sharply drive the striker downwards against the left-hand face of the tongue knife-edge; this forces the tongue over from the space to the mark contact to commence transmission of the first code element 20 ms after the beginning of the start element. The remaining code elements are transmitted in a similar manner, the pecker being unable to rise when it encounters unpunched paper for space elements.

When a space or mark element is followed by elements of the same polarity, the tongue is not moved from the contact on which it is resting until one of the opposite polarity has to be transmitted. The striker, however, continues to strike the tongue knife-edge at the appropriate instants corresponding to the terminations of the individual elements; these strikes occur on the same side of the tongue, which remains in the one position under the action of the jockey roller.

The step-by-step tape-feed mechanism is fitted in the recess at the back of the transmitter-head casting. A tape-feed cam is situated on the rear end of the transmitter-head driving spindle and a ratchet wheel is provided on the tape-feed spindle. A retention lever is used to locate the ratchet wheel so that the tape-feed wheel takes up the correct position relative to the peckers after feeding the tape forward. The tape remains stationary while the code elements are transmitted; during the stop signal the tape-feed cam moves the tape-feed pawl into engagement with the ratchet wheel to advance it by one tooth. The tape-feed wheel is rotated to feed the tape forward by one feed hole in readiness for transmission of the next character.

Automatic Transmitter No. 2E (Creed Model 6S/ 6M). This transmitter is similar to the No. 2D but includes an electromagnetic clutch-release mechanism to provide remote control of the transmitter head; it is suitable for pulsed operation to enable the transmitter head to be released character-by-character if required.

To complete the remote control a tight-tape mechanism operates a switch with change-over contacts to indicate when the supply of perforated tape (e.g. from a reperforator) is temporarily held up; and a tape-out mechanism operates a second micro-switch with change-over contacts. When the trailing edge of the tape has left the transmitter head the TAPE OUT switch is reset manually by a push-button when a fresh tape is inserted.

The detent for releasing and arresting the clutch is controlled by an electromagnet which has a roller mounted on its armature to engage with a sloping face of the detent. When the electromagnet operates, the roller engages the detent lever which releases the pawls and allows them to engage with the ratchets to start transmission.

The transmitter camsleeve has a flat section cut in its periphery at the front end. Tensioned against this track by means of a spring is a tape-out seeker-cam lever, which is attached to the tape-out pecker which protrudes through the pecker guide plate in line with the peckers and opposite the feed wheel. The tape-out pecker is capable of rising once during every revolution of the cam. If a tape is in position the pecker is prevented from rising, but when the tape runs out the pecker rises and causes its associated seeker-cam lever to rise. This lifts the tape-out latch-control lever extension, which projects through from the rear of the head. The latch-control lever rotates about its pivot and releases the latch to release the resetting lever and TAPE OUT micro-switch. The switch contacts, which are normally connected in series with the clutch electromagnet, are disconnected, the electromagnet is released, the detent engages the pawls and the transmitter mechanism is brought to rest.

When a new tape is loaded into the head the tape-out pecker will be depressed, but this action alone will not allow the electromagnet to re-operate to recommence transmission: the RESET button must be depressed to restore the latch and reset the micro-switch. The release of the latch-control lever allows the tape-out seeker-cam lever and its associated pecker to restore to normal, ready to detect a further tape-out condition.

Automatic Transmitters Nos 2E and 2F used with ARQ Systems. The character rate for an ARQ radio system is $411\frac{3}{7}$ char/min; a 7·5-unit machine has a nominal rate of only 400 char/min. For use on ARQ systems, transmitters Nos 2E and 2F are modified to transmit a 7·0-unit cycle which has an equivalent speed of $60/0·140 = 428·5$ char/min. The automatic transmitter is released character-by-character from 'call-in' stepping pulses, supplied by the ARQ equipment over a separate control circuit when the radio channel is open, at the maximum rate of $411\frac{3}{7}$ pulses/min.

On multiplex systems working over radio or submarine cable, 50-baud circuits at 400 char/min may be time-shared between two, three or four teleprinter installations, each supplied at a reduced character-rate of 100, 200 or 300 char/min. To avoid the need to provide separate control channels between the multiplex station and the teleprinter stations, automatic transmitters used for these services are modified to become self-pulsing at one of the reduced character-rates. The modification for this purpose consists of a 60-tooth gear driven at a speed of 100 rev/min from the transmitter driving shaft via a 15-tooth pinion. One, two or three small

conical studs can be fitted on the gearwheel to operate a spring set at 100, 200 or 300 times per minute; the springset causes the clutch electromagnet to release the transmitter head at the appropriate rate. Transmission at the selected rate is continuous, signals being stored in the multiplex equipment if necessary.

Ganged Automatic Transmitters. Tape transmitters with multiple heads are used in busy offices to maintain a high traffic loading on circuits. By means of a TAPE OUT contact, which operates when the trailing edge of a tape runs through the transmitting head, two or more transmitters kept loaded with tapes can be worked as a team so that a series of perforated tapes transmits continuously into a circuit which is automatically switched from one tape head to the other as each message tape runs out of the head. Another requirement is for a serial number to be inserted automatically prior to each tape transmission. This can be achieved by using a third transmitting head, loaded with an endless tape punched with a sequence of serial numbers and using automatic control circuits based on the presence of a group of LETTERS SHIFT signals in the message tapes, to interpolate a serial number between successive tape transmissions.

A universal motor with centrifugal governor drives all three heads, which can be brought into operation independently. The machines used (of U.S.A. origin) generate single-current 7·42 signals (i.e. a 1·42-unit stop signal) from six contact sets, including one which supplies the start and stop signals. The traffic speed is $60/(7·42 \times 0·020) = 404·3$ char/min. (This cannot be measured on a TDMS designed for a 150-ms character period.) Later models generate a 7·5-unit signal or a 7·0-unit signal (Models 3C, 4C and 5C) for multiplex time-division systems. On large installations serial-number interpolation is provided by electronic means.

Two types of transmitter machine are used. One has an additional pecker which rises and operates TAPE OUT contacts when the end of a message tape leaves the transmitter head; the other type has a seeker mechanism which closes a pair of contacts when all five code peckers rise simultaneously (i.e. on sensing the LETTERS SHIFT signal). These contacts are used to initiate switching operations.

Fig 6.42 Automatic transmitter No. 3 (*Courtesy of The Post Office*)

Automatic Transmitter No. 3. This machine (Fig 6.42) comprises three message heads, each controlled by an electromagnetic clutch, with step-by-step tape-feed mechanism and associated sequentially-operated transmitter contacts. The TAPE OUT contacts, which operate at the end of each message tape, remain operated until released by a manual-control lever. When this lever is depressed it disconnects the circuit of the transmitter-clutch electromagnet, which releases and brings the transmitting cam to rest; it also frees the tape-feed wheel so that it can be rotated to ease insertion or removal of tapes. TIGHT TAPE contacts are not provided; tape is fed from the front of the machine into a guide slot under the hinged tape-retaining plate and falls

away to the rear of the machine. For the right-hand transmitter unit a tape-guide chute directs the tape through the machine base. The transmitting units are individually geared to a common driving shaft.

One machine may be used to provide an automatic transmitter on each of three separate lines or two heads may serve one circuit and the third a second circuit. Two machines may be used on three heavily-loaded circuits to provide two message transmitters on each line, with automatic change-over from one to the other on completion of each tape transmission. Usually the No. 1 heads of the two machines serve circuit 1, and so on.

Automatic Transmitter No. 4. (Fig 6.43). This comprises three numbering units which are substantially the same as the message units of the Model 3. Contacts, which on the message units are used for TAPE OUT, are here actuated by a different mechanism which detects the LETTERS SHIFT combination in the tape.

Each transmitting unit is provided with a tape-storage reel to accommodate the serial-numbering tape before transmission, and a take-up reel to wind the tape as it is transmitted. Storage reels on the front cover plate are provided with spring-loaded jockey levers to cushion the initial drag on the feed wheel at the start of transmission A ratchet-driven shaft mounted above and to the rear carries the three take-up shaft spools which are coupled to it by friction clutches. To avoid damage to the feed holes, tension applied to the tape at the feed wheel is kept to a low value by introducing snubbing pins between the take-up spool and the feed wheel.

Fig 6.43 Automatic transmitter No. 4 (*Courtesy of The Post Office*)

References

1 Easterling, C. E., 'Improvements in the teleprinter No. 7', *POEE Journal,* 55, p.240 (1963).
2 Easterling, C. E. and Collins, J. H., 'The teleprinter No. 11', ibid. 46, p. 53 (1953).
3 Spurging, C. M., 'The new Creed teleprinter (444)', *Systems and Communications*, Jan. 1967, p. 1.
4 Wusteney, H., 'The new Siemens page-printer', *Siemens Review,* 25, p. 71 (1958).
5 Brader, C., 'Functional characteristics of the teleprinter T100', *Nachrichtentechn. Z.,* 14 p. 11 (1961).

7 Multi-circuit Voice Frequency Systems

The ability to use a common form of line plant for telephone and telegraph systems has important technical and economic advantages and greatly aids the flexibility of providing telegraph service. Transmission of telegraph signals by a.c. at voice frequencies enables these advantages to be secured and results in standardization of telegraph systems which give a highly stable performance — largely independent of distance — and readily interchangeable with similar systems when re-routing is needed. This is the universal method of operating long-distance telegraph circuits (in the United Kingdom those exceeding about 20 miles), both nationally and internationally, including those in inter-continental submarine telephone cables and in satellite systems.

These systems are known as *multi-circuit voice-frequency (MCVF)* systems; sometimes as *frequency-division multiplex (FDM)* systems; or simply as *voice-frequency telegraph (VFT)* systems, the fact of multiplexing being assumed.

Each VFT circuit normally carries one teleprinter communication (with duplex operation, if required), but a VFT circuit can be further multiplexed by TDM to carry two or more teleprinter transmissions on one voice-frequency circuit.

7.1 Principles of MCVF systems

The basis of the MCVF system is division of the available frequency bandwidth of a telephone channel into a large number (usually 24) of narrow-frequency bands, each used to provide a telegraph channel. For this application a 4-wire telephone circuit is allocated for telegraph use; when so used it is referred to as a *bearer* circuit. (Exceptionally, 2-wire circuits can be used.) The 2-wire/4-wire terminations, telephone supervisory signalling units and any echo suppressors are removed* from the 4-

*Recommendation R30.

wire circuit which now consists of two separate but associated 2-wire channels capable of transmitting signals in opposite directions.* The narrow-band telegraph channels are produced by connecting band-pass filters in parallel at both ends of each channel. Except for the use of different transmission bands and different carrier frequencies, all telegraph channels are virtually identical; the same carrier frequency is used in both directions for associated channels. The design of equipment is such that each VF circuit, except for its location in the frequency spectrum and for certain common supplies, is operated independently of the others: there is no requirement of synchronism between channels. Any modulation rate up to the capacity of the channel can be employed without regard for that of other channels, although normal teleprinter modulation rates and start—stop operation are generally used.

Non-linearity in the bearer-circuit phase characteristic can be considered separately over each of the narrow-band channels. For this reason a parallel mode of transmission can use efficiently a large part of the bandwidth and give a higher aggregate speed compared with a single-serial transmission system. Because of narrower filters, the liability to interference noise is less, and with longer signal elements reflections are less serious.

At the sending end a modulator is required to convert ±80-V double-current telegraph signals into appropriate form and frequency for a.c. transmission. At the receiving end a detector is required which will convert received VF signals back into ±80-V double-current mode. For impedance matching and isolation between unbalanced VFT equipments and balanced line conditions, MCVF equipments are connected to line via transformers: this results in the arrange-

*A *channel* is a means of one-way transmission; a *circuit* is a means of bothway communication between two points, comprising associated 'go' and 'return' channels.

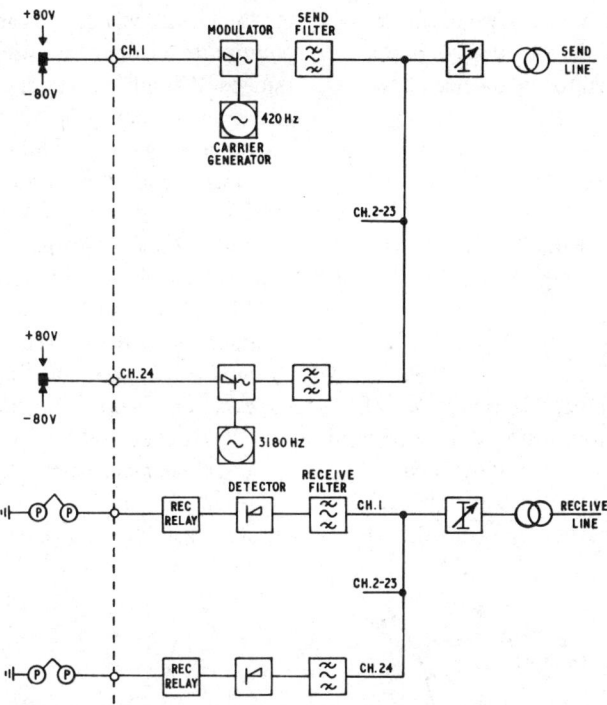

Fig 7.1 Principle of multi-circuit voice-frequency (MCVF) system

ments shown in the block diagram of Fig 7.1, which applies in principle to all forms of MCVF equipment. The design of the modulator at the sending end and of the amplifier—detector at the receiver depends on the form of modulation adopted.

When VFT channels are joined in tandem, whether by permanent 'jumper' wiring for leased circuits or by selectors for switched services, the connections are made at the ±80-V level — i.e. the receive-relay tongue of one channel is extended to the send modulator of another. The advantage arising from the presence of d.c. potential on relay and selector-bank contacts has been mentioned in Chapter 5. Apart from this, the aggregate distortion is much greater if channels are connected in tandem on an a.c. (VF) basis, i.e. RF OUT to SF IN. Measurements have shown that for two a.c.-connected channels, characteristic distortion is about 50% higher than for d.c.-connected channels, this adverse difference increasing rapidly if further channels are added to the tandem chain. The increase in characteristic distortion is due to the cumulative effects of bandwidth restriction and phase distortion through multiple pairs of narrow-band filters. With the a.c. method, interconnections would have to be restricted to channels using the same centre frequency; impedance matching between RF OUT and SF IN would also need to be incorporated. The d.c. interconnection is seen to be both very simple and flexible. If VFT circuits are to be picked up at random by automatic selectors for the purpose of joining several such circuits in tandem, the distortion contributed by each channel must be kept to a low value if five or six such channels are to be permitted in tandem connections.

7.2 Channel bandwidth

Information rate (modulation rate) is proportional to bandwidth.

At the transmitter, telegraph signals are square in waveform, containing an infinite series of *odd* harmonic frequencies, but before being transmitted over a d.c. line into a VFT modulator the waveform is modified by the low-pass filter 4B. When these signals, after modulating a VF carrier, are transmitted through the band-pass send filter, the effect of further curtailment of sideband frequencies is to delay the build up of the signal envelope. If f_1 and f_2 Hz are the *effective* cut-off frequencies of the filter, the time t required for a carrier current at mid-band frequency to build up to steady-state value is approximately equal to the inverse of the effective bandwidth:

$$t = \frac{1}{(f_2 - f_1)} \text{ s}$$

It follows that if a signal has a duration less than t it may not be transmitted properly, and the limiting telegraph modulation rate N is:

$$N = 1/t = (f_2 - f_1) \text{ bauds}$$

Double-sideband transmission is used in MCVF systems; single-sideband transmission cannot be used because the modulation products are the binary signal transients only. The frequency bandwidth required for a modulated carrier frequency is twice

50 bauds is equivalent to an a.c. signal at 25 Hz. The *minimum* channel bandwidth required for double-sideband amplitude-modulation of these signals with a carrier frequency f_0 will be $(f_0 \pm 25)$ Hz, i.e. a channel bandwidth of 50 Hz, numerically equal to the modulation rate of 50 bauds. The sideband frequencies $(f_0 \pm 75)$, $(f_0 \pm 125)$ ... resulting from the odd harmonic frequencies of the original squarewave will not be transmitted.

Fig 7.2 shows typical attenuation/frequency characteristics for the send and receive filters used for channel 2 ($f_0 = 540$ Hz) in MCVF systems. The *nominal* channel bandwidth has been standardized* at 120 Hz (480–600 Hz for channel 2). Due to practical characteristics of filters, the *effective bandwidth of the receive filter at the* 6-dB points is approximately 80 Hz, corresponding to a maximum

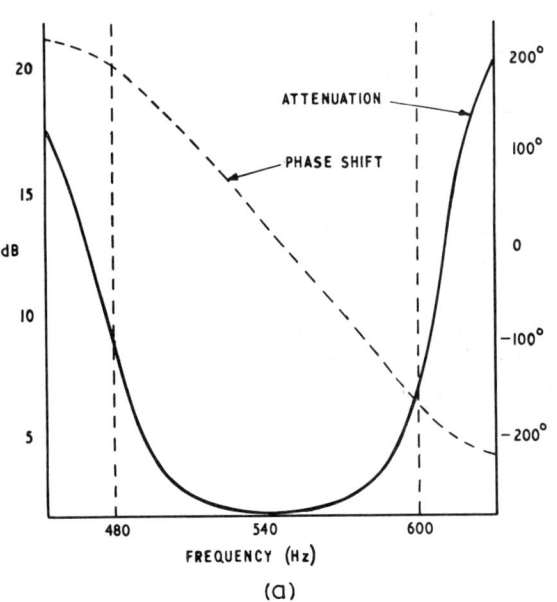

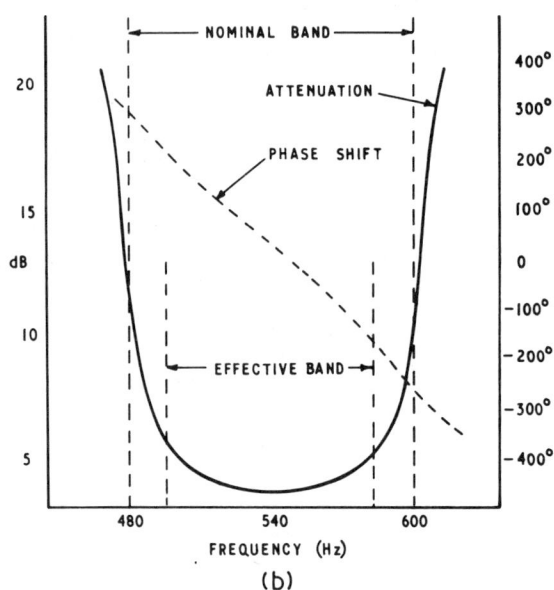

Fig 7.2 Attenuation/frequency and phase/frequency characteristics of channel-filters flanked on both sides: (*a*) send filter; (*b*) receive filter

the fundamental frequency of the original double-current signals. For a telegraph signal consisting of mark/space reversals at modulation rate N bauds, the modulating *frequency* is $N/2$ Hz. Combining the two statements, the minimum frequency band in hertz required for (double-sideband) modulated signals is numerically equal to the fundamental modulation rate in bauds.

Taking practical values, a standardized modulation rate for teleprinters is 50 bauds; sending reversals at

modulation rate of approximately 80 bauds. The fundamental frequency of reversals at 80 bauds is 40 Hz and if the carrier frequency $f_0 = 540$ Hz is amplitude-modulated by the telegraph signal at 40 Hz, sideband frequencies will be $(540 \pm 40) = 500$ and 580 Hz.

Although the bandwidth required is proportional to modulation rate, this relationship is tempered by

*Recommendation R31.

the ratio of effective to nominal bandwidth which can be realized in the practical filter. The question of performance stability also arises — the amount of falling-off in performance (due to such factors as changes in line-level, frequency-drift, supply-voltage variations, limitations of oscillator stability etc.) — which can be accepted. The narrower the channel frequency band, the greater the number of channels which can be operated on a transmission line of given bandwidth. In an economically-designed MCVF system, channel bandwidth will be no greater than is needed for the required modulation rate. At a modulation rate of 50 bauds and channel frequency spacing of 120 Hz, a balance is struck between achieving a high-quality service with adequate safety margin and due regard to economy in use of frequency spectrum, with some latitude for increasing modulation rate on certain special systems where other factors — e.g. signal/noise ratio — are favourable. At modulation rates higher than the design value, effects of interference, and amplitude and frequency variations become more onerous; also sidebands occur where phase characteristics are no longer linear and where amplitudes of the two sidebands may not be equal if filter attentuation characteristics are not symmetrical. This causes a significant increase in characteristic distortion and, in the limit, a modulation rate is reached where no sidebands are transmitted and transmission of information ceases. Due to the presence in the system of other sources of distortion it is normally unwise to attempt to work channels at a modulation rate too near the limiting value.

Although the above argument has been based on the use of amplitude modulation, when narrow-band frequency-modulated systems were later introduced, it was decided* to use the same bandwidth (120 Hz) and nominal mean carrier frequencies (those in Table 7.1) for these systems on the grounds that the FM system should provide an equivalent number of telegraph circuits from a bearer circuit.

7.3 Carrier frequencies
The principal factor affecting choice of channel-carrier frequencies is the need to prevent mutual interference between channels.

*Recommendation R35.

If the aggregate signal of an MCVF system is transmitted through a non-linear device, such as an amplifier with characteristics not completely linear, the waveform of the applied signal will suffer from amplitude distortion which results in the production of even harmonic frequencies not previously present. These harmonic frequencies beat together, introducing further spurious frequencies, some of which fall within the frequency range of one or other of the channel filters causing interference with signals in these channels and introducing fortuitous telegraph distortion.

If f_1 and f_2 are any two selected carrier frequencies, the even harmonic frequencies arising from non-linearity will be $2f_1, 4f_1, \ldots$ and $2f_2, 4f_2 \ldots$. Sum and difference products, introduced by the beating effect, will be $(f_1 \pm f_2), 2(f_1 \pm f_2), 2f_1 \pm f_2 \ldots$. The disturbing products are mainly caused by even harmonics and their resulting sum and difference products. The first-order products $2f_1, 2f_2$ and $(f_1 \pm f_2)$ have the greatest amplitudes; harmonics having frequencies which lie outside the range of the MCVF system are of no account. It is desirable that these first-order products shall fall midway between adjacent channel-carrier frequencies where filters offer maximum discrimination. This is achieved by selecting all channel-carrier frequencies to be *odd* multiples of a basic frequency equal to half the channel bandwidth. The even harmonic and the sum and difference frequencies then occur at midway between channel-carrier frequencies where least disturbance is caused.

Taking by way of example carrier frequencies 900

Table 7.1 Recommended carrier frequencies (AM) and nominal mean frequencies (FM) for MCVF systems operating at 50 bauds*

Channel number	Frequency (Hz)	Channel number	Frequency (Hz)
1	420	13	1860
2	540	14	1980
3	660	15	2100
4	780	16	2220
5	900	17	2340
6	1020	18	2460
7	1140	19	2580
8	1260	20	2700
9	1380	21	2820
10	1500	22	2940
11	1620	23	3060
12	1740	24	3180

*Recommendations R31 and R35.

and 1500 Hz, intermodulation will produce frequencies $(1500 \pm 900) = 600$ and 2400. These frequencies, being *even* multiples of the basic 60 Hz, will fall midway between other pairs of carrier frequencies where filters exercise maximum attenuation.

The standardized channel bandwidth based upon a modulation rate of 50 bauds is 120 Hz. The selected channel frequencies – see Table 7.1 – are odd multiples of 60 Hz, commencing with the seventh (420 Hz). Use of carrier frequencies in the table enables 24 telegraph circuits to be operated on a 4-wire bearer circuit which has the standardized bandwidth 300–3400 Hz.

7.4 Channel filters
A first consideration might suggest that, since signals from 24 channels are to be combined in the transmission line, the use of sending-end filters might be unnecessary since it is the function of the receive filter to separate out the channel signals from the aggregate signal. Though less critical in requirements than the receive filter (see Fig 7.2) the send filter has to fulfil the following important functions: (1) Restrict sideband frequencies of a modulated wave to the frequency band allotted to a channel. This prevents transmitting harmonic frequencies present in either the teleprinter modulating signal or in the VF carrier current which might otherwise interfere with other paralleled channels. (2) Present a high attenuation at the operating frequencies of other channels to prevent any of the parallel-connected channels from applying a significant load on individual channel-output terminals.

A receive filter must accept, with minimum attenuation, only energy proper to its pass band; it must offer high attenuation to signals at all frequencies lying outside its pass band, so that no appreciable energy is taken from other channels and interference is minimized. With parallel filters, the power loss at the midband frequency of any channel, due to the shunting effect of other filters, is about 2 dB. The receive filter needs sharper cut-off characteristics than those of the send filter and this is sometimes attained by using a two-section filter or by using a further stage of filtration within the channel-detector equipment.

Channel filters are usually of unbalanced type since this configuration is more economical in components compared with the balanced filter. For low characteristic distortion it is important that the phase characteristic be kept reasonably linear.

Channel filters delay the propagation of signals: for a 120-Hz spacing the delay due to the filter at each end of the system is $1/120 = 8.5$ ms. With other incidental delays added, the transmission time through the channel (excluding line-propagation time) can lie between 30–40 ms.

7.5 Power levels
With up to 24 teleprinter channels transmitting power into a common bearer channel it is clear that the amount of power permitted for each channel must be closely regulated. The power per channel must be high enough to maintain an adequate signal/noise ratio; on the other hand, power must not be so high as to overload the common amplifier in the line circuit, since this occurrence would produce waveform distortion and introduce interfering harmonic frequencies. The maximum permissible aggregate power at a given point in the bearer circuit having been established, the power per channel, allowing for all channels to transmit simultaneously, can be calculated.

If the maximum instantaneous power permitted at the given point is W watts and the system comprises N channels transmitting over a line of characteristic impedance Z_0, the maximum voltage V across the line is:

$$V = \sqrt{WZ_0} \text{ (from } W = VI = V^2/Z_0 \text{)}$$

The maximum permissible voltage V_c per channel is:

$$V_c = V/N = \sqrt{\frac{WZ_0}{N}}$$

and the maximum permissible power W_c watts per channel is:

$$W_c = \frac{V_c^2}{Z_0} = \left(\frac{V}{N}\right)^2 \times \frac{1}{Z_0} = \frac{WZ_0}{N^2} \times \frac{1}{Z_0} = \frac{W}{N^2}$$

For amplitude-modulated systems the CCITT recommends that the instantaneous maximum voltage produced by the whole of the telegraph signals on a single channel must not exceed that of a sinewave with a power of 5 mW at a point of relative zero level, calculated from the level diagrams of the telephone circuit. It is convenient to adopt the same value, in the main, for national systems.

For a 24-circuit AM VFT system the r.m.s. power per channel (measured at a reference zero point) is $5000/(24^2) = 8.6 \, \mu W = -20.6$ dBm0. (If a smaller number of channels is provided, it is permissible to use correspondingly greater power per channel — see Table 7.2.) The total average power allowed (at the zero point) for all the 24 channels is $24 \times 8.6 = 208 \, \mu W$.

When MCVF equipment was first introduced it was not possible to accommodate more than 18 channels on the line plant available. The permitted power level per channel was $5000/(18^2) = 15 \, \mu W$ per channel. In the national networks of the United Kingdom it was the practice to operate at slightly lower levels than those permitted for international circuits; it was found that the slight reduction in signal/noise ratio caused as a result of dropping the level from 15 to 10 μW (for 18-channel AM systems) was more than compensated by the resultant reduction in inter-channel interference due to cross modulation. When improved line plant enabled 24-circuit systems to be operated, the same power per channel, namely 10 μW equivalent to -20 dBm0,* was adopted, approximately equal to the CCITT value in Table 7.2. This power level was adopted for channels in 6-, 12-, 18- and 24- circuit systems with the advantage that if the number of circuits on a route were increased, say from 12 to 18, it was not necessary to re-adjust power levels as a consequence. The telegraph channel level is 20 dB below the corresponding telephone level.

With AM systems, power is not transmitted during space elements of a modulation nor while switched circuits are held disengaged in the free (space) condition. A diversity factor is included in the power consideration because mean power over a period is less than maximum power. With FM systems, the same amount of power is being injected into the line continuously whether the channel is in the mark or space condition. The total *average* power transmitted in the bearer circuit by all channels is limited to 135 μW at a point of zero relative level. This sets the limit to $135/24 = 5.6 \, \mu W$ per channel (-22.5

dBm0) for a 24-circuit FM system. As with AM systems, the British Post Office uses the same power per channel (-22.5 dBm0) irrespective of the number of channels in the system. Exceptionally in inter-continental submarine cables, on account of the higher proportion of MCVF traffic, special measures are taken to reduce the loading of common line amplifiers by working the MCVF systems (invariably FM systems) at a lower level, usually -24 dBm0; both directions of transmission may be handled by a common amplifier in each submerged repeater. There are exceptions to this value in specific cases.

The CCITT recommended power levels are shown in Table 7.2.

Table 7.2 Normal limits for the power per telegraph channel in MCVF telegraph systems

No. of channels in MCVF system	Power per channel at point of zero relative level			
	AM (R31)		FM (R35)	
	μW	dB	μW	dB
12	35	-14.5	11.25	-19.5
18	15	-18.25	7.5	-21.25
24	9	-20.45	5.6	-22.5

Note. By agreement between administrations slightly greater specified values are permissible.

In applying an MCVF system to a bearer circuit, the same relative power levels of the telephone circuit are maintained for telegraph operation. This procedure facilitates both provision and maintenance work; it also has the important advantage that a 'main' or working bearer circuit can be substituted at any time by an allocated, alternatively-routed 'reserve' circuit in the event of failure of the normal bearer circuit, without the need for any re-adjustment of power levels. Switching between main and reserve lines is effected at points A and B in Fig 7.3, where standard power levels of a 2-wire/4-wire telephone circuit having zero loss between the 2-wire ends are shown. When such a circuit is used as a bearer circuit for an MCVF system, removal of the 2-wire/4-wire terminations releases a power loss of approximately 4 dB at each end of the circuit. The telegraph system is applied to the 'go' terminal amplifier at a point -4 dB reference to zero telephone level; and the output from the receiving terminal amplifier is at a point $+4$ dB relative to zero telephone reference level. Each channel has a gain of 8 dB. The power per channel at the input to the sending terminal ampli-

*See Appendix C.

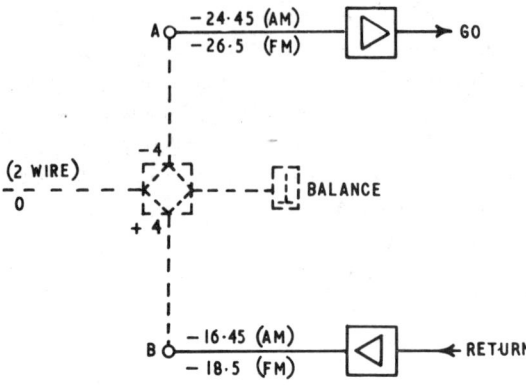

Fig 7.3 Power levels (dBm0) for MCVF systems

fier must in all cases be 4 dB *lower* than the power values quoted above for the relative zero point, e.g. −24 dBm0 (AM) and −26·5 dBm (FM systems). (The normal value for submarine cables would be −28 dBm0.) Input values to MCVF receive equipments will be 8 dB higher than send-line input values.

The foregoing considerations apply specifically to MCVF systems spaced at 120 Hz. For systems with nominal channel bandwidths of 240 Hz (110 bauds) a maximum of 12 channels can be accommodated on a bearer circuit. The allowed power per channel at the zero relative level point of the telephone circuit is 11·25 μW, equivalent to an absolute level of −19·5 dBm0. When such equipment is connected to the −4 dB relative level point of a bearer circuit, the output to line for each telegraph channel is adjusted to −23·5 dBm0.

For 6-circuit (200-baud) FM equipment with a nominal channel bandwidth of 480 Hz, a power of 22·5 μW per channel (−16·5 dBm0) is allowed at the zero relative level point of the bearer circuit; the individual telegraph-channel output level to the line is adjusted to −20·5 dBm0.

The range of adjustment on MCVF equipment enables output levels to be achieved with a tolerance of ± 0·5 dB.

7.6 Amplitude-modulated systems

Apart from the '2-tone' systems used exclusively if AM systems are required on radio-teleprinter circuits, landline AM systems, in which VF tone is sent for a mark signal and space signals are represented by no

tone, are essentially single-current systems. They are subject to as much as 10% bias distortion by a 1-dB change in line level and one of the main problems is to maintain a constant value for the demodulated mark signal in comparison with the fixed bias value which inserts space signals at the receiver. To compensate for line-level changes which result from such causes as valve ageing, variations in battery voltage at repeater stations and changes of cable attentuation with temperature, some form of automatic gain control (AGC) for the amplifier detector is essential in AM systems.

The simpler type-3 amplitude modulator introduced an effect known as 'carrier beat' into the lower channels, which increased distortion of transmitted signals. If the harmonic components of the squarewave signal from the transmitter, of fundamental frequency f_M, are present with significant amplitude, when these modulate the carrier, f_C, the modulated sidebands include the products:

$$f_C \pm f_M, f_C \pm 3f_M, f_C \pm 5f_M \ldots f_C \pm (2n-1)f_M$$

When n has a value such that $(2n-1)f_M$ is approximately equal to $2f_C$, the difference frequency $f_C - (2n-1)f_M$ lies close in value to f_C, i.e. in the middle of the send-filter pass band. For example, the 17th harmonic of 25 Hz is 425 Hz. Its presence, together with the 420 Hz carrier frequency of channel 1, produced a beat frequency of (425 − 420) = 5 Hz. Because of their position in the frequency spectrum these interfering frequencies (such as 425 Hz) cannot be eliminated by the band-pass filter and their presence caused 'cyclic' or 'beating' form of distortion having a value of a few percent. This effect was pronounced on channel 1, and to a less extent on channel 2. The more complex type-4 modulator (AM) was designed to prevent the incidence of carrier beat.

If the local S-wire (an earth-return d.c. circuit) becomes disconnected due to a fault, the modulator loses its control. If the line disconnection occurs towards the sending end of a line of appreciable length, the unterminated line is sensitive to induction from signals in other wires sharing the same cable; these induced e.m.f.s may be of sufficient magnitude to operate the modulator and transmit false signals into the channel. To prevent this, the modulator is biased with a 4-mA current (in the space direction) which will take control of the modulator should the line become disconnected; it is found that 4 mA gives immunity against the effect of induced signals. This bias control is important on VFT circuits used in switched networks to overcome the effects of brief disconnections and ensure the release of a call which might otherwise be held by a faulty line condition. Leased circuits not used for automatic switching need this 4-mA bias in the mark direction.

of the modulated signal. Further rounding of the envelope due to the transient effects caused by the receive filter is shown at (d). Signals from the output of the receive filter are rectified to operate the receiving instrument; these signals must operate a receiving relay in such a way that time intervals between instants of restitution conform as closely as possible to corresponding time intervals between

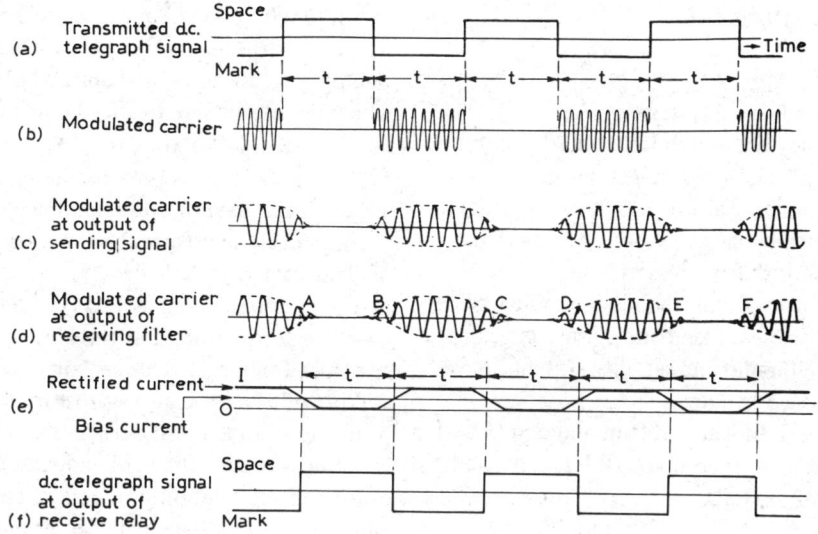

Fig 7.4 Signal waveforms in an AM VFT channel

This bias path introduces some bias distortion, which increases with the length of the d.c. line: due to the low-pass filter 4B and the line characteristics, the instant at which the modulator registers a transition is no longer the zero point of the waveform (except when the waveform is square, as with a test-signal generator applied at this point). For a fixed-input condition the channel bias can be neutralized by adjustment of the detector circuit.

Fig 7.4 indicates the nature of typical telegraph signals at various stages of transmission in an AM VF telegraph channel. At (a) are shown d.c. reversals from a transmitter. The resulting amplitude modulation of the channel-carrier frequency by d.c. signals in the modulator is shown at (b). For simplicity 100% modulation is shown, i.e. the carrier completely suppressed during space signals: suppression is usually about 35 dB. At (c) the form of modulated-carrier signal at the output of the send filter shows the effect of build-up time of the filter on the envelope

instants of modulation of the original signal at the transmitter, i.e. with minimum distortion.

At (d) the signal envelope is symmetrical about points A, B, C, D, E and F which correspond to instants at which the amplitude of the signal envelope is 50% of its steady value; in the ideal case, time intervals t between these points correspond to time intervals t between the original instants of modulation of the d.c. signals transmitted. The aim in design of the receiver is to cause the receive relay to restitute signals at these points; for this purpose an amplifier-detector circuit operates into a polarized relay. Signals from the output of the receive filter are first amplified and then passed to a detector which rectifies the modulated-carrier signals to produce d.c. signals which closely follow the shape of the envelope of the modulated-carrier signal. The rectified current is a single-current signal and the polarized relay has to be biased in opposition to received signals to an extent equivalent to half the amplitude of received signals. The relay will then operate at points A, B, C, D, E and F of the received signal. In the ideal case shown, signals are restituted without distortion; this is not attainable, but a high degree of performance is obtained in practical design.

Although many AM MCVF systems are still in operation, equipment for new installations employs FM exclusively. Space does not permit description of the older AM systems.*

7.7 Frequency-modulated systems

Although frequency-shift modulation has been in use for many years on radio-telegraph systems, this method of modulation was for some time considered suitable only for wide-band systems. In later years, results and predictions from research investigation into the use of narrow-band frequency-modulated MCVF systems for use on land and submarine-cable circuits, and on HF radio and satellite systems, have been well established by experience.

It is found that frequency-shift modulated (FSM) systems, with channels spaced at 120 Hz, can give a very good ratio of modulation rate to bandwidth. With care and experience in design, but without undue complexity, FSM systems have low inherent characteristic distortion and will maintain good performance at modulation rates well in excess of basic design rate.

Under comparable conditions, an FM circuit can show a signal/noise advantage of 6 dB over an AM circuit. This ensures a good margin of performance in face of noise inherent in very long submarine-cable circuits which are used to provide MCVF telegraph systems in intercontinental routes. The same factor renders FM systems particularly suitable for radio-telegraph circuits on which high noise levels occur.

FSM systems are insensitive to wide changes in line level; abrupt level changes cause only small increases in signal distortion and these only when the level change occurs at or near to a significant instant of modulation. FSM systems give a performance superior to AM systems on bearer circuits subject to abrupt level changes. On the other hand, FSM systems are inherently sensitive to unwanted

*The types-3 and 4 AM MCVF systems are described in *Telegraphy* (J. W. Freebody), Pitman.

frequency drift during transmission; a frequency drift of 1 Hz, which may result from frequency translation on an HF bearer circuit, will introduce a signal bias distortion of 2·5% at 50 bauds and 5% at 85 bauds with a 30-Hz shift. High-frequency stability in all carrier-frequency supplies is vital. For relatively slow changes in frequency which occur on cable systems, this disadvantage can be overcome by automatic-compensation devices which rely on pilot frequencies suffering the same frequency drift as the traffic channel frequencies. The pilot frequencies used are 300 and 3300 Hz which lie just outside traffic-channel frequencies.

The fact that VF tone is always present on an FM channel increases power loading on the circuit; although this necessitates reducing channel power it does not detract from the superior performance of these systems.

For systems spaced at 120 Hz, channel frequencies have been standardized at the same values as for AM systems (Table 7.1). To a certain extent this choice still provides some guard against intermodulation products although in the rest condition the frequency is not now that of the midband, and the first-order intermodulation products of channels at rest will not occur halfway between nominal channel frequencies. These centre frequencies are *virtual* frequencies – they are not actually present in the channel because the channel frequency is normally at one of the characteristic deviated frequencies which represent the mark or space signal.

A frequency deviation of ±30 Hz is used, this being the greatest permissible value to prevent significant sidebands being generated outside the channel range. With frequency modulation, a theoretically infinite number of side frequencies can be produced. Mathematical analysis of the composition of a frequency-modulated wave is a matter of some complexity; amplitudes of the side frequencies at various values of deviation ratio are illustrated graphically in Figs 4.19 and 4.20. This value of 30 Hz also has the advantage of keeping to a minimum the interchannel interference which results from intermodulation.

On this account the attenuation of receive filters over the range of adjacent channels must be increased by some 15 dB beyond that of AM systems.

For a *mark* signal the *centre frequency* is *lowered* by 30 Hz; for a *space* signal it is *raised* by 30 Hz. The centre frequency is equal to half the sum of these two characteristic frequencies. The frequency-shift modulation system is a balanced double-current method. For a modulation rate of 50 bauds (25 Hz) the deviation ratio or modulation index m is:

$$m = \frac{f_C}{f_M} = \frac{\pm 30}{25} = 1 \cdot 2$$

where f_C is the carrier frequency shift and f_M is modulating frequency; at higher modulation rates the value of m is reduced. With an effective channel bandwidth of about 100 Hz and a modulation rate of 50 bauds (25 Hz) the second side frequency can be transmitted but the third will be considerably attenuated. The chosen value of m is the maximum for which amplitudes of third and higher side frequencies are negligible (30 dB below unmodulated carrier level). Although lower values of m would also result in smaller amplitudes of side frequencies, it is desirable to use the greatest frequency deviation permissible from consideration of the effects of frequency instability on bias distortion and for improvement in noise immunity.

Group modulation is often included in the design of a 24-circuit system. When the lower sideband is selected after group modulation, characteristic mark and space frequencies become *inverted* in the line. If F is the group-modulating frequency and f_M and f_S are the characteristic channel mark and space frequencies respectively, the result of modulation is:

$$F \pm (f_M \text{ to } f_S) = F + (f_M \text{ to } f_S) \text{ and } F - (f_M \text{ to } f_S)$$

Taking the lower sideband, $(F - f_M)$ to $(F - f_S)$, since $f_S > f_M$, $(F - f_M) > (F - f_S)$ i.e. the characteristic frequencies are inverted. For those channels which are to be group modulated, arrangements are made in the telegraph-channel modulator for the effect of the d.c. control signal to be reversed, i.e. to produce in the modulator a mark frequency of $(f_0 + 30)$ Hz and a space frequency of $(f - 30)$ Hz. The line frequencies of all channels then follow the standard i.e. $(f_0 - 30)$ Hz for mark and $(f_0 + 30)$ Hz for space signals.

Should either the d.c. modulator-control circuit or the bearer send channel become disconnected, the modulator or the detector respectively would be left without control and the channel would become sensitive to noise. Guard circuits are included to ensure that under either of these fault conditions an auto-matic control will function to hold the channel positively in one of its binary states. For international circuits and those used in automatic switching systems the channel reverts to the space condition; for other requirements reversion to the mark condition can be arranged by simple strapping.

FM and AM systems are not compatible but they can be linked in tandem at their d.c. ends.

The term frequency-shift modulation implies that a signal transition produces a phase-continuous change from one characteristic frequency to the other. In some FM systems used for telegraphy, the frequency changes abruptly from one characteristic to the other causing signal distortion. These may be distinguished by using such terms as frequency-shift modulation (FSM) or frequency shift (FS).

Because variations in resonant frequency shift (FS) circuits caused by temperature changes are proportional to the frequency, the error (in hertz) resulting from a given change in ambient temperature is higher for higher-frequency channels and also for the group modulator—demodulator, e.g. for the 3180 Hz channel the effect is 7·5 times as great as for the 420 Hz channel. At the same time each channel, regardless of its virtual frequency, is deviated by ±30 Hz about its centre frequency and a frequency error of 1 Hz will produce the same degree of distortion on all channels. To maintain the same standard of performance on all channels, the frequency-temperature stability of higher-frequency channels must be made considerably greater than that of lower-frequency channels. This problem is sometimes approached by splitting the 24 channels into two groups so that the highest channel centre frequency is 1740 Hz; the upper group of channels 13—24 is produced by group modulation. The group-modulation circuit is given very high frequency stability.

In addition to traffic channels, an unmodulated pilot frequency of 300 Hz ('channel 0') may be transmitted, and also a second pilot of frequency 3300 Hz ('channel 25'), the latter obtained from modulating 300 Hz in the 3600 Hz group modulator.

In a frequency modulator the frequency of an oscillator is deviated by ±30 Hz from nominal mid-frequency, under control of the modulating signals; the frequency deviation is linear over a limited range (Fig 7.5). The modulator circuit is designed to prevent injection of modulating signals into the

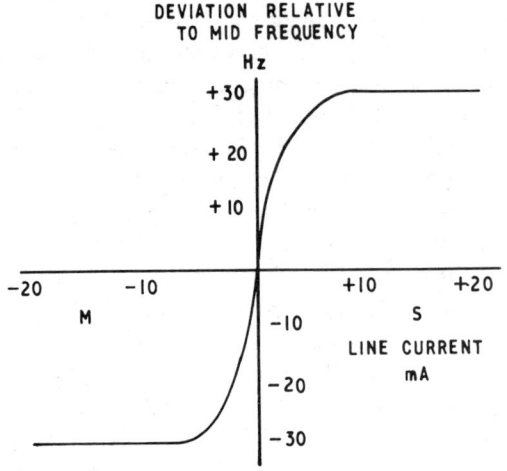

Fig 7.5 FM modulator characteristic

and indicates typical signal waveforms at points in the FM receiver. Received signals are first amplified and then fed to a symmetrical diode voltage-limiter which restricts positive and negative peaks of the VF waveform equally to a predetermined level for all input levels within the design range. In designing the detector, any variation in signal amplitude in the discriminator (equivalent to amplitude modulation) would cause distortion. Signals are then fed to a conventional discriminator consisting of two tuned circuits which resonate at ±45 Hz from the channel centre frequency, just beyond the deviated frequencies. The resultant output-voltage characteristic of the discriminator (Fig 7.7) passes through zero volts at mid-band frequency and is approximately linear for 40 Hz on either side of this. The a.c. outputs from the discriminator are rectified by diodes, and the two outputs are fed in series opposition to the output stage. Preceding the output stage is a second stage of filtration — a low-pass filter which eliminates carrier ripple from the rectified waveform and also improves channel filtering by rejecting interference from adjacent channels.

Any bias produced by drift in the resonant frequency of the tuned circuits in the discriminator

carrier-frequency portion of the circuit; by this means the carrier-beat effect is avoided.

The FSM receiver comprises an amplifier limiter, a discriminator to convert the frequency-modulated VF currents into direct currents, and an output stage; a low-pass filter is usually inserted between the discriminator and the output stage. Fig 7.6 shows a block diagram of the general arrangement

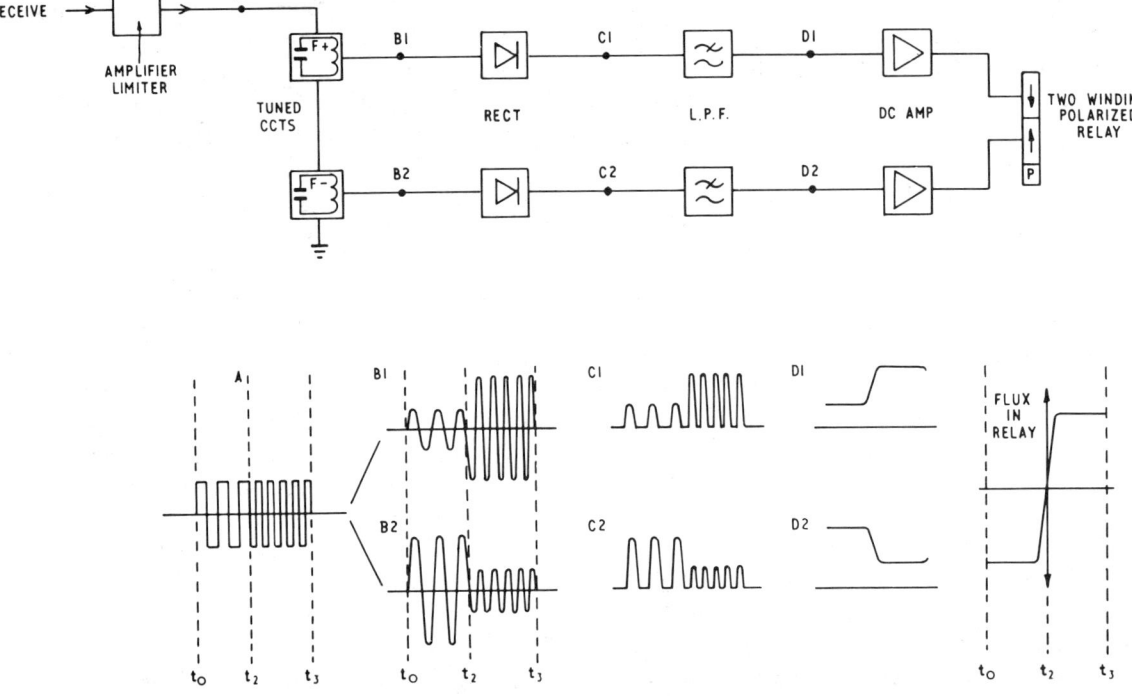

Fig 7.6 FM receiver — block diagram and waveforms

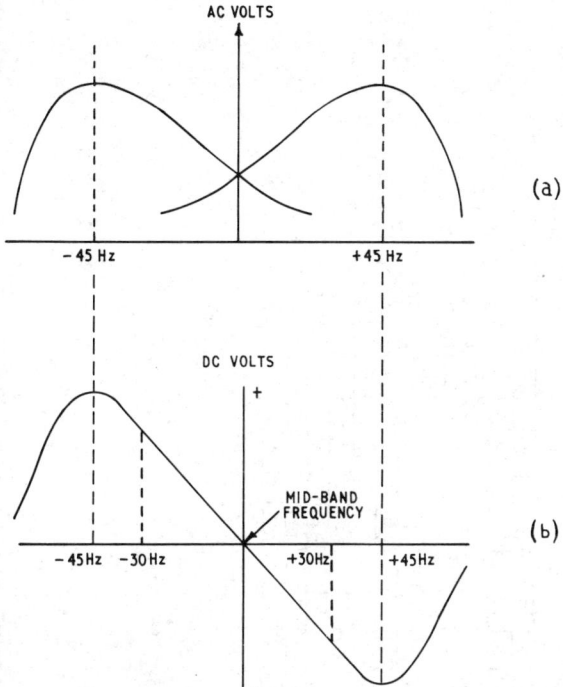

Fig 7.7 FM discriminator: (a) tuned-circuit characteristics;
(b) discriminator output

as a result of changes in ambient temperature is individual to each channel and can be counteracted by introducing into the output amplifier an equal and opposite bias which varies proportionately with temperature.

Bias introduced by frequency drift due to frequency translation of the bearer circuit is corrected by the presence of the pilot channel which undergoes the same frequency drift as traffic channels. The pilot frequency may be received on a detector so arranged that any error in the received frequency produces a proportional change in a d.c. voltage. The voltage produced by the pilot detector can then be fed to the output amplifier of each channel detector and is such as to produce a bias to nullify that produced in each channel due to a frequency drift of its own VF signals. There is no voltage output from the pilot discriminator unless a frequency drift has occurred; any such output may be positive or negative depending upon a rise or fall in drift. If the 3300 Hz pilot has been group modulated and demodulated, it compensates not only for frequency drift within the bearer channel but also for any error introduced by

frequency difference between carrier oscillators of the group modulator and demodulator at the ends of the circuit.

7.8 56-type FSM system (120-Hz spaced channels)

Although designed nominally for a modulation rate of 50 bauds, it is inappropriate to refer to a modern 120-Hz-spaced system as a 50-baud system, since performance at modulation rates even up to 100 bauds is acceptable for service using synchronous methods.

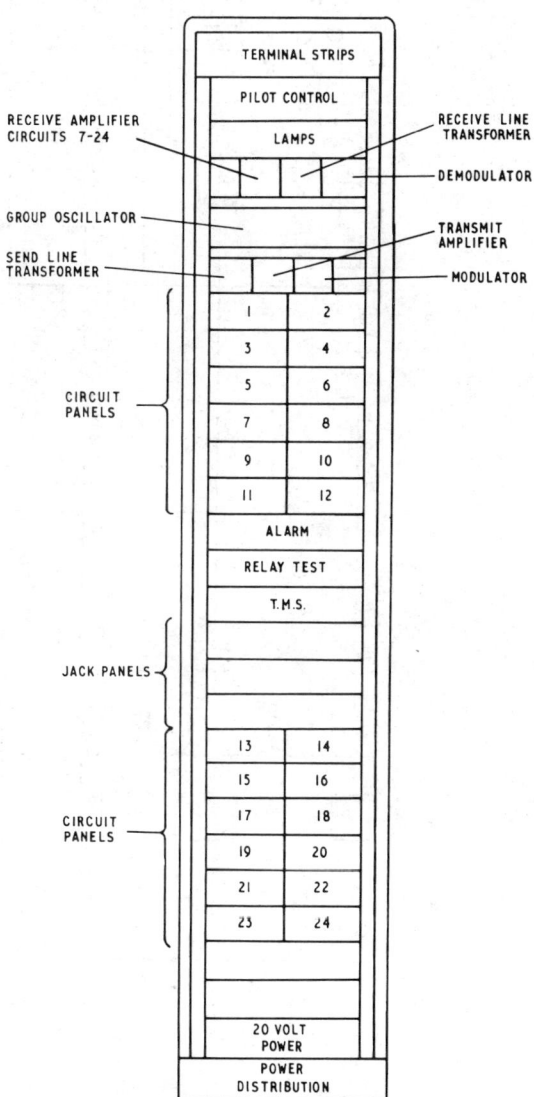

Fig 7.8 24-circuit FSM system (TF3A) — circuit rack layout

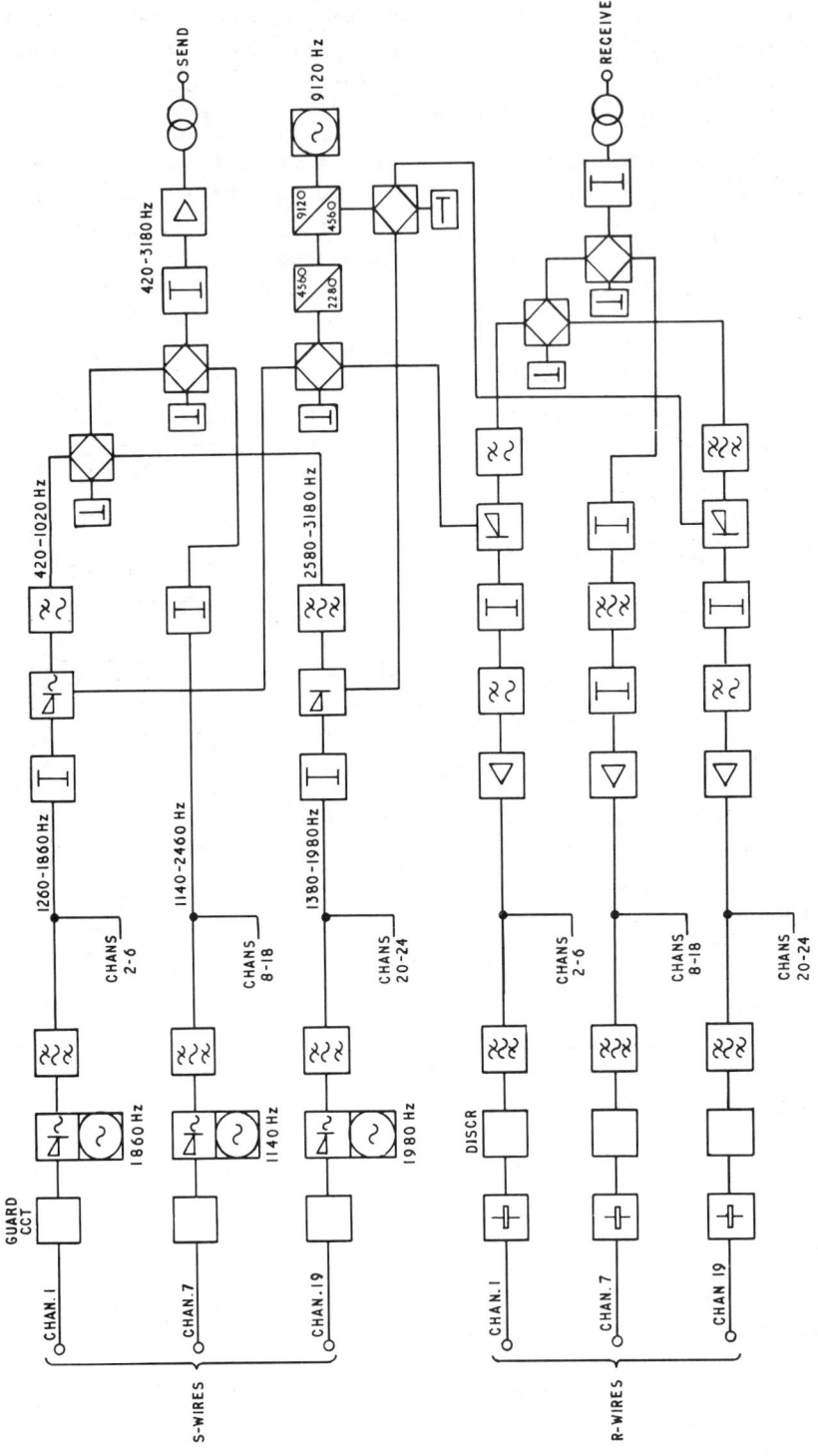

Fig 7.9 24-circuit FSM system (TF3A) — block diagram

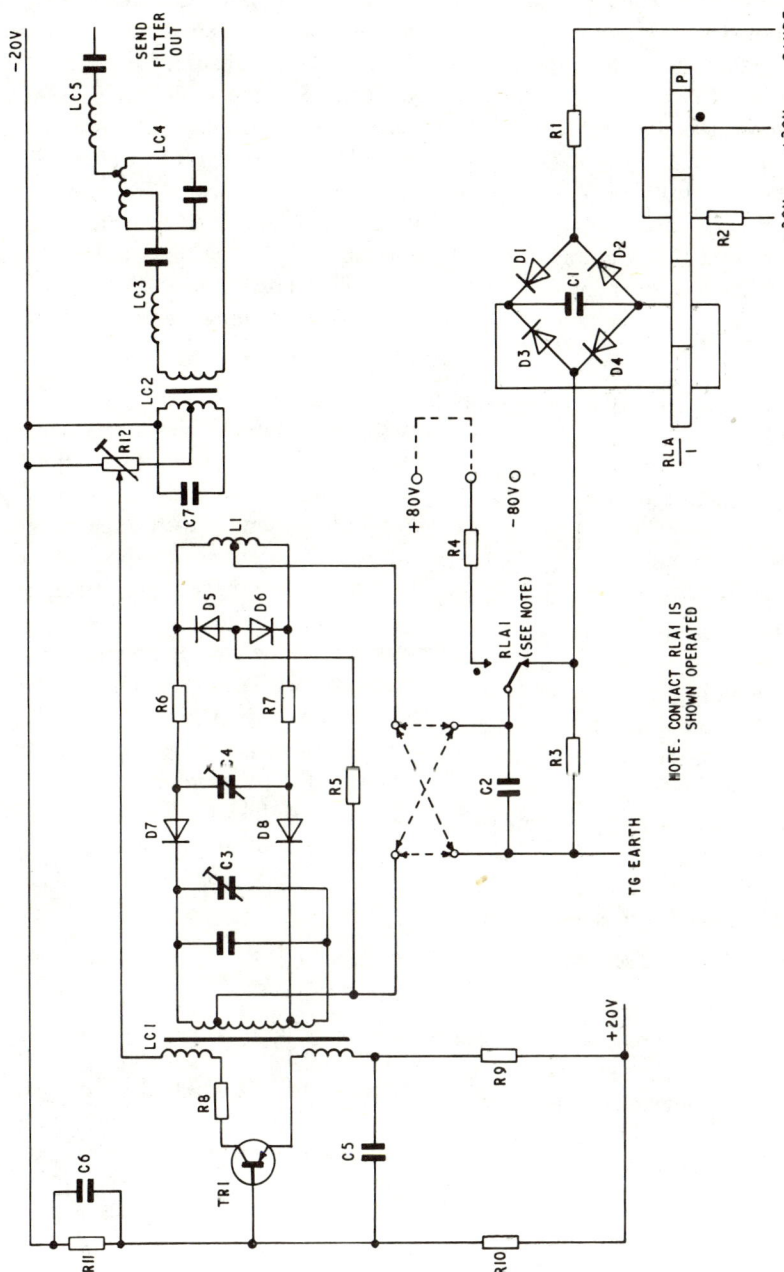

Fig 7.10 Oscillator-modulator circuit (TF3A)

In this early transistor system, designed by STC as the TF3A system, 56-type rack equipment practice is used. The rack layout is shown in Fig 7.8: two circuit equipments side-by-side are accommodated in one shelf-unit. The rack houses equipment for 24 circuits, together with pilot control, transmission-measuring set (TMS), relay test panel, test-jack fields and power packs; not all these auxiliary panels are required on every rack. Semiconductors are used throughout, together with a polarized relay (4199). The equipment may be fed either from a.c. mains or from a 24-V battery; in the latter case the supply to the rack is regulated at -20 ± 0.3 V.

Channel Frequencies. Standard channel frequencies, 420–3180 Hz with deviation of ±30 Hz, are used. Two pilot frequencies, 300 and 3300 Hz, are available for automatic correction of frequency drift in the bearer circuit*.

A block diagram of the equipment is shown in Fig 7.9. Line frequencies for channels 7–18 are directly derived from individual oscillators for each channel; those for channels 1–6 and 19–24 are obtained by group modulation. Group carrier frequency supplies are 2280 and 4560 Hz, obtained by frequency division from the same crystal-controlled basic frequency 9120 Hz; this process avoids the need for providing two separate stabilized carrier-frequency oscillators. Line frequencies for channels 1–6 are obtained by group modulating the outputs of channel equipments identical with those of channels $13 \frac{1}{N} 8$ by 2280 Hz and selecting the lower sideband; those for channels 19–24 are obtained by group modulating the outputs of channel equipments identical with those of channels 14–9 by 4560 Hz. On the receive side, individual channel frequencies are recovered by a complementary process of demodulation.

The Oscillator Modulator (Fig 7.10). Diodes D5–D8 are the active elements of a conventional switching network. Negative (mark) potential from the S wire to L1 causes D5 and D6 to conduct and D7 and D8 to become high impedance. Capacitor C4 is isolated from the oscillatory circuit of transistor TR1 and the higher of the two characteristic frequencies is

transmitted through the send filter LC2–LC5. On installation, the frequency under this condition is adjusted by varying C3, and the output level by R12. Positive (space) potential from the S wire to L1 causes D7 and D8 to conduct and D5 and D6 to become high impedance. Capacitor C4 is now connected into the oscillatory circuit and the lower characteristic frequency is transmitted through the send filter; adjustment of C4 allows an accurate setting of this frequency. The alternative strapping adjacent to C2 permits the switching network to be reversed to invert the characteristic frequencies – required in those channels whose line frequencies are obtained by selecting the lower sideband following group modulation.

Modulator Control. Telegraph modulators must be insensitive to currents less than 4 mA in the S wire to be inoperative to line noise when no modulating current is present: in these circumstances the signal sent to line may be steady space or mark as desired. This facility is provided by polarized relay RLA; while normal conditions exist on the S wire, the relay is held operated by double-current line signals which are rectified by the full-wave rectifier D1–D4 connected in series with the S wire. The contact of this relay (shown operated) maintains the S-wire connection to the modulator. When line curent falls to about 5 mA, relay RLA is controlled only by its bias winding and changes over its contact to apply +80 V (or −80 V if desired) to the modulator via R4.

The Detector (Fig 7.11). Input to the detector is fed via the receive filter to an amplifier limiter; each receive filter is designed to have a mid-band insertion loss of 15 dB and the normal input level to the receive filter is −25 dBm. Since it is not possible to make an individual channel-level measurement at the input to the receive filter, this is assessed by measurement at the output of the receive filter where the normal level is −40 dBm. Successive stages of amplification (not shown in detail) are used as a limiter to enable a signal of constant amplitude to be applied to transformer T1 whenever the input level to the channel band-pass filter exceeds −40 dBm. This represents the lowest input level, as the required operating range is +10 to −15 dB about the nominal design input level of −25 dBm.

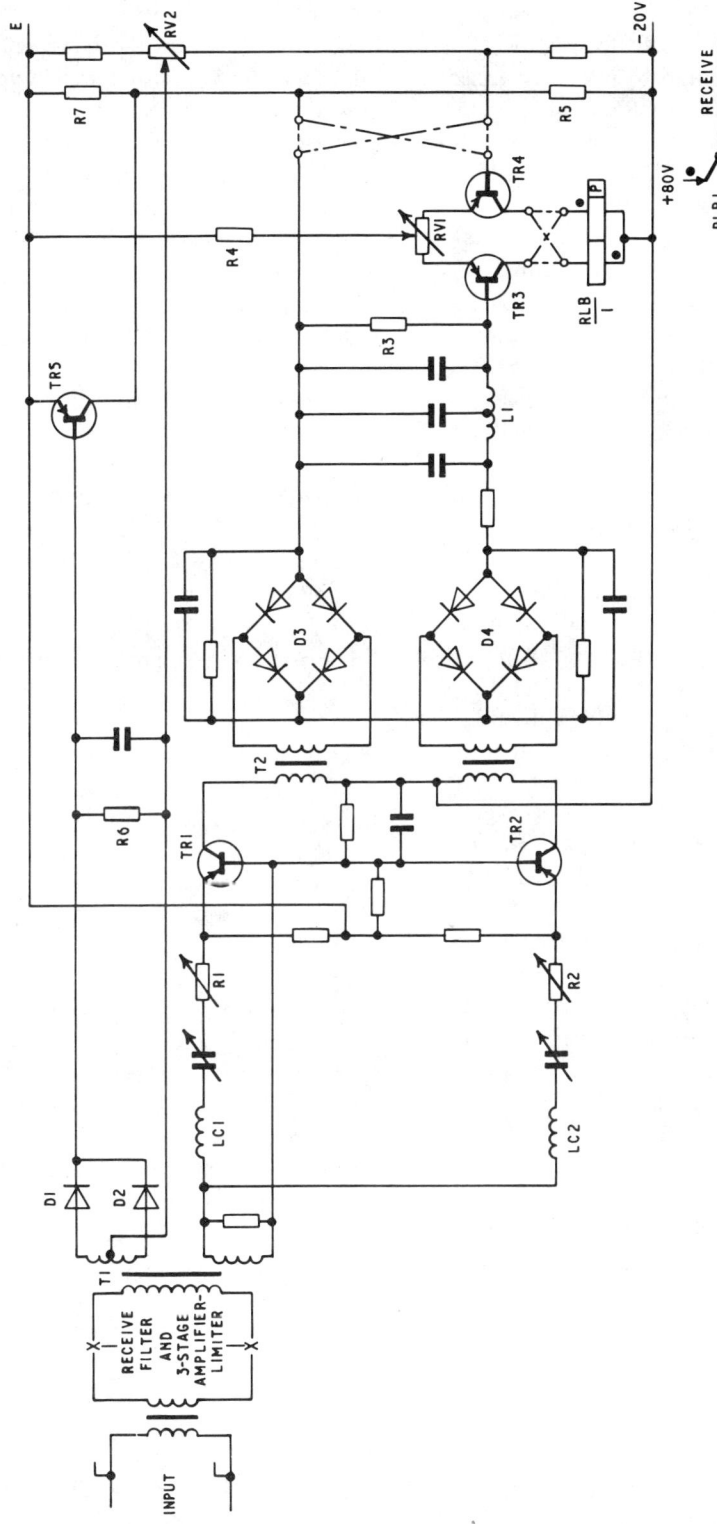

Fig 7.11 Detector circuit (TF3A)

Two series-resonant circuits are connected in parallel across one output of transformer T1. One resonant circuit, LC1 and its associated components, is tuned to the higher characteristic frequency; the other, LC2 etc., is tuned to the lower. Characteristics of these tuned circuits are similar to those of Fig 7.7 (*a*). Outputs from these two circuits are individually amplified in TR1 and TR2 and rectified by D3 and D4. The two outputs are differentially connected and applied through the post-discriminator filter, L1 etc., to load resistor R3. The combined characteristic at this point is as shown in Fig 7.7 (*b*). The output relay is controlled by a bi-stable trigger circuit of TR3 and TR4 which have a common resistor R4 in their emitter circuits. Potentiometer RV1 is in this circuit and, by its use, operation of the trigger can be adjusted to give a bias-free output from the relay.

When the higher characteristic frequency is being received, the base of TR3 is driven relatively positive and is cut off, TR4 conducts and the polarized output relay RLB operates to the space contact. For the lower characteristic frequency, the base of TR3 is driven negative and conducts; TR4 is cut off and relay RLB is operated to the mark contact. The optional strapping associated with relay RLB decides, for any channel, which way the relay is operated for a given modulation condition. This allows adjustment on the receive equipment complementary to the inversion sometimes required on the send equipment. The impedance which TR3 presents to the discriminator output when cut off is different from that presented when it conducts so that operation is not entirely electrically symmetrical; in the original setting up of the circuit, R1 and R2 are adjusted to compensate for this asymmetry.

Line Failure. When channel input falls to 6 dB below threshold-operating value for the detector (−40 dBm), relay RLB must be operated firmly to the space contact (or mark contact, if desired) to avoid the relay chattering under line-failure conditions. This facility is provided by TR5 and its associated auxiliary circuit.

Under normal operation, the VF signal from a second output of transformer T1 is full-wave rectified by D1 and D2 to develop a smoothed potential across R6. This voltage is applied in series opposition

with the voltage developed across potentiometer RV2, which is negative with respect to earth, to hold the base of RT5 positive for all input levels in the normal operating range. RT5 under these conditions is cut off, and the potential at the junction of R5 and R7 is appropriate to the normal operation of the output trigger circuit of RT3 and RT4. When the incoming signal level falls to the point where the voltage developed across R6 is less than that across RV2 (which is adjustable to allow some discrimination in the level at which this occurs) the base of RT5 becomes negative, and it conducts; more current is drawn through R5 and the junction of R5 and R7 becomes more positive. Coincident with this action, the line-signal voltage applied to the base of RT3 becomes virtually zero. Under these conditions, the increased potential at the junction of R5 and R7 will cause either RT3 and RT4 to be cut off, depending on which of the optional strappings associated with the base of RT4 is used. The output relay RLB is operated to the appropriate contact.

Pilot-frequency Control. This equipment corrects the frequency of signals entering channel band-pass filters and will compensate for frequency errors up to ±15 Hz.

The basis of the method employed is comparison of frequency of the 300-Hz pilot tone with a standard reference frequency — the local pilot tone. Any disparity between the two frequencies results in an equivalent inverse correction being applied to the aggregate-system frequencies. Referring to Fig 7.12 assume that the aggregate of channel frequencies f_X and pilot frequency f_P have suffered a change of d Hz during transmission. A band-pass filter in path B separates the incoming pilot frequency from traffic-channel frequencies.

Signal frequencies in path A and pilot frequency in path B are separately modulated in M1 and M2 respectively by an intermediate frequency f_1. In path B the lower side frequency $f_1 - (f_P + d)$ of this modulation is further modulated in M4 by the local standard frequency f_P; selection of the upper side frequency from this process gives a frequency $(f_1 - d)$. This frequency is then used in M3 to demodulate the lower sideband of the translated signal frequencies in path A, as a result of which the original signal frequencies are applied to the telegraph equipment.

The accuracy of this method of compensation depends upon the stability of pilot-frequency oscillators at each end of a system. There are advantages in using oscillators of the same stability as channel oscillators and this has been found adequate.

Performance. The isochronous characteristic distortion on mixed signals at 50 bauds does not exceed 4% over a range +10 to −15 dB about normal input level; at 85 bauds the comparable figure is within 10%, signals being present in adjacent channels in both cases.

tional rackside is replaced by a lightweight framework made from pressed-steel sheets which, when bolted together, form a free-standing, open-fronted rack which is used to hold a number of self-contained equipment shelves. Circuit components are mounted on printed wiring boards fitted within steel frames carrying multiple connectors which plug into sockets at the rear of the shelf assembly, connected directly to the station cabling without need for the traditional terminal blocks at the top of the rack. This design not only simplifies installation and

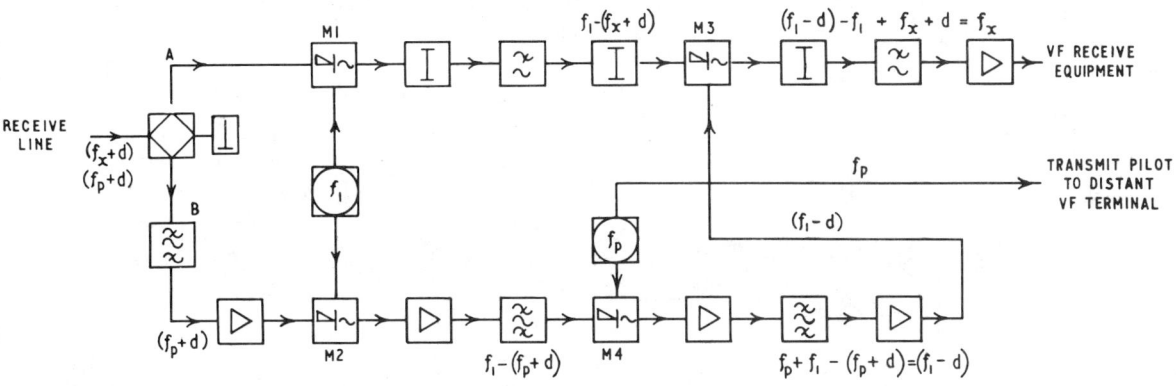

Fig 7.12 Pilot-control circuit (TF3A)

The TF3 equipment was replaced by the later TF8 system, basically similar but with improvements in design detail. In this system, channels 7–24 are all directly modulated; channels 1–6 are group modulated with a carrier of frequency 2280 Hz, supplied from a crystal-controlled oscillator at 9120 Hz via two frequency-dividing stages. This arrangement avoids 'carrier-beat' effect.

7.9 62-type FSM system (120-Hz spaced channels)

Considerable growth in new routes and in traffic which have occurred in recent years necessitated a new approach to transmission-equipment design with a view to economy in accommodation requirements. A degree of miniaturization was made possible by availability of new devices and materials such as the transistor, printed wiring cards and the mercury-wetted relay.

This led to a new form of equipment construction, known as 62-type. In this design the conven-

maintenance but also improves reliability by the reduction in soldered joints.

An exploded view of the rack assembly is shown in Fig 7.13, and of the shelf assembly in Fig 7.14. The numbered items are:

1 Rackside – right-hand	10 Panel (rear footplate)
2 Rackside – left-hand	11 Mounting bracket
3 Base	12 Blanking plate
4 Top	13 Rear cover
5 Dust cover	14 Front cover plate
6 Tie bar	15 Shelf
7 Cable support	16 Card lock
8 Earthing strip	17 Card guide
9 Panel (front footplate)	18 Shelf runner

The rack dimensions (to a CCITT standard) are width 20·5 in., depth 17·7 in. and height in a range of values up to 10 ft 6 in. (520 x 450 x 3200 mm). The rack consists of two side pieces bolted to a box base section and a top section; side sheets are stiffened by channels running the complete length

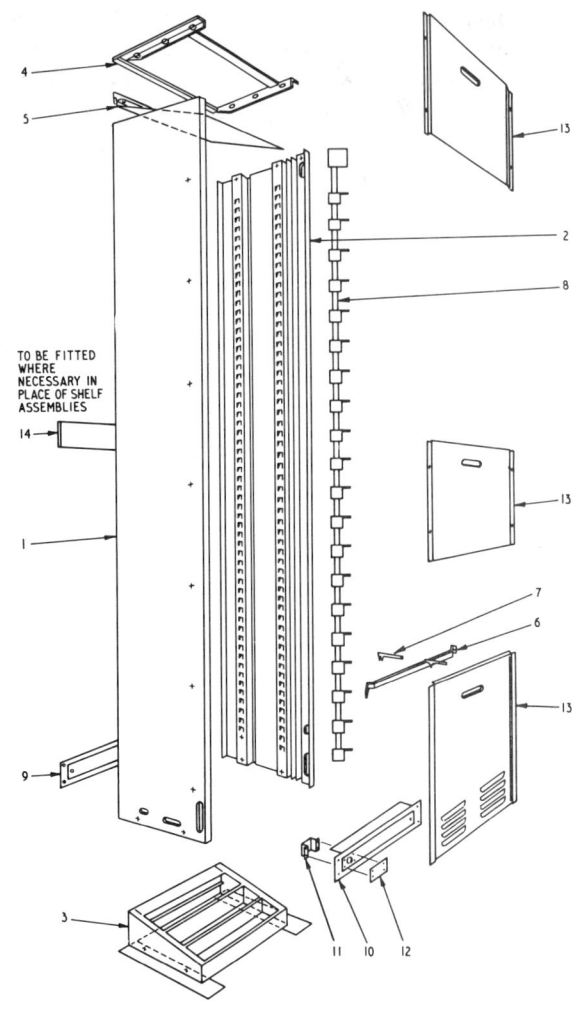

Fig 7.13 62-type rack construction

which are slotted and accept shelf supports, inclined at 15° to the horizontal. This sloping provides an easy airflow over the circuit equipment from rear to front and improves ventilation. A space of $2\frac{3}{4}$ in. (70 mm) is available at the rear of the rack over the full height and width to accommodate cabling; the top of the rack is angled to allow space for cable turning. Cable tie bars and supports are fitted and rear covers protect the cabling. Except at empty shelf positions, front covers are unnecessary, the front panel of each unit giving a finished appearance.

The shelf consists of a corrugated-metal base on which are mounted front and rear stirrups. When fitted into the rack (15° slope) the shelf rests on angled supports and is secured in position by front

flanges bolted to the front of the rack sideplates. The shelf is fitted with metallic screens bolted to the base and to the front and rear stirrups. These screens improve rigidity, are shaped to provide guides into which the equipment cards slide, and provide thermal and electrical shielding between units. Drillings for the assembly of screens into the shelf are provided at intervals of 0·2 in. (5 mm) and allow maximum flexibility in accommodating units of various widths. Sockets to accept card connectors are mounted on floating assemblies to ensure accurate homing of the inserted units. A gravity-operated locking bar across the full width of the shelf front locks all units in place.

The circuit card comprises one or more printed wiring boards enclosed by a metal frame which carries the multiple gold-plated plug connectors at the rear. The frame has a front panel equipped with a lipped handle for withdrawing a card unit from its shelf guide; the front panel mounts such items as controls, test sockets and access points to allow for setting up and maintenance testing.

An outrigger, a dummy card with a plug at the rear and a matching socket at the front, may be inserted in the shelf guide in place of any individual circuit card. This serves to extend any card clear of the rack; it is hinged left and right so that all components are accessible, while still in operation, for inspection and testing. An outrigger is seen in use in Fig 7.15 which shows part of a large installation of T24P MCVF equipment developed by the Telephone Manufacturing Co. Ltd, and which provides up to 24 duplex telegraph circuits. With 62-type construction, an MCVF system is readily extendable from 1 to 24 circuits. Eight circuits are accommodated on one shelf, i.e. a 24-circuit system requires three shelves and five 24-circuit systems (120 circuits) can be mounted on one rack; this may be contrasted with but three circuits per rack side in the pioneer types 1, 2 and 3 AM VFT systems. At the top of the rack are fitted alarm equipments, fuse cards and power units; the bottom is occupied by a mains unit. A test and monitor jack field can be included at mid height of the rack; at large installations, a centralized jack field known as a test-access frame (TAF), similar to an ECB, is used.

In the installation shown, 80- and 24-V supplies from a centralized power plant are fed to the racks through distribution fuses shown at the upper end

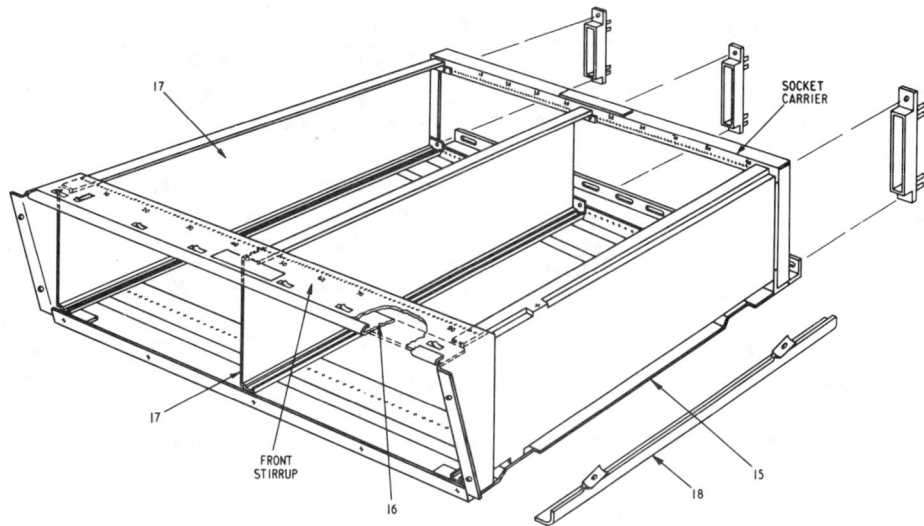

Fig 7.14 62-type shelf construction

of the suite of racks. The illustration also shows a part of the IDF serving the MCVF equipment.

The equipment for each VF circuit is mounted on one card, seen opened up in Fig 7.16 to reveal the filter and other components. A mercury-wetted output relay is seen at the rear of the vertical card (the small unit bearing a vertical arrow to indicate the mounting direction). From the front panel projects the double-filament barretter, and the single control – a receiver bias potentiometer. Test links and access points (S wire, R wire, SF OUT, RF IN, and DET IN) are available without need to remove the card.

All carrier frequencies are directly derived from transistor oscillators with frequency stability of ±1 Hz and frequency adjustment range of ±6 Hz. Alternative inputs are available for telegraph circuits (4 kΩ for ±80 V) or data circuits (1 kΩ for 5–25 V). A miniature, sealed guard relay is included to transmit unmodulated mark or space frequency (as required) if the d.c. input circuit fails. Test links enable +80 or −80 V to be applied to the modulator for test purposes.

Fig 7.15 Suite of T24P FSM MCVF equipments (*Courtesy of Telephone Mfg Co. Ltd*)

Fig 7.16 VFT circuit card in open position (T24P) (*Courtesy of Telephone Mfg Co. Ltd*)

If the level of received signal for a channel falls by 21 ± 2·5 dB from the input planning level, the receiver ceases to respond to incoming signals and the output is clamped to the mark or space condition, as required; a hysteresis range is included to guard against random operation when the threshold level has just been reached, and also against random operation to line noise under fault conditions. The mercury-wetted relay gives output signals at ±80 V or at ±6 V for data circuits.

A separate circuit card houses the main line transformer and attenuation pads common to the 24-circuit system. Test links give access to both input and output of the go and return line transformers.

A frequency-control unit is available for correcting effects of frequency drift in an HF bearer circuit, based on the use of an unmodulated 300-Hz pilot frequency (stability ±0·2 Hz). This control will automatically correct for any frequency-translation error of the bearer circuit over a drift range of ±10 Hz.

A system-fail alarm operates to failure of the pilot frequency; alternatively, if pilot equipment is not fitted the system-fail alarm operates to indicate a level drop of more than 21 ±3 dB in the aggregate signal level.

Performance. Telegraph-signal distortion can be measured and quoted under a variety of conditions — on reversals or mixed signals, in local or over a line, with input levels raised or lowered, with or without adjacent channels in operation, with or without frequency drift in the bearer circuit, in the presence of negligible or significant noise, at various modulation rates, over a range of ambient temperatures, and so on.

With a circuit looped in local (via amplifier and attenuator), after the bias distortion on 2 : 2 reversals at 50 bauds has been eliminated by adjusting the bias control, the isochronous characteristic distortion on mixed signals will not exceed 5%, while adjacent channels are being modulated by signals not in synchronism with the text-test signals. This value will not rise beyond 7% when the level of received signals is varied between +10 and −15 dB with respect to the planning level.

At 85 bauds, the maximum values of isochronous characteristic distortion are 14 and 18% at nominal level, and over the range +10 to −15 dB about the nominal level respectively. At 96 bauds the equipment will produce not more than 25% distortion, including the effects of adjacent channel interference at the planned level.

Transmission time through the looped circuit does not exceed 40 ms; as far as the MCVF equipment is concerned, the spread of this value is within 3 ms for all circuits of a system.

Performance of MCVF channels spaced at 120 Hz. Details of the CCITT Recommendation R35 for performance of single-link 120-Hz-spaced MCVF channels at 50 bauds (without line) are quoted in Table 7.4. Some examples of distortion measurements at 50 bauds and higher modulation rates made on 120-Hz-spaced channels in service over long international circuits (direct single links) routed in transoceanic cables are shown in Table 7.3. These circuits include national landline-cable extensions; the Pacific routes include the trans-Canada microwave system.

The figures below are approximate average values from measurements made on a number of VFT

Table 7.3 Typical performance of FSM MCVF circuits (120 Hz channel-spacing) on long submarine-cable routes

Route	Approximate distance (miles)	Isochronous distortion (%) with Q9S signals			Noise (dBm0 in 3 kHz)	Loop transmission time (ms) in VFT channel
		50 bauds	85 bauds	96 bauds		
London–Montreal	4000	6	7	18	−45	140
London–New York	4000	6	7	18	−45	142
London–Sydney	16 000	6	12	20	−42	375
London–Auckland	14 000	6	12	35	−42	310
London–Tokyo	16 000	6	15	24	−38	340
London–Hong-Kong	26 000	8	14	24	−38	476
London–Singapore	28 000	8	14	28	−38	490

circuits using equipments of differing origin, and measured on more than one bearer circuit for each route. The distortion is measured while adjacent channels are being modulated.

7.10 62-type FSM system (240-Hz spaced channels)

In recent years a considerable demand has arisen for data-transmission circuits suitable for modulation rates in the range 75—110 bauds. This need has been met by provision of 12-circuit MCVF systems with channels spaced at 240 Hz using the standard frequencies and deviations shown in Table 7.4. These installations are of 62-type equipment, identical in all main respects with those described for 120-Hz-spaced channels.

In the T12P2 equipment of the Telephone Manufacturing Co. Ltd a 12-circuit system is provided for operation over the normal 4-kHz-spaced bearer

circuit. The equipment for two such systems takes up the space of three shelves. Fig 7.17 shows the 240 Hz circuit unit opened up. All channel-carrier frequencies are directly modulated and means are provided to set transmitted power per channel to the correct value (see Table 7.2). Guard circuits in transmitter and receiver ensure maintenance of the unmodulated mark or space condition in the event of line failure. A system-fail alarm indicates a fall in aggregate-signal level of more than 21 ± 3 dB. The input circuit is designed for either earth return or loop signalling. Mercury-wetted relays are used for outputs at ± 80 and ± 6 V.

Performance at 100 bauds is similar to that quoted for 120-Hz spaced channels at 50 bauds. Modulation rates in excess of 100 bauds can be used with proportionately higher values of signal distortion. Transmission time through a circuit looped in local does not exceed 20 ms.

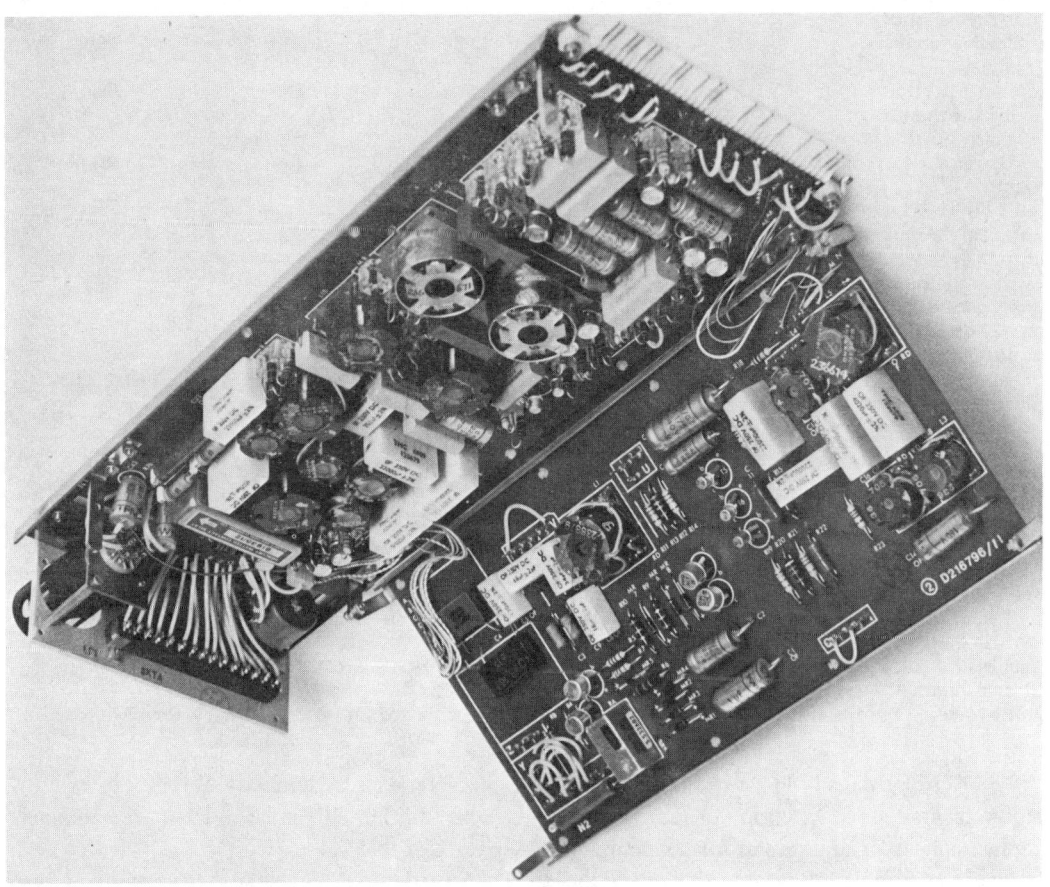

Fig 7.17 VFT circuit card in open position (T12P2) (*Courtesy of Telephone Mfg Co. Ltd*)

Table 7.4 Summary of CCITT recommendations for FSM MCVF systems

CCITT Recommendation	R35	R37	R38A	R38B*
Channel spacing (Hz)	120	240	480	360
No. of circuits†	24 (22)	12 (11)	6 (5)	8 (7)
Modulation rate (bauds)	50	100	200	200
Nominal centre frequency (Hz) n = channel number $f_0 = \dfrac{f_A + f_Z}{2}$	420+ $(n-1)\,120$	480+ $(n-1)\,240$	600+ $(n-1)\,480$	540+ $(n-1)\,360$
Tolerance on mean sent frequency f_0 (Hz)	±2	±3	±4	±3
$f_A - f_Z$ (Hz)	60±2‡	120±4	240±6	180±4
Sent power per channel (μW at zero point)	5·6§	11·25	22·4	19·2
Difference between mark and space power (dB)	±1·7	±1·7	±1·7	±1·7
Condition without modulator control	space ± 5 Hz	space ± 10 Hz	space ± 20 Hz	space ± 10 Hz
Regulated level range (dB ref. nominal input level)	−17·4	−17·4	−17·4	−17·4
Operation to space if S-channel open (dB below nominal input level)	−23·5	−23·5	−23·5	−23·5
Maximum distortion in local with adjacent channels operating	5%	5%	5%	6%
Distortion as above but over level range +8·7 dB to −17·4 dB ref. nominal input level	7%	8%	7%	8%
Distortion as above but with single-frequency 12% interference (f_A or f_Z at −20 dB S/N)	12%	12%	10%	15%
Max. distortion from frequency drift Δf not exceeding 5 Hz	$(5+2·5\,\Delta f)\%$	$(5+1·3\,\Delta f)\%$	$(5+0·7\,\Delta f)\%$	$(6+1·2\,\Delta f)\%$
Pilot frequency (if used) Hz (tolerance ± 1 Hz)	300 preferably or 3300	Same	Same	Same
Pilot level (dBm0)	−22·5	Same	Same	Same

Notes; In all cases the space–condition is the higher and the mark is the lower of two characteristic frequencies.
*For use only on long intercontinental bearer circuits spaced at 3 kHz (360-Hz-spaced channels are not used by the British Post Office).
† Channels 23–24, 12, 6 and 8 respectively cannot be used if a 3-kHz bearer circuit is used.
‡ 70 Hz by mutual agreement is permitted.
§ 9μW by mutual agreement is permitted; also, higher powers are permitted for systems with less than 24 channels.

7.11 FSM MCVF systems (480-Hz spaced channels)

For requirements such as extension of radio-telegraph circuits over landlines or for data communication requiring modulation rates in this range, the CCITT has recommended standards for the main characteristics. A summary of these, together with those for 120, 240 and 360-Hz-spaced channels is given in Table 7.4.

A further recommendation* suggests preferred

*Recommendation R36.

methods for the best use of bandwidth for composite systems comprising channels spaced at 120, 240 and 480 Hz.

An example of the use of 100- and 200-baud systems is for landline extensions of radio-teleprinter systems. For reasons concerned with sites, radio propagation and radio interference, radio stations serving long-distance routes are sited well away from urban areas. Radio-transmitting and receiving sites are some distance apart and both may be some distance from the central 'overseas' MCVF terminal station. Radio stations are connected to the MCVF terminal station by VFT channels over cable routes and a typical arrangement is shown in Fig 7.18. The send-channel equipment (*a*) represents *either* the MCVF equipment at a radio-receiving station for relaying signals over a landline towards the 'overseas' telegraph office; *or* conversely, for serving a teleprinter transmitter at an 'overseas' central telegraph office (e.g. in London) for extending signals by landline towards a radio-transmitting station. The lower diagram (*b*) shows the complementary

receiving equipment at either the MCVF station serving the teleprinter receiver at an 'overseas' telegraph office, or the equipment at a radio-transmitting station. These multiplex systems carry signals for ARQ systems operating at modulation rates of 96 or 192 bauds and data-communication signals at various modulation rates, in a group of channels ready assembled to fill a 3-kHz independent sideband (ISB) radio transmission. These radio landline connections are channels as distinct from circuits.

The following description refers to the CT6A system designed by GEC/AEI to provide six channels spaced at 480 Hz. Block diagrams of the send- and receive-channel equipment are shown in Figs 7.19 and 7.20 respectively. Channels 4, 5 and 6 are modulated; 1, 2 and 3 are group modulated using a group-carrier frequency of 3600 Hz and selecting the lower sideband. Maintaining a high ratio between channel-carrier frequencies and group-modulating frequency produces a smooth frequency change with little discontinuity of amplitude or phase and avoids a source of distortion which is present when this ratio

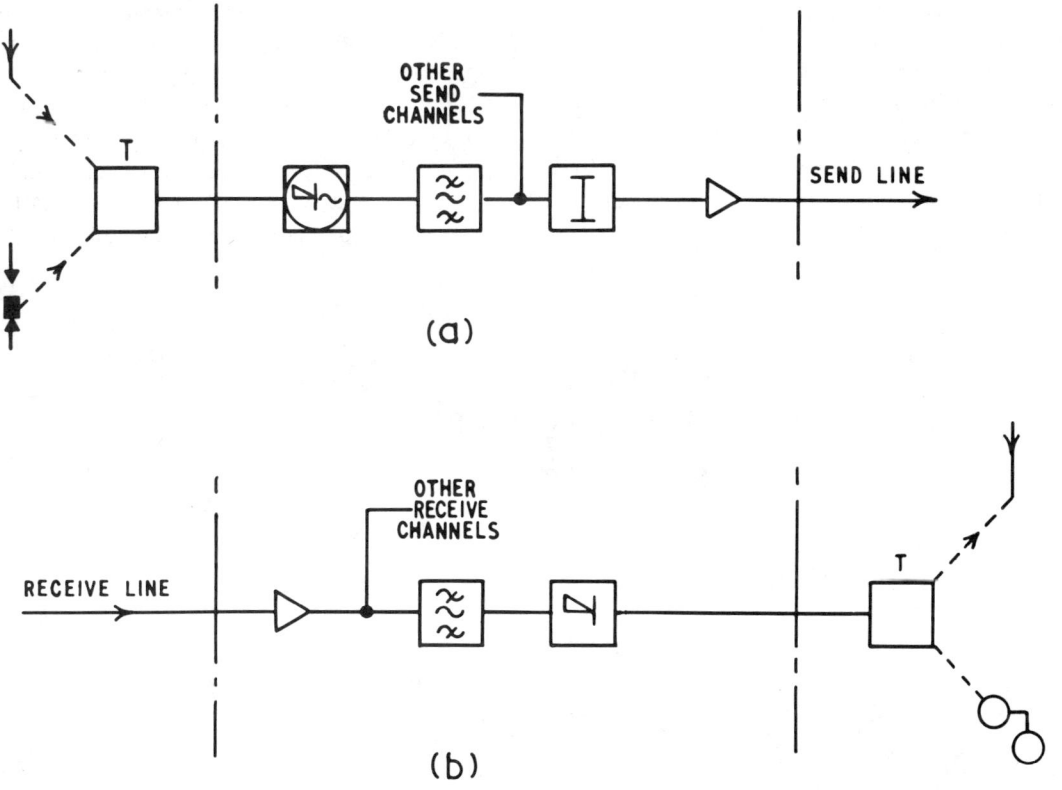

Fig 7.18 Radio-channel landline extensions: (*a*) send-channel equipment; (*b*) receive-channel equipment

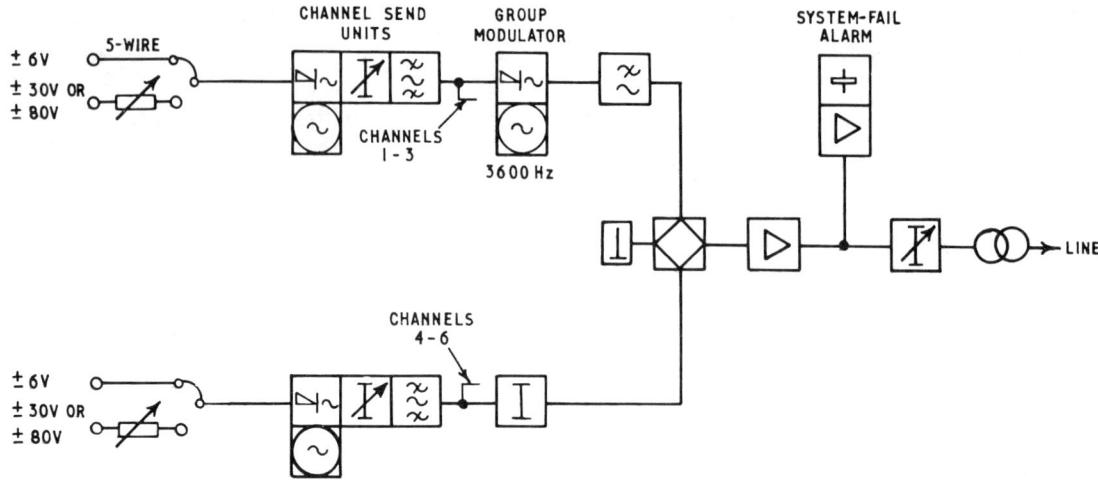

Fig 7.19 6-channel 200-baud send equipment for radio-landline extension

is of low value. The line frequencies and frequency shift are in accordance with Table 7.4.

The Send-channel Equipment. The principle of the modulator is shown in simplified form in Fig 7.21. A bi-directional transistor is used as a switch across the primary winding of a transformer. With negative potential on the S wire this transistor conducts, causing a low resistance to be reflected into the transformer secondary winding which then acts as a low resistance connection. With positive potential on the S wire the primary winding is open-circuited and the secondary acts as a disconnection. For

directly-modulated channels (Fig 7.21(a)) the LC network is set to produce the upper channel frequency f_A with positive input voltage; when input voltage is negative an additional capacitance CA is connected in parallel with the LC network, lowering the frequency. The value of CA can be adjusted to give a frequency shift of exactly 240 Hz.

For group-modulated channels (Fig 7.21 (b)) the opposite conditions must apply. With positive potential on the S wire, the LC network is adjusted to produce the lower characteristic frequency f_Z. When input voltage is negative, inductor LA is switched into circuit and by adjusting the value of LA the frequency of oscillation is raised by 240 Hz. The

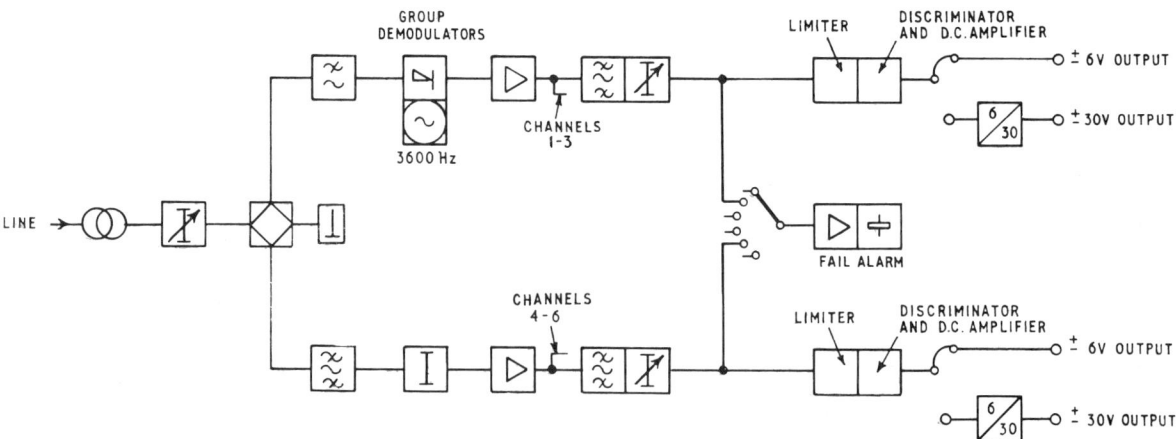

Fig 7.20 6-channel 200-baud receive equipment for radio-landline extension

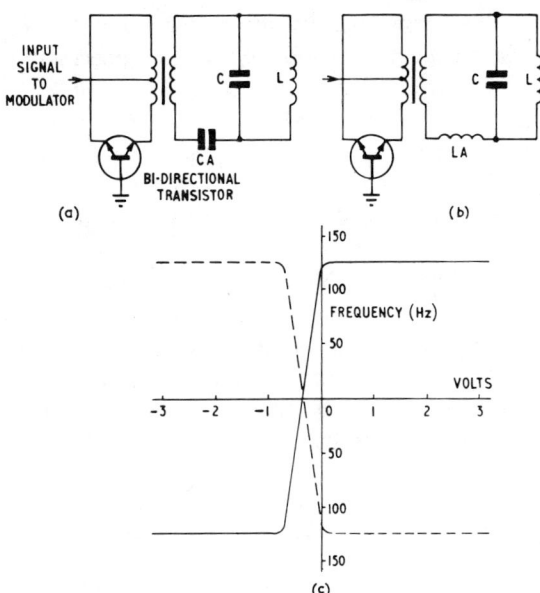

Fig 7.21 Principle of 200-baud modulator: (a) erect channel; (b) inverted channel; (c) voltage/frequency characteristic

collector oscillator TR2. The bi-directional transistor switches either LA or CA into circuit, depending on the channel number; the modulated signal is amplified by TR3. Zener diodes D1–D2 in the feedback path of the emitter of TR2 limit the signal at this point and restrict change in output level with temperature variations. The send filter has $1\frac{1}{2}$ prototype sections.

Receive-channel Equipment. This comprises the usual band-pass receive filter, amplifier limiter, discriminator and d.c. output stage. Send and receive filters are each of 600-Ω characteristic impedance and have an insertion loss which is approximately symmetrical about the channel mid-frequency; the receive filter has $2\frac{1}{2}$ prototype sections. The limiter maintains signal level to the discriminator at constant value, despite level changes in the range +10 to −15 dB about nominal level. Output signals from the limiter are of square waveform.

The discriminator (Fig 7.23) is of unusual design, which gives good sensitivity, low distortion and extreme simplicity. Circuit LC is tuned to the channel centre frequency f_0 and the bi-directional transistor short-circuits the input to a low-pass filter for half a cycle of every cycle of channel frequency. Network LC delays switching action by 90°, exactly a quarter

voltage/frequency characteristic of the modulator is shown in Fig 7.21(c). The modulator is arranged to send the upper-channel frequency f_A to line if the S wire becomes disconnected.

The send unit, including modulator, oscillator, amplifier and send filter is shown in Fig 7.22. Network L2C2 is the resonant circuit for the tuned

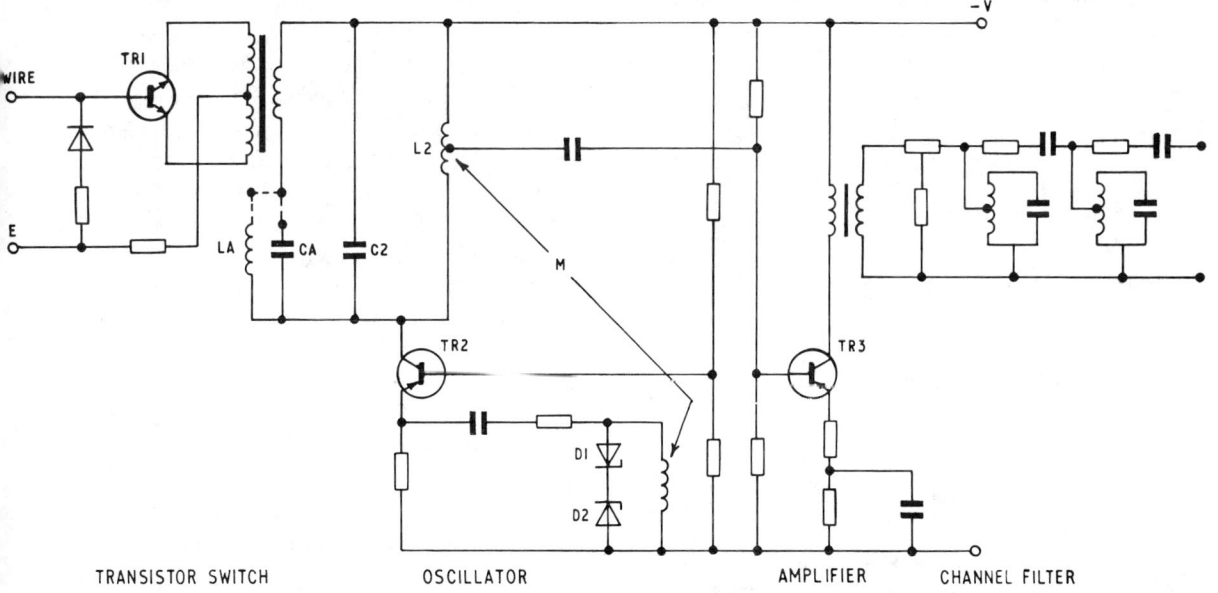

Fig 7.22 Send unit: 200-baud system

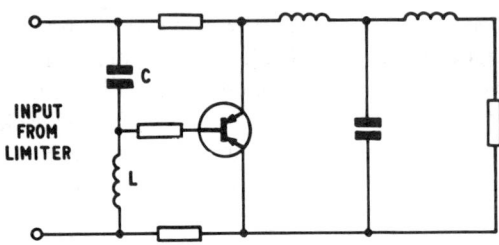

Fig 7.23 Discriminator: 200-baud system

of a cycle of the channel centre frequency and, as seen in Fig 7.24, the two quarter-cycles of conduction (shown as shaded areas in Fig 7.24(b)) into the filter are equal and opposite, hence the filter output is zero. For higher frequencies switching action occurs later in the cycle (Fig 7.24(c)), and the overall difference in conducting areas results in a positive-potential output signal. Conversely, with lower frequencies the width of the negative-potential areas of conduction increases at the expense of positive areas, producing a negative-potential output (Fig 7.24(d)). Output from the post-discriminator low-pass filter is the difference between the shaded areas. The resulting p.d. drives the d.c. amplifier stage which includes a bias control and is arranged to give an output of nominally ±6 V into a 1000-Ω load. Polarity of the discriminator output can be reversed by means of soldered straps, a facility needed to give correct output polarity for both group-modulated and directly-modulated channels. The bias control corrects any bias distortion present in incoming signals, including a small amount originated by the bias circuit in the send modulator.

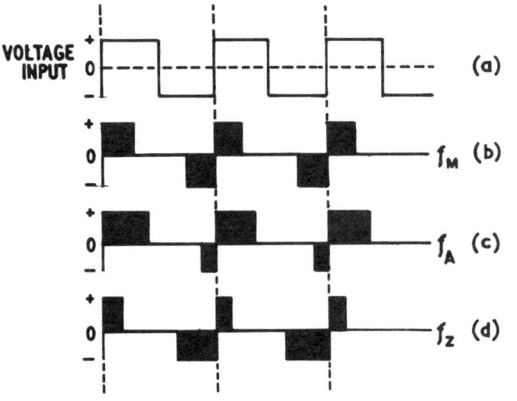

Fig 7.24 Action of 200-baud discriminator: (a) squarewave-input signal; (b) mean characteristic; (c) upper characteristic frequency f_A; (d) lower characteristic frequency f_Z

The channels are the landline extensions of HF radio channels. To avoid the need for unnecessary signal regeneration it is important that signal distortion should be low. Isochronous distortion on any channel is less than 7% at modulation rates up to 240 bauds, when all other channels in the system are carrying signals and with signal-level variations of +10 to −15 dB about normal level.

Channels of this MCVF system each carry signals of a time-division multiplex ARQ system. In view of the importance of this high traffic loading, immediate recognition of equipment failure is necessary. Alarms indicate when there is failure of either of the ±12-V stabilized supplies; or fall in level of 15−20 dB from normal level of the aggregate signal at the send terminal; or a fall of 9−14 dB below normal level at RF OUT of any selected receive channel.

7.12 Maintenance of MCVF equipment

Maintenance work of a VF terminal station is mainly concerned with lining up and fault clearing on MCVF circuits. This involves co-operation with renters of leased circuits and with telex and TAS exchanges to deal with faults on trunk lines. To facilitate this, all d.c. extension circuits from VF circuits are led through a jackfield in the VF terminal station.

In earlier types of equipment, test-access jacks were on the MCVF-equipment racks. In 62-type MCVF stations the arrangement has been simplified by installing centralized test-access frames (TAF) − jackfields in which each circuit is represented by four jacks for test access, wired as shown in Fig 7.25. A 2-jack appearance of a circuit does not lend itself to easy patching, but the 4-jack arrangement enables a faulty circuit to be temporarily replaced or patched by a spare VFT circuit. All d.c. testing is centralized on the TAF. Apart from the teleprinter S & T positions* the testing equipment (e.g. the TDMS) is portable. All circuits on the TAF are at d.c. points; the VF-access points are on the MCVF-circuit equipment.

Fig 7.25(a) shows that the INST jack gives access on tip and ring to the d.c. or teleprinter side; the VF jack gives access on tip and ring to the VF circuit. Using a crossed double-ended cord (Fig 7.25(e)) the VF circuit may here be intercepted for patching, e.g.

---

*Speak & Test (S & T) positions are described in *The International Telex Service* (R. N. Renton), Pitman.

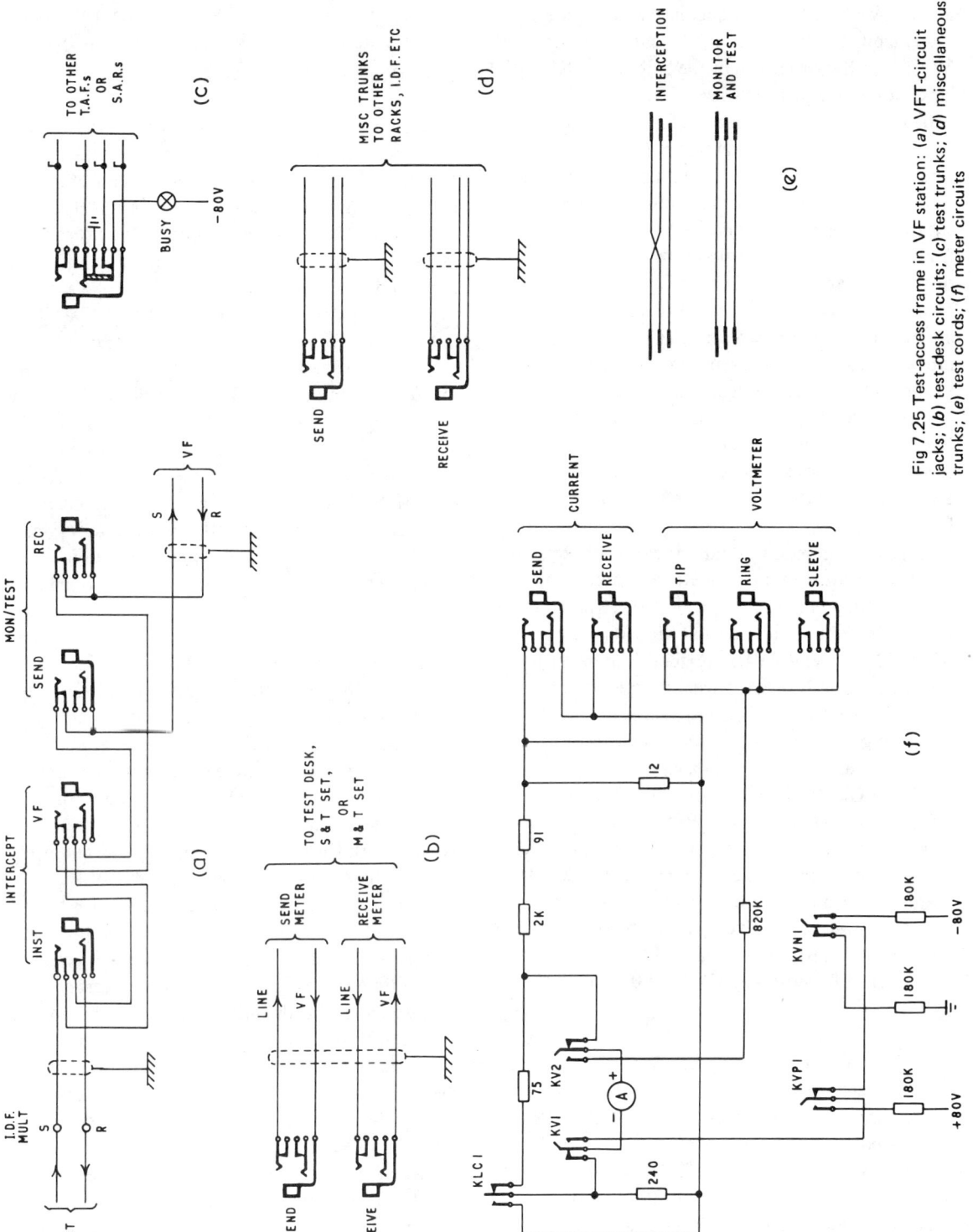

Fig 7.25 Test-access frame in VF station: (a) VFT-circuit jacks; (b) test-desk circuits; (c) test trunks; (d) miscellaneous trunks; (e) test cords; (f) meter circuits

a faulty VFT circuit can be replaced by a spare one by patching the INST jack of the faulty circuit to the VF jack of the spare. Use of the SEND and RECEIVE jacks, (a) with a pair of straight 3-way double-plug-ended cords (e) also intercepts a circuit, at the tip and sleeve, for monitor and test purposes. Provided that one end of each plug is first inserted into the test-set jacks (b) – where the S and R wires are each separately looped through a meter – a circuit which is intercepted is not interrupted when the other ends of the cords are plugged in the SEND and RECEIVE MONITOR/TEST jacks. Test trunks to other TAFs are provided as at (c); when such a circuit is taken into use, auxiliary jack springs extend earth potential to the multipled lead to light the BUSY lamp at the points where the jack is not in use. Test trunks to the IDF and other miscellaneous points are provided as at (d). Other jacks, not shown, give access to ±80 V (via barretters) and to the signalling earth lead.

D.C. measurements of current and voltage can be made using the centre zero meter as a milliammeter or as a voltmeter (Fig 7.25(f)). For these measurements, the circuit under test is patched to the CURRENT or VOLTMETER jacks. The meter is normally on its 100-mA range, but operation of the LOW CURRENT key KLC introduces the 75-Ω shunt to give full-scale deflection at 20 mA.

The meter may be used as a voltmeter by operating the VOLTMETER key KV and patching the circuit under test to the TIP, RING or SLEEVE jack, according to which of the jacks at (a) is used. One side of the meter is connected to earth and the meter will read positive or negative voltages from the line under test; alternatively, it may be connected to +80 or −80-V instead of to earth, by operating the VOLTMETER +80-V key KVP or 80-V key KVN.

The Transmission-measuring Set (TMS). The TMS is an essential item of test equipment in every VF station and is used for checking VF power levels at various points in the equipment.

The TMS comprises an oscillator whose frequency is adjustable over the range 300–3400 Hz or more, and whose output can be accurately adjusted by a variable attenuator to a known level (e.g. 1 mW in 600 Ω); together with a decibelmeter, the latter being a high-impedance voltmeter calibrated in deci-

bels referred to 1 mW in 600 Ω, i.e. the meter reads 0 dB for a power of 1 mW in 600 Ω.

Lining up a VFT Channel. The process of checking and adjusting voltages, currents, power levels, distortion and generally ensuring that a VFT channel is performing according to specification is traditionally termed 'lining up the channel'. This procedure takes place on first installation or after clearing a fault, and as a routine at infrequent intervals.

For this purpose the channel must first be taken out of service temporarily, substituting another if need be. To economize on effort, it is usual first to line up each end of the VFT circuit in local loop by connecting SF OUT to RF IN via an amplifier and attenuator; after the send- and receive-channel ends have been properly lined up in local, an end-to-end test over the two channels forming the circuit may be made by co-operation with the distant VF terminal station. Communication between stations while the circuit is being lined up is established by a teleprinter 'speaker' circuit over an MCVF system.

If necessary, supply voltages are first checked to verify that they are within permitted limits, and the receive–output relay is either cleaned and adjusted or replaced by one which has been so treated.

The sent power level of each channel will be measured (usually at SF OUT, according to the location of the test access point, e.g. SF OUT = −17 ±0·25 dBm). On FM systems, channel output level at both characteristic frequencies is measured; on AM systems the modulator-insertion loss for the space condition would be measured to ensure that it is at least −33 dB with respect to the mark level.

For lining up a channel, distortion measurements are made at 50 bauds (or other appropriate modulation rate) using 2 : 2 reversals as the test signal; on VF channels distortion measurements are made in the isochronous mode. It is found that a properly line-up channel produces substantially the same bias distortion with 2 : 2 reversals as with mixed signals. There is a pronounced difference in bias distortion which results from 2 : 2 reversals and 1 : 1 reversals owing to the occurrence of side frequencies at different points in the attenuation/frequency characteristic of the channel filter. After adjusting the receiver for minimum bias distortion (virtually zero) the distortion on Q9S signals may be measured with the input level adjusted over the specified range.

Characteristic distortion on Q9S signals, measured isochronously, will rarely exceed 3% in local or 5% over the line circuit (at 50 bauds on channels spaced at 120 Hz) on modern equipment.

After lining up the channel in local, the threshold operating level of the receiver is measured in local, using an adjustable attenuator in the loop. Reversals 2 : 2 are then applied and the loop attenuation is reduced until the receiver just responds. If the receiver were out of limits the gain of the receiver would be adjusted.

Since they have a high standard of reliability the frequency of oscillators is checked only at infrequent intervals.

If an MCVF circuit or system becomes faulty, since it is not possible to 'busy' the trunk-circuit multiple at the VF station, co-operation over a speaker circuit between the VF station and the telex or TAS exchange is necessary.

If an MCVF system bearer circuit becomes faulty the 4-wire reserve circuit is brought quickly into use by means of U-links (points A and B in Fig 7.3) in co-operation with the controlling telephone-repeater station and the telephone-traffic staff.

Polarized Relays. Testers used in VF stations for checking the performance of the polarized output/ receive relays are described in Chapter 5. On 62-type equipment the sealed mercury-wetted relays used do not require any maintenance attention; they have a life in excess of 5 years and are discarded and replaced if they become faulty.

7.13 MCVF systems for radio-telegraphy (170 and 340-Hz spaced channels)

Radio-telegraph systems are classified according to the frequency (or wavelength) used for transmission, since propagation characteristics vary over different frequency bands.

VHF radio systems, used for short ranges (about 50 miles), useful for bridging stretches of water where the cost of a submarine cable may not be justified, are not needed to any extent for commercial multiplex telegraphy in the United Kingdom. Provided the distance is not excessive, signal/noise ratio is

adequate and conditions sufficiently stable for operating standard MCVF systems using frequency modulation.

SHF radio-relay systems also maintain a good signal/noise ratio, and bearer circuits are in use for carrying standardized FM MCVF systems.

Communication satellites also use SHF propagation. Bearer circuits on a synchronous satellite system are of very high quality and stable performance. Circuits on these systems are now a regular means of providing intercontinental circuits, using standard MCVF narrow-band systems with frequency modulation.

HF radio systems, which relay for long-distance propagation upon reflection from the varying ionized layers surrounding the earth, are used extensively for radio-telegraph circuits on routes where neither submarine cables nor communication satellites are available. Propagation on HF radio circuits is characterized by fading — deep and rapid falls in signal level, accompanied by high noise levels — and also multi-path propagation which results in the presence of two or more appearances of the received signal, these arriving at different instants, having traversed different paths. Fading which occurs over a single path is termed *flat fading* and propagation over two or more paths produces *selective fading*; at any instant it produces different signal levels in adjacent receiving aerials, and also in similar channels with a frequency separation of only a few hundred hertz. Under selective-fading conditions, fades on the mark and space frequencies generally occur at different times. Although termed rapid, relative to telegraph modulation rates fading is nearly always very slow. Reception on HF systems is also adversely affected by interference from other HF transmissions in this very congested frequency band.

The effect of multi-path propagation places some restriction on maximum telegraph-modulation rates. Path-time differences of 1–2 ms are common and values up to 4 ms are experienced. A path-time difference of 4 ms could cause up to 20% signal distortion at 50 bauds (20 ms elements), but at 200 bauds (3-ms elements) the distortion could be as much as 80%. At 100 bauds (10-ms elements) the same multipath differential delay could cause 40% signal distortion, tolerable in synchronous telegraph systems. For long-distance communication a practical limit to telegraph-modulation rate is about 200 bauds, although for general application about 100 bauds is preferred. This limitation is brought about only by the multipath effect. Exploitation of transmission facilities by time-division multiplexing is severely restricted and generally frequency-division multiplex methods are employed on radio channels. For this reason, the use of 2-channel ARQ systems (96 bauds) is favoured in preference to 4-channel systems (192 bauds) on routes where the multipath differential delay is high. Although a 4-channel system is capable of carrying twice as much traffic as a 2-channel system under stable circuit conditions, if the margin of satisfactory operation is lowered the time efficiency of the system may well be reduced by the greater call for repetitions, and the increased potential capacity is not fully realized.

In HF radio transmission, economy of radio bandwidth is of the utmost importance because transmission bandwidth has to be shared according to international regulations. Maximum power per transmission must also be limited to avoid causing interference with other transmissions in the same band.

Radio-telegraph transmission systems in the HF band use either frequency-shift modulation of the radio carrier or independent sideband transmission.

The FSM systems (classified as F1 radio emissions) have a non-linear amplifier characteristic and normally can carry only one telegraph channel at a modulation rate up to 200 bauds; this may be a 2-channel or 4-channel time-division telegraph system. In FS systems, mark and space conditions are distinguished by transmission on two frequencies, typically a few hundred hertz apart; these systems are fairly economical in bandwidth.

A variant of the FS system, known as *twinplex* (or 4-frequency diplex), classified as an F6 radio emission, enables two telegraph channels to be carried simultaneously using four radio-shift frequencies, only one of which is emitted at any instant. This method does not increase modulation rate as with time-division systems, nor does it necessitate reduction in power per channel as with frequency-division systems, since only a single tone is emitted. At any instant, a pair of telegraph channels can be in only one of the four modulation conditions shown in Table 7.5. Four frequencies are used to represent these four conditions.

Signals for the two channels are separated in the radio receiver. The space-diversity system is used in conjunction with the F6 emission.

The ISB systems have linear characteristics which enable two 3-kHz independent transmissions to be carried on each sideband. These 3-kHz bands are used for telephony or as bearer circuits for multi-channel telegraph systems. A reduced level of radio carrier is available for demodulation and this reduces the need for extremely high frequency stability.

Power transmitted by each channel of a multi-channel frequency-division telegraph system must be as high as practicable to load the radio transmitter fully, but not so high as to overload it and cause intermodulation. The aggregate signals in a frequency-division telegraph system may be regarded as the sum of a number of equal-amplitude vectors having a more or less random phase relationship. If there are C carriers of amplitude V which may be transmitted simultaneously, the maximum instantaneous amplitude that the aggregate signals can attain is CV. To transmit this signal without introducing intermodulation distortion, the peak power-handling capacity P of the transmitter must be proportional to $(CV)^2$. The power available for each channel is P/C^2.

Assuming that noise present in the common path is uniformly distributed in frequency, the noise in each telegraph channel is proportional to its bandwidth. Noise N in each telegraph channel before demodulation is equal to:

N = Total noise power in the common path $\times$

$$\frac{\text{bandwidth of telegraph channel}}{\text{bandwidth of bearer channel}}$$

Before demodulation, the average signal/noise ratio for a channel calculated on the foregoing basis would be proportional to:

$$\frac{P}{C^2} \times \frac{1}{N}$$

Table 7.5 Twinplex-modulation frequencies

Condition of Channel 1	Condition of Channel 2	Designated frequency	Deviation of radio-carrier frequency (Hz)
M	M	F4	+600
M	S	F3	+200
S	M	F2	−200
S	S	F1	−600

For a given transmitter peak power the signal/noise ratio in each channel deteriorates as the number of channels is increased. As the number of channels is increased, the chance of the aggregate signals approaching closely the peak amplitude quoted above for a given percentage of time becomes statistically less, and this enable a higher power to be used for each channel. Table 7.6, based on a CCIR Report, indicates typical values employed. Up to four channels, powers quoted correspond to voltage addition; beyond this number they correspond to power addition.

Table 7.6 Maximum power levels for radio emissions

No of simultaneous tones	Maximum level per tone (dB) relative to transmitter peak-power rating
1	0
2	−6
3	−9·5
4	−12
6	−13·8
8	−15
12	−16·8

When a common radio link is employed simultaneously for telephony and telegraphy, the level of the telephone channels is maintained at its usual value, but in the channel used for telegraphy the level of the emission is reduced by up to 3 dB if one telephone channel is also present, and up to 6 dB for two telephone channels.

Frequency Diversity. Fading is frequency selective — sometimes one channel of a multiplex systems may be temporarily affected by fading while another in the same system, operating on a frequency differing by only a few hundred hertz, is still working satisfactorily. By combining two telegraph channels, separated by 600 Hz or more, to transmit in parallel, and selecting the received signal from the channel which is instantaneously delivering the higher-level signal, at the same time suppressing both the weak signal and the noise from the channel subjected to the deep fade, a considerable improvement in reception is obtained. If need be, three channels can be used to transmit the same signal in parallel, the received signal of highest level always taking control of the receiver output to the exclusion of noise in channels undergoing deep fading. This improvement

in reception is obtained only at the expense of uneconomic use of valuable radio-frequency spectrum; it also necessitates a reduction is permissible power per channel and in signal/noise ratio.

Spaced-aerial Diversity. An alternative form of diversity reception is spaced-aerial diversity; in this system two, or possibly three, receiving aerials are spaced a distance apart equivalent to a few wavelengths. Fading rarely occurs simultaneously at the different aerials. An MCVF equipment is associated with each diversity aerial and outputs are combined as for frequency diversity, the signal of highest instantaneous level being selected to control output of the receiver by a built-in diversity switch. With spaced-aerial diversity, channel equipments of identical frequency are used for diversity channels. This method conserves both bandwidth and power and is used regularly at radio-receiving stations provided that somewhat extensive additional site space is available.

Space diversity and frequency diversity may be employed on the same transmission, the latter usually within the group of telegraph channels contained in the same 3-kHz band. The AGC of the radio receivers will stabilize the mean level of the aggregate telegraph signals, but the shorter-term variation between the channels due to more rapid and selecting fading must be handled by individual telegraph-channel equipment. Diversity combination or switching is effected after complete demodulation of the telegraph channels.

When amplitude-modulated MCVF systems were in general use for landline routes, '2-tone' systems were used on HF radio circuits. A 2-tone channel comprised two adjacent AM VFT channels, one of the modulators being suitably inverted so that mark signals were transmitted by VF tone on one channel and space signals by a different VF tone on the other. This had the advantage that tone was always present on the joint channel to facilitate the operation of a rapid-response AGC device in the receiver. If frequency diversity were required, a further pair (or even two more pairs) of similar channels were used for simultaneous transmission, resulting in a 4-tone (or 6-tone) system. These remained AM systems and were subjected to the effects of severe level changes in received signals.

In an AM system, due to its single-current characteristics the receiver has to distinguish between tone and no tone at its input. In the no-tone condition the receiver is particularly susceptible to noise interference. The necessity for deriving bias from the received-signal level narrows the difference which can be accommodated between maximum and minimum input levels. Amplitude modulation is not very suitable for radio operation. By using FM, in which binary conditions are represented by two frequencies, a double-current effect is obtained by detecting the FM signals and adding them differentially. The presence of a continuous tone in the transmission path improves the signal/noise ratio. With the development of narrow-band FSM systems, the original 2-tone method became obsolescent since, apart from its inferior performance, a 2-tone AM system occupies twice the bandwidth compared with an FM system for a given modulation rate.

Recently the 2-tone AM system (classified as a *frequency-exchange system*, to distinguish it from a frequency-shift system) has been brought back into use with a fundamental change in the method of detection using an assessor technique associated with the name of H. B. Law. Each telegraph channel can be regarded as comprising a pair of AM channels with complementary modulation, separate detection and additive combination of the detected signals to produce a frequency-diversity improvement. For modulation at 100 bauds, the modular spacing of frequencies is 170 Hz. The two frequencies used for a channel may be separated by a multiple of 170 Hz, to give a measure of frequency diversity, being interleaved with frequencies of other channels. The separation is chosen according to the most likely multipath propagation-time difference, optimum frequency being equal to half the inverse of this time difference in seconds, e.g. for a path-time difference of 1 ms, the half-inverse value is $1000/2 = 500$ Hz. The system may be used in conjunction with spaced-aerial diversity reception. The inherent frequency-diversity characteristic of this system gives an improvement in performance with selective fading, which is optimum if propagation times differ by the amount appropriate to the separation of the two modulating frequencies and if the activity of the paths is equal.

Investigations suggested that conventional frequency-diversity-detection methods failed to exploit completely the information contained in received signals, some of it being suppressed in the limiting action, because information presented by *absence* of a signal could be equally significant to that due to presence of a signal. For example, during a period of selective fading when mark signals are comparatively strong and space signals are weakened by the fading, the weak signals are detectable as space signals *because they would have been strongly received had they been mark signals*. Errors in detection then need arise only when fades occur simultaneously on both frequencies. This reasoning led to design of equipment in which received signals are assessed in terms of the expected amplitudes of mark and space signals, derived from the preceding elements.

The arrangement of such a radio receiver for spaced-aerial-diversity reception is shown in Fig 7.26. Mark and space signals in each aerial path are processed in separate diversity branches. With a mark branch, for example, after suitable frequency-changing, signals are applied to a filter responsive to the mark frequency and having a build-up time approximately equal to the duration of a signal element. The build up is linear and at an instant corresponding to the end of a signal element; the filter output is a measure of the mark-frequency energy applied to it during that element. The filtered mark signal is rectified and then applied to an assessor, which gives a quantitative indication of presence or absence of the mark signal by comparing the rectified-signal voltage with a judgement level about half-way between the voltage to be expected in the presence of a mark signal and that to be expected in its absence; these reference voltages are obtained from preceding signals and noise by RC storage circuits of suitable time constant. In effect, the assessor subtracts the voltage corresponding to the judgement level from the detector-output voltage and the result is a voltage of which the sign depends on whether the mark signal is judged to be present and the magnitude depends on the mark-signal strength. Absence of a mark signal, which would be strong if it were present, gives a strong indication of space. The space branches operate similarly, giving outputs of opposite polarity. Outputs of all branches are combined on a voltage basis, limited and fed out as square telegraph signals. Theory shows that it is better to combine outputs than to select the largest

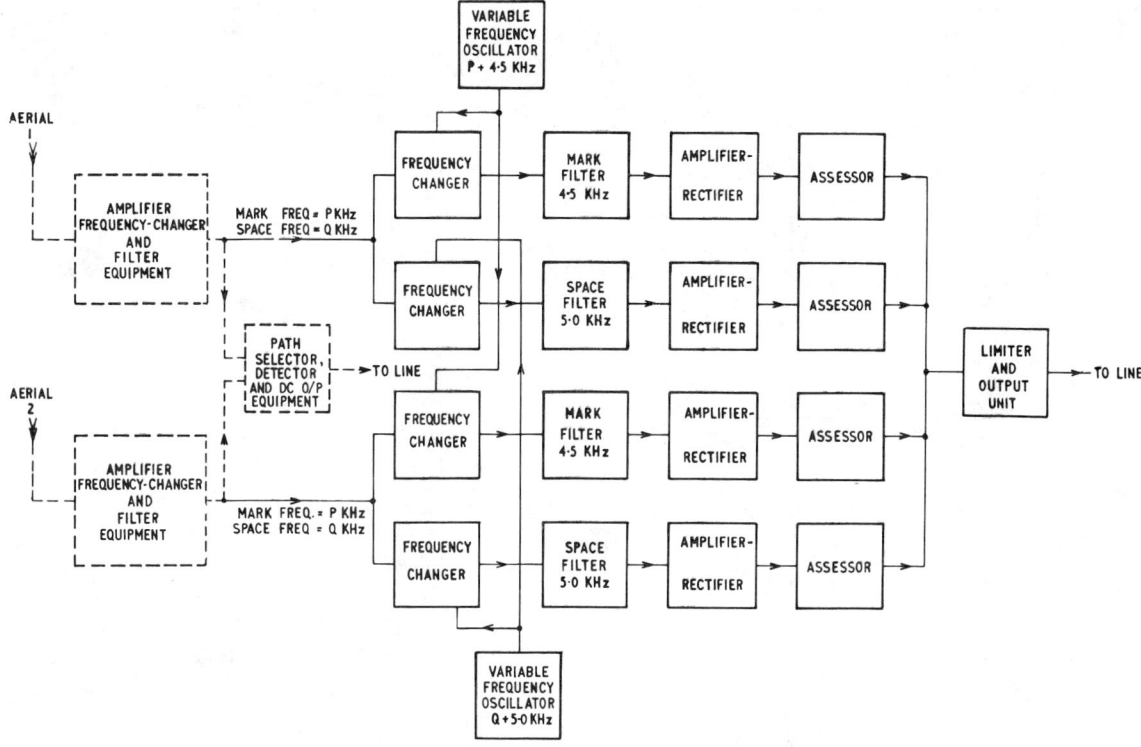

Fig 7.26 2-tone radio-telegraph detector using assessor technique

as is more usual; ideally, combination should be made on a power basis.

With flat fading, very good performance is obtainable with FSM systems but, when multipath propagation is present, performance deteriorates rapidly with increase in path-time difference, particularly at lower signal/noise ratios, and the benefit of dual space diversity may be desirable. In conditions of multipath selective fading, the frequency-exchange system using the assessor type of detector gives much superior reception; on the other hand, it requires twice the bandwidth compared with the normal FS system. This comparative performance is demonstrated by Fig 7.27 which shows the result of a practical simultaneous test carried out over a long-distance HF radio ISB system using a 96-baud ARQ system with channel separation of 240 Hz in the 2-tone equipment. Dual space-diversity reception was used for both AM and FM transmissions, the latter using ±40-Hz deviation in a 170-Hz band.

Bandwidth is controlled by filtering the modulated signal; the filtered signal will in general be to some extent amplitude modulated. Investigation showed that high error rate and poor diversity action experienced with narrow-band FSM radio-telegraph systems in the presence of fading multipath signals are due to a combined effect of the frequency modulation and spurious amplitude modulation of the signal which together cause severe phase abberations during signal transitions at troughs in fading. The spurious amplitude modulation is due to removal of high-order side frequencies by normal channel filters. A useful reduction in residual error rate and improved diversity operation may be achieved if second-order sideband frequencies reach the output of the channel-receive filter at correct level. Some increase in channel spacing in relation to modulation rate would be necessary to restrict inter-channel interference.

Channel Spacing. Channel spacing and modulation index standardized for FM systems on landline circuits are not ideal for HF radio systems on account of selective fading. It is found advisable to work each system with a deviation designed to cover completely the available bandwidth with significant

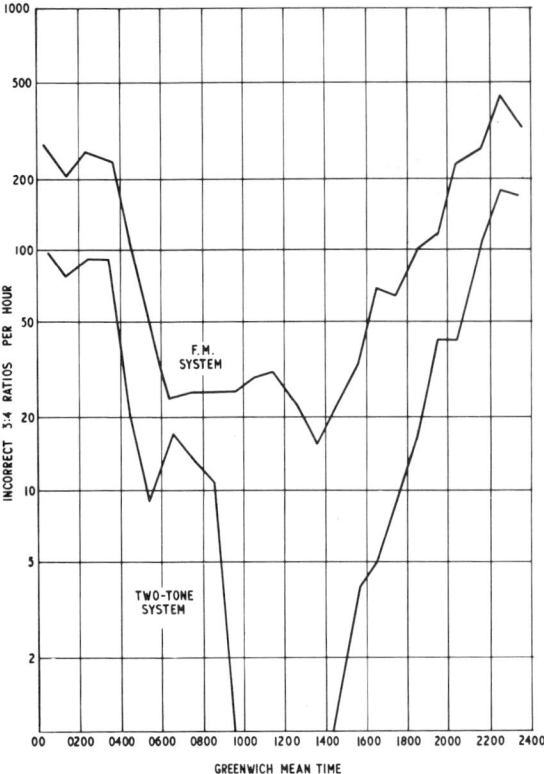

Fig 7.27 Comparative errors for simultaneous reception by FM and 2-tone assessor systems

Table 7.7 Centre frequencies for FSM VF systems with channel separation of 170 Hz and modulation index of approximately 0·8*

Channel	Centre frequency (Hz)	Channel	Centre frequency (Hz)
1	425	9	1785
2	595	10	1955
3	765	11	2125
4	935	12	2295
5	1105	13	2465
6	1275	14	2635
7	1445	15	2805
8	1615		

* Recommendation R39.

station and receive filter followed by limiter, discriminator and output stage at the receiving station. Lower channels may be derived by group modulation, for reasons outlined earlier. High frequency stability in the equipment is required since a frequency error of 1 Hz will introduce a bias distortion of 1·5%. A particular requirement of the receiver is rapid response to wide variations in signal level (e.g. ±15 dB) tolerated at the receiver. Arrangements also need to be made for modulator and detector to assume the space condition if control fails due to a fault. If these systems are extended over landline circuits, power levels quoted in Table 7.2 apply. Characteristic distortion at 100 bauds (measured isochronously) would be within 8%.

If dual frequency diversity is required it is usual to use a 12-channel FM system, coupling certain of the channels to provide a 6-channel dual-diversity system. The coupled channels are spaced at 680 Hz (i.e. 4 x 170 Hz) and joint channels so formed are distinguished by using reference letters instead of channel numbers. Diversity channels A–F are formed by pairing respectively channels 1 and 5, 2 and 6, 3 and 7, 4 and 8, 9 and 13, 10 and 14 (11 and 12 are not used for pairing).

Fig 7.28 shows a block diagram of the arrangement adopted by STC for their TF6 system in which channels 1–8 are formed from channels 18–11 by group modulation and demodulation, using a group-carrier frequency of 3740 Hz and selecting the lower sideband; channels 9, 10, 13 and 14 are directly modulated. At the sending end the input is applied to the pair of modulators in parallel.

At the receiving station a diversity-combining circuit is associated with each pair of channel detectors (Fig 7.29). Each diversity pair of channels is fed

sidebands otherwise, in the presence of interference, unoccupied bands at the ends of the channel bandwidth produce noise voltages which increase distortion.

For radio transmission at high frequencies in the 3–30 MHz band the CCIR has issued Recommendations concerning the main characteristics* for telegraph frequency-division multiplex equipment. The preferred frequencies are based upon the use of odd multiples of a basic frequency of 85 Hz, resulting in a range of centre frequencies spaced at 170 Hz; a modulation index of about 0·8 is suggested. These frequencies, listed in Table 7.7, permit up to 15 channels to be worked on an ISB radio emission with a bandwidth of 3 kHz; the frequency deviation of the telegraph channels is 42·5 Hz and the modulation index 42·4/50 = 0·85.

The design of an FSM system to this standard follows the general principle already described — oscillator modulator and send filter at the transmitting

*CCIR Report No. 199; also CCITT Recommendation R39.

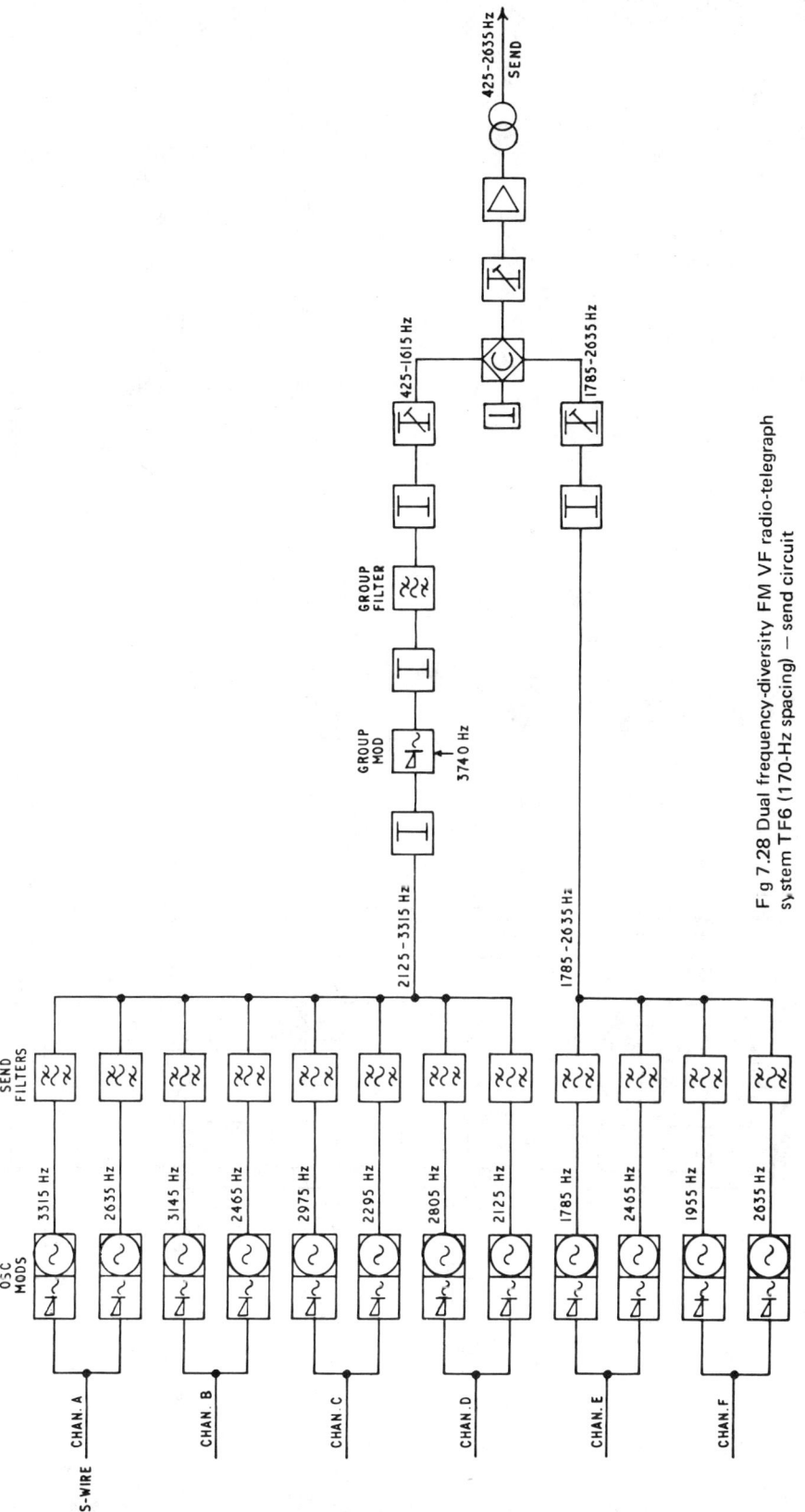

F g 7.28 Dual frequency-diversity FM VF radio-telegraph
system T F6 (170-Hz spacing) — send circuit

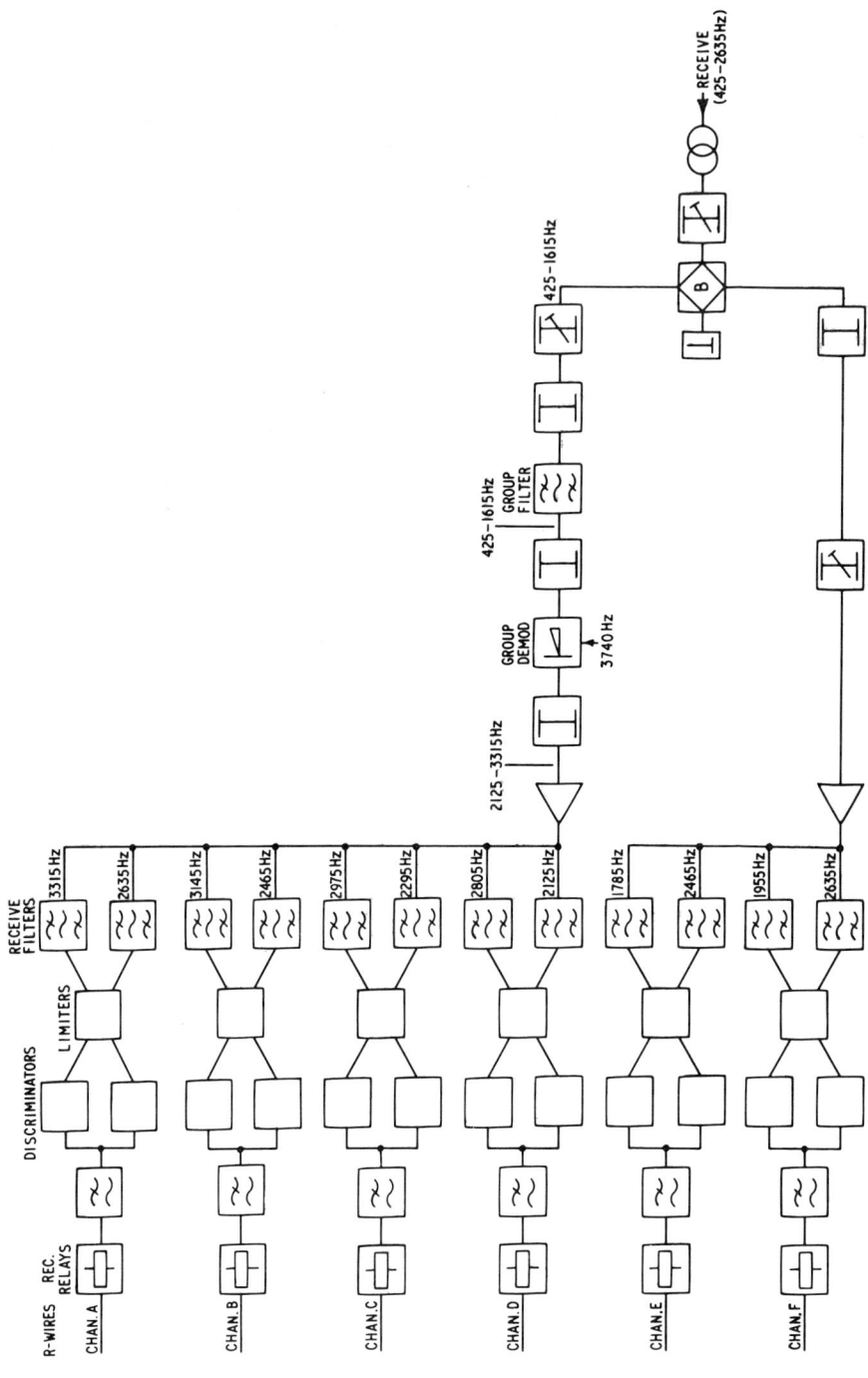

Fig 7.29 Dual frequency-diversity FM VF radio-telegraph system TF6 (170-Hz spacing) — receive circuit

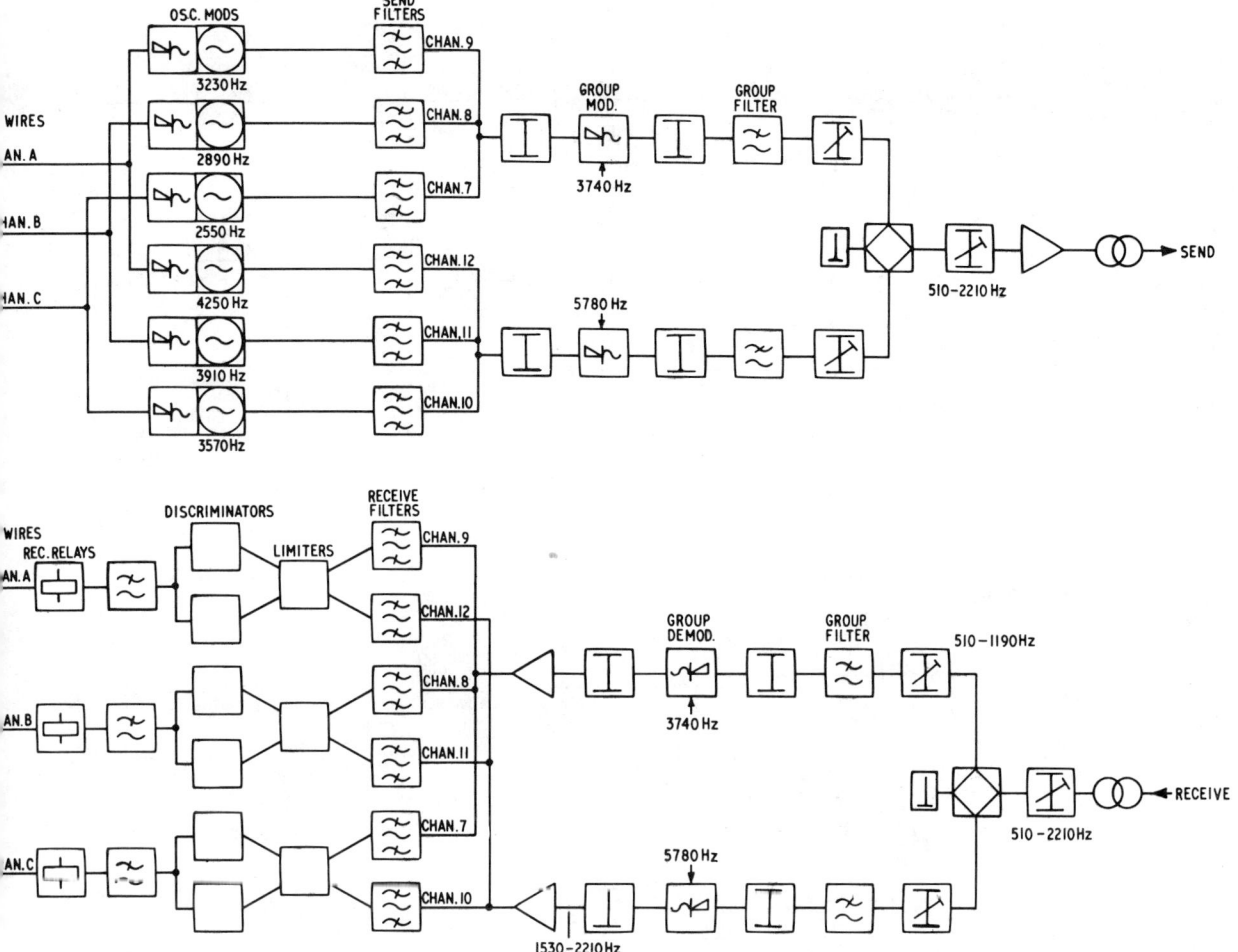

Fig 7.30 Dual frequency-diversity FM VF radio-telegraph system (340-Hz spacing) — Send and receive circuits

to a common limiter whose output is applied to two discriminators; outputs from discriminators are added in the receive-relay circuit. The effect of the common-limiter action is that the weaker signal makes a negligible contribution to the final output in the event of the signal from one diversity channel strongly predominating over the other. The phase of the output approximates to that of the stronger diversity signal, but if both signals are of nearly equal strength the phase of the output corresponds to the mean of the receive delay in the diversity paths. The receivers are designed to accept signals in the range −10 to −20 dB about the nominal level.

For higher modulation rates, up to 200 or 240 bauds, channels spaced at 340 Hz are used with a frequency deviation of ± 85 Hz, i.e. a modulation index of 0·85. Channel centre frequencies are $f_0 =$ 510, 850, 1190, 1530, 1870 and 2210 (and 2550) Hz.

In the STC 6-channel system TF4 (100 bauds), channels 1, 2 and 3 are obtained by group modulating channels 9, 8 and 7 (3230, 2890 and 2550 Hz) with a carrier frequency of 3740 Hz; channels 4, 5 and 6 are formed from group modulating channels 12, 11 and 10 (4250, 3910 and 3570 Hz) with a carrier frequency of 5780 Hz, the lower sideband being selected in each case.

For spaced-aerial-diversity reception the 6-channel system is available. If frequency diversity is used, three channels A, B and C are formed by paralleling channels 1 and 4, 2 and 5, 3 and 6 respectively — each diversity pair of frequencies being

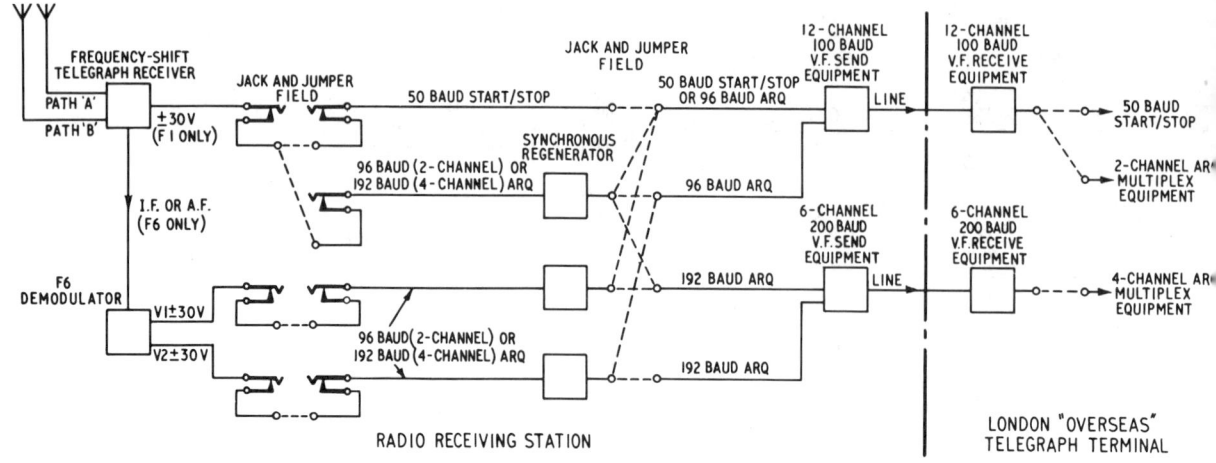

Fig 7.31 Radio-telegraph-receiving station — FS system

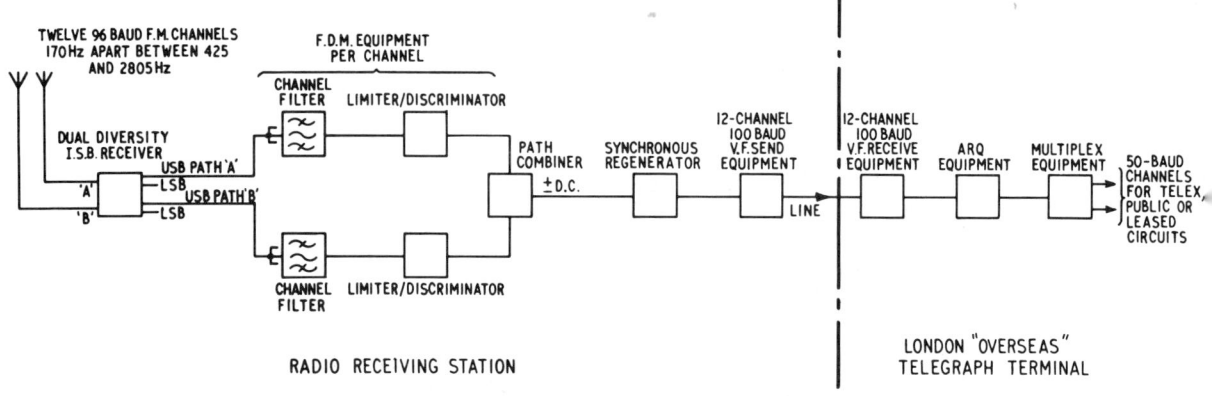

Fig 7.32 Radio-telegraph-receiving station — ISB system

separated by 1020 Hz. This layout is depicted in Fig 7.30 which shows the use of a common limiter for each pair of diversity channels at the receiver. The limiter output is applied to two discriminators whose outputs are added in the receive-relay circuit.

Figs 7.31 and 7.32 show outlines of two typical radio-telegraph-receiving systems. In Fig 7.31 for the FS system, the combination of signals from the A and B spaced-aerial diversity paths is made in the radio receiver. For 50-baud start—stop F1 channels (or F6 if two such channels are combined) the output of the receiver is connected directly to the input of a unidirectional MCVF system for transmission over a landline circuit to the central 'overseas' VF telegraph-terminal station. For synchronous F1 or F6 channels working at 96 or 192 bauds, signals are regenerated in a synchronous regenerative repeater

before retransmission over a landline circuit.

Fig 7.32 shows the arrangement for ISB MCVF systems. Audio-frequency outputs from two spaced-aerial diversity paths of the radio receiver contain similar channel-aggregate signals which are applied to the FDM equipment to separate the channels, demodulate them and combine the two diversity paths for transmission over the landline circuit, after synchronous regeneration. The MCVF equipments feeding into the landline circuits will be suitable for modulation rates of 100 or 200 bauds.

7.14 Bandwidth economy
Attention has been continuously directed throughout the history of telegraphy towards methods of deriving more circuits from a single bearer circuit for better

utilization of the transmission path. For HF radio services in particular, the limited number of radio channels that can be provided in the crowded spectrum and the ever-increasing traffic impose requirements of bandwidth economy. The choice and adoption of any such system is mainly a question of economic balance between relative costs of terminal equipment and of the transmission medium in all its aspects. Some methods which have received recent consideration are briefly discussed below.

Frequency-multiplex Systems. The standard MCVF narrow-band (120-Hz) system providing 24 teleprinter circuits each capable of at least 6·6 char/s from a 4-wire circuit with a bandwidth of about 3 kHz, may be already regarded as highly economical, not only in bandwidth but in equipment costs, power consumption, accommodation needs and maintenance cost.

The excellent performance of present-day MCVF systems with 120-Hz spaced channels enables standard channels to be exploited for use at modulation rates much higher than the 50-baud rate for which they were designed. The graph in Fig 7.33 shows the performance which can be expected from such a

channel and indicates that rates up to 100 bauds may be considered. In graph (a) all other channels are being modulated at the rate being measured; in (b) all channels are held at the upper characteristic frequency.

Bandwidth Limitation. If regeneration of signals at every junction of channels connected in tandem were an acceptable feature, much higher signal distortion per link could be tolerated. Channel spacing could be decreased to about 60 Hz for 50-baud operation permitting channel distortion due to narrow filters, and inter-channel interference to rise to a limit in the range 30–40% for a single channel. The advantage from doubling the number of channels would have to be weighed against the cost of supplying and maintaining regenerative repeaters.

Synchronous Systems. By dispensing with the need for start and stop signals, synchronous systems make more efficient use of bandwidth. They are not readily adaptable for general use by asynchronous instruments — the teleprinter — for which the VF channel provides a very flexible transmission medium.

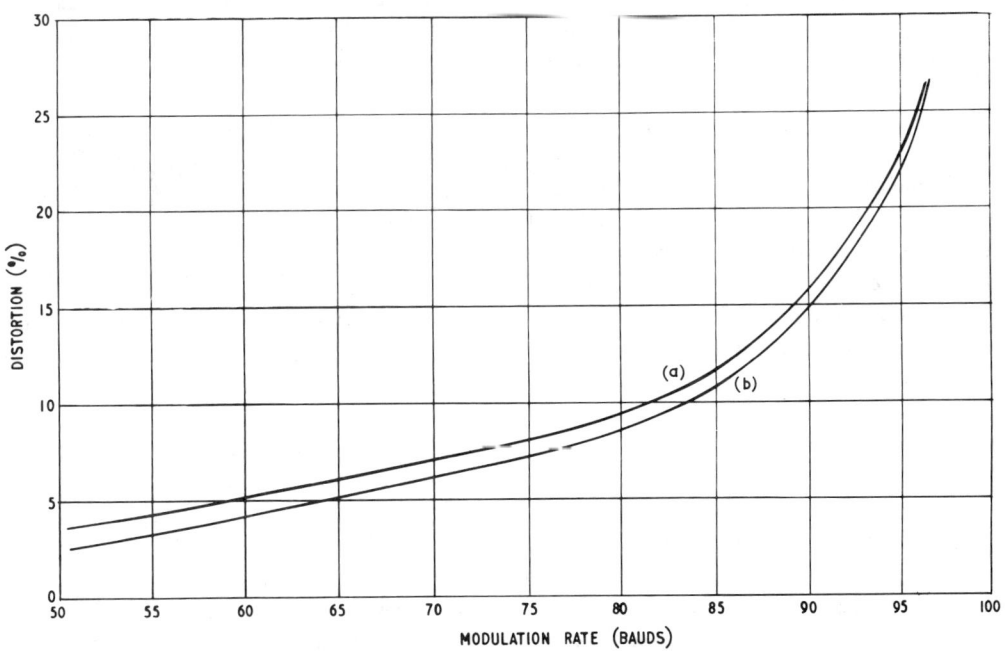

Fig 7.33 Performance of 120-Hz FM channel at modulation rates greater than 50 bauds

Combined FDM and TDM. Later designs of MCVF systems with 120-Hz spacing enable one VFT channel to carry a 2-channel 6-unit TDM system, operating at $82\frac{2}{7}$ bauds; or a 2-channel 7-unit TDM (ARQ) system at 96 bauds; or even a 3-channel 6-unit TDM system at $123\frac{3}{7}$ bauds by compensating characteristic distortion with the CDC unit described below. These methods of bandwidth conservation are widely exploited on HF radio circuits and on costly transoceanic cables where bandwidth needs to be used with utmost efficiency.

Duobinary and CDC have similar standards of performance at 125 bauds and are equally degraded by the presence of adjacent-channel interference — equivalent to a reduction of signal/noise ratio of 2–3 dB. The duobinary system requires additional equipment at both ends of the transmission channel whereas CDC requires about the same total quantity of equipment at the receive end only.

Channel capacities, in terms of 50-baud start—stop 5-unit teleprinters, of some possible systems are summarized in Table 7.8. The different methods of

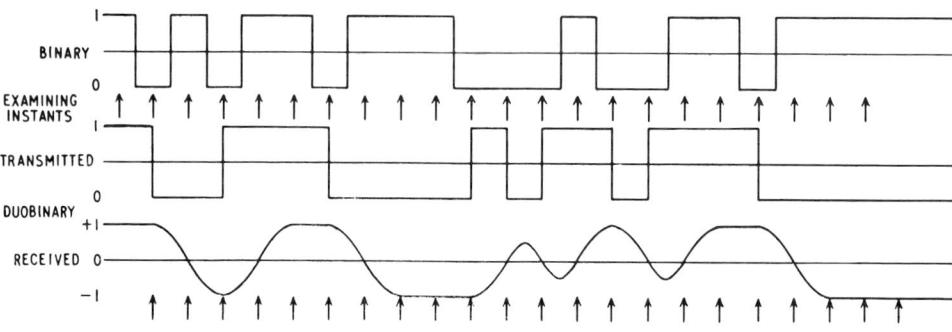

Fig 7.34 Duobinary coding

Duobinary Transmission. Duobinary transmission involves a form of coding with the object of offering an increase in modulation rate of about 50% relative to normal binary transmission in a given bandwidth. Of the two forms of duobinary transmission which have been published, the more significant is briefly described below.

Using the binary symbols 0 and 1, the duobinary signal is a differentially-coded binary signal derived from the original binary signal by *changing its condition if a 0 occurs but not changing it if a 1 occurs*. An example of duobinary coding is shown in Fig 7.34. This shows that for reversals the number of transitions in a given period and hence the modulation rate is halved. The performance of the system at high speed results from the method of detection.

At the input to the transmission circuit a scale-of-two counter operates under control of clock pulses which are inhibited when the input is 1. In the receiver, the detector has two thresholds so that +1 and −1 become 1 and the zero value becomes 0 when the waveform is examined at appropriate instants.

modulation and combination will have different degrees of susceptibility to noise, and the tolerable level of uniform—spectrum random noise in the bearer circuit (nominally 300–3400 Hz) for a basic error rate of 10^{-9} when the aggregate mean power of all the carriers is −10 dBm0, is estimated. For comparison, the possible teleprinter-circuit capacities which might result from time division over the full 3 kHz band are also given.

7.15 Characteristic distortion compensation
If a means were provided for separately detecting all *single* elements of telegraph-code signals, remaining elements would have a duration of at least two unit intervals and the modulation rate would be virtually halved. This would enable, for example, 100-baud signals to be transmitted through a channel capable of only 50-baud transmission. A 'suppressed-singles' telegraph system was used many years ago in which single elements were re-inserted artificially by the receiving equipment, from recognition of a third condition midway between the levels of mark and

space signals. The receipt of such a condition signified that the element to be inserted — a single element which had insufficient time to build up to normal level — was one of opposite sign to that last received.

This principle is employed in revised form in the characteristic distortion compensation (CDC) system which enables signals at a modulation rate of 96 or $123\frac{3}{7}$ bauds to be detected after transmission over a

In an MCVF channel, build-up time is approximately the inverse of the modulation rate. At 50 bauds the build-up time of a signal transition is almost 20 ms; maximum amplitude is just accomplished within the duration of a unit element. As modulation rate increases and the period of the unit interval decreases, characteristic distortion rises and the amplitude of a unit element at the discriminator

Table 7.8 Comparison of telegraph systems for bandwidth economy

System (Aggregate power −10 dBm0	No. of FDM circuits	No. of teleprinter circuits	USR noise (dBm0 in 3 kHz)
1 Conventional 24-circuit AM	24	24	−43
2 Conventional 24-circuit FM	24	24	−37
3 24-circuit (60-Hz spaced) AM with regenerative repeaters	48	48	−43
4 24-circuit (60-Hz spaced) FM with regenerative repeaters	48	48	−37
5 24-circuit (120 Hz) FM with 2-circuit TDM	24	47	−25
6 24-circuit (120-Hz) FM with 3-circuit TDM and CDC	24	71	−32
7 Kineplex-type (49 Hz) with 2-phase modulation	58	114	−26
8 Kineplex-type (49 Hz) with 3-phase modulation	58	171	−32
9 4-phase or vestigial sideband-suppressed carrier	1	57	−30
10 Vestigial sideband suppressed carrier, 8-level	1	174	−42

Notes:
1 Start—stop performance — distortion not greater than 10%.
2 Synchronous performance — element error rate $1/10^6$.

nominal 50-baud channel in an intercontinental submarine cable. By this means, 2-channel 7-unit ARQ systems (96 bauds) or 3-channel 6-unit time-division multiplex systems ($123\frac{3}{7}$ bauds) can be operated over a channel nominally designed for a 50-baud single transmission, so increasing the channel capacity by a factor of two or three.

The purpose of using an ARQ system on such a channel is not primarily for signal protection but for more efficient bandwidth utilization. It also enables half-rate and quarter-rate (200 and 100 char/min, both at 50 bauds) leased circuits to be provided at reduced tariffs over long transoceanic circuits. Moreover, it provides a ready means of standby against emergency should a section of submarine cable become faulty; during the emergency the ARQ system can more readily be temporarily re-routed over an HF radio system.

output in an FM receiver becomes greatly reduced. At a limiting value of about 120 bauds the discriminator-output signal has insufficient amplitude to cross the detection level and single elements cannot be accurately restituted. At modulation rates up to about 90 bauds, signal distortion can be corrected by regeneration, but as signal amplitude is very low the detector is more susceptible to noise interference. The CDC unit provides the regeneration required to eliminate telegraph distortion and, in addition, it produces a compensation signal which, when fed back into the discriminator, adds to the amplitude of the single element and raises it above the detection level. Between transitions, the feedback must be of opposite polarity to the detected signal so that it can anticipate a single element of opposite condition. Immediately a change of condition is detected the condition of the compensation signal is changed. The

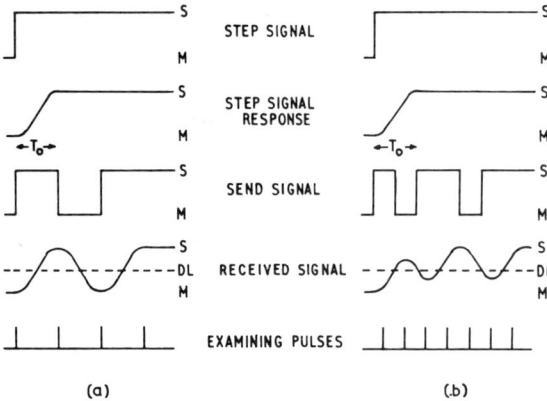

Fig 7.35 Signal transmission at (a) normal rate, (b) higher modulation rate

system necessarily entails the use of synchronous operation.

The build up of current resulting from an abrupt or step change in significant condition is shown in Fig 7.35(a) together with a typical received waveform at 'normal' modulation rate in which each signal element attains full steady-state amplitude. If modulation rate is considerably increased (Fig 7.35 (b)) the amplitude of the shortest (single) element is materially reduced. The essential problem at the receiver is the determination, at uniform time intervals, of whether the signal is nearer mark or space condition, this being effected by recording whether the signal lies above or below the decision level DL midway between the two steady-state mark and space levels. The phase of the periodic signal-examination instants is arranged so that they coincide with the peak values of the shortest elements.

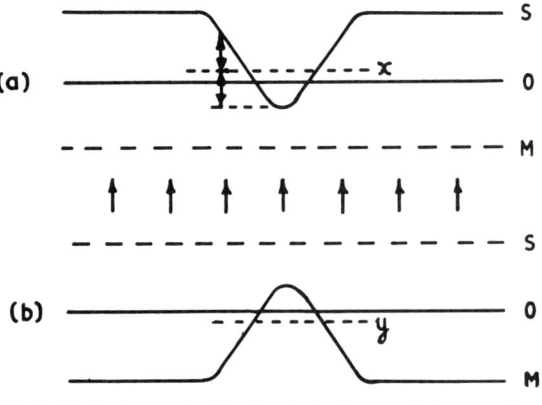

Fig 7.36 Optimum decision levels for isolated high-speed signal elements

In Fig 7.35(a) the margin against mis-interpretation (e.g. the amount of perturbation by interference which could be withstood) is the same for all units of the signal no matter where they lie in a sequence, but when the modulation rate is materially increased as shown in Fig 7.35(b) this margin is greatly reduced for some signal units, depending upon their position in the sequence. Since the reliability of the signal depends on the weakest element, security of reception at the higher speed can be improved by increasing the margin for interpretation of the signal units of smaller amplitude even at some expense of the margin for those of larger amplitude.

This may be illustrated by considering a modulation which includes a single element as in Fig 7.36; x and y (midway in the rise time) are better decision levels than the neutral (zero) level for distinguishing

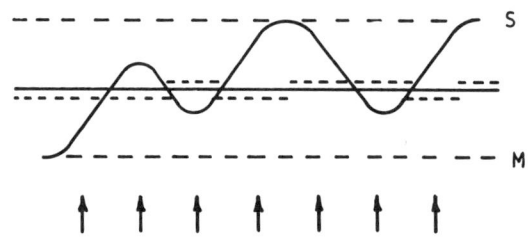

Fig 7.37 Reception with two decision levels

the isolated unit. There should be two levels of decision, choice of one or the other depending on the sequence of mark and space elements. It is sufficient, in the simple example considered, to choose x or y according to whether the preceding unit-signal element has been interpreted as space or mark. Fig 7.37 shows the signal waveform of Fig 7.35(b) repeated with the x and y decision levels superimposed. The effective minimum margin of discrimination is now considerably enhanced. Even if the received-signal waveform should exhibit no modulations at all resulting from mark—space reversals at maximum modulation rate, the alternation of the decision level effectively re-inserts them at the receiver. Since this compensating waveform is dependent upon the last interpretation of the signal, its modulation will follow that of the regenerated output of the system. The compensating signal can be directly derived from the system output, amplitude and absolute level being suitably adjusted to match the received signal.

The compensating process is equivalent to subtracting a fraction of the regenerated signal from the incoming signal; the same end could be achieved by inverting the regenerated signal, transforming the level and adding it to the received signal before presentation to a restituting device with a fixed decision level. Although the minimum margin for discrimination is improved, the margin is not constant for all signal units, but is influenced by the following unit in the sequence and may be influenced by more of the preceding elements than the immediate one.

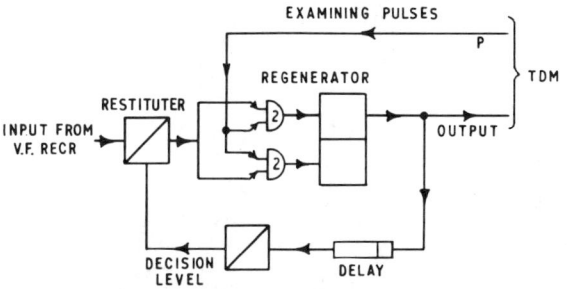

Fig 7.38 Characteristic distortion compensator

The logic diagram of the basic characteristic distortion compensator is shown in Fig 7.38. The restituter is part of the normal channel termination (e.g. the receiving relay) and its output controls a regenerator (which may be part of a multiplex synchronous receiving device). At signal-examining instants, a pulse P energizes the gates so that the binary stage is set in accordance with the input signal. The output from the binary stage is the regenerated signal which is fed back after a short delay via a level-converting device to provide the compensating-decision level in the restituter (i.e. the equivalent of bias current for a receive relay).

In synchronous systems, examining instants are maintained in correct phase with respect to the signal by comparing the instants of restitution with the examining-pulse instants, using the mean error to control the receiving timing source. Consideration of Fig 7.37 shows that when the incoming signal comprises alternate single mark and space elements, instants of restitution will be considerably influenced by the local timing source itself. Measures must be taken to reject these in the synchronizing system and to accept only those which are produced by the incoming signal following two or more consecutive units of a similar nature.

Application of CDC depends on fore-knowledge of the waveforms and particularly the relative amplitudes of elements of various duration. The compensating signal must be pre-determined, so it is essential that the signal to be compensated is stable. It is very desirable that the telegraph-transmission system should not be influenced by normal variations in line attenuation, or else adequately compensated against them. FM telegraph channels are eminently suitable provided that the amplitude limiters in the channel receivers compress the level variations sufficiently (±0·5 dB for the regulated-level range) and that the system is compensated or protected against frequency-translation errors of bearer circuits. With AM channels, the usual type of AGC circuit is likely to present difficulties for CDC. Transient reponse of the channel as observed in the demodulated signal before restitution is all-important: it should exhibit a smooth transition from one significant condition to the other without appreciable oscillation, according to the characteristics of Fig 7.39.

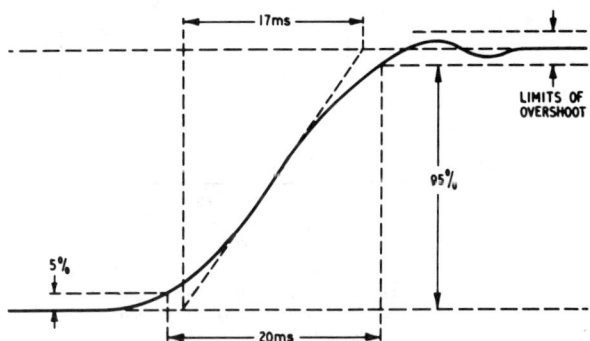

Fig 7.39 Transient-response characteristic required for CDC

Timing of the regeneration is controlled by a pulse at element rate supplied by the TDM equipment. Since propagation time may vary slightly from one VFT channel to another, an adjustable delay circuit is provided to enable the regenerator-examining pulses to be off-set accordingly and permit control of the TDM equipments on a number of channels from the timing source. The restituted signal from the VFT channel, which has been produced by comparison of the demodulated signal against the decision level set by the compensating signal according to the preceding regenerated element, is examined in the input gates of the regenerator circuit by the element-timing pulse P and the output circuit is set accordingly. The output circuit feeds the regenerated signal to the TDM equipment and also the feedback circuit which, after a short delay caused by the receive relay, produces the compensating signal.

To ensure that the synchronizing waveform passed to the TDM equipment has only those transitions of the restituted signal which reliably indicate the phase

with an ideal system having an error-probability law: $P_e = \frac{1}{2} \exp(-R)$, where R is signal/noise power ratio in the 80-baud bandwidth of the 120-Hz spaced channels. With such an increase in modulations rate there is a penalty to pay as regards the tolerable

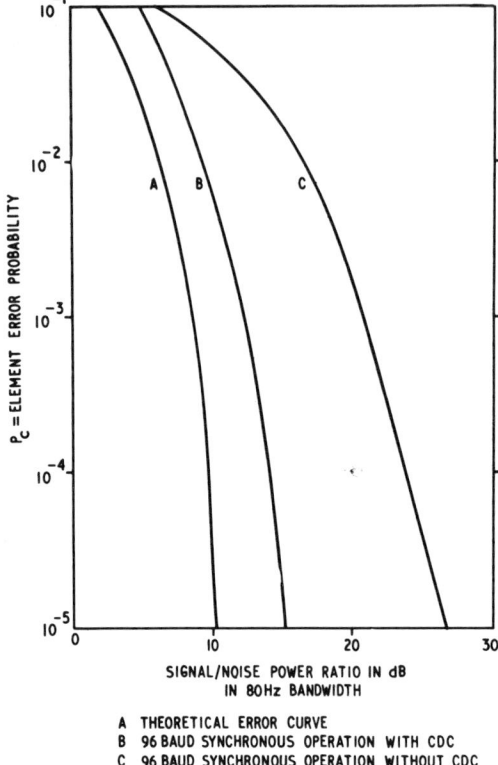

Fig 7.40 Performance of 96-baud synchronous channel in URS noise

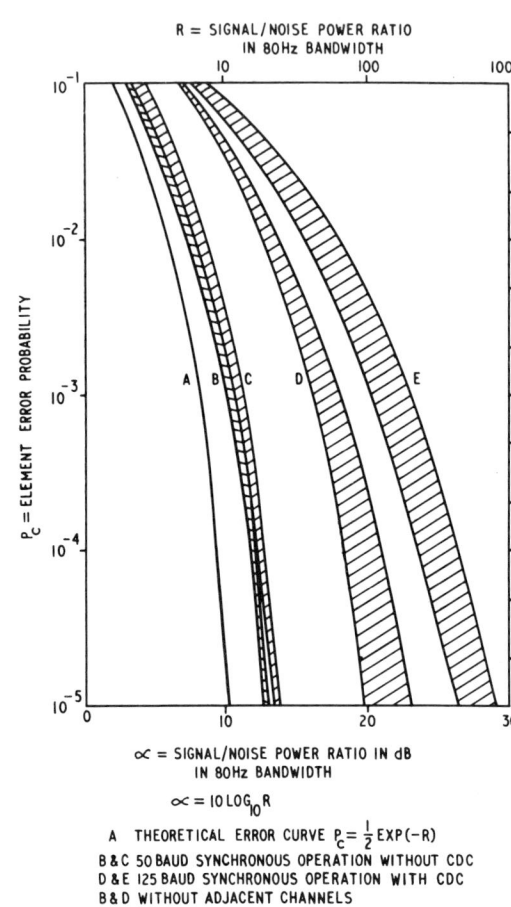

Fig 7.41 Performance of 50- and 125-baud synchronous channels in URS noise

of the received-signal modulation, circuits are provided to reject the transitions following others at unit interval and to produce a synchronizing signal from those which are preceded by at least two units of the same polarity; these circuits are known as the synchronization separator. A variable-delay circuit is included so that a predetermined phase shift may be inserted to suit the requirements of the TDM equipment.

In Figs 7.40 and 7.41 the performance of the CDC channels at 96 and 125 bauds (with and without inter-channel interference) is compared with the same channels operating synchronously, and also

circuit-noise level; the measure of this is shown by the displacement of any curve from the ideal.

Although synchronous operation at 96 bauds on channels spaced at 120 Hz is possible without CDC, the demodulated waveforms of single mark and space elements reach only 0·5–0·63 of the steady-state excursion. Use of CDC gives an additional 10-dB immunity against effects of noise at this modulation rate.

An advantage of the CDC unit is that it can readily be added to a suitable FDM system; slight modifications to an existing VFT channel equipment may be necessary at the receiver, but no changes are required to the send equipment.

For successful CDC operation the performance of the VFT channel should be such that the steady mark- or space-signal output from the post-discriminator filter should not vary by more than 5% of the mark- space difference when the line attenuation changes smoothly over the regulated level range. The transient response should be such that the build-up (or decay) time at the mid-amplitude rate (see Fig 7.39) should be less than 17 ms; and less than 20 ms from 5–95% of steady-state amplitude. Overshoot in the wave- form should not exceed 5% of the mark–space level difference. In a well-designed FSM MCVF equipment no difficulty would be encountered in meeting these requirements.

7.16 Speech-plus-telegraph circuit
A leased circuit is sometimes required to provide both speech and teleprinter facilities. This can be met by a simple change-over switch giving alternative use of the line to the telephone or to the tele- printer; VF transmission is used for the latter in

order that signals may pass through line amplifiers and the transmission bridge of a private telephone exchange (PBX). The fact that the circuit may often be found engaged when needed for the alternative form of telecommunication has led to the provision of facilities giving *simultaneous* telephone and tele- printer service over a single circuit. Two disym- metrical narrow-band (120-Hz-spaced) teleprinter channels are provided and, although normally used for two-way simplex working, the circuit is often known as a speech-plus-duplex (S + D) circuit.

To derive a simultaneous telegraph circuit from the telephone circuit, a band-stop filter is used in conjunction with a pair of band-pass filters; the latter can be the normal MCVF send and receive fil- ters. By this arrangement, use of the frequency range 1600–2000 Hz (approximately) is withdrawn from the original telephone circuit to be made available for telegraph purposes. These telegraph frequencies are selected in order to avoid frequency bands below 1000 Hz which are used for VF supervisory signalling on long-distance telephone circuits; and also to avoid the range 1200–1400 Hz where effects of cross-talk are more severe. On older line plant, heavy loading

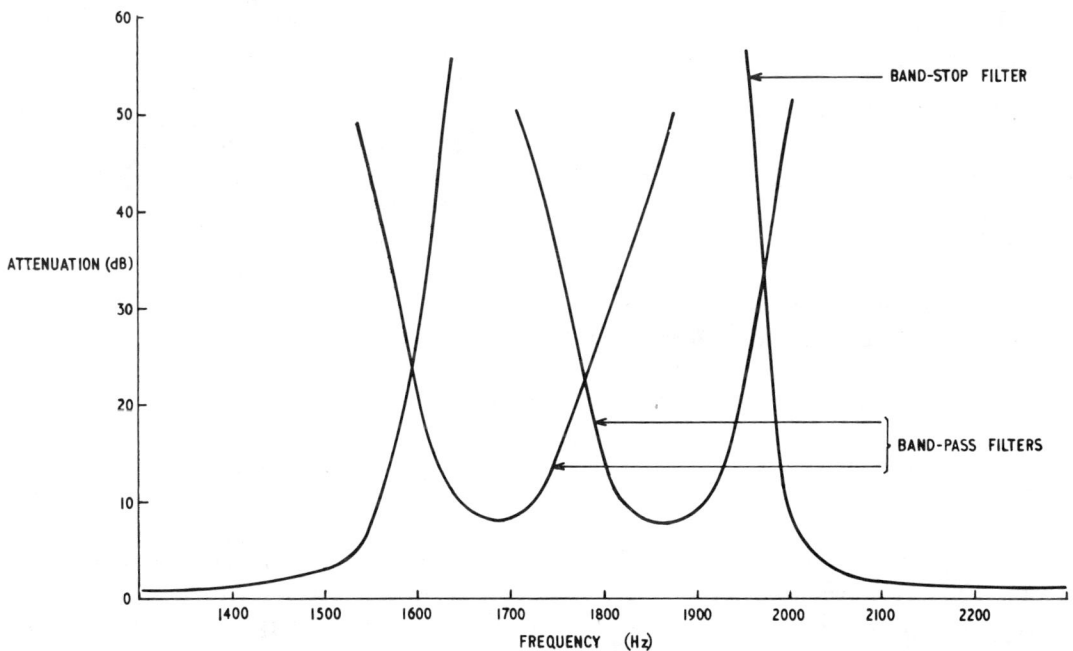

Fig 7.42 Filter attenuation/frequency characteristic for simultaneous speech and telegraph circuit

(with inductance coils) may be present which prevents useful transmission of any frequencies above 2000 Hz. The centre frequencies used for the telegraph circuit* are 1680 and 1860 Hz (two frequencies from the MCVF series) to provide a telegraph channel in each direction. Since the telegraph frequencies used for transmission and reception are inverse at the two stations, these must be designated as A and B stations for identification purposes. The telegraph bandwidth allocation is also chosen so as not to cause excessive degradation of speech quality in the telephone circuit; the main speech energy is concentrated at the lower frequencies and the loss in quality of speech reproduction will mainly affect the harmonic frequencies and consequently articulation.

Typical attenuation/frequency characteristics for the set of filters are shown in Fig 7.42.

A block diagram for the dual circuit is shown in Fig 7.43. For the telegraph circuit FSM is used with a frequency deviation of ±30 Hz — the lower frequency for the mark condition. The general principles of the FSM circuits are as already described. The power level for the telegraph channel is −13 dBm, measured at the send-line terminals. Unlike the MCVF systems, which operate over +8 dB channels, the attenuation of the line circuit may be as much as 30 dB and the telegraph receiver must be designed for operation over a wide range of input levels which may be as low as −43 dBm. Signal distortion in the telegraph channel could be in the range 2–5% at 50 bauds, rising a little at the lowest input levels. Within the available bandwidth, higher modulation rates than 50 bauds could be used, with some increase in signal distortion.

The use of supervisory signals on the telephone circuit needs careful consideration in order to avoid interference to the teleprinter circuit. The normal 500/20-Hz telephone ringers cannot be used in the usual way since this arrangement disconnects the return channel and applies 17-Hz ringing current towards the 2-wire circuit. A modified ringer arrangement maintains continuity of the 4-wire circuit during ringing and applies 17-Hz ringing current only into the telephone termination. A fourth harmonic from the 500/20-Hz ringer supply (4 x 480 = 1920) falls within the pass range of the band-pass receiver filter at the 'B' station and it is

necessary to eliminate this harmonic frequency which would otherwise cause interference with teleprinter reception every time ringing occurred on the speech circuit. Care is also necessary to ensure that line impedance is not changed by application of the 500/20-Hz ringing-generator supply to the 'go' channel, otherwise a sudden change in level of telegraph signal could occur. Dialling from the telephone circuit will also cause interference with the teleprinter circuit unless the dialling circuit is arranged to avoid this.

When a telephone circuit of standardized bandwidth (nominally 300–3400 Hz) is provided, it is possible to increase the number of VF telegraph circuits up to ten at 120-Hz spacing or, alternatively, five at 240-Hz spacing, leaving the band 300–2000 Hz for speech. Other alternative arrangements include a speech band of 300–2400 Hz plus either six VF telegraph circuits spaced at 120 Hz or three spaced at 240 Hz; or a speech band of 300–2600 Hz plus five VF telegraph circuits spaced at 120 Hz (or two spaced at 240 and one at 120 Hz). In fact, pairs of 120-Hz spaced VF telegraph-circuit cards can be replaced by 240-Hz-spaced circuit cards. The VFT centre frequencies (and deviations) are those standardized for channels 15–24 (120-Hz spacing) or channels 8–12 (240-Hz spacing).

The wider speech bands may be used either for speech with data communication at speeds up to 2400 bits/s, using suitable modems, or for phototelegraphy. A block diagram of such an arrangement is shown in Fig 7.43. The speech, phototelegraph or data-input signal is passed through a limiter which ensures that the power level will not exceed −10 dBm at the line terminals. The bandwidth is limited by the low-pass filter and the attenuator and amplifier enable the correct output level to be achieved. Telegraph input signals (from one or more send channels) are combined with speech, data or phototelegraph signal in a hybrid termination and fed to the go line.

On the receive side the combined signal is distributed by a hybrid termination. Each telegraph receive filter selects the appropriate signal for demodulation. Mercury-wetted relays feed the ±80-V output signals. In the speech/data/phototelegraph receive line, telegraph-channel frequencies are rejected by the low-pass filter and an attenuator and amplifier enable the correct signal level to be obtained. Ringing/signalling facilities for the telephone circuit are provided according to requirements.

*Recommendation R43.

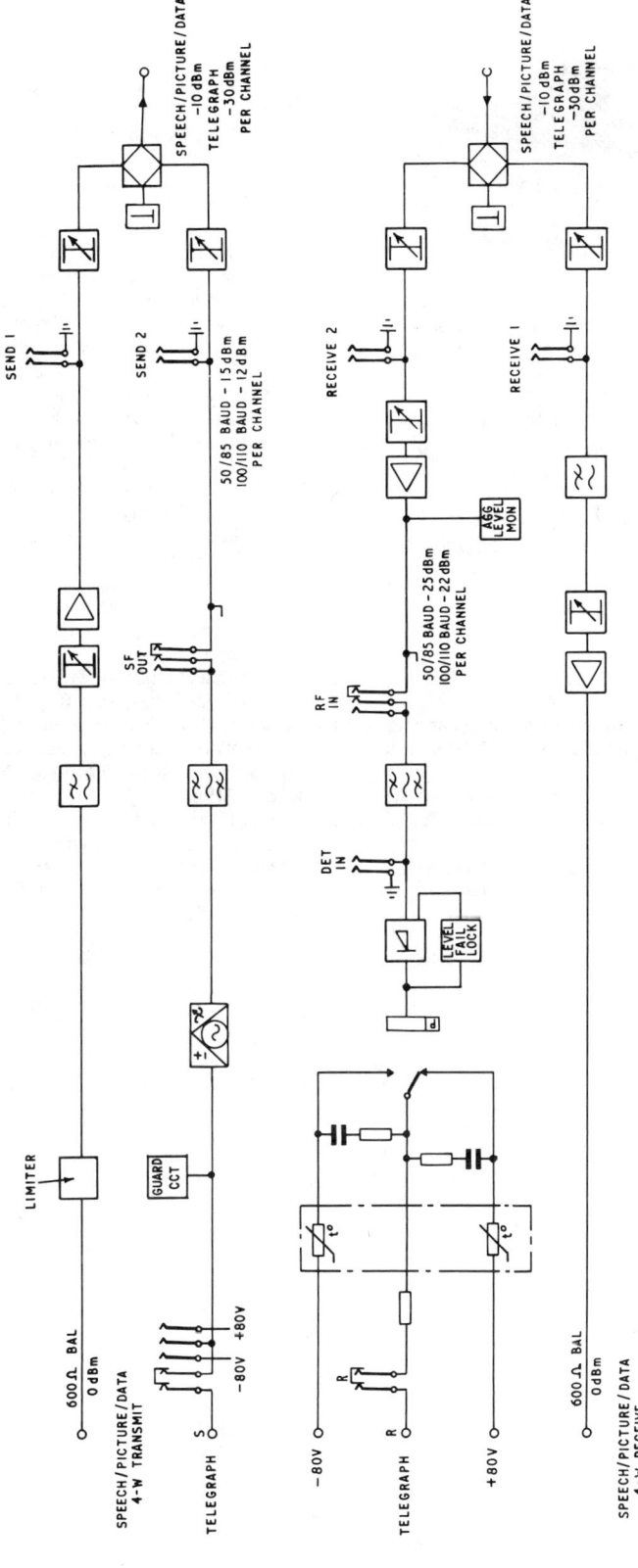

Fig 7.43 Speech/data/phototelegraph circuit plus VFT telegraph circuit

Fig 7.44 Equipment for speech/data/phototelegraph plus telegraph circuit (*Courtesy of Telephone Mfg Co. Ltd*)

Fig 7.44 illustrates a speech and telegraph equipment providing three VFT circuits using 62-type equipment. From left to right are the three VFT circuit cards, a telephone signalling circuit, 2-wire/4-wire hybrid termination, test and miscellaneous card and a mains-power unit.

References

1 Chittleburgh, W. F. S., 'Some modern developments in telegraph transmission equipment', *Electrical Communication*, **35**, p. 230 (1959).
2 Chittleburgh, W. F. S., Green, D. and Heywood, W. A., 'A frequency-modulated voice-frequency telegraph system', *POEE Journal,* **50**, p. 69 (1957).
3 Croisdale, A. C. and Harris, E. T. C., 'Transatlantic cable: special equipment – telegraph channels' *ibid.* **49**, p. 443 (1957).
4 Sallis, R. T. G., 'A multi-channel voice-frequency telegraph system using transistors', *ibid.* **57**, p. 95 (1964).
5 Hunt, C. S., 'A wideband multi-channel voice-frequency telegraph system', *ibid.* **60**, p. 10 (1967).
6 Ridout, F. N. and Wheeler, L. K., (i) 'The choice of multi-channel telegraph systems for use on HF radio links', (ii) 'A comparison of binary signalling methods', *Proc IEE*, **110**, No. 8, p. 1402 (1963).
7 Bronsdon, E. G., 'The new Brearley radio receiving station', *POEE Journal*, **61**, p. 75 (1968).
8 Watt-carter, D. E. *et al.*, 'The new Leafield radio transmitting station', *ibid.* **59**, p. 130 *et seq.* (1966).
9 Cook, A and Hall, L. L., 'The Rugby-B HF transmitting station', *ibid.* **50**, p. 15 (1957).
10 Bray, W. J. and Morris, D. W., 'Single-sideband multi-channel operation of short-wave point-to-point radio links', *ibid.* **45**, p. 97 (1952).
11 Allnatt, J. W., Jones, E. D. J. and Law, H. B., 'Frequency-diversity in the reception of selectively-fading binary frequency-modulated systems', *Proc. IEE,* **104**, part B, p. 98 (1957).
12 Law, H. B., 'The signal/noise performance-rating of receivers for long-distance synchronous radio-telegraph systems using frequency modulation', *ibid.* **104**, Part B, p. 124 (1957).
13 –, 'The detectability of fading radio-telegraph signals in noise', *ibid.* **104**, Part B, p. 130 (1957).
14 Hayward, H. L., 'Comparison of 170-Hz spaced FM and 2-tone voice-frequency telegraph systems on a practical HF radio circuit', *Point-to-point Telecommunications,* **8**, p. 5 (1964).
15 Groves, K. and Ridout, P. N., 'Effect of multipath propagation on the performance of narrow-band radio-telegraph systems', *IEE Paper* 5120E.
16 Walkden, M. R., 'Construction Practice for Transmission Equipment', *POEE Journal*, **66**, p. 25 (1973).

8 Time-division Multiplex Systems

The first part of this chapter deals with the ARQ type of error-correcting synchronous multiplex telegraph system, associated with the name of Dr H. C. A. van Duuren. The system operates in the synchronous mode, with multiplex facilities, and uses an error-detecting code from which a detected error results in an automatic call for repetition of transmission.

The system is used on HF radio circuits where error rate at times of poor propagation would otherwise be uneconomically high. For telex service it is essential for subscribers to have error-free copy using normal operating procedure.

For leased circuits the ARQ system confers great benefits in operation, avoiding the need for users to be frequently calling for repetitions of characters which appear to be mutilated. It enables a circuit to be subdivided for time sharing between two, three or four renters and puts long-distance teleprinter circuits at their disposal at reduced tariffs even though at reduced character rate. The system is essential for message-relay systems in which the originator of a message and the addressee are not in direct teleprinter communication on a 'conversation' basis.

Though circuits of the general overseas telegram service have been, and could be, operated by trained telegraphists without use of the ARQ system, there is no doubt that its use reduces operating costs and increases efficiency of circuit utilization by calling for repetitions with automatic processes which can take place much more rapidly than by comparable manual methods. For gentex service, HF radio circuits could not be used for 'dialling' connections without the security of the ARQ system.

In the latter part of the chapter, multiplex systems used for bandwidth economy on VF telegraph channels, used in transoceanic cables, are described.

8.1 Principles of synchronous systems

In a synchronous system, transmitting and receiving equipments operate continuously — ideally at identical speeds and with constant phase relationship. This state cannot be achieved without some complexity in equipment which, though well justified on expensive routes where the economics of multiplexing can be exploited, cannot bear comparison with the simple effectiveness of the start–stop system for normal subscriber-station use.

When 5-unit start–stop signals are to be sent over a synchronous system there is no purpose in transmitting start and stop signals — 5-unit code signals can be sent instead of 7·5-unit start–stop signals, with consequent economy in circuit time or in bandwidth.

In addition to greater code efficiency, there are other important advantages in using a synchronous system when compared with a start–stop system. In a 50-baud start–stop system, each character received is timed from the leading edge of the start element, and examination of the polarity of code elements takes place ideally at the midpoint of each code element, i.e. at $t = 30, 50, 70 \ldots$ ms counting from $t = 0$ for the leading edge of the start signal. Since this leading edge may itself become distorted, as much as any other transition in the character signal, during transmission the examining instants may not occur at the midpoints of code elements but are delayed or advanced according to the distortion of the start signal. In a start–stop receiver the maximum receive margin (e.g. ±40%) of any signal element must in fact be shared with that of the start element. In contrast, a synchronous receiver is timed by the average of many transitions so that the examining instant takes up an optimum position which always occurs nominally at the element midpoint; an element can tolerate distortion up to a maximum value which approaches ±50% — much greater than the effective margin of a start–stop receiver. A further advantage of the synchronous system is that it cannot readily be thrown out of synchronism as a result of excessive distortion. In the start—stop system, poor radio-propagation conditions could cause mutilation of the stop signal and result in signals from an automatic transmitter remaining out of synchronism for several characters. Another advantage in a synchronous system is that all signals are inherently regenerated at the output of the system. Compared with a start—stop system, a synchronous system will withstand a worse level of noise for a given error rate.

Synchronism is achieved by operating transmitter and receiver from highly stable oscillators and integrating incoming signal elements to determine their average phase. An ARQ system using 7-unit code with alphabet No. 3 requires a duplex circuit so that requests for repetition can be sent over the backward channel; for this system transmitters and receivers on both channels need to be in a locked phase relationship. The two stations connected are, for control purposes, designated *master* and *slave*

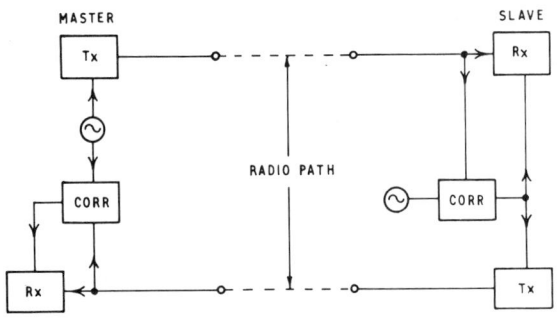

Fig 8.1 Principle of synchronous control on duplex circuit

stations (Fig 8.1). At the master station the transmitter is controlled entirely from the local oscillator; the receiver at the master station is *driven* by the same local oscillator but it is *corrected* from incoming signal transitions. At the slave station, receiver and transmitter are locked together via a device which can introduce an adjustable delay up to 14 signal elements into the transmitter timing compared to the receiver; both receiver and transmitter are driven by the local (slave) oscillator but jointly corrected from incoming signal transitions.

It is important that the instant for transmitting a character shall not coincide with identification of a received character, which could be the request-for-repetition RQ signal. Otherwise there would be difficulty in determining whether to transmit the next character or inject an RQ signal; this could result in lost or duplicated characters. This danger period is avoided by the time delay inserted between operations of the receiver and transmitter at the slave station. The ideal setting for this delay is such that at the master station the receiver is half a multiplex cycle (7 aggregate elements on a 2-circuit system) behind its transmitter. A 14-position switch is used at the slave station to set the number of elements by which the transmitter lags the receiver; at the master station an indication of the relative receiver—transmitter delay is given as an element count.

In telegraph systems, synchronism and phase have often to be considered separately. The term *element synchronism* is used to denote *the condition in which an element of the local timing coincides completely with an element of the received signal.* The oscillators at transmitter and receiver cannot run at precisely the same frequency; eventually they are bound to drift slightly out of synchronism. A synchronous system must transmit signals continuously; to effect synchronism, pulses generated at element rate by the local oscillator are compared with incoming transitions and (unless coincident) the local pulses are assessed as early or late. Timing differences will also arise from occurrence of small changes in propagation time of the radio path, and from other sources of fortuitous distortion. It is undesirable to correct or synchronize to every early or late transition and it is usual to employ an electronic counter which will integrate over about eight elements, either early or late, or preferably to use a differential counter which will reset to the centre position when a synchronizing step is taken after an excess of about eight early (or late) transitions has been registered. When a synchronizing step is taken the amount of correction is normally 1% of an element.

Under certain conditions of poor radio propagation it is desirable for the synchronizing device to be temporarily switched off automatically. This requirement could occur at a master or slave station which is receiving errors on all channels and it is then necessary to prevent a receiver from driving itself out of a good synchronous (and also eventually, phase) position when receiving random noise, or possibly when receiving via a synchronous regenerative repeater which may be off-speed. This requirement could also arise at a master station which is receiving errors on all channels, in which case it is desirable to prevent the master receiver from driving out of a good synchronous position by following a slave transmitter operating as described above.

To reduce unnecessary re-synchronizing and re-phasing operations after the radio path has failed for a prolonged period, the CCITT recommends* that control oscillators should be stable within 1 part in $\pm 10^6$. To examine what this means in practical terms, consider a synchronous system operating at 192 bauds (4-circuit system) and assume that one station is +1 part in 10^6 off frequency and the other is −1 part in 10^6 off frequency. The total relative drift must not exceed half an element, i.e. one-quarter element at each station, or phase will be lost. The time for this to occur with the two oscillators free-running is:

$$\frac{10^6}{4 \times 192} = 1300 \text{ s or } 21\cdot6 \text{ min}$$

This figure must be adjusted for the overall receive margin: taking this as 90%, the stability period becomes about 20 min. For a modulation rate of 96 bauds (2-circuit system) the period would be about 40 min: higher-stability oscillators will give correspondingly longer periods.

The input to the transmitting station of the synchronous system originates from a telex or equivalent station using start–stop equipment; similarly, the output from the receiving station of the synchronous system feeds into the start–stop equipment of the ultimate receiving station. There are three distinct links in such a connection, the central one being operated synchronously, the other two in the start–stop mode. It is essential to ensure that traffic shall not be offered to the synchronous link at a higher rate than it can be cleared, otherwise loss or mutilation of characters will occur.

The character-clearance rate of the synchronous link must be either equal to or greater than the character rate of the incoming start–stop signals; similarly, the receiving station of the synchronous link must not offer traffic to the start–stop receiving station at a greater rate than it can be accepted. Allowing for the specified speed tolerances of the terminal teleprinters, the synchronous system is designed to clear traffic at a character rate about 3% higher than the nominal input rate. For this reason, the character period of the synchronous link is fixed* at $145\frac{5}{6}$ ms when transmitting signals which arise from a standardized 7·5-unit 50-baud start–stop source. This period corresponds to a character rate on the synchronous path of $60\,000/145\frac{5}{6} = 411\frac{3}{7}$ char/min, compared with the nominal rate of 400 char/min for the teleprinter. It follows that the start–stop signals sent out by the receiving synchronous station (character period = $145\frac{5}{6}$ ms) will be less than 7·5-unit signals, i.e. the stop signal is $25\frac{5}{6}$ ms in duration.

A synchronous transmitting station 'calls-in' characters one by one from the originating source. To deal with the random arrival of signals at the transmitting station, a one- or two-character store is used so that each character is presented to the input of the system at the instant when it is required for processing.

8.2 Multiplex principles

In a 2-circuit TDM system, signals from two sources are time-interleaved (corresponding to frequency interleaving of the FDM system). To maintain the original character-clearance rate of the separate channels the duration of all elements must be reduced by 50% and this necessitates doubling the bandwidth requirement and appears to introduce no overall advantage. In a TDM system each contributory channel has the brief but exclusive use of the main channel and can operate at full power — there is not the need to share the power with other channels as in the FDM system, and maximum signal/noise ratio is obtained. An FDM channel can be operated at a modulation rate much higher than its nominal rate if increased distortion is acceptable; also the synchronous system has twice the receive margin of the start–stop system. The fact that start and stop elements can be discarded on the synchro-

*CCITT Recommendation S12; CCIR Recommendation No. 242.

*Recommendation S12; CCIR Recommendation No. 242.

nous system would result in a modulation rate lower by the ratio 5/7·5 per channel (for the case being considered, i.e. 33·3 bauds for a single-channel and 66·6 bauds for a 2-channel system). However, in the ARQ system advantage is taken of dispensation from start and stop signals to use a 7-unit error-detecting code — the alphabet No. 3.

For a 2-circuit 7-unit system (character period = $145\frac{5}{6}$ ms) the duration of one element is

$$\frac{145\frac{5}{6}}{7 \times 2} = 10\cdot4 \text{ ms}$$

approximately.

The modulation rate of the aggregate signal (the multiplexed 7-unit signal) in the 2-channel synchronous path is 1000/10·4 = 96 bauds (approx.). For a corresponding system, but with four circuits, the element duration is

$$\frac{145\frac{5}{6}}{7 \times 4} = 5\cdot2 \text{ ms}$$

approximately. The modulation rate is 1000/5·2 = 192 bauds (approx.). Slightly different modulation rates apply to U.S.A. start—stop standards, such as 45·5 bauds.

Due to multipath differential delay the modulation rate for an HF radio-telegraph circuit is limited to 200 bauds. This limits the TDM system to a maximum of four circuits (192 bauds) or preferably to two circuits (96 bauds). In a TDM system all channels forming the same aggregate must be operated at the same modulation rate.

Channel Marking. The multiplexed channels are designated A and B (for the 2-circuit 'diplex' system), or A, B, C and D (4-circuit system).

In any multiplex system it is vital to ensure that circuits cannot become crossed, i.e. traffic originated on channel A (or B) must not be received on channel B (or A). In FDM systems the danger is safeguarded by channel band-pass filters. In TDM systems, signals in each channel must be given a unique character marking to permit channel recognition. This is achieved simply and effectively by inverting the

signal conditions on certain channels. The recommended inversions are: *

2-circuit system	*4-circuit system*
Channel A erect	Channel A erect
Channel B inverted	Channel B inverted
	Channel C inverted
	Channel D erect

Even with this precaution it is possible to cross channels in 2-circuit systems if the aggregate signal is accidentally inverted. As a refinement a special *marked character-cycle* pattern has been introduced. With the inversion scheme the aggregate signal is now balanced — there are equal numbers of mark and space elements in both 2-circuit and 4-circuit systems, since the 3:4 Z/A signals become 4:3 Z/A ratios in the inverted channels. Signal inversion provides reversals for synchronizing even when all channels are idle.

Interleaving. In a TDM system, signals may be interleaved on an element basis, character basis, or a combination of both. Character interleaving has the advantage over element interleaving that it allows earlier injection of the RQ (request for repetition) signal.

Character interleaving is used for 2-circuit systems, i.e. characters for the two channels are transmitted consecutively. In 4-circuit systems a combination of character and element interleaving is used and here, again, characters from channels A and B are transmitted consecutively (character interleaving). Elements of channel C characters are interleaved with those of channel A; elements of channel D characters are interleaved with those of channel B. Elements of channel A precede those of channel C and elements of channel B precede those of channel D.

Aggregate signals have the following form (A, B, C and D refer to channels, suffix numerals 1—7 are elements of the 7-unit code, and main numerals 1, 2, 3, . . . are successive characters):

2-channel: $A1_1 \ A1_2 \ \ldots A1_7 \ B1_1 \ B1_2 \ \ldots B1_7 \ A2_1$ 　　　. . . etc.

4-channel: $A1_1 \ C1_1 \ A1_2 \ C1_2 \ \ldots A1_7 \ C1_7 \ B1_1$ 　　　　$D1_1 \ \ldots B1_7 \ D1_7 \ A2_1 \ C2_1 \ \ldots$ etc.

*CCITT Recommendation S13; CCIR Recommendation No. 242.

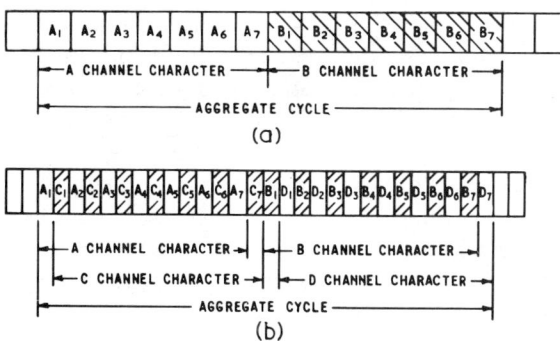

NOTE :— THE POLARITY OF THE SHADED ELEMENTS IS INVERTED

Fig 8.2 Channel interleaving: (a) 2-circuit; (b) 4-circuit

Interleaving and inversion are shown in Fig 8.2.

Sub-divided Circuits. Mentioned has been made of time-sharing circuits to enable lower-speed circuits to be made available to renters with lighter traffic.

Sub-division of a main circuit must bear a simple relation to the duration of the correction (RQ) repetition cycle; the basis for sub-dividing a main circuit is one-quarter. The renter of a quarter-rate circuit would clear traffic at one-quarter the normal rate, i.e. at 100 char/min compared with the nominal 400 char/min. For the system being described the actual rate would be one character every $583\frac{1}{3}$ ms (= $4 \times 145\frac{5}{6}$ ms). The three intervening character periods are filled by traffic from three other renters of reduced character-rate circuits. During the 'blank' intervals of the sub-circuits the start—stop 5-unit links are held to the mark condition.

Possible combinations for sub-divided circuits are: $4 \times \frac{1}{4}$ rate; or $2 \times \frac{1}{2}$ rate; or $2 \times \frac{1}{4} + 1 \times \frac{1}{2}$ rate; or one each at $\frac{1}{4}$ rate and $\frac{3}{4}$ rate. The last-mentioned allocation is rarely used.

Sub-divided circuits also must be protected against inadvertent crossing and this has been achieved by inverting one character in four. The sub-channels are numbered 1—4 and the first of any main channel is inverted, the remainder remaining erect.* This inversion overrides the channel inversions illustrated in Fig 8.2; in some cases, an inverted element will be re-inverted, i.e. made erect.

*Recommendation S13.

Phasing. Character phasing is the condition in which a character cycle of the local timing completely coincides with a character cycle of the received signal.

In addition to achieving element phase (synchronism) and character phase, it is essential that the two stations are also in *system phase,* i.e. the condition in which a character cycle of the local timing completely coincides with the *related* character cycle of the received signal.

Once correct phase has been found it is maintained during good radio conditions by the synchronizing action. If the radio path fails and this condition persists for longer than the period covered by the oscillator stability an eventual loss of phase will occur. This can also occur when the radio frequency has to be changed if this involved a change in propagation time, e.g. a possible routing through a radio-relay station. As when first setting up the system, correct re-phasing may be achieved at any time by one of three procedures:

(1) Manually by a push-button, each operation of which retards the receiver or transmitter by one element. This method is useful when working to an unattended remote terminal, e.g. to a slave station where automatic phasing is not provided or is switched off. (2) Semi-automatically, i.e. automatic phasing, manually initiated. This has little application except to force automatic phasing during testing or fault localizing. (3) Automatically. This is the preferred method for maintaining a high-grade service. Automatic phasing relies on the 'RQ repetition cycle' in a special pattern known as the 'marked character cycle.' Briefly, this cycle comprises four characters (sometimes eight), made up of the special request-repetition signal RQ together with the last three (sometimes seven) characters just transmitted and stored at the sending end against the possible need for repetition of transmission. On reception, this RQ cycle can be fully tested for authenticity.

Whenever the tested-RQ-cycle criteria are not met on both circuits for a set period (10 s or 5 min, by choice) re-phasing is initiated automatically. The equipment then retards the local timing (except for the transmitter at the master station) in steps of one element per detected error received, with the qualification that only one error can be detected per aggregate character cycle. Phasing steps are made in one direction only. The aggregate character cycle referred to here is the period in which each channel of a TDM transmission has completed one character in the synchronous path. When the tested repetition criteria are met on both circuits, phase hunting stops. On a 4-circuit system, phase hunting is initiated by failure to meet the tested-repetition-cycle criteria on either channels A and B or on channels C and D and it is continued until satisfactory repetition cycles are received on all four circuits.

circuits. If character phase is regained, but the repetition cycles at the two terminals do not also achieve the same relationship as before, then one or more of the characters in store will be lost in one direction of transmission, while in the other direction an equal number of characters will be printed twice. Fig 8.3 illustrates the loss and duplication of characters which result when, following a loss of phase in traffic, character phase is regained without system

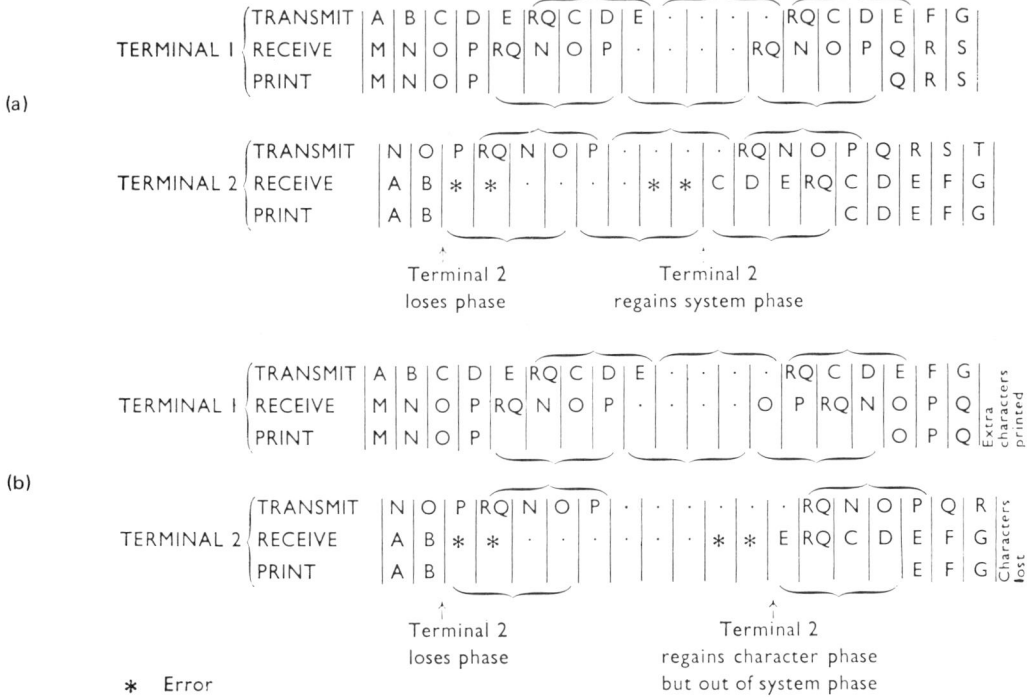

Fig 8.3 Regaining (a) system phase, (b) character phase only

The Marked Character Cycle.* While automatic phasing can be established successfully with the interleaving and inversion patterns of Fig 8.2, certain residual disadvantages remain which prevent printed errors from being reduced to a minimum. When a circuit is initially phased at the time of installation, character phase is achieved and a relationship is established between the repetition cycles at each of the terminals. When for any reason phase is subsequently lost, each terminal equipment repetition-cycles and the characters held in store are re-inserted continuously; only when phase has been re-established are these characters passed on to the 5-unit output

*CCIR Recommendation No. 342.

phase. The correct repetition cycles in the two directions are RQ CDE and RQ NOP.

If an acceptable in-phase condition is found only once per repetition cycle (i.e. every 56 elements instead of every 14 for a 2-channel 4-character repetition-cycle system; or 112 elements for a 4-channel 4-character or 2-channel 8-character cycle; and 224 elements for a 4-channel 8-character-cycle system) when phase is re-established according to this criterion, it will give not only character phase but also repetition-cycle phase or system phase. Printing will resume at exactly the point at which it ceased when loss of phase occurred; no additions or omissions will occur in the printed copy.

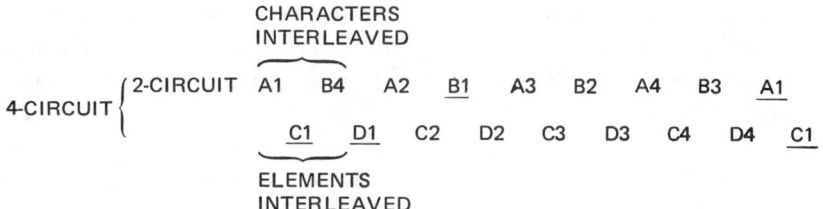

Fig 8.4 The marked repetition cycle applied to 2-circuit and 4-circuit systems

System phase can be made possible by marking the aggregate signal with an inversion pattern, over and above that shown in Fig 8.2, in such a way that the composite marking pattern is unique over as many characters as are transmitted in the period of a repetition cycle. A suitable marking pattern is obtained by the inversion of one character in four on each channel (for a 4-character repetition cycle). If, in addition, the relationship between the inverted characters in each channel is chosen so that the pattern is unique despite any inadvertent inversion of the resultant aggregate signal (as sometimes happens after a change of radio frequency), phasing to a signal with reversed aggregate polarity will not be possible.

The marked cycle applied to 2-circuit and 4-circuit systems for a 4-character repetition cycle is shown in Fig 8.4. Underlined characters have their polarities reversed *in addition* to the inversion of B- and C-channel characters as already shown in Fig 8.2.

When a signal having this marking pattern is received, if the receiver is in phase there will be one character in every four (for a 4-character repetition cycle) which is inverted at the transmitter and is re-inverted at the receiver. During phase hunting, if the receiver is in character phase but not yet in system phase, two acceptable characters (with

correct 3:4 ratio) and two unacceptable characters will be received because the inversion at the transmitter is not matched at the receiver. Fig 8.5 shows how, by the use of the marked cycle and tested repetition cycle, the out-of-phase terminal is forced to regain system phase.

One of the acceptable characters may be the RQ signal; recognition of this is not in itself a sufficient criterion for system phase. Phase hunting continues until the tested repetition-cycle criteria are met in both channels; these criteria can be satisfied only when system phase is regained. In this way, the equipment uses the marking pattern to ensure completely unambiguous phasing.

Having reached the point where system phase can be established, equipment can be designed to operate automatically without the possibility of losing or duplicating characters, or causing incorrect channel routing when re-phasing in traffic. Once loss of phase has been recognized (by setting an arbitrary time limit to continuous cycling, or by other means) re-phasing can be initiated automatically and system phase regained. When this has occurred, and only then, traffic is passed forward and printing will resume precisely where it left off before the interruption. This is an advance over earlier systems which could obtain phase automatically but only

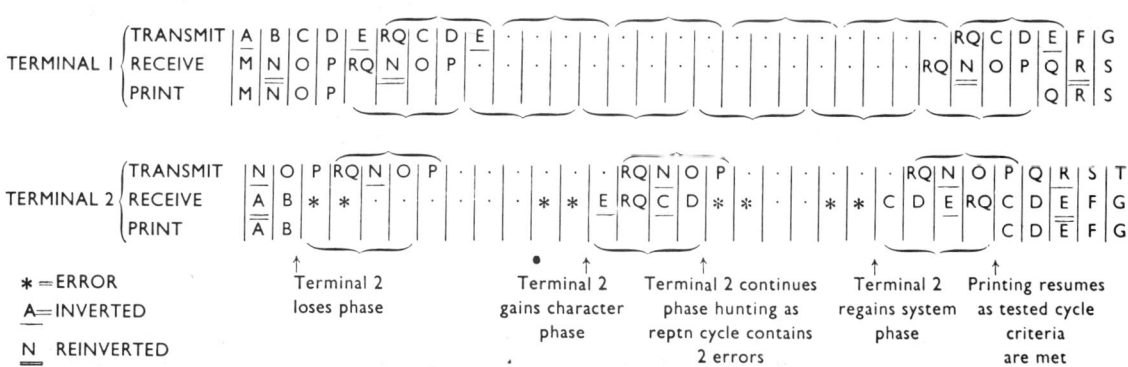

*=ERROR
A=INVERTED
N REINVERTED

Fig 8.5 System phasing with the marked cycle

with the possibility that errors could occur if it were regained automatically in traffic.

The marked character-repetition cycle requires the inversion of one character in every four per main channel, regardless of the presence of sub-divided channels. This pattern ensures correct system phase and simultaneously provides the marking necessary for subdivision, for which no separate marking is otherwise required. In fact, if all channels are subdivided the marking pattern previously described for the sub-channels is the same as that of the marked character cycle.

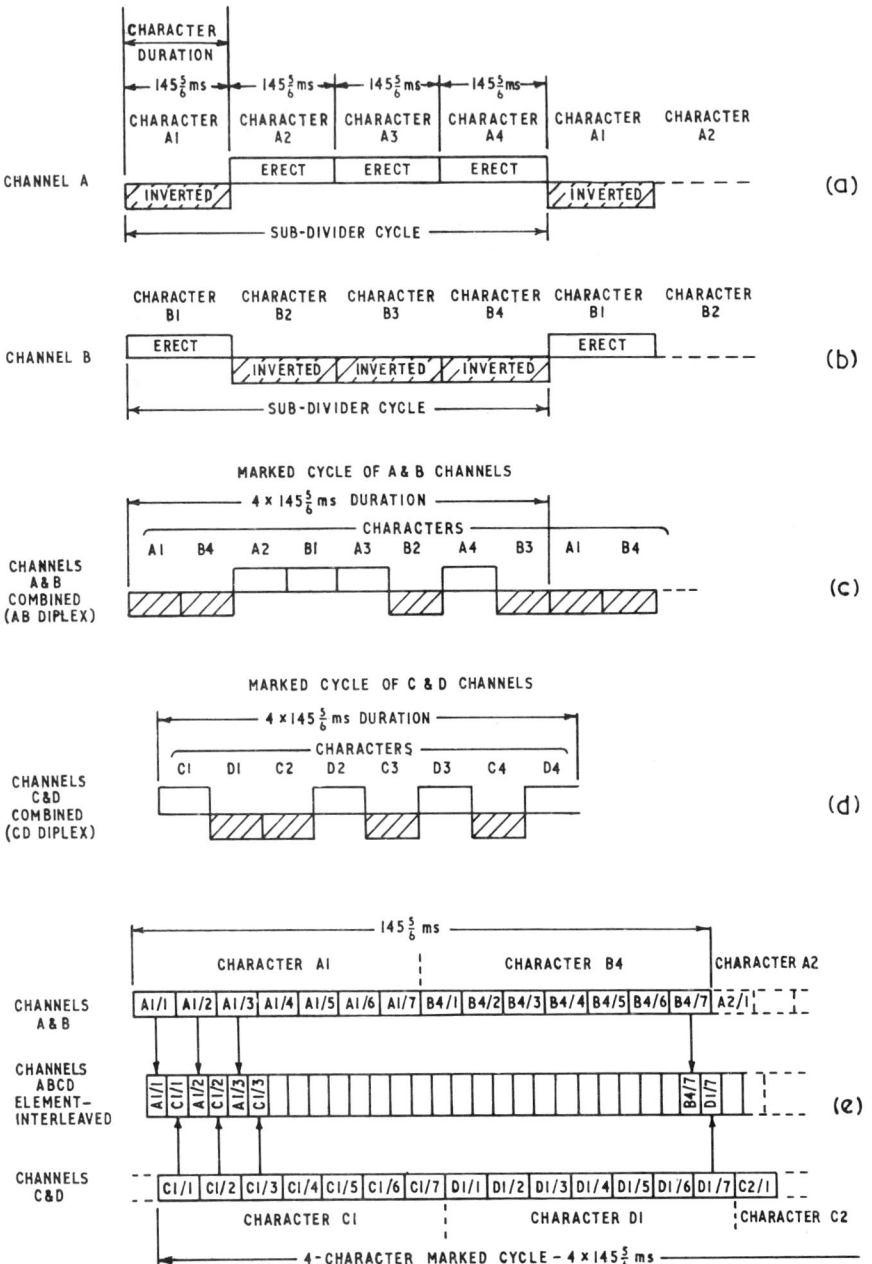

Fig 8.6 Formation of 2-channel and 4-channel marked cycle of four characters: (*a*) marked cycle on channel A; (*b*) marked cycle on channel B; (*c*) marked cycle interleaved on AB diplex (B4 follows A1); (*d*) marked cycle interleaved on CD diplex (D1 follows C1); (*e*) part of 4-character aggregate signal

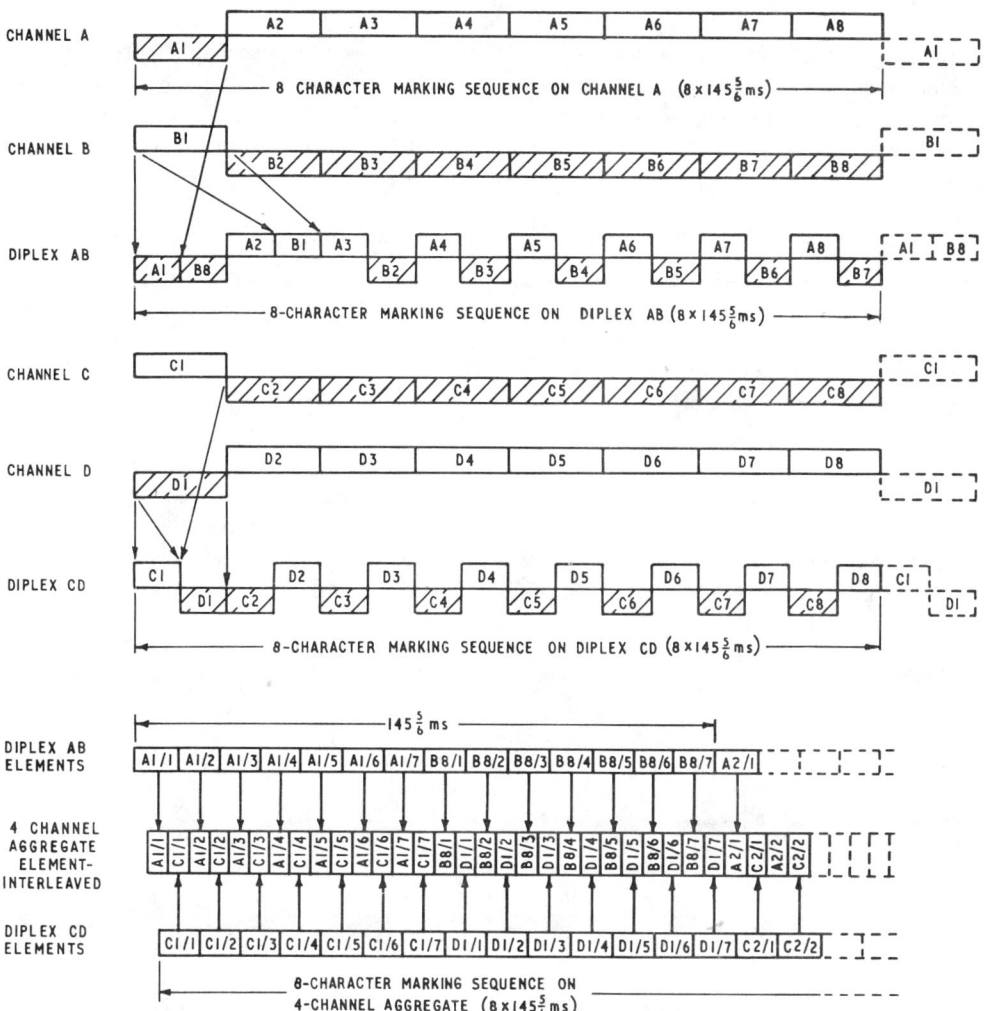

Fig 8.7 Marking pattern for 8-character repetition cycle

When an 8-character marked character-repetition cycle becomes necessary the first character in the group of *eight* of the cycle is marked by inversion; in other respects operation is the same as described for the 4-character cycle.

Operation of the marked cycle is summarized in Fig 8.6 for the 4-channel repetition cycle and in Fig 8.7 for the 8-character repetition cycle.

In Fig 8.6(*a*), A1 is a character on sub-channel 1, A2 is a character on sub-channel 2, and so on. At (*b*), B1 is a character on sub-channel 1, B2 is on sub-channel 2, and so on; all these sub-channels of channel B are inverted which implies that the marked character B1 is re-inverted, i.e. erect. At (*c*) is shown the interleaving of signals from (*a*) and

(*b*) with consequent halving of the separate character periods. Note that character A1 is followed by character B4. At (*d*) is shown the character interleaving on channels C (inverted) and D (erect); C1 being re-inverted becomes erect. At (*e*) is shown the first part of the element interleaving of channels A, B, C and D. The relative polarities may be deduced from charts at (*c*) and (*d*).

In Fig 8.7 are shown separately the marking pattern for the character interleaving of each channel (A, B, C and D) and of each diplex (AB and CD). B1 and C1 becomes erect by re-inversion; B8 follows A1 in the AB diplex, while D1 follows C1 in the CD diplex. Finally, the element interleaving of the two diplexes and of the 4-channel aggregate signals are shown. In these three lower charts the polarities may be deduced from the upper timing charts.

With the marked-cycle method of automatic phasing it may be necessary to step for the number of elements of the whole repetition cycle less one element, i.e. a 2-channel system operating a 4-character repetition cycle may have to step a maximum of $(2 \times 7 \times 4 - 1) = 55$ elements; on average only 27 steps will be necessary and this may be achieved in about 5 s. Twice this period will be required for a 4-channel system, since the rate of stepping can only be decided on the A and B channels, because the C and D channels are element interleaved and so overlap the character periods of the A and B channels.

8.3 Principle of error-correction

Incoming 7-unit signals at the receiver are examined for the correct 3:4 Z/A ratio. If correct, the signal is decoded and transmitted as a 5-unit start–stop signal towards the receiving teleprinter; if incorrect, an RQ signal is sent back to the transmitting station to invoke a further transmission of the faulty signal. For this purpose, the last three characters transmitted are always held in a cyclic store. In the 8-character repetition cycle, seven characters are held in store; or, alternatively, only the last four characters are stored and they are supplemented by three 'packing' signals when the RQ cycle is in operation.

Correction of a Mutilated Character. The passage of traffic on a 2-circuit multiplex system is represented in Fig 8.8. In one direction the characters AaBbCc . . . are being sent; capital and small letters are used to distinguish the two channels. In the reverse direction, characters LlMmNn . . . are being sent on

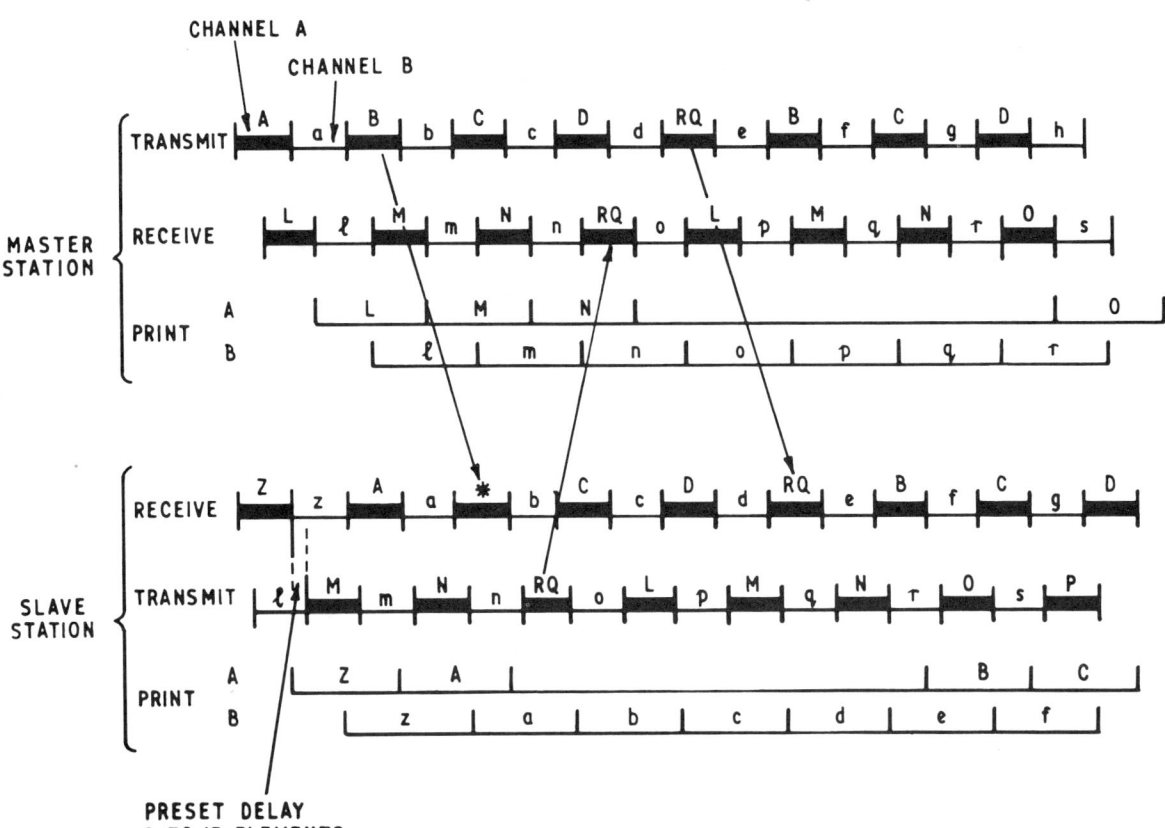

Fig 8.8 Error correction by 4-character repetition cycle

the two channels. Corresponding signals at transmitter and receiver are shown displaced by a duration approximately equal to that of two characters (master to slave) and one character (slave to master) to represent the overall transmission times over the two paths; the preset receiver–transmitter delay at the slave station is also shown.

Suppose that character B sent from the master transmitter is mutilated. At the slave receiver an error in the 3:4 ratio is detected. The following chain of events occurs on circuit A:

(1) At the slave receiver, the 5-unit output path is blocked by applying Z signals for four characters; the faulty character and the three following characters are not printed.

(2) At the slave station, intake of further characters for transmission over the backward channel is suspended and the next character is not immediately sent.

(3) At the slave station, transmission is interrupted at the earliest timing opportunity to allow the RQ signal to be injected and transmitted, followed by a retransmission of characters LMN which are in storage.

(4) At the master station, on receipt of the RQ signal, the 5-unit backward path is blocked by Z signals for four characters.

(5) At the master transmitter, further intake of characters is suspended; character E is not yet sent.

(6) At the master station, the RQ signal is transmitted followed by retransmission of the preceding characters BCD from the cyclic store. These retransmitted characters are again stored in the normal manner.

(7) At the slave station, on receipt of the RQ signal followed by characters BCD, all four characters are checked: provided that there is no further mutilation, the 5-unit output path is re-opened for transmission of characters BCD. . . . Suspension of the backward path is ended and normal operation continues with transmission of characters OP . . . if no further errors are received.

(8) At the master station, the 5-unit backward path is re-opened four characters after the arrival of the original RQ signal. The intake of characters by the synchronous channel is resumed.

Had the second transmission of character B again

resulted in faulty reception the whole RQ procedure described would have been repeated, indefinitely, until character B was received with the correct 3:4 ratio.

The above operations occur only on the circuit affected by the mutilation; the other circuit of a diplex is unaffected and continues to print bcd . . . and opq . . .

The total content of the RQ cycle is four characters; although in certain circumstances an 8-character RQ cycle may be used.

The possibility exists that the falsified character B could have been an RQ signal from the master station and it is for this reason that repetition from storage takes place in the backward direction. At the master station, since no RQ was in fact originated, this repetition is ignored.

Interruption of the backward channel does not cause any loss of traffic since characters are always held in a cyclic store prior to transmission over the synchronous path. Each character is 'called' into the channel sender and, by suspending the pulse which calls in the character, it is possible to inject the RQ signal instead of the next character (O in the case quoted).

During the RQ cycle four characters are suspended from transmission on each channel of the duplex circuit; the effect of the 3-character cyclic store is to enable a mutilated character to be repeated four character periods later; no extra characters are printed and none are lost during the RQ cycle. This is true for various propagation times up to the limit set by the store capacity.

The receipt at the slave station of the confirmatory RQ signal (along with characters BCD) from the master station has no effect other than to unblock the onward 5-unit path to the receiving teleprinter and allow printing to be resumed at the end of the 4-character cycle. Had the radio channel caused this confirmatory RQ signal to be also mutilated, its detection as a mutilated 3:4 signal would have invoked a further request for repetition. By the use of this 4-character RQ cycle *in each direction of transmission* the system is immune from breakdown by mutilations to any character of the repetition cycle. The injection of four characters into the return path when only the RQ signal is effective appears on first sight unnecessary; but it is possible for an RQ signal to be produced –

though with low probability — by a transposition* in another character. This would produce a repetition cycle and, if this did not block the succeeding characters, the repeated characters would be printed twice. A deterioration of the radio path in one direction of transmission may be accompanied by a deterioration in the other and in no circumstances may characters be either lost or printed twice.

station is opened for service in time for M to be printed.

Any mutilations received are taken care of by this 4-character repetition cycle at each terminal with the minimum interruption to traffic. During any repetition sequence a genuine RQ signal must be received before printing is resumed at the end of the 4-character cycle. Each character received after the teleprinter circuit is restored is tested for

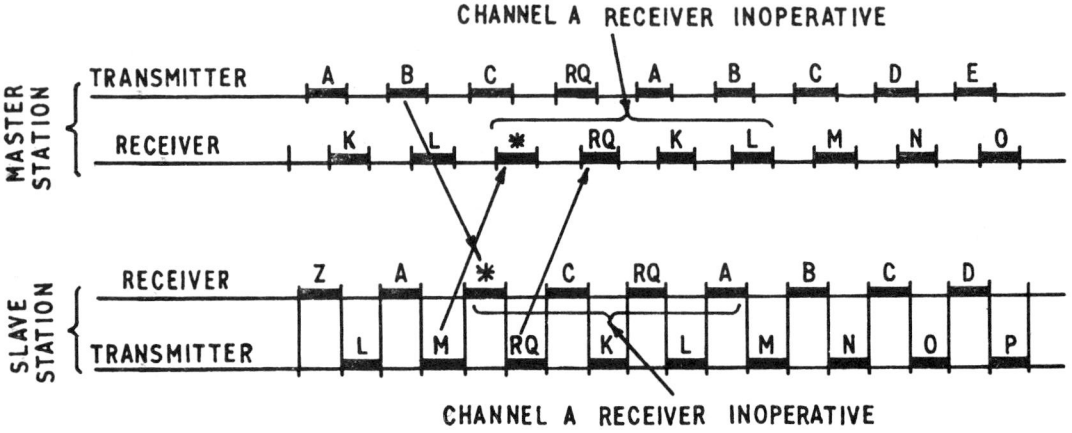

Fig 8.9 Correction of simultaneous errors in both channels of a circuit

Correction of Mutilations Occurring Simultaneously in Both Directions. This case is illustrated in Fig 8.9. If character B from the master station is mutilated on arrival at the slave receiver, the RQ signal is injected into the return channel. However, a mutilation has occurred on the character M just transmitted from the slave station and this, on arrival at the master station, causes a repetition cycle to commence by the injection of an RQ signal followed by the previous three characters ABC.

At this stage there are repetition cycles running at each station. The characters from the master station (RQ signal followed by ABC) will be received at the slave station, and B arrives just after the receiving teleprinter connected at the slave station has opened for service. In the other direction (slave to master), the characters RQ KLM are received and the receiving teleprinter connected at the slave

the 3:4 Z/A ratio before it is printed. If an RQ signal were mutilated on receipt at either station it would be treated as an error, initiating an RQ signal and calling for a retransmission.

The cases described cover all the sequences of errors possible in transmission but, in poor radio conditions, it may happen that a number of RQ cycles may elapse before printing is resumed; delay inevitably arises but no characters are lost, and the received message is printed 'clean' and error-free.

Error Indication. If a return channel is not available the system can be used for error indication only. In this case one of the characters from alphabet No. 2 (usually a special symbol allocated to combination No. 32* for this purpose) is transmitted over the 5-unit start—stop output circuit to the receiving teleprinter, which must have its typehead modified to print the special signal.

*Transposition — one A element received as Z, together with one Z element received as A, within the same 7-unit signal. This maintains the correct Z/A element ratio, and an undetectable error results.

Testing the Repetition Cycle. Three methods have been used for checking the authenticity of the RQ signal to reduce the incidence of undetected errors caused by transpositions. RQ 'cycling' occurs on receipt of an error or when the system is phase hunting. It is necessary to make certain checks to ensure that conditions will be satisfactory before re-opening the circuits for traffic and with the advent of automatic phasing these checks have become more stringent.

In 'gated-RQ' (GRQ) the correct *sequence* of A and Z elements for the RQ combination is checked.

In 'tested-RQ' (TRQ) a check is made of all the elements of the RQ signal together with a 3:4 ratio check of the three (or seven) characters that follow the RQ signal within the repetition cycle.

In 'tested-repetition-cycle' (TRC) the whole repetition cycle is tested for all characters having the correct 3:4 ratio, and for the presence of an RQ signal. This test includes the two or more characters which follow the mutilated character but precede the arrival of the RQ signal — e.g. characters CD at the slave receiver in Fig 8.8 — and ensures that each printed character is preceded by at least three correct characters. Errors on ARQ systems tend to occur in groups and it is unlikely that an undetected error (a transposition) will occur unaccompanied by detectable errors. The application of a tested repetition cycle during automatic phasing is very desirable for finding unambiguous phase. In some systems, TRQ is operated during normal traffic but TRC during phasing.

Circuit-propagation Time. It is clear from Fig 8.8 that the characters stored at the transmitter must be held at least for the time in which the return of an RQ signal is possible. This time is the sum of the two character periods involved (the signal which gets mutilated and the RQ signal which is returned as a consequence) and the overall loop-transmission time, which includes not only the propagation time of the radio circuit but also the fixed transmitter—receiver phase displacement at a slave station, and the response times to the RQ conditions by the equipments (working to a fixed time cycle) at both stations. The loop-propagation time over the radio path (at 3×10^8 m/s or 186 000 mile/s) between any two radio stations rarely exceeds 100 ms

except for such longer routes as United Kingdom—Australia (115 ms) or New Zealand (125 ms) — both loop times. Propagation delays also occur in any landline extensions. Taking very general figures for different types of plant, something like 10 ms for filters at radio-receiving stations, 10 ms for a regenerative repeater, 10 ms for landline propagation and 20 ms for MCVF equipment, might result in as much as 50 ms for one joint transmitter—receiver landline path, i.e. up to 100 ms for both landline ends of the overall circuit. In total, the overall loop-delay time could attain something like 200 ms, with higher values for the longer radio paths and much higher values if a radio-relay station is involved (i.e. two radio links in tandem) with the inherent additional landline extensions and associated equipment, possibly including a regenerator. Long loop delays are also incurred for the special case of ARQ systems operated on transoceanic telephone cables, for which the propagation velocity is much lower (about 100 000 mile/s) compared with radio propagation.

The use of a 4-character RQ cycle is related to this loop-transmission time. It should be capable of accommodating a loop delay of approximately 290 ms, i.e. the difference between the time for the four characters and the time occupied by the mutilated and RQ characters themselves.

On routes which have longer propagation times it is necessary to use a longer RQ cycle. On account of the use of subdivided channels and the marked character cycle the RQ cycle must comprise a multiple of four character periods. The 8-character RQ cycle which is used when long propagation times are present may consist either of (RQ + seven stored characters) or (RQ + three β + four stored characters) which allows for loop-propagation times up to 868 ms.

8.4 The ARQ radio-teleprinter system

The following description refers to the principles of the 2-circuit equipment. Although interleaved sequential signals are delivered to the synchronous system, the input signals are taken in simultaneous form as this is more convenient for the storage and encoding processes. This necessitates receiving the incoming signals on a staticizer.*

*Staticized = made static or simultaneous, i.e. converted to the simultaneous mode for temporary storage.

Leased circuits, and circuits of the overseas telegram service, which operate from automatic transmitters are terminated on sequential-to-simultaneous converters or *extensor* units. Characters from tape in the automatic transmitter are called in one by one by clutch-release pulses whenever the ARQ equipment is ready to deal with them; these release pulses are suspended during an ARQ cycle.

For telex traffic it is not feasible to extend the additional control circuits through the network. Telex circuits terminate upon a buffer store which enables characters from either a teleprinter keyboard or from a tape transmitter to be sent into the ARQ equipment without restriction during ARQ cycles.

Tape readers can be used at telegraph-operating positions if they are in the same building as the ARQ equipment. Since in this case the signals are fed direct to the ARQ distributor which interleaves the signal elements, they need to be 'gated' or timed so that their arrival corresponds to the appropriate instants for transmission. This requires seven wires per circuit from the instrument room to the ARQ equipment and this arrangement is rarely used.

The Sequential-to-Simultaneous Converter. As signals from the automatic tape transmitter are sequential, only the S wire is needed for transmission, but in addition to the R wire for teleprinter reception a separate control circuit for the automatic-transmitter clutch is required between the ARQ equipment and the renter's premises. This connection may be via one or more MCVF channels and there will be a delay, due to the loop-propagation time of the inland circuit, between the emission of the clutch-release pulse at the ARQ terminal and the arrival of the corresponding start–stop signal at the same point. This involves storage in the converter, which is designed to accommodate delays up to 290 ms (= 2 x 145-ms characters).

The automatic tape transmitter is the Model 2F (or equivalent) fitted with a clutch magnet which responds to the release pulses, provided a perforated tape is loaded ready to pass through the transmitter head. The release-control pulse is a +80-V pulse of 35–50 ms duration. Since the release pulses will arrive at the $145\frac{5}{6}$-ms intervals of the ARQ period, a normal 7·5-unit tape transmitter would not come to rest before the next pulse

arrived. Therefore, it is necessary to use a transmitter with a shorter stop signal to give an overall character period of less than $145\frac{5}{6}$ ms. The Model 2F transmitter has a 7-unit cycle at 50 bauds which results in a cycle time of 140 ms (a potential speed of 428 char/min); the automatic transmitter is at rest for about 6 ms between characters.

At the output from the converter, five wires correspond to the five code elements (+6 V = A and −6 V = Z). In addition, it is necessary to indicate to the ARQ equipment when traffic is awaiting transmission. The condition on a sixth wire indicates 'traffic' (−6 V) or 'idle' (+6 V), and a seventh wire indicates whether the idle condition is the β (Z or call = −6 V) condition, or the α (A or clear = +6 V) condition.

Referring to Fig 8.10, under normal traffic conditions, release pulses for the automatic transmitter are applied from the ARQ transmitting equipment (Fig 8.11) to the auto-drive pulse circuits from where they are extended to the automatic transmitter. Each character called in is presented in sequential form to the input trigger which reshapes the input signals. The leading edge of the start element of a character starts the multivibrator and, after a delay of half an element, the distributor-drive circuit produces pulses at element rate to drive the distributor. To guard against spurious signals from the line, a short-start rejection circuit is incorporated to switch off the multivibrator if the input does not remain at A polarity for at least half an element; under this condition the distributor-release pulses are not produced.

The distributor produces output pulses timed to the midpoint of each element of the incoming character, and these are applied to both registers A and B. Alternate traffic characters are routed to registers A and B by gates which are switched at character rate by the distributor. In the registers, characters are converted from sequential to simultaneous form under control of the distributor; the latter resets after the seventh output pulse, switching off the multivibrator.

When the ARQ equipment is not cycling it examines the simultaneous-signal wires and then sends out a clutch-release pulse to the automatic transmitter. The leading edge of this pulse clears the store and sets the supervisory detector to its rest position (the output corresponding to 'no

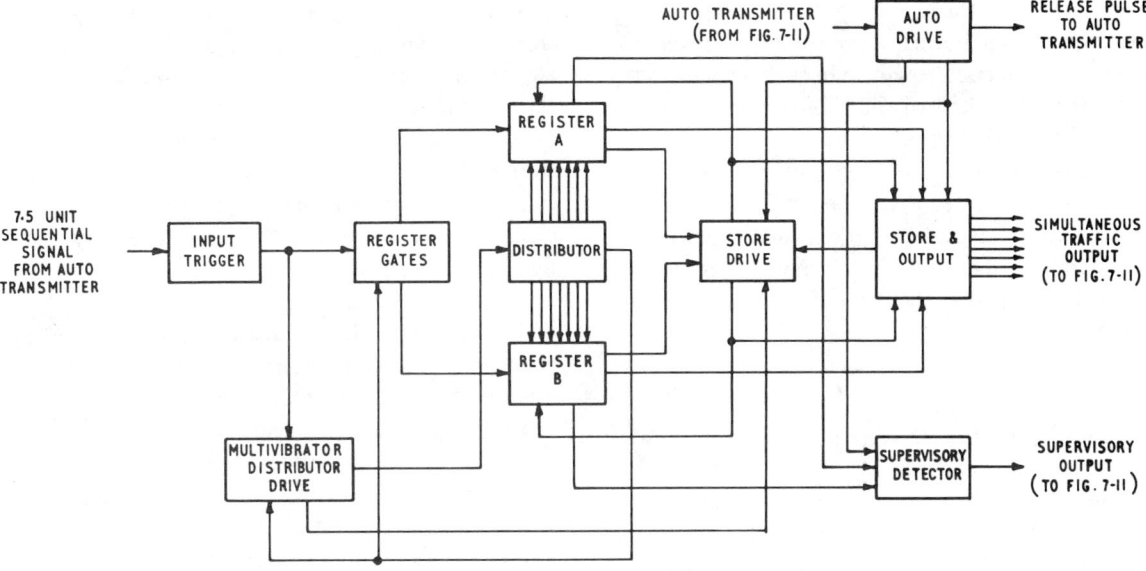

Fig 8.10 Sequential-to-simultaneous converter

traffic'). The store, in clearing, provides one of the required input conditions for the store-drive circuits. Other input conditions for the latter are provided by the registration of the start element of a character in one register and the absence of a start element in the other; a character being registered thus routes the read-out to itself. When the multivibrator is stopped, the distributor-drive circuit produces a transition which is passed by the store-drive circuit and causes the transfer of the character from the register to the store, setting up the appropriate conditions on the traffic-output wires. At the same time the supervisory detector is set to produce an output corresponding to the

'traffic' condition for a character, or the 'no traffic' condition for either of the supervisory signals.

When the ARQ equipment starts cycling it neither reads in the character held in the converter store nor sends out the clutch-release pulse. With a long loop delay on the line between converter and automatic transmitter it is possible for two characters to be on the line when cycling commences. These characters will be read into the converter and held, one in each register. As soon as cycling ceases, the simultaneous output corresponding to the character held in the store is examined, and a clutch-release pulse is sent, the leading edge of which clears the store in the normal way.

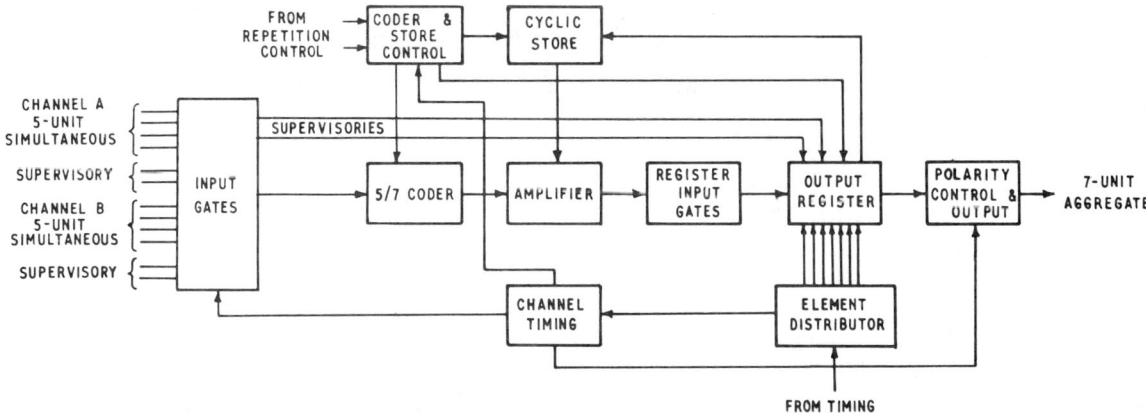

Fig 8.11 ARQ transmitter

Since, due to the line delay, the first character after cycling will not have arrived at the converter, the store-drive circuits will be in such a condition that the first of the two characters registered is transferred to the store. The trailing edge of the first release pulse after cycling is used to switch the store-drive circuits so that, when the store is next cleared, the character from the second register is transferred to the store. Traffic then proceeds normally as the first character after cycling is read into the converter.

2-channel Transmitter. Referring to Fig 8.11, 5-wire input signals are applied to the input gates of each channel. Under the control of the channel timing the selected input is applied to the coder. Its function is to convert signals from alphabet No. 2 (5-unit) into corresponding signals in alphabet No. 3 (7-unit with the 3:4 Z/A ratio). Immediately prior to transmission of a clutch-release pulse, the previous character held on the output store of the sequential-to-simultaneous converter is transferred into the 5/7-unit code converter. The code-conversion speed is such that it is available to both channels of the diplex. The code converter comprises a matrix either of semiconductor diodes or of ferrite cores. The application of appropriate potentials on the 5-wire input circuits activate the corresponding 3-out-of-7 wires on the output.

Output pulses from the coder, after amplification, are used to set up the required 7-unit code-combination in the output register, and the character held in the register is read sequentially by means of pulses from the element distributor. Before the next character (from the other channel of the diplex) is applied to the register the existing character is transferred to the cyclic store. Provision is made here for the storage of three characters as required for the 4-character repetition cycle. Where the total loop-propagation time requires an 8-character repetition cycle, seven characters are held in the store (sometimes four only, to be supplemented by three 'packing' β characters). The last three or four characters are stored here at all times in readiness for retransmission. The store consists of a 3- (4- or 7) - position 7-line shift register, elements being stored on capacitors, semiconductor bistable triggers or ferrite cores as 7-unit coincident signals.

In normal traffic a character is put into the cyclic store immediately prior to being read out sequentially by the main distributor for transmission towards the radio transmitter. When this character is read into the store, the characters already held are each shifted along one position, the ultimate one at each shift stage being discarded. The output from the cyclic store is normally disconnected and only used during an RQ cycle.

To prevent falsification due to a supervisory signal being applied while a 5-unit character is being coded, the supervisory signals are retimed by the input gates and are only effective prior to normal coding on the channel to which they are applied. During supervisory conditions the normal coding action on the corresponding channel is stopped and the required supervisory signal is set up in the output register where it is read in the normal manner. The character-interleaved aggregate signal is applied to the polarity-control circuits where the required channel inversion and marked character-recognition inversion are introduced.

During repetition cycling, on instruction from the repetition-control circuits, normal coding action is stopped and the store read-out timing for the appropriate channel is delayed until the register-input gates are open. Stored information is retransmitted, and after this is again returned to the store. On receiving instruction from the repetition-control circuits to transmit the RQ signal, both the store and the coder are stopped and the RQ signal is set up in the output register. The RQ signal is not stored after transmission but is inserted when necessary on instruction from the repetition-control circuits.

The input circuit of the sequential-to-simultaneous converter has the usual 4-kΩ input impedance for ±80-V working. The 7-unit aggregate signals, probably at the ±6-V level, pass to the modulator at the input of the MCVF system (with 240- or 480-Hz spacing) which conveys the signals by landline to the radio-transmitting station.

The ARQ Receiver (Fig 8.12). The receiver performs the converse functions of those at the transmitter and, in addition, checks the 3:4 Z/A ratio of all incoming signals and applies the RQ signal over the backward channel whenever an error is detected.

At the receiver the 7-unit aggregate signal from

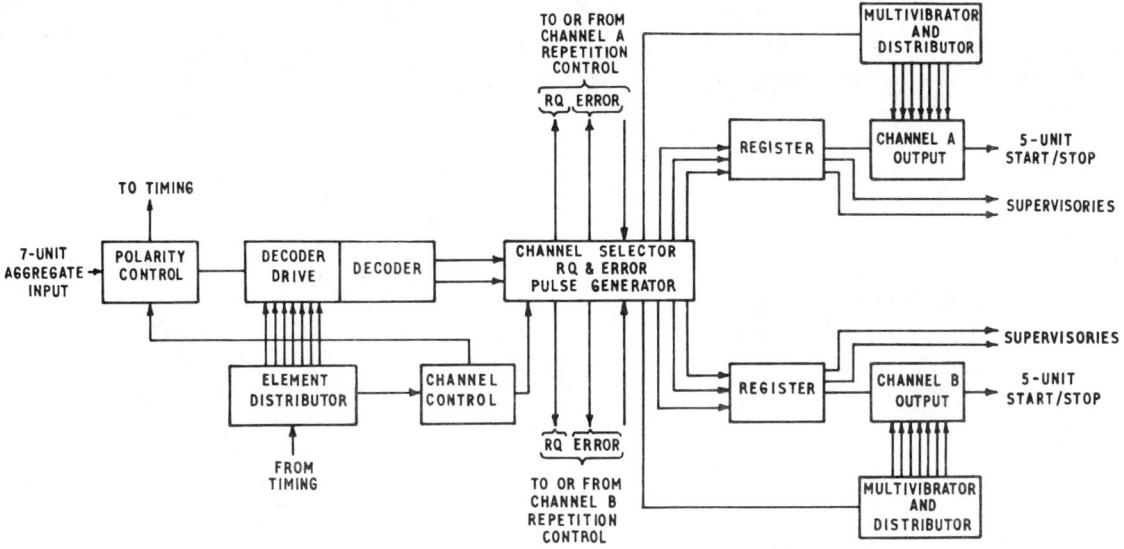

Fig 8.12 ARQ receiver

the radio circuit is applied to the polarity-control circuits, where the signal is reshaped and the channel inversions and marked character-recognition inversions are removed. The resultant squared signal is applied to the decoder-drive circuits together with pulses from the element distributor; the received signal is also applied to the synchronizing section of the timing circuits.

The 7/5-unit converter is a complementary equipment to the 5/7-unit converter at the transmitter; the code converter is usually common to both channels of the diplex. In addition to producing the 32 characters of alphabet No. 2, in 5-wire form, the α, β and RQ signals appear on separate output wires.

Error detection may consist in checking for the presence of three Z elements; any count other than three will initiate RQ action; the counter may be individual to a channel or shared by a diplex pair. Alternatively, an OR gate in the 5-wire output from the 7/5-unit code converter detects the presence of a signal for any of the 32 permitted combinations for alphabet No. 2. There are additional outputs for the α, β and RQ signals and the special case for combination No. 32. No output results from 7-unit signals which have not the 3:4 Z/A ratio. Absence of a condition on any of these nine wires indicates no recognizable output, i.e. a detected error, and the RQ action is commenced.

The decoded output is applied to the channel selector where the A and B channels are separated and their signals fed to the appropriate channel register. From the register, 6th and 7th wires provide signals indicating additional information on traffic/no traffic and α/β conditions. This information is required by the telex interface panel.

In the register, start and stop elements are added to the 5-unit character, which is then converted to a sequential output by means of a multivibrator and distributor. The accuracy of the modulation rate supplied from the multivibrator must conform to the standard of $\pm0.5\%$. The character cycle being $145\frac{5}{6}$ ms in duration, the character transmitted out to the teleprinter receiver has a stop signal of $25\frac{5}{6}$ ms. The output unit may provide a ±6-V signal to drive a relay and provide ±80-V signals. If a supervisory signal is received the start—stop output remains at stop polarity for the corresponding character period and the appropriate supervisory condition is indicated on the supervisory output wires.

An unacceptable 7-unit signal (one not containing the 3:4 Z/A ratio) is not translated by the decoder and as a result a channel-error pulse is produced by the error-pulse generator. Under normal ARQ operating conditions, error and RQ pulses are routed direct to the repetition-control circuits and the 5-unit start—stop output remains at stop polarity.

Under error-indication conditions, error signals may also be routed to activate the teleprinter so that it prints a special error symbol or other required character.

During repetition cycles the channel selector, on instruction from the repetition-control circuits, inhibits the passage of traffic to the output register corresponding to the channel that is cycling.

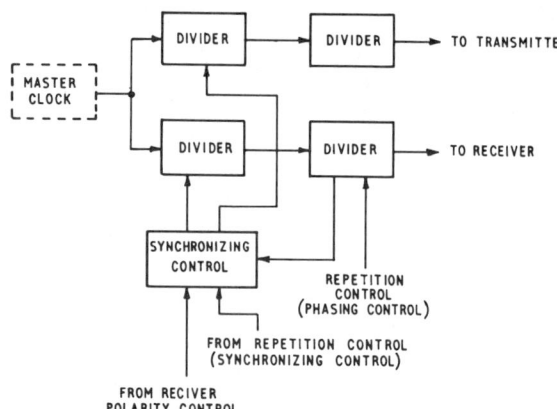

Fig 8.14 ARQ-timing circuit

CHARACTER

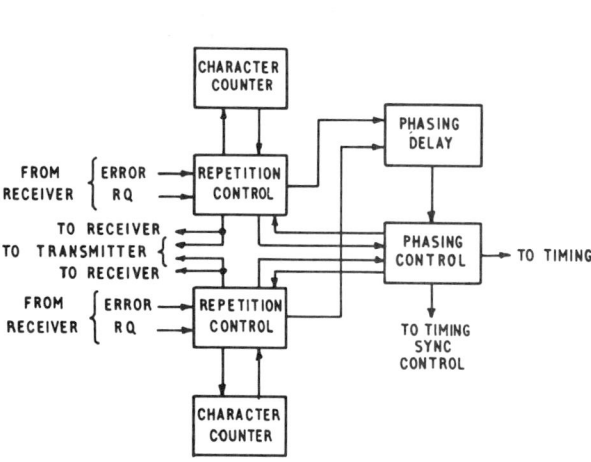

Fig 8.13 ARQ repetition-control circuits

Repetition Cycling (Fig 8.13). Under normal conditions the repetition-control circuit is inactive and comes into operation only on receipt of an error or an RQ signal. Application of an error or RQ pulse from the receiver channel selector to the repetition control starts the character counter on the corresponding channel. The repetition control also instructs the transmitter to insert the RQ signal on the required channel and informs both transmitter and receiver that repetition cycling is in progress. If the cycling is initiated by receipt of an RQ signal the cycle will terminate automatically at the end of the repetition unless errors have been received in the intervening period. However, if cycling is initiated by receipt of an error it is necessary that an RQ signal (with no errors) is received back before cycling ceases (i.e. the tested repetition cycle).

Timing (Fig 8.14). Both the transmitter and the local receiver of the ARQ equipment are controlled from

the same timing source. All waveforms derive their timing from a crystal-controlled master clock whose output, at 9·6 kHz, is applied to separate divider chains for the transmitter and the receiver. These dividers, which act effectively as ring counters, drive the transmitter and receiver-element distributors. Switches enable the overall division to be adjusted to provide aggregate modulation rates of 96 bauds (for 2-circuit operation) or 192 bauds (4-circuit operation), or other rates.

Synchronizing. Under normal conditions, synchronism is maintained by continually modifying the divisor of the first counter in the receiver–divider chain. Pulses from the receiver-polarity control, derived from transitions of the aggregate received signal, are applied to the synchronizing control. Here the relative timing of the pulses is compared with that of similar pulses derived from the output of the receiver–divider chain. Signal transitions that occur before or after the expected time – early or late – are used by the synchronizing control to apply the necessary correction. After receiving eight early or eight late transitions the synchronizing control resets and in doing so changes the number of input pulses that must be applied to the first divider during one cycle of its operation before an output pulse is produced. In this way the timing of the output of the receiver–divider chain is adjusted in steps of the order of 1% of an element.

Phasing. When the two repetition controls indicate that cycling due to receipt of errors is occurring simultaneously on both channels of a diplex, the phasing delay is brought into operation. This circuit, after a pre-determined delay, initiates automatic phasing via the phasing control.

Should either of the repetition-control circuits cease to indicate error cycling during the delay period, the delay circuit is restored to the rest condition. Once re-phasing has been initiated it is maintained by the phasing control until satisfactory repetition cycles are received simultaneously on both channels. A satisfactory cycle is one that contains no errors and includes a single RQ signal.

While automatic phasing is in progress the transmission of the RQ signal is suppressed at the slave equipment until the slave receiver is in phase with the master transmitter and is replaced by a character which does not have the correct 3:4 Z/A element ratio. This is recognized as an error by the master receiver and prevents the master equipment from phasing before the slave is in phase.

If both channels are cycling due to receipt of errors, information from the phasing-control section of the repetition-control circuits inhibits synchronizing. This ensures that the equipment is not pulled out of synchronism as a result of a failure of the incoming radio path. At the master station, if the outgoing radio path fails, the equipment will continuously receive repetition cycles containing the RQ signal. Under these conditions, synchronizing is applied to both the transmitter and the receiver and enables the master terminal to maintain system-synchronism even though the radio path to the slave terminal is interrupted and the slave equipment is drifting relative to its true timing. This facility is especially valuable when operating to a slave terminal with poor frequency stability or which does not have the facility of automatically switching off synchronizing when receiving errors.

In response to signals from the phasing-control section of the repetition-control circuits during automatic phasing, the final divider in the receiver—divider chain is caused to omit one element-drive pulse each time an error is received. This moves the relative timing of the receiver distributor and the elements of the aggregate input in steps of one element. Phasing steps are taken at the rate of one for every detected false character received since, when the tested repetition-cycle facility is in operation, all characters must be in correct ratio.

This criterion enables system phase to be established when the marked cycle is in use. During automatic phasing, the synchronizing control operates normally, i.e. synchronizing is applied to the receiver only.

The Marconi ARQ equipment HU121 is shown in Fig 8.15.

Fig 8.15 Autoplex ARQ telegraph equipment HU121 (*Courtesy of the Marconi Co. Ltd*)

7-unit Monitor. As the 7-unit aggregate signals contain the information from two or four channels, and they also have a complex marking pattern, no useful purpose is achieved by a straightforward monitoring of these signals other than to confirm their presence.

To enable useful information to be gained from monitoring the aggregate 'line' signals, a special 7-unit monitor is used. The monitor equipment must include most of the basic features of the ARQ system since it must be able to select any channel or sub-channel from the aggregate, effect synchronizing and phasing, and remove marking inversions. There is no need for a separate decoder: a special Siemens—Halske 7-unit teleprinter associated with the monitor responds direct to the signals of alphabet No. 3. The monitor presents a high impedance (40 kΩ) to the circuit being monitored, has a polarity-reversing switch and can respond to modulation rates of 96 or 192 bauds (or other rates).

Automatic synchronizing is provided. To simplify the phasing equipment the criterion of 'minimum errors' is adopted. An adjustable error

count per ten characters received is controlled by a
switch to enable one phase step to be taken when
1–5 errors (as selected) occur; a lamp indicates a
detected error. False phasing could occur to signals
of a repetitive nature (e.g. RQ cycle, α or β) but
this is prevented by using detectors for these signals.
A phasing button is also available for step-by-step
phasing under manual control. Provision is made
for operation to marked or non-marked character
cycles and to the 4- or 8-character-repetition cycle.

The 7-unit signals of the selected channel are
displayed on lamps and presented in 7-wire simultaneous form for decoding by the teleprinter which
prints the characters of alphabet No. 2, together
with special symbols to indicate the receipt of
errors and RQ signals. The printer does not recognize the α and β signals and it is necessary to rely
on the lamp display to identify them. For sub-
channel monitoring, four switches select for printing one separate character in every four received; a
full-rate channel would be monitored with all four
switches thrown.

The equipment is designed for mounting on a
test trolley or for fitting into a control desk.

Performance. The plots of Fig 8.16 show the
relationship between element-error rate and
character-error rate for the ARQ system, assuming

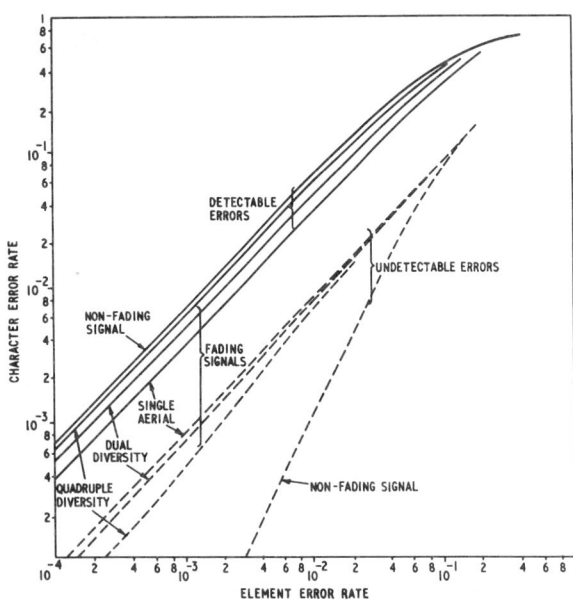

Fig 8.16 Performance of ARQ circuit

random occurrence of element errors within each
character. The graphs relate to propagation under
both non-fading and fading conditions, with single
aerial and dual and quadruple aerial-diversity
systems. Undetectable errors are due to transpositions which maintain a 4:3 Z/A ratio but produce a false character. Such errors can be detected
only by comparison of sent and received 'copy'
under laboratory or field-testing conditons.

The time efficiency of an ARQ circuit falls under
poor propagation conditions as the number of
repetition cycles increases. The Time Efficiency
Factor (TEF) of an ARQ circuit over a given period,
expressed as a percentage, is:

$$\text{TEF} = \frac{\text{Number of characters printed}}{\text{Theoretical clearance rate}} \times 100\%$$

For the system described the character-clearance
rate per circuit is 411 char/min on each channel;
the circuit efficiency would be 100% if no repetition
cycles occurred in either direction during the stated
period and 411 characters were printed.

The plots in Fig 8.17 show time efficiency factor
plotted against detectable character-error rate on
a system which uses a 4-character repetition cycle.
Character errors are assumed to occur randomly and
to show equal mean error rates in both channels of
a circuit. The graphs show the progressive improvement in TEF achieved by using gated RQ and, more
particularly, TRC, especially at the higher character-
error rates.

The relationship between signal/noise ratio and
TEF is shown by the graphs in Fig 8.18, for equal
mean-error rates in both channels of a circuit. The
graphs relate to a modulation rate of 96 bauds on
channels spaced at 170 Hz and are taken under conditions of 40 fades/min using dual aerial diversity.

The CCITT suggests* that for a leased, gentex or
telex circuit which includes a radio path, the
maximum tolerable character-error rate is 10 in 10^5;
this may be compared with a similar recommenda-
tion† that for landline circuits the maximum tolerable error rate is 3 in 10^5 alphabetic telegraph signals. The figure of 10 in 10^5 is at least one order
higher than that achieved by normal radio-telegraph
systems. On typical radio circuits the use of ARQ

*Recommendation F11.

†Recommendation F10.

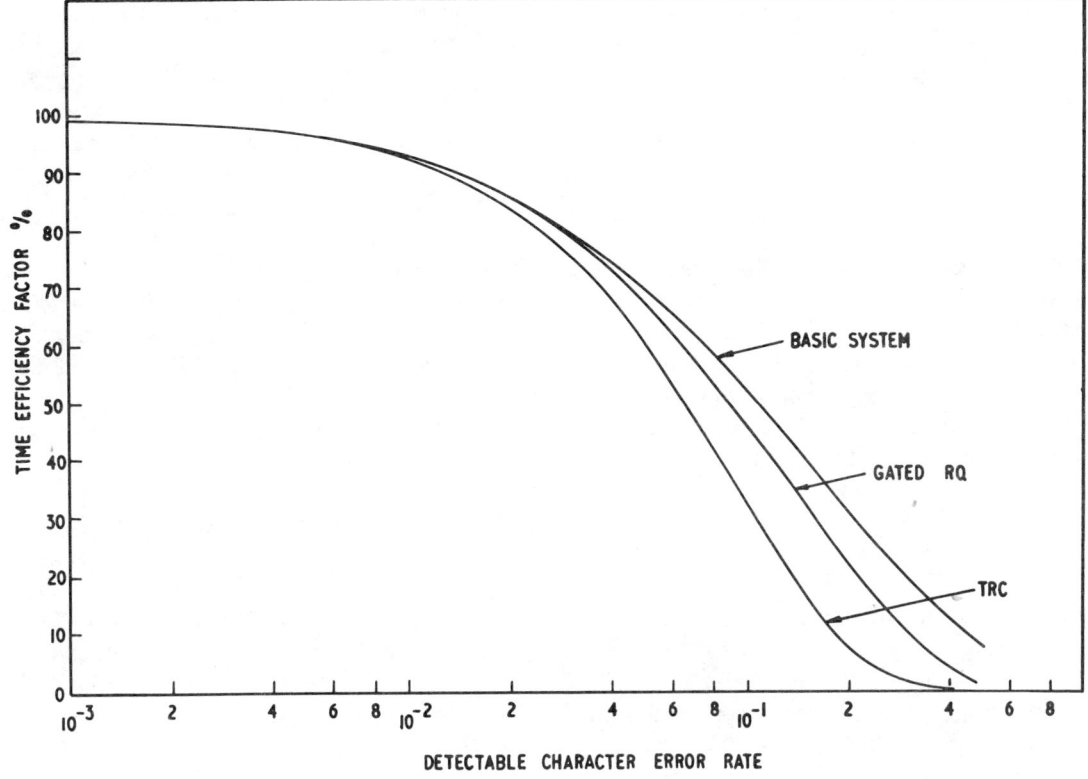

Fig 8.17 Time efficiency factor of ARQ circuit

equipment provides an improvement in error rate of approximately 100 to 1.

8.5 The telex buffer store

Since it is not feasible to control the transmitter at a telex station from the ARQ centre, character storage is provided for each radio-telex circuit at the ARQ centre: this enables a telex subscriber to continue transmitting independently of delays due to RQ cycling. Use of a buffer store permits the arrival of signals at random.

This buffer store originally took the form of the electromechanical reperforator—tape reader combination in which characters are stored in perforated paper tape which can accumulate in a long tape loop during RQ cycling. With a normal reperforator— tape reader pair there is always a residual length of perforated tape joining the two machines to which the tape reader has not present access due to the form and dimensions of these machines. A special form of 'fully-automatic reperforator-transmitter-

distributor' — the FRXD — was used, in which a pivoted tape-reader head could step along the tape and read the latest perforated character. These

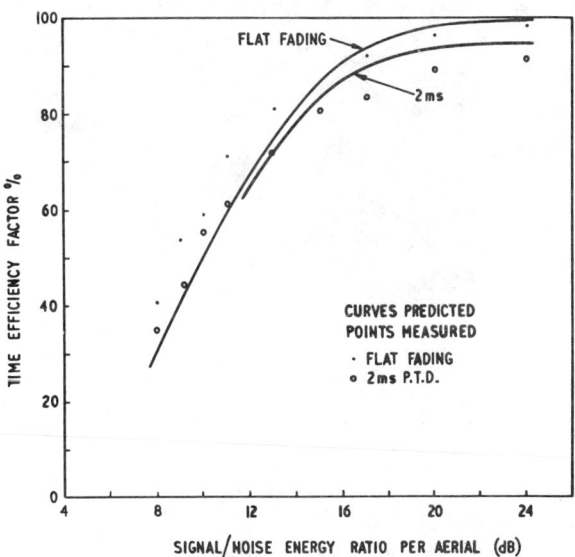

Fig 8.18 Time efficiency factor for various signal/noise ratios

machines met the traffic requirement, but maintenance problems such as risk of tape jamming, and costs of machines which were heavily worked during peak traffic and consumed a large quantity of paper tape, led to more convenient media for character-signal storage.

Magnetic devices in various forms have become available for signal storage — the magnetic drum or disc, magnetic tape, magnetic (ferrite) cores and magnetic thin-film stores. The drum has a large, fixed capacity; the tape has access and other problems; the ferrite core, a static device, is convenient but becomes costly if large amounts of information have to be stored but, on the other hand, access time is of the order of a few μs so that sharing a common store between several circuits offers no fundamental difficulty.

The Magnetic-drum Buffer Store. Each magnetic drum is used as storage for six circuits and together with its control equipment is housed on a pair of standard racks. Fig 8.19 shows two such equipment: the left-hand set with doors removed reveals the enclosed magnetic-drum compartment on the upper part of the left-hand rack, the remaining equipment consisting of printed-circuit logic cards; the right-hand equipment shows supervisory panels for each of the six stores.

The storage capacity required depends on the state of the radio channel. Sequential characters from a telex station can arrive at a maximum rate of 400 char/min; if the efficiency of the radio channel drops to 50% it will clear 411/2 char/min. The difference (400−205 = 195 char/min) has to be absorbed by the buffer store. With the drum it was convenient to allot storage of 4000 characters to each telex circuit, allowing for a call duration of 4000/195 = 20 min (approx.) under the conditions quoted.

The main requirements of the store are that it shall accept and store characters arriving at random at speeds up to 400 char/min; any character must be capable of being read out immediately (within one character period) or, alternatively, after being held for any time up to 4000 character periods. Outputs signals are presented in 5-wire simultaneous form when called out by character-release pulses from the ARQ equipment, which also receives a traffic-in-store supervisory signal from the drum equipment.

A sequential output is available for monitor purposes.

The magnetic drum (Fig 8.20) is a 9-in. metal

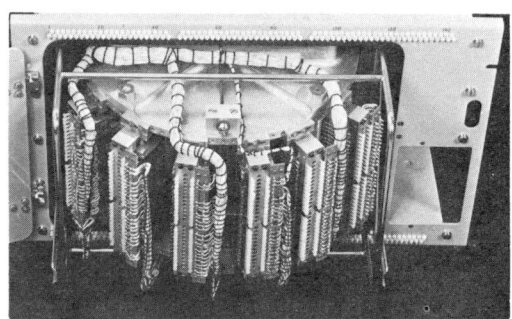

Fig 8.20 Magnetic-drum store showing writing and reading heads (*Courtesy of the Plessey Co.*)

cylinder sprayed with iron oxide particles suspended in a resin base to form the recording medium. Reading and writing heads are fixed to the drum mounting, separated from the cylindrical surface by a small airgap. A current waveform through a winding on a writing head impresses a magnetic-flux pattern on the drum surface as it sweeps past, the flux pattern remaining in the drum surface until it is

Fig 8.19 Telex buffer store using magnetic drum (*Courtesy of The Post Office*)

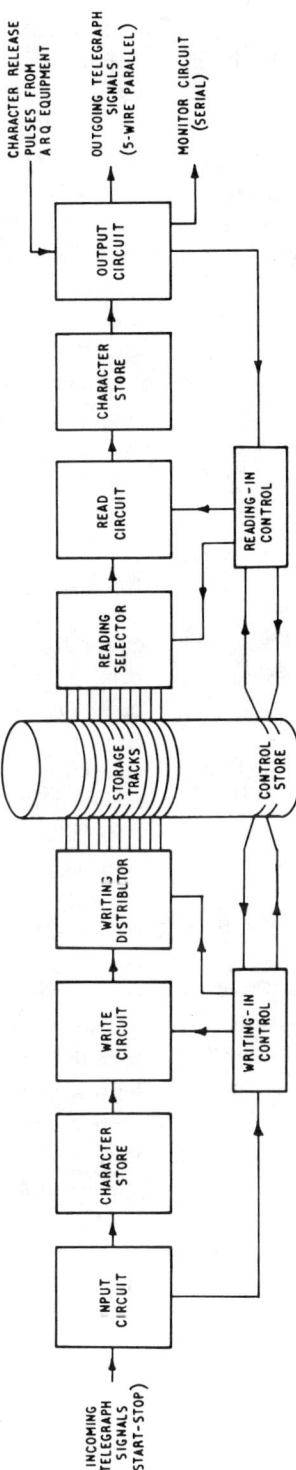

Fig 8.21 Block diagram for one channel of the buffer store

either erased or over-written by applying a further current waveform to the writing head. As the magnetic-flux pattern produced by the writing head of a particular track passes under a reading head associated with the track it induces a voltage in a winding on the reading head. By suitably controlling the phase of the current waveform in the writing head it is possible to store binary information on a drum track; the two flux patterns recorded on the track produce different output waveforms from a reading head and enable digits 0 or 1 to be detected.

The drum is driven by an induction motor in the drum, at a speed maintained at 1800 rev/min by eddy-current braking. The drum is 5 in. high and can provide up to 100 circumferential tracks, each of which can store 466 6-bit characters, a bit occupying approximately 0·01 in. along the track. To guard against possible failure, six telex circuits only are served by each drum; each circuit is allocated ten tracks which allows storage up to 4600 char/circuit. Other storage positions and tracks are required for control and supervisory purposes.

Writing in. An explanatory diagram of the equipment for one channel is shown in Fig 8.21; each control unit serves three storage channels.

Incoming messages arrive as a succession of start—stop signals and are received by the input circuit which assembles each character, element by element, and temporarily stores it pending the arrival of the allocated drum-track position at the writing head. An instruction from the writing-in control circuit causes the character to be transferred into the appropriate storage location on the magnetic drum.

The transfer of a character from the input circuit to the drum takes place as follows:

(1) A signal from the input circuit indicates to the writing-in control circuit that a complete character is held.

(2) The write circuit is switched on when the next free position in the currently-used storage track reaches the write head; the character is then written into this storage position and the write circuit is switched off.

(3) A record is kept of occupied and free positions on storage tracks by means of a control-store track on the drum. Although each channel has ten storage tracks, writing-in takes place only on one at a time, and the control track maintains the 'occupied/free' record for the particular storage track in use. As each character is inserted in a storage-track position the corresponding position on the control track is marked engaged accordingly.

(4) As each storage track is filled the write circuit is switched to the next track in the group of ten by means of the writing distributor — a 10-state ring counter. This is stepped to successive tracks in cyclic rotation by a signal from the writing-in control circuit. The stepping signal occurs when the final character position on a storage track is occupied; at the same time the control track is cleared ready to provide the 'occupied/free' record for the new storage track.

Reading out. This process is basically the reverse of writing in. Each character is transferred to the output circuit from whichever storage track is currently being used for reading out. When a character-release pulse arrives from the ARQ equipment the character is transferred from the output circuit to the ARQ equipment, where it appears on five wires in simultaneous form. As each character is called in to the ARQ equipment the next character to be read out is transferred from the storage track to the output circuit. Each outgoing character is also reproduced in start—stop form and this allows a monitor teleprinter, or other device requiring sequential drive signals, to be used for test purposes.

Functional units used in the writing-in process have their reading-out counterparts, e.g. the particular storage track from which reading out is taking place is determined by the reading selector. The latter is stepped on from track to track, in the same cyclic order as the writing distributor, by signals from the reading-out control circuit. The same control track is used for both writing-in and reading-out functions, separate parts of it being allocated to each purpose. Writing in and reading out are independent processes and, at any one time, may be taking place on the same or different storage tracks.

At the end of a telex call the store will have cleared all its traffic for this call and the writing in will have reached a random point in the store. For subsequent traffic, writing in will commence from the point reached for the last call.

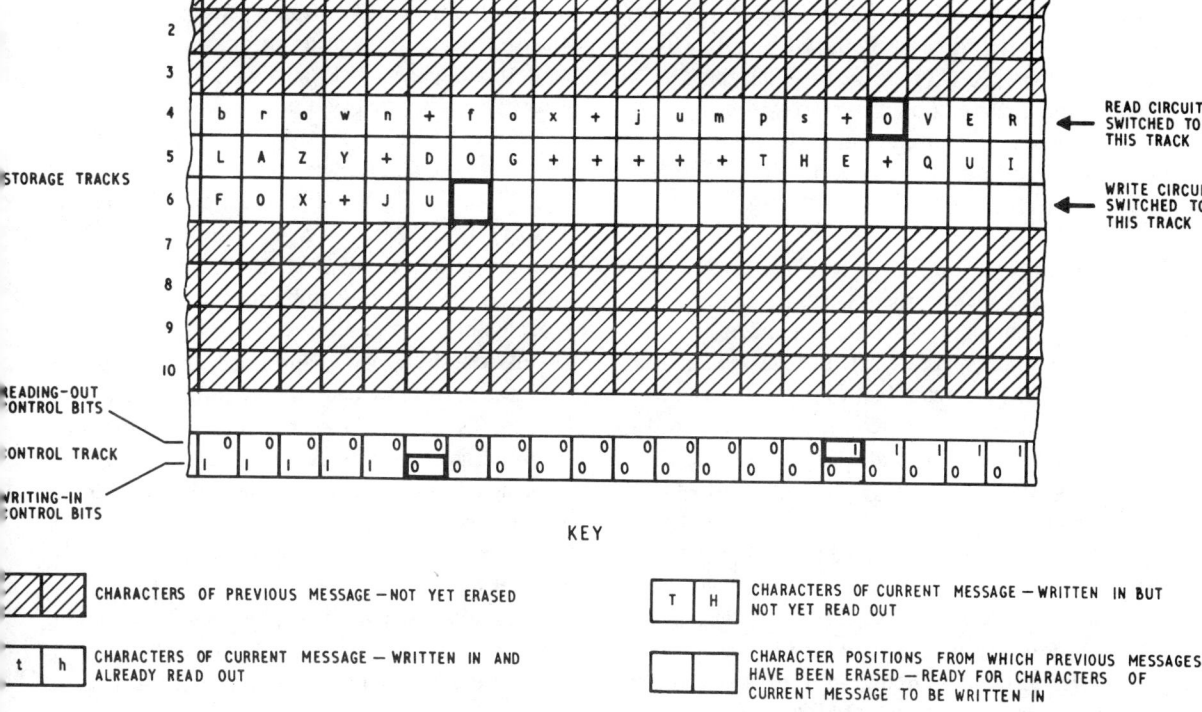

Fig. 8.22 Writing-in and reading-out control

Storage and Control Tracks. The storage and control tracks occupy the complete drum circumference. Fig 8.22 shows portions of the ten storage tracks and the control track associated with one channel, individual character positions being denoted by squares.

On storage tracks, each character is stored in binary form as a 0 and 1 pattern. Only five out of the six bits in each character position are used (the sixth is for supervisory purposes).

Each character position on the storage tracks is associated with a character position on the control track. On the latter, one of the six available bits per position controls the writing-in operation and another the reading-out operation. Starting with an empty storage track and digit 0 in all writing-in control positions, as each character is stored digit 1 is inserted in the writing-in bit position of the corresponding control-track character position. Similarly, in reading out from a previously-filled storage track (initially with digit 1 in all reading-out control bit positions), as each character is read digit 0 is inserted in the appropriate reading-out control-bit position.

The character positions on the control track are in advance of the corresponding ones on the storage tracks. This is illustrated in the diagram where the next positions to be used for writing in (on storage track 6) and reading out (on storage track 4), together with the corresponding control-track character positions, are denoted by heavy outlines.

For explanatory purposes it is assumed that the message currently being handled on the channel to which Fig 8.22 relates consists of a series of repetitions of the sentence THE + QUICK + BROWN + FOX + . . . Storage of this message has started at the beginning of storage track 4. As each storage track in turn is selected by the writing distributor, all earlier information on that track is erased in one revolution of the drum and the writing-in bits of all character positions on the control track are at the same time set to 0. The storage track is then filled, character by character, and on the control-track character positions digits 1 are correspondingly substituted for 0 as the writing-in indication. In the diagram, the next character to be written in will be letter M in the word JUMPS; when this is

done the 0 shown in the corresponding writing-in
control-bit position will be changed to 1. Similarly,
the next character to be *read out* will be letter O
of the word OVER. On the control track the
corresponding read-out control bit 1 will be changed
to 0, a 1 having been inserted in all read-out positions
at the start of reading out a new track.

Input Circuit. Incoming characters are sampled at
the midpoint of their element periods. As the
polarity of each code element is recognized it is
recorded in a shift register. When the last element of
a character has been recorded, filling the shift
register, the complete character is transferred to the
incoming character store and a signal is extended to
the writing-in control circuit, indicating that a complete character is available for transfer to the drum.

Sampling times are determined by a stable multivibrator which is triggered at the leading edge of each
true start element. A short-start rejection circuit
guards against effects of line interruptions.

The standard input circuit, with receive margin
of ±47%, accepts double-current 50-baud signals.
Following the sampling process all functions within
the input circuit are performed by standard bistable
trigger circuits and gates.

Control Track. The control track and delay track
together form a regenerative loop, the writing process being continuous. Since each channel requires
only one writing-in and one reading-out control bit
(and there are six bits within the character position)
it is possible to share a control track between three
channels.

Strobe and Clock System. One track on the drum —
the 'strobe' track — contains a permanently-stored
sequence of 2796 zeros occupying the complete
circumference. The signal induced in the associated
read head, after amplification and shaping, is used to
generate strobe pulses. These are the basic timing
signals, synchronized with the drum rotation, which
drive the clock system and determine the exact
instants at which the various logical operations occur.

Another track — the clock track — contains a
permanently-stored sequence of 2795 zeros and a

single 1 bit. This 1 bit provides a reference point for
locating the first bit position on the storage and
control tracks. Character positions on these tracks
are defined by TB clock signals generated by two
separate clocks; one cycle of TB clock signals
(TWB1–TWB466) is phased with the writing (W)
process on the storage and control tracks, the other
cycle (TRB1–TRB466) being phased with the
reading (R) process. The single 1 on the clock
track initiates both the TWB and the TRB clock
cycles, as it is read in turn by read heads individual
to each clock.

Only the first, second and final character
positions need to be distinguished; the TWB and
TRB clocks are simple, each containing only four
bistable trigger circuits. A TA clock produces
clock signals TA1–TA6, defining individual bits
within the character positions. The phase
difference between TWB and TRB clock signals is
an exact multiple of six bit times, permitting the
same TA clock signals to be used in both writing
and reading operations. Fig 8.23 illustrates the
TA/TB bit and character-timing arrangement.

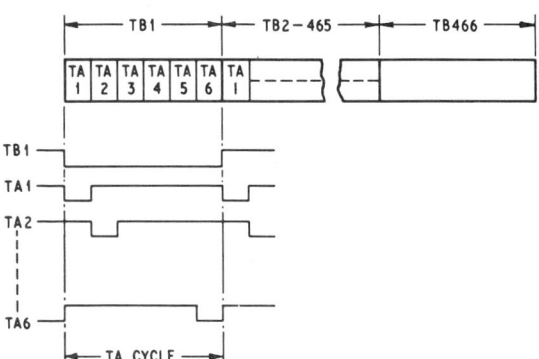

Fig 8.23 Storage-track timing on magnetic drum

The output of the clock track has a frequency
directly proportional to drum speed. This signal
is amplified and applied to a frequency-sensitive network whose output is rectified and used to control
the drum speed.

Supervisory Features. Facilities on each channel
include a lamp display of characters held in the
incoming and outgoing character stores; a lamp
display of the channel occupancy (derived from

the track count); and local and remote STORE CLEAR and RESET controls. A counting circuit registers the number of storage tracks in use at any time: since there are ten storage tracks for each channel the proportion of channel-storage capacity currently in use is displayed in 10% steps.

To prevent any portion of an earlier message being read out, even under fault conditions, all previously-stored information is erased before a newly-selected storage track is used for writing in. This is done by writing check codes into spare positions at beginning and end of the storage track during the erasing operation; these are subsequently read and if found incorrect any further writing in or reading out is inhibited and an alarm given.

Alarms indicate when the percentage occupancy of the channel has reached a predetermined limit; separate alarms can be arranged at any two 10%-step levels.

A key, suitably protected against accidental operation, is provided for erasing all message information from the drum.

Magnetic-Tape Buffer Store. The Philips TMBO tape buffer store uses a tape deck with an endless tape loop and has a capacity of 4000 characters. A tape unit is associated with each outgoing telex radio circuit.

Writing and reading are performed with the tape stationary: it moves 2 mm for each character stored. Code elements are stored in the tape in 5-unit parallel form on five tracks; a synchronizing signal is recorded on a sixth track. A unique feature is the inclusion of a 1-character 'bridging' store using transistor bistable triggers to overcome the delay while the tape travels between input and output heads. This enables a character to be re-transmitted immediately once it has been received into the store.

If the radio path is good and no storage other than that on account of the random arrival of characters is required, characters are accepted in sequential form at the input, transferred in parallel form by the receiving distributor and forwarded to the electronic bridging store (Fig 8.24). Within 10 ms of the time the first character starts entering the store, a 'traffic-in-store' signal is sent from the control circuit to the ARQ telex panel.

When a release pulse is received from the ARQ equipment the character is passed on via the output store to the parallel output — or, if sequential output is required, to the transmitting distributor, in which case the information is not recorded on the tape.

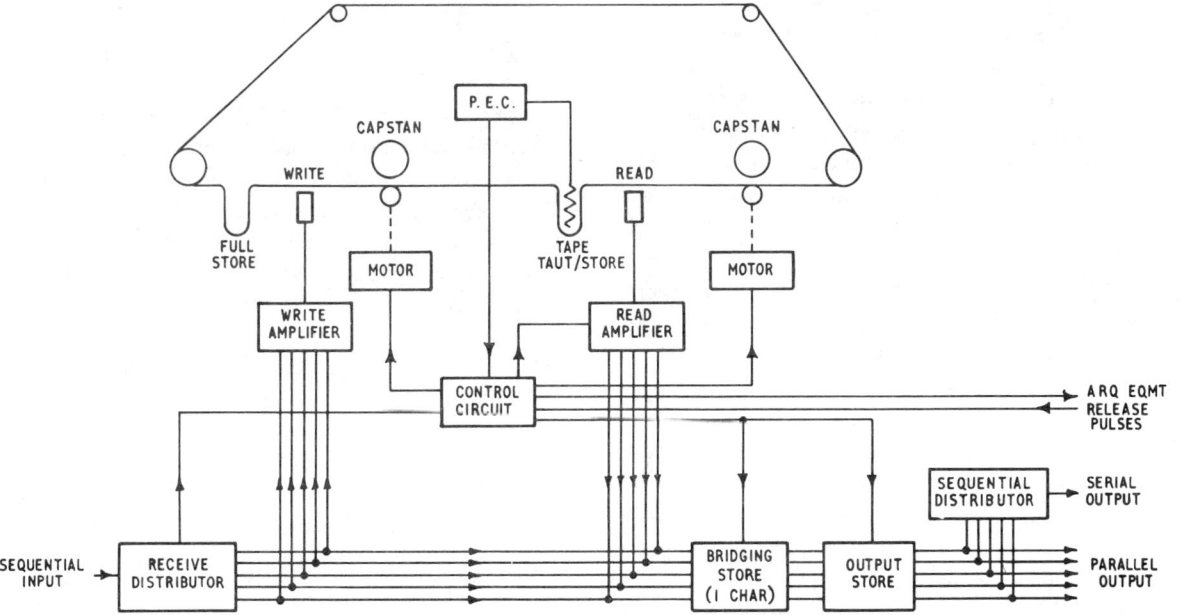

Fig. 8.24 The Philips TMBO magnetic-tape buffer store

If the radio path is cycling and characters have to be stored, the traffic-in-store signal is not followed by receipt of a release pulse; the first character then remains in the transistor 1-character store. Subsequent characters have to be stored in the tape; the control circuit ensures automatic change-over to the tape store.

The information is now passed on to the writing amplifier which forwards the five code elements, together with a synchronization element, to the writing head; at the same time, the tape starts running from the writing to the reading head. The writing head records the information on the fast-moving tape which transports the stored characters to the reading head, and there it will stay until a release pulse has been received. As soon as the written information exceeds the read-out information, so forming a tape-storage loop, the tape changes from fast-running to step-by-step motion.

The tape is fed by means of capstans and rubber rollers driven by stepping motors, the motion of which is determined by the control unit on the basis of incoming information and the tape condition. The condition of tape-taut, small-loop or full-store is indicated by lightweight brackets which can be moved by the tape to operate a photoelectric-cell unit.

After receipt of a release pulse the character which is still in the bridging store is passed on to the output store and thence to the output. In the meantime, subsequent characters which are under the reading head are read out and passed on to the bridging store via the reading amplifier. The electronic 1-character storage device and the output store ensure that no gaps can occur in the outgoing traffic flow.

For the output circuits, use is made of mercury-wetted relays; alternatively, low-voltage, low-power output can be provided by electronic relay.

The tape unit provides the normal supervisory signals to the ARQ telex panel. On the front of the unit panel are lamps which indicate when (1) traffic is in store; (2) more than five characters are in store; or (3) traffic is being emitted by the store. A teleprinter can be plugged in for monitoring purposes using a sequential output from the tape unit. All recorded information can be erased if necessary by a signal from the telex panel or from a switch on the unit, adequately protected against accidental operation.

The receive margin is ±45%; output distortion is less than 5%.

Stored-program Shared Ferrite-Core Buffer Store.

The Hasler buffer store is a stored-program information processor, static in operation and requiring very little maintenance attention.

The processor is programmed to carry out required functions after instructions in an 8-unit machine code have been read into its program store. The heart of the equipment is the processor control (Fig 8.25) – a ferrite-core store with instruction counter for continuous processing of the program and information-processing loops which are called in by the program. Input and output units provide an interface between the processor and 15 traffic circuits. An additional circuit is provided to a teleprinter for printing in additional operating instructions and receiving information from the equipment. A separate input is used for feeding in the program for a tape reader.

The ferrite-core store has a capacity of approximately 24 000 characters, shared between a maximum of 15 telex circuits, which is equivalent to a storage of about 1500 char/circuit ($3\frac{3}{4}$ min at 400 char/min) if all circuits demand this storage simultaneously. At a large ARQ station this is an unlikely event since the storage load can be spread by judicious selection of radio routes with peak demands occurring at different times of the day. Each circuit is allocated a fixed minimum storage of 512 characters; at the other extreme, if only one or two circuits are demanding storage beyond this figure, each can store up to 8000 characters by drawing on the store capacity which is common to the group of circuits. Alternatively, the store is available in capacities of 8000 or 12 000 characters which may be sufficient in view of a CCITT Recommendation* that telex calls on ARQ circuits should be forcibly released if the circuit efficiency falls below 80% for a period of 20 s, or when a comparatively limited store capacity (e.g. 700–800 characters) has been filled.

*Recommendation U23.

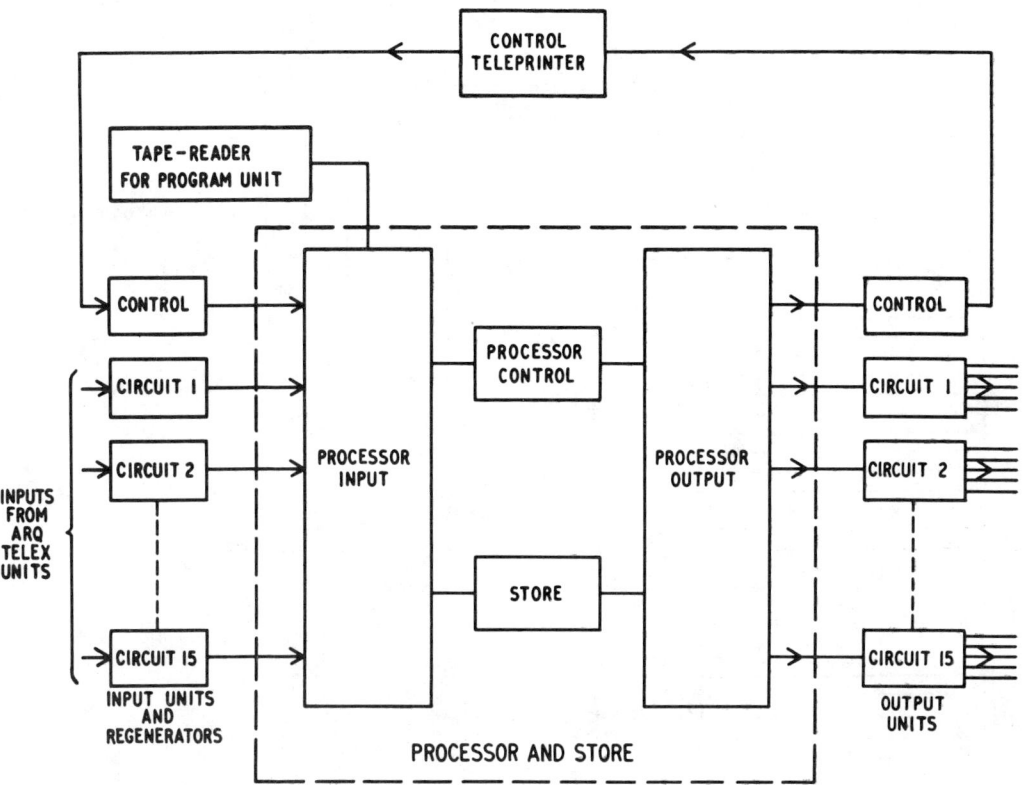

Fig 8.25 General arrangements of stored-program buffer store

The Core Store. The store consists of a series of ferrite-core planes, each comprising 64 x 64 = 4096 cores. There are 46 planes in two groups (Fig 8.26). For storing program instructions an 8-bit code is used; two zones each of eight planes are required, giving a capacity of 8192 eight-bit computer words. In the section of the store which caters for traffic information in 5-unit code, six zones of five planes each provide a capacity for 6 x 4096 = 24 576 words; this gives a total of 32 768 (= 2^{15}) word positions requiring addresses. A code containing 15 bits is necessary for addressing – six bits for the X-axis, six for the Y-axis, and three for the zone (Z). These are incorporated in one 8-bit and one 7-bit word as follows:

$$\underbrace{e_7 \quad e_6}_{Y\text{-axis}} \quad \underbrace{e_5 \quad e_4 \quad e_3 \quad e_2 \quad e_1 \quad e_0}_{X\text{-axis}} \quad IX \text{ address}$$

$$\underbrace{e_6 \quad e_5 \quad e_4}_{Z\text{-axis}} \quad \underbrace{e_3 \quad e_2 \quad e_1 \quad e_0}_{Y\text{-axis}} \quad IY \text{ address}$$

To address the program store, one bit only is required as there are but two zones.

The information store is divided into 96 blocks of 256 five-unit traffic characters each, and each circuit is given a minimum storage capacity of two blocks, or 512 characters. Traffic information is stored consecutively in the first block, the second serving as a standby; once the first is filled, traffic continues and is stored in the second block. At the same time a search is made to earmark a third and this process can continue, providing that there are blocks unallocated, up to a maximum of 32. Stored characters are transmitted consecutively from the first block, and when transmission from the second commences, the first is released into the common pool. From 1–256 characters can be stored in each block and the maximum storage capacity one circuit can obtain is 7937–8192 characters. All circuits together can store a total of 20 496–24 576 characters.

The Stored Program. The program store is supplied with a series of instructions for the central processor in 8-bit computer words. A circuit block which

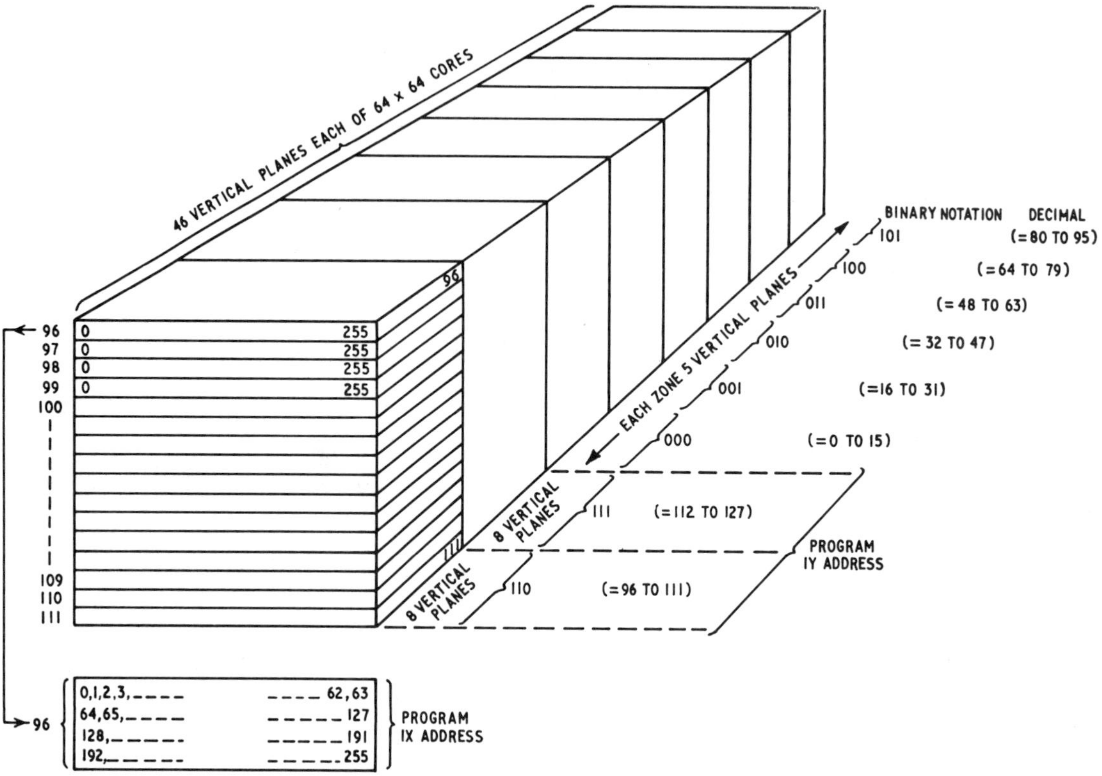

96 VERTICAL PLANES EACH OF 64 x 64 CORES

	BINARY NOTATION	DECIMAL
101		(= 80 TO 95)
100		(= 64 TO 79)
011		(= 48 TO 63)
010		(= 32 TO 47)
001		(= 16 TO 31)
000		(= 0 TO 15)

EACH ZONE 5 VERTICAL PLANES

96	0	255
97	0	255
98	0	255
99	0	255
100		
109		
110		
111		

8 VERTICAL PLANES   III   (= 112 TO 127)

8 VERTICAL PLANES   110   (= 96 TO 111)

PROGRAM IY ADDRESS

0,1,2,3, ____	____ 62,63
64,65, ____	____ 127
128, ____	____ 191
192, ____	____ 255

PROGRAM IX ADDRESS

96

Fig 8.26 The Elstore core assembly showing addressing method

delivers information (the 'source') is specified by three bits in binary code since eight transfers of information are required. A circuit block which receives information (the 'sink') requires five bits to specify any one sink among 28. Many combinations between source and sink can be effected by the 8-bit program word of which the first three bits specify the source and the last five the sink.

The transfer of instructions takes place on two 8-wire 'highways' (Fig 8.27) — a 'collecting' highway which connects all sources and a 'distributing' highway which is connected to all sinks and distributes information from the sources to the sinks. Use of 8-wire simultaneous signal paths reduces the time required for routing and transferring information. An instruction is normally carried out in one storage cycle (6·5 μs) but some operations, such as writing and reading, require two cycles (13 μs). The cycle is the sequence in which the store accepts an address, reads out the corresponding cell and either rewrites the old information or writes in new information.

The program of instructions is stored in the pro-

gram zone of the core store in the form of sequences of instructions having consecutive addresses, a succeeding instruction usually being located in the next higher storage address. All instructions are given a specific program-store address (designated the PX

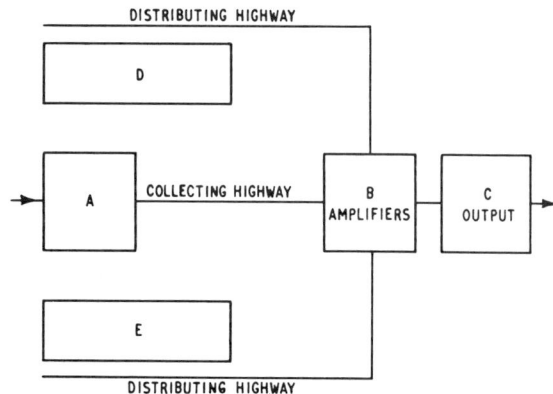

Fig 8.27 Stored-program core store — outline of central processor

and PY addresses); during each cycle the address is automatically stepped on so that a succeeding cycle causes the extraction and execution of the next instruction. To execute a specific program it is necessary only to set the program-store address of the first instruction for this program, and during normal operation this is performed by the preceding program. A jump from one instruction sequence to another may take place either as a result of reading into the program-store address the start address for the succeeding program sequence, or as a result of a combination of decision circuits.

The series of functions is related to the incoming traffic signals and to the calling-out pulses from the ARQ equipment. The periodicity of the program is based upon the 20-ms element of the input signals, and that for the transfer of information on a character basis occurs every 140 ms; other programs occur at longer intervals. A storage program transfers characters from receive-buffer stores to the information store, and from there to the send-buffer store as required.

The control program allocates the storage position to each channel, registers the allocated position and the number of characters stored, and responds to instructions received on the control channel. The program information is fed in from a tape reader at 250 char/s; an equipment can be completely programmed in $2\frac{1}{4}$ min. The information on the tape is in groups of four characters:

(1) Synchronizing character (binary 27).
(2) The Y-address.
(3) The X-address.
(4) The required instructions to be inserted in that address.

The Processor. The program repertoire contains instructions necessary to execute all logical functions. To do this, information is transferred between sources and sinks. The source outputs are normally closed, and the designated one is opened at the correct instant in the program cycle, releasing the source information on to the collecting highway. Likewise, the sink inputs are normally closed, the designated one opening 3.5 μs after the source opens. As an example of the operation, to transfer information from register R1 (binary notion 010) to the

store SR (binary 10100), the 8-bit instruction will be in the form:

$$\underbrace{0 \; 1 \; 0}_{\substack{\text{source} \\ \text{designation}}} \quad \underbrace{1 \; 0 \; 1 \; 0 \; 0}_{\substack{\text{sink} \\ \text{designation}}}$$

In the program flow charts, a repertoire sheet shows the eight sources with binary and equivalent decimal notation, and similarly the 28 sinks and their notations. In addition there are five further instructions used for special functions not involving source-to-sink transfers.

The processor may be divided into five sections designated A–E (Fig 8.27):

Section A is composed of input-circuit units with associated signal regenerators and the input matrix or address register for selection of the desired channel output. There is also the core store with its division into two parts, the information store, and the program store containing the control program. An instruction decoder, with its associated buffer register, receives the instruction word from the source and puts it into the specified sink. It also directs the special operations not involving a source/sink transfer and controls most of the processor operations.

Section B contains amplifiers and connecting circuits which act as interfaces between the collecting and distributing highways.

Section C contains channel-output circuits, address register and address decoder for the selection of any channel output.

Section D is equipped with address registers containing addresses of program and information-store positions. The addresses are used to extract stored instructions and to read information from core-store locations assigned with their specific addresses. They are also used for writing instructions or information in the store locations with the assigned addresses. Five registers are used for addressing program-store locations, while two more are used for addressing the whole of the core store, although they are used mainly for addressing information-store locations.

Section E contains units for special functions such

as an accumulator register, the contents of which can be added to a word during its transfer from a source to certain sinks; also comparator circuits in which an 8-bit constant is put in and compared with an unknown 8-bit word. If the two are equal, a condition is put out to cause a required function. This section also contains a single-bit evaluating circuit in which it is possible to shift the position of a specified bit an arbitrary number of positions, thus influencing any desired output.

The Equipment. The incoming-line unit contains the signal-voltage-changer devices and inspects the incoming line every 20 ms; a regenerator corrects any signal distortion present. All inputs are combined in an input matrix, consisting of semiconductor elements, through which passes all the incoming flow of information to the central processor. The 16 inputs are divided into two groups of eight, and the state of them is read into the processor by a scanning instruction, each group being dealt with in turn but both within the 20-ms cycle. Similarly, all outputs from the equipment are combined in an output matrix, where they are also dealt with in two groups of eight alternately. Each output is equipped with a bistable trigger circuit to maintain the signal state between changes of condition, and provide signal-output-voltage conversion. Signals, pulses, supervisory signals and control pulses for indicator lamps are all (apart from the monitor output) transmitted via the output matrix.

All pulses are derived from a 7·68-MHz oscillator, with a frequency stability not worse than 10^{-4}. A series of divider chains provides pulses of the periodicities required.

A view of a single 15-circuit rack, together with the control position, is shown in Fig 8.28. There are seven shelves of plug-in 'book' units: one contains the core store and associated amplifier units; another is for control and monitoring; and three are occupied by mains-operated stabilized power units, which include a nickel-cadmium floating battery to maintain continuity during short breaks in mains-power supply.

Silicon transistor modules, some in integrated form, are used throughout. Signal output is at ±6 V to match the requirements of the ARQ equipment.

Fig 8.28 Elstore buffer store with control position (*Courtesy of Hasler AG*)

Commands and interrogations can be initiated by the control teleprinter; printed confirmation or answers are given from the equipment. Lamps and audible alarms indicate equipment failures, some of which are advised by automatic print-out on the control teleprinter. A control panel provides facilities for maintenance operations.

8.6 The ARQ-telex interface
Signals are transmitted over the synchronous ARQ system at specific instants; there is normally some delay – of variable amount, which may include ARQ repetition cycling – before signals are sent over the radio path. For telex service, provision has to be made for extracting and repeating supervisory signals* such as the call (Z) and clear (A); these can be transmitted over the radio path only in 7-unit form as α and β signals. Information must be passed to the switchboard on operator-controlled outward calls to indicate periods when the radio circuit is cycling so that the subscriber is not charged for these.

The ARQ Telex Panel. Between the telex exchange and the ARQ equipment, a control unit, the 'telex panel', is required, in addition to the buffer store, to provide control signals and interface conditions;

*These signals are described in *The International Telex Service* (R. N. Renton), Pitman.

this panel is part of the ARQ equipment. The relationship between units is illustrated in Fig 8.29.

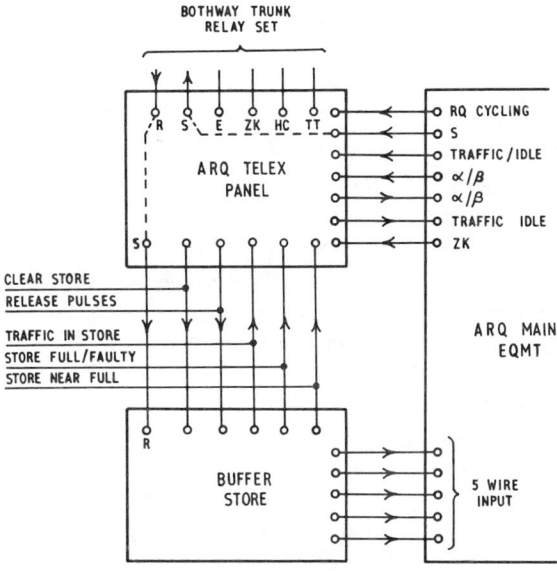

Fig 8.29 Intercommunication between exchange, telex panel, buffer store and ARQ equipment

ARQ panel to telex bothway trunk relay set

S and R, are the normal send and receive wires operated by 50-baud start–stop 5-unit signals at ±80 V.

E, indicates to the exchange equipment that the radio circuit is in use for an incoming or outgoing call; it is at earth potential on receipt of a β combination over the radio channel, or on receipt of Z polarity on the R wire, otherwise it is open-circuit. If, during the free-line condition, receipt of a β combination is not followed by a call signal, earth potential is maintained until two consecutive α combinations are received. Under all other conditions, earth potential, once applied, is maintained until the telex panel has transmitted the clear signal and detected the clear-confirmation signal.

CR, monitors the transmission of ungated character-release pulses from the telex panel; earth-potential pulses are coincident with the transmission of ungated character-release pulses on the ZK wire.

HC, indicates to exchange equipment that the radio circuit is to be busied for maintenance purposes or for certain radio circuit-fault conditions;

negative potential is applied when the BUSY key is operated or when the buffer store signals that storage is approaching exhaustion; earth potential is applied when the buffer store is out of order or the BUSY key is normal, otherwise open-circuit.

LB, earth potential when the BUSY key is operated, otherwise open-circuit.

MI, monitor signals received over the radio circuit at the panel-output S wire.

TC, earth potential when the TEST TELEPRINTER key is operated.

TE, indicates to the exchange when the telex circuit is engaged on incoming or outgoing call; earth potential is applied whenever the radio circuit is engaged (incoming or outgoing).

TT, indicates when the buffer store is in use; earth potential is applied when the store has one or more characters stored.

ZK, character-release pulses relayed from the main ARQ equipment; earth-potential pulse (44 ms) whenever an α, β or teleprinter character has been accepted for transmission.

ARQ panel to buffer store

CE, effects high-speed clearance of characters in storage; +6 V is applied when a clear signal is received over the radio channel and (if necessary) when a clear signal is received from the exchange. The condition is maintained until –6 V is removed from the CS wire; the rest condition (–6 V) is then restored on CE wire.

CM, causes the buffer store to release the next character; rest condition is –6 V. A +6 V pulse (44 ms) is applied for each character to be released.

CS, indicates to the telex panel when there are one or more characters stored; –6 V is applied when characters are stored and there is no fault, otherwise +6 V.

MF, indicates to telex panel that the buffer store is out of order. During normal conditions –6 V is applied, changed to +6 V or open-circuit when the store is out of order, not available to traffic or storage capacity is exhausted.

S, the signalling path from the telex panel to the buffer store.

TL, indicates when storage capacity is approaching exhaustion (+6 V – otherwise –6 V).

ZA, ZB, ZC, ZD, ZE, 5-wire simultaneous-signal

paths for code elements from the buffer store (±6 V).

ARQ panel to ARQ main equipment
Five-wire simultaneous input for code elements (±6 V).

Transmitter, traffic/idle conditions.

Transmitter supervisory signals, α and β.

Sequential 5-unit output wire.

Receiver, traffic/idle conditions: −6 V during normal traffic; +6 V pulse (44 ms) for an α or β combination received.

Receiver supervisory wire: −6 V on receipt of β combination, and +6 V pulse (44 ms) on receipt of α combination.

Automatic transmitter drive pulse: +6 V for 44 ms once per character; rest condition −6 V.

External cycling conditions: +6 V for the first half of each repetition cycle; −6 V rest condition.

Provision is made for a test teleprinter to send or receive into either the exchange or radio equipment.

When an outgoing call is received from the exchange, the telex panel sends a signal to the ARQ equipment which causes the appropriate transmitter-channel supervisory signal to change from α to β. This results in a call signal being received at the distant exchange, the latter responding by changing its supervisory condition from α to β. Receipt of β combinations at the originating terminal station applies the answering call-confirmation signal to the exchange equipment. Normally, two consecutive β combinations must be received as call confirmation at the originating ARQ terminal before a traffic circuit can be established, but traffic can follow a call signal without waiting for call confirmation.

A call originating at the distant terminal results in supervisory signals received over the radio path changing from α to β. Two consecutive β combinations must be received before the calling signal is extended to the exchange. The circuit ensures that at least four consecutive β combinations, without repetition cycling, are sent as call confirmation.

Telex signals from the subscriber are relayed by the telex panel to the buffer store, where they are held until the radio path is available for their transmission; the store provides an indication to the telex panel that traffic is held there. Subject to the radio path being in a condition to transmit the informa-

tion, characters from store are presented to the ARQ equipment under the joint control of the ARQ main equipment and the telex panel. Character storage is not necessary for received traffic over the radio path: this is relayed direct to the exchange equipment by the telex panel.

On completion of an outgoing call the clear signal from the exchange is registered in the telex panel, which either continues to control transmission of stored traffic or else rapidly erases it. When the store is empty the telex panel causes the ARQ equipment to send consecutive α combinations as a clear signal. At the distant terminal, receipt of two consecutive α combinations is accepted as a clear signal, confirmed by transmission of consecutive α combinations. Receipt of two of these consecutively at the originating terminal causes a clear-confirmation signal to be sent to the exchange equipment. The circuit enables α combinations to be transmitted as a clear signal when there is still traffic in store but the storage equipment is faulty.

The telex panel does not restore to the free-line condition until it has (1) caused the sending of seven or eleven α combinations (without repetition cycling) as a clear signal, and (2) received two consecutive α combinations as a clear-confirmation signal. The reason for sending seven or eleven α combinations before clearing down is to ensure that the distant terminal will receive two consecutively to complete its own clearing of the call. Seven α combinations are sent when the ARQ equipment is operating to a 4-character RQ cycle, or eleven are required for an 8-character RQ cycle. If a clear signal is received from the distant terminal during an outgoing call, no more traffic is presented to the buffer store and the latter is cleared by an erasing signal from the telex panel.

Circuits provided by communication satellites are now a normal means of providing commercial international circuits; they are very stable and do not require error-detection or correction facilities. HF radio circuits provide service on routes not accessible by a more stable medium.

8.7 Integrated Programmed ARQ System
A computer can be programmed to provide all the operational facilities of the normal ARQ system — multiplexing, synchronizing, phasing, 5/7-unit

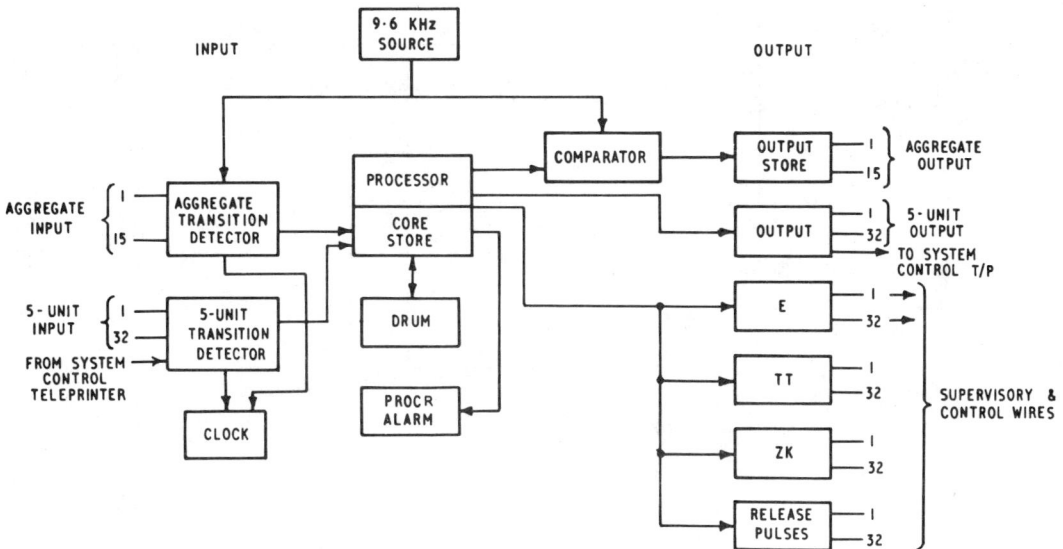

Fig 8.30 Block diagram of integrated ARQ system

encoding, 7/5-unit decoding, error detecting, RQ cycling, sub-channelling, controlling remote automatic transmitters, telex signalling, buffer storage, etc. The integrated ARQ equipment will work to any other form of standardized ARQ equipment as 2-circuit or 4-circuit systems, with or without marked-character cycle, sub-channels, etc. Such facilities as sequential-to-simultaneous conversion, telex control (excluding supervision), or sub-channelling have no need for physically-separate equipment units; the facilities are provided entirely as the result of programming.

The main equipment of an integrated ARQ system comprises magnetic-core storage units, magnetic-drum storage, line input and output access units for start—stop and aggregrate circuits, supervisory-control circuits for leased lines and telex lines, and a teleprinter for subsidiary program-control purposes. A simplified block diagram of the equipment is given in Fig 8.30.

To enable a single computer-type equipment to control a large number of telegraph circuits, both 5-unit and aggregate, advantage is taken of the high speed at which the computer operates. In a further application of the time-division principle, the condition of each incoming-line circuit is examined repeatedly for extremely brief intervals; according to the condition found on each line, and dependent upon the program written into the computer prior

to installation, the processor causes the required condition to be applied to the outgoing-line circuits.

Characters from line, both 50-baud start—stop and 96/192-baud aggregate, are accepted bit by bit for examination by the transition detector which determines, against the clock and counter, the instant at which a transition takes place and its direction. This information is placed in core storage. A program determines the polarity of elements received and passes these to a character-assembly program. Short-start rejection and protection against noise are provided.

Fig 8.31 illustrates how each aggregate line is scanned by the common equipment. The processor is controlled from a 9·6-kHz crystal-controlled oscillator, so that an element at 96 bauds can be divided into 100 equal parts. At these intervals the counter and all transition detectors jointly supply signals to the core store to indicate the instant when any transitions occurs, and also its sense (0 to 1, or 1 to 0); the passing of the counter from 100 to 0 is also recorded as a time reference. The sketch illustrates the form in which these signals are recorded and how they are formed into a separate table for each line circuit. This process enables the correct midpoint sampling instant to be determined and synchronism to be automatically adjusted. The 5-unit incoming lines, leased or telex, are similarly scanned and their signals recorded.

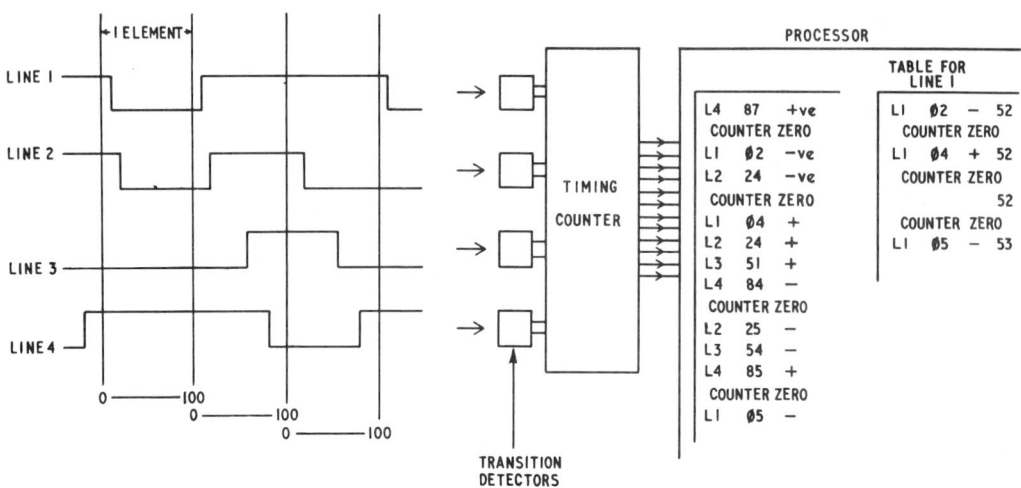

Fig 8.31 Integrated ARQ system — line scanning

The Synchronization Program. When the core store contains entries showing a list of aggregate lines and the times at which transitions have occurred on those lines, the synchronization program investigates, for each line, whether they occur at times which are earlier or later than expected. The time at which a transition is expected will be an exact number of elements after the previous transition if there is no distortion, and the program makes the decision as to polarity of the input exactly half an element of time after the expected time of the transition; in this way each element is sampled at midpoint. If reversals do not occur at the expected instants the program causes the amount and direction of the displacement to be stored but does not alter the sampling point until the direction of discrepancy is maintained for six transitions without there being a displacement in the opposite direction; the program then makes a change to the sampling point to bring it again to the midpoint of the element. Each of the instructions in the synchronizing program can be performed in approximately 1·5 μs. The effective sampling point for each line is stored in the processor, adjustments being made as necessary from time to time.

The Phasing Program. A new element arrives every 10·4 or 5·2 ms (96 or 192 bauds) and this is examined by the phasing program which takes seven elements at a time and decodes them to ascertain whether the character signal is erect, inverted, an RQ signal or an error. Unless an error is received, the information is stored awaiting the arrival of the next seven elements. If an error is detected, subsequent action depends upon whether the system is in traffic or is in process of automatic phasing. If the former, an RQ signal is initiated; if the latter, or if further program checks show the circuit to be out of phase, a single phase step is taken. The program waits for the next element to arrive and attempts to decode the combination formed by this and the six preceding elements. This process, which takes about 10 μs, continues until good characters are received. The type (i.e. whether erect, inverted or RQ) of these good characters is examined and stored in a table. The phasing and sub-channel decoding program is constantly looking at this table to fit it to the standard reference table held in the core store. This will show the pattern for the particular aggregate input being considered and will take into account the channelling, sub-channelling and length of RQ repetition cycle. When a fit is obtained the phasing process is complete and the program now knows to what channel each incoming character is allocated.

Phasing can take place automatically or under operator control. For manual phasing, a specific typed instruction is made by the attendant, who then depresses the space bar of the teleprinter once for each phasing step desired. A selected channel on the circuit being manually phased is automatically monitored on the teleprinter so that the operator can be certain when correct phasing has been achieved.

The RQ Cycle Program. This is called in if there is an error in a received character. The processor causes the RQ character to be transmitted. Any further characters entering the system from a telex circuit for that radio path are stored; if a leased circuit is concerned the clutch-release pulses are temporarily suspended. The RQ + 3 or RQ + 7 cycle is initiated. At the conclusion of RQ cycling, characters from the telex or leased circuit are again transmitted.

If error indication only is being given, the computer is programmed so that the receipt of an error will not initiate an RQ cycle, but will cause an error symbol to be transmitted to the receiving teleprinter.

Aggregate Outgoing Signals. Characters received into the unit from telex or leased circuits are interleaved according to channel requirements, and then converted to 7-unit code. Under control of the RQ cycle program the multiplexed signal is presented, element by element, for transmission from the unit. Although the program may have an element ready immediately after multiplexing and processing, it must not be sent immediately but must wait so that the transition will be timed to occur at the correct relationship to the received transition times on that channel of the synchronous system. The element is temporarily stored in the aggregate-output device, and the program places into a register the time relative to an accurate clock at which the element can be transmitted; this is obtained from knowledge of transition times of the synchronization program. This clock is the same one used for the incoming circuits, and when the selected instant occurs the element is extracted from the temporary store and sent to line; this process occurs on all aggregate channels.

5-unit Output Signals. The 7-unit signals from the incoming aggregate are decoded and allocated to their respective channels by the phasing and channel-decoding programmes. Conversion to 5-unit code is achieved and elements are now presented to the output circuits for transmission. Under the control

of a crystal clock running at a multiple of the 50-baud rate, elements are sent to line sequentially; start and stop elements are inserted automatically.

The Telex Panel. The telex-control functions are carried out by program to control the wires E (engaged), ZK (call-duration timing pulses for characters transmitted) and TT (traffic in store) extended from the integrated ARQ unit to the telex switching equipment. These wires are connected to a series of registers which are loaded by program whenever a change to a condition is indicated. For a leased circuit the ZK lead will be conditioned to provide the automatic transmitter-release pulses.

Buffer Storage. This comprises a magnetic drum and a magnetic-core store. The core is used for taking in characters from line, processing them and holding the part of the program currently being used. The drum is used to keep some of the less frequently-used programmes and for temporarily storing characters that have been received and are waiting processing and transmission.

The core store is an integral part of the computer and the contents of an address can be retrieved in $2-3$ μs. Access to the drum store is via the core store which acts as a buffer for the transfer of information on and off the drum. Average access time to a word is 10·35 ms.

In the STC Adex system, illustrated in Fig 8.32, each integrated unit makes provision for 15 duplex

Fig 8.32 Integrated ARQ system (*Courtesy of STC*)

aggregate circuits and 32 five-unit start—stop circuits which can be programmed as either telex or leased circuits. Additional units can be readily added to increase the line capacity.

The drum storage is for 65 000 computer words (12-bit) of which some 12 000 are used for program store leaving 53 000 words (=106 000 telegraph characters) freely available to all telex circuits as buffer storage. A 10-in. diameter drum is used, rotating at 2900 rev/min. Data is stored at a density of 466 bits/in.

The equipment uses semiconductors on a modular basis throughout, apart from reed relays and mercury-wetted relays on the control wires.

Operator Control Position. This consists of a teleprinter with tape-perforating and transmitting facilities, and alarm indications. By use of programming from the teleprinter the operator can control any 5-unit or 7-unit circuit and interrogate the computer for operational or statistical information. The teleprinter copy (printed in black and red for messages into and out from the computer, respectively) provides a running log of the state of the system. Programming by the control operator is used to provide supervisory features. For example, changes to 2- or 4-circuit working, phasing, sub-channelling, marked-cycle, 4- or 8-character repetition cycle etc. can all be effected selectively by specific programming changes acceptable within the main computer program.

Alarm conditions, if not provided on lamps, are automatically printed out on the control teleprinter: these cover such items as excessive cycling or pre-set error rate on a particular circuit being exceeded. Certain statistical information is also readily obtained from the computer. In addition, diagnostic programmes enable a measure of fault localization to be achieved.

Reliability. It is a fundamental requirement for telecommunication plant of every type that reliance must never be placed upon common equipment feeding a large number of circuits without adequate provision for rapid change-over to spare plant in emergency. For this reason an identical spare complete unit would be installed to serve from one to five integrated units. This spare unit would be programmed in advance so that it could take over in a matter of minutes by changing over patching cords and, if necessary, feeding in a perforated tape to update the program on any special items in current use. When not in emergency use, the spare unit can be used for monitoring on any selected 5-unit or 7-unit line circuit (using a 5-unit teleprinter).

8.8 Cable multiplex system

This system* is designed to make economical use of bandwidth in costly transoceanic telephone cables where the number of bearer circuits is necessarily limited. This it does by interleaving two 50-baud circuits, or three such circuits using CDC in the bandwidth normally occupied by a single start—stop MCVF teleprinter circuit with 120-Hz spacing. Error correction is unnecessary but in other respects the system shares some features of the ARQ system — automatic synchronizing and phasing, provision of $\frac{1}{4}$- and $\frac{1}{2}$-rate sub-circuits. In the cable multiplex system alphabet No. 4 with 6-unit code is used, and provision made for transmitting dial pulses and international telex signals types A, B and C. Circuits in this multiplex system can be used without restriction for the overseas telegram service, gentex or telex services (using dial or keyboard selection), or leased circuits, all at 50 bauds.

Modulation Rate. This is determined from the character period of $145\frac{5}{6}$ ms used for the ARQ system. An ARQ circuit may be connected in tandem with a circuit on the cable multiplex system and it is important that their character-clearance rates should be compatible.

Using a 6-unit code for 2- and 3-circuit multiplex systems, aggregate element durations are:

$145\frac{5}{6} \div (2 \times 6) = 12\frac{11}{72}$ or 12·153 ms (2-circuit)
$145\frac{5}{6} \div (3 \times 6) = 8\frac{11}{108}$ or 8·102 ms (3-circuit)

The respective aggregate modulation rates are:

$1/12·153 = 82\frac{2}{7}$ bauds (2-circuit)
$1/8·102 = 123\frac{3}{7}$ bauds (3-circuit)

*Recommendation R44.

Multiplexing. Element interleaving is adopted because this is beneficial to telex signalling in that a change in condition is registered with minimum interleaving delay. This alleviates the problem of double seizure of a bothway telex circuit from either end and is also advantageous in passing dial pulses over the system.

The multiplex circuits are designated A and B (diplex system) or A, B and C (triplex system). The main channel interleaving has the following pattern, the suffix numerals referring to code elements 1–6:

2-circuit operation: A_1 B_1 A_2 B_2 . . .

3-circuit operation: A_1 B_1 C_1 A_2 B_2 C_2 . . .

Provision is made for sub-circuits to operate at $\frac{1}{4}$-speed (100 char/min) and $\frac{1}{2}$-speed (200 char/min); there are no $\frac{3}{4}$-speed sub-circuits. The interleaving pattern for sub-circuits 1, 2, 3 and 4 is:

2-circuit operation: A1 B1 A2 B2 A3 B3 A4 B4 A1 B1 . . .

3-circuit operation: A1 B1 C1 A2 B2 C2 A3 B3 C3 A4 B4 C4 A1 B1 C1 . . .

Note: One $\frac{1}{2}$-rate circuit will comprise the two $\frac{1}{4}$-rate circuits 2 and 4 together. Two $\frac{1}{2}$-rate circuits will comprise the $\frac{1}{4}$-rate circuits (2 and 4) and (1 and 3).

The multiplex system is built in a unit serving six VFT circuits, numbered from 1–6. When circuits are subdivided, the four $\frac{1}{4}$-rate circuits are designated 1–4 following the multiplex-circuit letter, e.g. 2A1 is the first $\frac{1}{4}$-rate circuit on multiplex circuit A of MCVF circuit 2. Half-speed circuits are equivalent to a pair of $\frac{1}{4}$-rate circuits and so carry the sub-circuit designation either 1/3 or 2/4, e.g. in full, 1A1/3 or 1A2/4. The full designation for the circuits available in a 6-circuit group is shown in Table 8.1.

Either 11 (diplex) or 17 (triplex) full-rate circuits plus three $\frac{1}{4}$-rate circuits are available from a 6-circuit group. As an alternative there could be up to 23 x $\frac{1}{2}$-rate or 47 x $\frac{1}{4}$-rate circuits; or any appropriate composite arrangement of full-rate, $\frac{1}{2}$ rate and $\frac{1}{4}$ rate circuits.

Bearer circuits on submarine-cable routes are usually spaced at 3 kHz, allowing a 22-circuit MCVF system to be operated. Three multiplex groups, each of six VFT aggregate circuits, can be operated using 18 of these VFT circuits; the remaining four are used as separate 50-baud start–stop circuits. With the aid of multiplexing, the MCVF system can be progressively loaded with a total of 37 (diplex) or

Table 8.1 Designation of multiplex circuits and sub-circuits

Full-rate circuits					
1A	2A	3A	4A	5A	6A
*	2B	3B	4B	5B	6B
1C	2C	3C	4C	5C	6C

Quarter-rate circuits											
1B1	2B1	3B1	4B1	5B1	6B1	1C1	2C1	3C1	4C1	5C1	6C1
1B2	2B2	3B2	4B2	5B2	6B2	1C2	2C2	3C2	4C2	5C2	6C2
1B3	2B3	3B3	4B3	5B3	6B3	1C3	2C3	3C3	4C3	5C3	6C3
*	2B4	3B4	4B4	5B4	6B4	1C4	2C4	3C4	4C4	5C4	6C4

Half-rate circuits					
1B1/3	2B1/3	3B1/3	4B1/3	5B1/3	6B1/3
*	2B2/4	3B2/4	4B2/4	5B2/4	6B2/4
1C1/3	2C1/3	3C1/3	4C1/3	5C1/3	6C1/3
1C2/4	2C2/4	3C2/4	4C2/4	5C2/4	6C2/4

*Notes: (1) Circuit A is not available for subdivision. (2) The $\frac{1}{4}$ rate circuit 1B4 is used for phasing control only and is not available as a full-rate circuit.

55 (triplex) full character-rate circuits plus nine $\frac{1}{4}$-rate circuits; these figures include the four start–stop circuits.

Channel Marking. Signals on all channels are transmitted with 'erect' polarity, except for the whole of that channel B (i.e. channel 1B) which carries the phasing control (channel 1B4), i.e. sub-channels 1B1, 1B2, 1B3 and 1B4 have their six element polarities *inverted*.

Protection Against Crossed Circuits. If the system falls out of phase it would be possible for characters to be received on incorrect circuits. To avoid occurrence of crossed circuits a special marking pattern is introduced by transposing certain code elements in each possible sub-circuit. Any fault which would otherwise cause crossed circuits will now result instead in the reception of mutilated signals. The transposition pattern, which is carried into effect by suitable wiring connections at the staticized signal mode, is:

Circuit	A	1	2	3	4	5	6	
	B	1	3	2	4	5	6	Sub-circuit 1
	C	1	2	4	3	5	6	

	A	1	2	3	5	4	6	
	B	1	2	3	4	6	5	Sub-circuit 2
	C	1	4	3	2	5	6	

```
A 1 2 5 4 3 6 ⎫
B 1 2 3 6 5 4 ⎬ Sub-circuit 3
C 1 5 3 4 2 6 ⎭

A 1 2 6 4 5 3 ⎫
B 1 6 3 4 5 2 ⎬ Sub-circuit 4
C 1 6 5 4 3 2 ⎭
```

Full character-rate and half character-rate circuits take the sequence allocated to their lowest-numbered sub-circuit, i.e. a full-rate circuit takes the sequence for its sub-circuit 1, a $\frac{1}{2}$-rate sub-circuit using sub-circuits 1 and 3 takes the sequence for its sub-circuit 1, and a $\frac{1}{2}$-rate sub-circuit using sub-circuits 2 and 4 takes the sequence for its sub-circuit 2.

Synchronizing. This is from a common frequency source which controls the 6-circuit aggregate group. The six VF telegraph circuits working as a group will have slightly different propagation times (within a range of ±3 ms); the correct sampling or examining instant for one channel would not necessarily be the optimum instant for the other five. For synchronizing, the aggregate channel which has the mean value of propagation time is designed the Group Control Channel and the remaining five are dependent channels. The receiver of the 6-circuit aggregate system is synchronized to the received aggregate signal of the group-control channel. Examination pulses for each dependent aggregate channel bear a fixed time relationship to the examination pulses of the group-control channel, but the time relationships for sampling instants may be varied by ±3 ms in steps of 0·5 ms.

Synchronizing is achieved by examining transitions of successive received elements and integrating the early and late elements. A synchronizing step is equivalent to 0·1% of an element per element.

When CDC is employed, only those transitions of the received signal of the group-control channel which are preceded by two or more unit elements are used for synchronizing purposes. In this case the synchronizing signal to the group timer will be the binary output from the 'synchronizing separator' of the CDC.

If all channels were idle the only transitions received would be those of the occasional phasing signal. For this reason the signals of the phasing-control channel B are inverted so that transitions

for synchronizing are available at all times, even under idle conditions.

Phasing. To provide automatic phasing, irrespective of whether circuits are in traffic or idling, it is necessary to have a marked-character signal which can serve at the receiver to identify a particular channel or sub-channel. The $\frac{1}{4}$-rate circuit 1B4 of each 6-circuit aggregate group is allocated for this. The phasing signal ZZAAZZ is sent continuously over this channel and it is received on a phase-detector circuit. This provides automatic identification of the phase-control channel and of all the other multiplex circuits of the system.

If the phasing signal is not recognized on three consecutive occasions the phase detector examines the contents of the receive shift register of each of the channel registers of the group-control channels. Provided that the aggregate signal is being received without mutilation, the phasing signal must appear in one of the channel registers within a period of four characters, i.e. in $4 \times 145\frac{5}{6}$ ms.

When the phasing signal is detected in one of the channel registers, three of the frequency dividers are reset, thereby adjusting the timing of the receive-pulse generators to the correct phase position in relation to the incoming signal; this operation takes a few seconds only, even under adverse conditions. After a phase adjustment is made, a check for reception of three correctly-received phasing signals is carried out before the group recommences transmission of traffic.

The multiplex system will fall out of phase only if a VFT system fails or an equivalent emergency occurs. A VFT system-failure alarm operates if the line-power level falls by 10 dB for a period of 500 ms. The multiplex system will tend to fall out of phase slowly (the system-control-oscillator stability is better than one part in 10^6), but re-phasing during MCVF system failure is inhibited. The A condition is applied to all circuits while this MCVF system failure persists and until re-phasing is completed after restoration of the MCVF system; alternatively, Z condition can be applied, e.g. to leased circuits.

Rephasing takes place automatically when three successive phasing signals have failed recognition. It can also be initiated by operating a press-button,

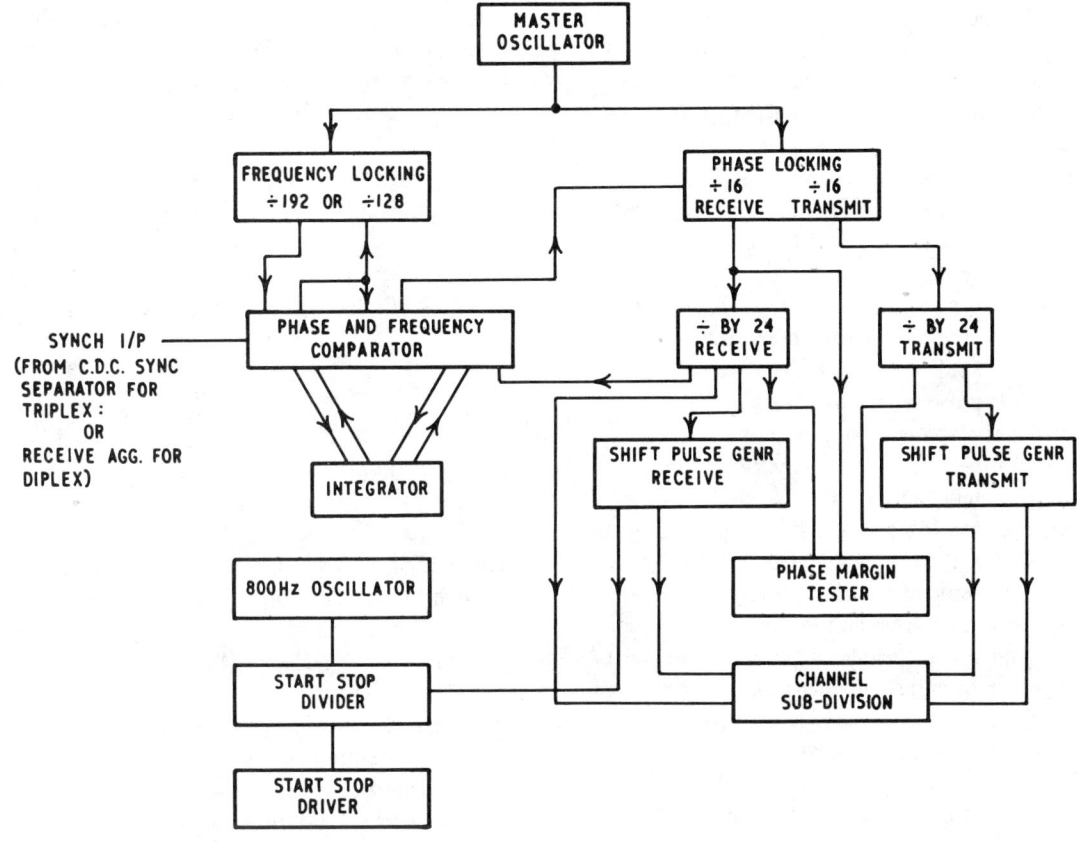

Fig 8.33 Cable multiplex system — pulse generation

or the system can be rephased manually by single-element steps for each operation of a manual-phase press-button.

Pulse Supplies. Except for the 800-Hz oscillator used for timing the start—stop signals, all timing pulses are derived from a common oscillator whose frequency is 15·798 kHz. A frequency stability of better than 10^{-6} is assured by enclosing the oscillator in a temperature-controlled oven. This master oscillator feeds both the frequency-locking and phase-locking units as shown in Fig 8.33.

The frequency-locking unit consists of a 1/16 divider stage feeding a further stage set to divide by 8 or by 12, depending on whether the group-control channel is a 3- or 2-channel multiplex. Output pulses of the frequency-locking unit have a period of approximately 8 or 12 ms and are used to drive the phase and frequency-comparator unit. The latter is also driven by the input-synchronizing signal from

the aggregate receive line to produce a phase comparison between the synchronizing signal and the output of the frequency-locking unit, so determining whether the frequency of the receiver is greater or less than the frequency of the synchronizing signal. The comparison pulses are then integrated in the integrator unit, the output of which is fed via the phase and frequency-comparator unit to the frequency-locking and phase-locking units. The outputs of these latter two units are thus adjusted to the same frequency as that of the synchronizing signal, in steps of approximately 0·06 ms, the frequency of steps being proportional to the difference in frequency of the two signals being compared. For operation at $82\frac{2}{7}$ bauds the group-control channel aggregate is used as the synchronizing signal, but for operation at $123\frac{3}{7}$ bauds the synchronizing signal is produced in the CDC equipment.

The phase-locking unit consists of two 1/16 dividers, the output of one being connected to the receive 1/24 divider and the other feeding the transmit 1/24 divider. The output from the receive divider is compared in phase with the synchronizing signal in the phase and frequency-comparator unit, the result integrated and fed via the comparator unit to the phase-locking unit. The phase-locking 1/16 divider, which feeds the receive 1/24 divider, is thus advanced or retarded in 0·06-ms steps, enabling the receiver timing to be kept in element phase with the received aggregate.

Various outputs of the receive 1/24 divider are connected to the receive shift-pulse generator and to the receive side of the channel subdivider. The receive shift-pulse generator consists of a number of converters producing stagger pulses with a repetition time of approximately 24 ms, and a 1/6 divider producing an output pulse with period of one character ($145\frac{5}{8}$ ms) which is used to drive the receive side of the subdivider. These pulses are used in the receive side of the channel register, the receive side of the telex bypass and the timing and polarity unit. A similar arrangement of 1/24 divider, shift-pulse generator and subdivider is used in the pulse generation for the transmit side of the system.

Receive aggregate examination pulses are provided by the receive 1/24 divider for 3-circuit operation, and by the phase-margin tester for 2-circuit operation. The phase-margin tester consists of a shift register driven by 0·5-ms pulses and synchronized by the receive 1/24 divider. Outputs of the phase-margin tester and receive 1/24 divider are also available to determine the optimum timing for the aggregate examination pulses.

The start–stop divider is driven from the 800-Hz oscillator and produces outputs which are combined in the start–stop driver to produce 20-ms pulses, these being used in the start–stop terminal to produce the 50-baud output. The 1/16 start–stop divider is synchronized to the receiver timing so that a character received on the aggregate during one receiver character cycle is transmitted during the next character cycle. The output of the 1/16 divider is fed to a 1/7 divider to produce the 20-ms timing pulses.

The Transmitter. The block diagram of Fig 8.34 shows the equipment for one multiplex circuit. For each full-speed channel the transmit path consists of a read-out unit, a 1-character store and a channel register. The store is needed because, although the synchronous channel clears traffic faster (411 char/min) than the maximum input rate (400 char/min normally – 411 char/min if from an ARQ system), the characters arrive at random and it is necessary to store a character to present it to the multiplex equipment with the correct phase timing. A short-start rejection circuit precedes the read-out to absorb false start signals.

Receipt of a start signal starts a 1/16 divider chain in the 800-Hz supply to produce sampling pulses at 20-ms intervals (50 bauds) for testing the polarity of the incoming signals; the code elements are staticized on a shift register. On reception of the stop signal the divider-chain cycle is stopped and

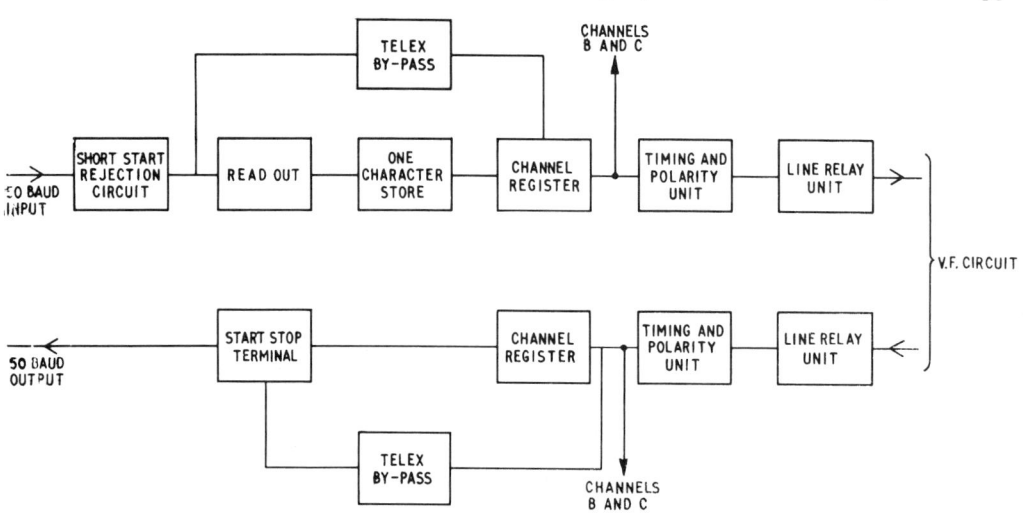

Fig 8.34 Cable multiplex system — transmitter and receiver

the character is transferred to the character store where it is held pending the instant when it can be transferred to the channel register. This occurs when the sixth element of the previous character is being transmitted from the channel register into the aggregate equipment, where a delay of up to one character period ($145\frac{5}{6}$ ms) can take place.

In the channel register a series of pulses moves the stored elements along a shift register, the output of which is read off into the timing and polarity unit. The outputs of the channel registers are continually examined at intervals of $12\frac{11}{72}$ ms (diplex) or $8\frac{11}{108}$ ms (triplex) and passed forward on an interleaved-element basis to a multiplex terminal unit. The interleaved output signals from this unit may be at ±6 or ±80 V as required.

Since common pulses are used to feed the channel registers and timing and polarity units, characters and elements are transmitted simultaneously on all six aggregate circuits.

For leased circuits, clutch-release pulses are sent to the sending station to control the call-out character rate. These pulses are inhibited during re-phasing or during MCVF system failure.

The Receiver. On the receive path, incoming aggregate signals from the VFT system are converted, if necessary, from ±80 to ±6 V in the multiplex terminal unit and then passed to the timing and polarity unit. Here the aggregate signal is sampled by examining pulses and elements proper to the two or three channels are separated and passed to their respective channel registers. From here the simultaneous signal is read out by sequential-timing pulses from the 800-Hz divider chain at the start–stop terminal unit. At this point, start and stop signals are added, except when received signals are the α and β characters. The 5-unit signals are transmitted at ±80 V.

Each sub-channel equipment consists of a read-out, a 1-character store and a start–stop terminal unit. The channel registers handle characters from each sub-channel in turn so that only one channel register per full-speed channel is required. In the transmitter, subdivision pulses are used to gate a character from the appropriate 1-character store into the channel register. Subdivision pulses in the receiver are applied to the start–stop terminals to allow the received character to be transmitted

by the appropriate sub-channel. Sub-channel outputs are held to Z condition during intervals between their printing-character periods.

Telex Circuits. Due to random arrival of character signals, these must be held staticized in a 1-character store until the time comes for the character to be sent out in the interleaved aggregate signal. This delays the moment of transmission and to avoid this delay to the calling signal (which would increase the possibility of the double seizure of a bothway circuit) the 1-character store is bypassed to reduce the time required for the call signal to be transmitted over the synchronous path. This bypass circuit is normally switched out shortly after the channel is seized (between 97–121 ms) but for type B dial selection it remains in circuit until completion of dial pulsing. It is switched back into circuit by receipt of the clear signal.

At the bypass unit, signals are sampled at 1-ms intervals. The condition found is transferred at 24-ms (= $145\frac{5}{6} \div 6$) intervals to the output of the channel register, so passing on to the timing and polarity unit a change of condition without the normal delay.

The bypass unit counts five Z elements to switch itself out of circuit, and two characters of the clear signal to switch it back into circuit ready for a following telex call. For type B signalling, reception of one β character (following dialling) is required additionally on the receive side before switching out the bypass circuit.

On the receive side, use of the telex bypass unit avoids the channel register. The receive side of the bypass unit is switched out on receipt of one β character and switched in, ready for the next call, by receipt of two α characters of the clear signal.

Dial pulses may be passed in either direction through the bypass unit. At the receiver, dial-pulse duration is in multiples of 24 ms; in sampling and reforming the signals, they are regenerated by the multiplex equipment to have minimum periods of 44 ms (A) and 36 ms (Z).

The Haslar SYNTOC equipment, which uses semiconductor logic circuits on printed-wiring cards, is shown in Fig 8.35. It includes a fixed proportion of units for telex and leased circuits. A patching panel enables subdivided circuits to be provided as required (the latter cannot be used on telex service).

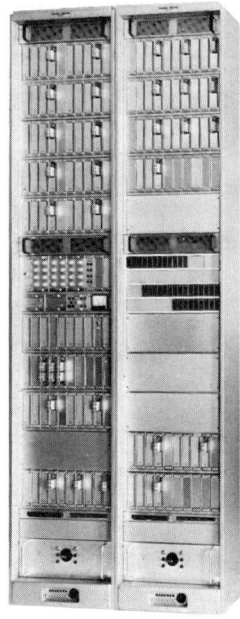

Fig 8.35 SYNTOC cable multiplex equipment (*Courtesy of Hasler AG*)

Each full-rate circuit on which subdivision is possible has three appearances on the patching field of the auxiliary rack, and each sub-circuit position has one appearance. Connection of sub-circuit appearances to full-rate circuit appearances automatically subdivides the full-rate circuit to appearances selected.

A control panel with lamps draws attention to failure of timing or power supplies, or to rephasing operations; push-buttons are provided for semi-automatic and manual phasing. Any circuit can be operated at either modulation rate — $82\frac{2}{7}$ or $123\frac{3}{7}$ bauds; the receive margin is ±45%, distortion on output signals does not exceed 3%.

8.9 Bandwidth-economy systems

Both the ARQ and the cable-multiplex systems give considerable economy in use of bandwidth. Synchronous systems can withstand a worse level of noise — see comparative figures in Table 8.2

Table 8.2 Tolerable noise levels for equal error rate

Noise level* (dBm0)	Mode of operation
−40	50-baud, start–stop
−29	83-baud, synchronous
−34·5	123-baud, synchronous (with CDC)

*Uniform spectrum random noise in 3-kHz band, weighted.

relating to the tolerable noise level before an element error rate of 10^{-6} is introduced by the noise.

The real value of TDM systems in transoceanic cables lies in the fact that the number of telegraph circuits per kHz can be increased two- or three-fold without increasing power loading to the common amplifiers in submerged repeaters.

In conventional FDM systems, inter-channel interference is controlled by the combined action of the send and receive filters to restrict the overlap of the spectra of the individual channels. Interchannel interference can also be effectively avoided in the transmission path, provided that the signals are orthogonal and matched filtering is used for reception. (Two signals $S1$ and $S2$ of duration T are orthogonal if $\int_0^T S1.S2.dt = 0$.) Orthogonality results in all the unwanted signals integrating to zero when the wanted signal achieves its maximum. It can be achieved to a close approximation by ensuring that, at the receiver, the modulations of all channels are both isochronous and synchronous, that the carrier frequencies have a separation which is the reciprocal of the received signal-element duration and that the lowest carrier frequency is greater than about eight times the modulation rate. This is one of the principles of the Kineplex and Rectiplex systems. Ideally the carrier spacing would be $\Delta f = 1/B$, where B is the modulation rate, but to accommodate difference in propagation times of individual channels the aggregate signal is admitted by a gate to the channel receivers for a period slightly shorter than the sent-element duration, to ensure that the signal accepted does not contain the end of an adjacent element which has been displaced by difference in propagation time. Since the effective element length is reduced, the spacing must be accordingly increased. If an allowance of about 3·5 ms is made for delay variation, synchronous teleprinter channels could be accommodated by a carrier-frequency spacing of 48 Hz; this would produce 57 channels (plus a synchronizing channel) in the conventional voice-frequency bearer band. The relatively high bandwidth utilization stems from the permissible overlapping of the spectra of the channels. Additional clearance must be allowed for the end channels where there is no overlap. Because of this and the proportionately greater effect of delay variation, higher modulation rates entail a reduction in the aggregate capacity of the system. The basic principle is applicable to amplitude modulation but, in practice, differential phase modulation is used.

The principle of orthogonality can be pursued further by the use of carriers of identical frequency but in quadrature (with a phase difference of 90°, i.e. carriers of $e_1 = \sin \omega t$ and $e_2 = \cos \omega t$) to provide two independent channels. The aggregate signal appears as a 4-phase modulation. This leads to consideration of higher orders or phase multiplexing: for eight or more phases, which would provide three or more channels, mutual orthogonality of channels cannot be achieved and unequal performance could result. If eight or more phases are used it is better in principle to recode before transmission to a higher radix by combining binary elements from the same channel and then time-dividing the aggregate. This, however, is counter to the desirability of interleaving the channels element by element.

The Kineplex System. The Kineplex system (developed in U.S.A. by Collins Radio Co.) is a telegraph- or data-transmission system which uses a combination of FDM and phase multiplexing to provide 40 telegraph circuits operated at a modulation rate of 75 bauds on 20 VFT telegraph carriers, spaced at 110 Hz in a 3-kHz bearer circuit. Differential phase modulation is used, the carrier being changed in phase by $\pm\pi/4$ or by $\pm3\pi/4$, relative to the phase in the previous signal element, to transmit the information from two start–stop input channels. Reception is by means of a matched-filter technique embodying gated resonators. At the transmitter all channels are modulated isochronously and in unison and a pilot channel synchronizes the receiver to the transmitter. Start–stop-to-synchronous converters are provided to receive teleprinter signals and regenerate them in the time scale of the transmission system. Since the modulation rate of the Kineplex channels is that of the nominal value (75 bauds) for the teleprinters with which the system is designed to be used, time-scale conversion is not provided, but the stop element of character signals will be sometimes only one unit and sometimes two. This high bandwidth utilization is obtained chiefly by the phase multiplexing, together with the narrower channel-frequency bands made possible by the form

of reception employed, which necessitates immunity against delay distortion. Provision is not made for subdivided nor for telex circuits.

FDM Arrangements. The Kineplex system uses VF telegraph-carrier frequencies which are the odd multiples of 55 Hz, from 605–2 695 Hz, the channel frequencies being $f_0 = 605 + (n-1)110$ Hz. A pilot channel at a frequency of 2915 Hz (spaced 220 Hz above the highest traffic channel) is used for synchronizing.

Modulation coding applied to each channel is quaternary and is derived from the combination of two start–stop channel inputs (the coding is binary if one only of the start–stop channels is required). Information is transmitted by differential phase modulation, i.e. by the phase of the carrier relative to that during the preceding element period and not with respect to a fixed reference.

The signals from two start–stop telegraph circuits after conversion into the synchronous time scale of the Kineplex transmitter are used to control the modulation of a single VF carrier to produce the equivalent of two separate binary phase-modulated carriers of the same frequency, but with a quadrature phase relationship. Modulation is such that information on each channel is represented by a vector in one of two possible phases 180° apart, and the vectors of the two channels are separated in phase by

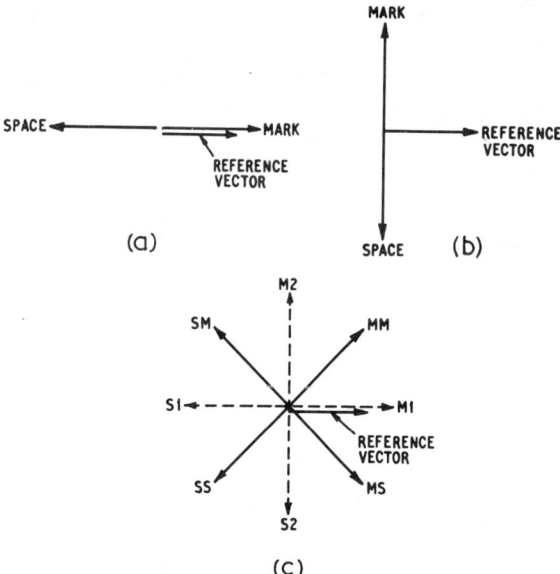

Fig 8.36 Carrier vector diagrams: (a) channel A; (b) channel B; (c) channels A and B combined

90°. The aggregate signal vector in the VF telegraph channel has four possible phases (see Fig 8.36 (*a*), (*b*) and (*c*)). To simplify the problem of detecting the signal at the receiver, the phases of the transmitted signal are not generated from a constant datum but with respect to the resultant of the previous element.

All carriers are modulated at the same instants. At the receiver, the signals at each carrier frequency are integrated over the same time interval, but this is made appreciably shorter than the duration of the transmitted unit element signal, although this results in a proportionate loss of signal energy. The integrating period is nominally centred within the element period so that differential delay (of less than a pre-determined amount) between channels will not result in parts of successive elements being integrated; similarly, overlapping of successive elements in the same channel (which would result from multipath radio propagation) will have no effect provided it is contained within the marginal interval.

Each integrator comprises a mechanical resonator to which positive feedback is applied to produce a very high Q-factor. Two of these filters are provided for each VF channel, and are used during alternate element periods for integrating the incoming signal and then for storing it as an a.c. waveform. The 2-element cycle can be divided into three periods:

(1) a receiving or sampling period during which the signal is integrated in the filter;

(2) a storage period during which the integrated signal is preserved;

(3) a period primarily to allow for possible differential-delay distortion during which the filter is quenched in preparation for the next 2-element cycle. These operations are demonstrated in the diagram of Fig 8.37.

There are two phase-comparison demodulators for each VF channel. To one are directly applied the new signal to be interpreted and the stored previous element; to the other, one of the two components is subjected to a 90° phase shift (see Fig 8.38 (*a*) and (*b*)). This effectively separates signal components due to the two signals which were combined at the transmitter. Outputs from the demodulators are inspected and signals are regenerated immediately before a resonator is quenched.

Channel send filters are not used but harmonics

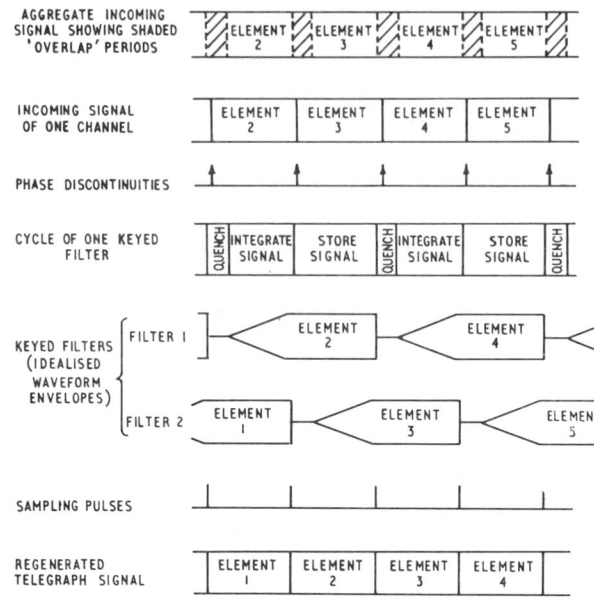

Fig 8.37 Kineplex system — operation of keyed filters

of the carriers are removed in the transmitter. To eliminate inter-channel interference, carriers are spaced at intervals which are the reciprocal of the sampling period, so that by the end of each sampling period the interference from the other channels has cancelled itself out.

Performance. The Kineplex equipment does not use the standardized character period for synchronous systems (145⅝ ms) and is not compatible with other TDM systems. The synchronous margin of the system is ±25%; distortion of output start—stop signals is about 10%. Approximately half of the signals will have but a single-element stop signal.

For the same error rate, the Kineplex system provides 50% more information-carrying capacity compared with the 2-circuit (83-baud) cable multiplex system, in return for an increase in aggregate

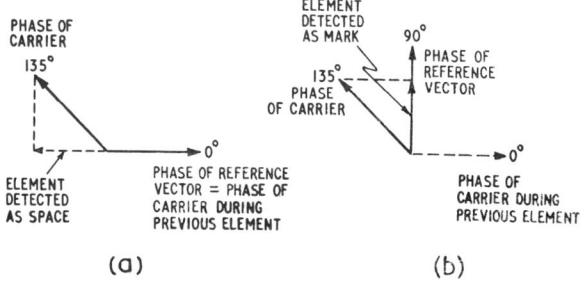

Fig 8.38 Vector diagrams for detection of SM signal

signal/noise ratio of 4 dB. On radio-telegraph circuits subject to fading and with path-time delays up
to 2 ms, the performance of a Kineplex channel in
its normal mode is very similar to that of a good
2-tone system.

Spacing of carriers at 110 Hz is a suitable compromise for operation at a modulation rate of 74·2
bauds. For a modulation rate of 50 bauds, a channel
spacing around 64 Hz would be appropriate.

For operating a given number of synchronous
telegraph circuits on a cable or radio system, Kineplex offers a very slight advantage in the tolerable
noise level but a considerable advantage in bandwidth (about 2:1 compared with an FM MCVF
system, or 4:1 compared with a 2-tone system) for
a prescribed error rate and peak signal power. The
error rate/noise characteristic is shown in Fig 8.41.

The Rectiplex System. In this system (by KDD of
Japan) up to 108 circuits for 50-baud teleprinter
operation can be obtained from a 3-kHz bearer
circuit. Alternatively, $\frac{1}{4}$-rate and $\frac{1}{2}$-rate subdivided
circuits, or data-communication circuits for
modulation rates of 600, 1200 or 2400 bauds, may
be provided.

This extreme utilization of bandwidth is achieved
by a combination of FDM, using multi-phase modulation, and channel filtering by integrator circuits,
together with TDM principles.

FDM application. An 18-circuit MCVF system
forms the basis, with standardized centre frequencies for circuits 2–19 (540–2580 Hz). In addition,
a synchronizing carrier is used, corresponding to
circuit 21 (2820 Hz), since the MCVF channels
must all operate in synchronism.

Multi-phase Modulation. This provides multiplex
facilities by phase division without increasing
modulation rate; one VF carrier accommodates two
or three channels using 4- or 8-phase modulation.

The modulation scheme is shown in Fig 8.39. As
in the Twinplex HF radio system, each of several
modulation conditions is used to transmit information concerning the instantaneous significant conditions present on two or three teleprinter channels
(Table 8.3). For 4-condition modulation, phase
shifts of +90° are used; for 8-phase modulation, a

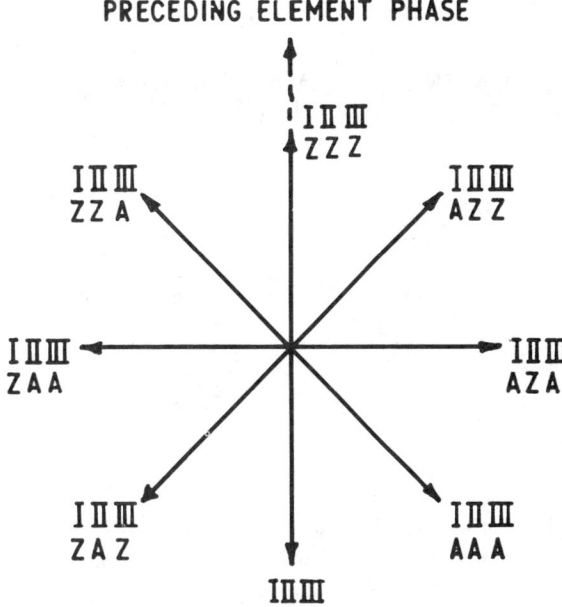

PRECEDING ELEMENT PHASE

Fig 8.39 Rectiplex modulation scheme

phase-shift of +45° is necessary. By way of example,
suppose that the input for the previous element was
$Z_I Z_{II} Z_{III}$ and the carrier phase was left at 0°. If the
next input is $Z_I A_{II} A_{III}$, a phase shift of 90° (from
0°) takes place. If the next input is $A_I A_{II} A_{III}$, the
phase of the carrier will shift 225°, i.e. a 315° phase
shift with respect to 0° phase of the carrier. The
absolute phase of the carrier is of no significance —
only the phase shift with respect to the phase of the
preceding element indicates the information.

Demodulation. For detection, the reference phase
is the phase of the preceding element of carrier;
this is differential phase modulation, not dependent
on a knowledge of the absolute phase of the signal.

Demodulation of a differential phase-modulated
carrier is made firstly by detecting the phase shift
of the incoming signal by reference to a locally-
generated carrier which has stored the same phase as
that of the preceding element of the incoming signal.
This stored-phase condition is shifted into four
phases and supplied to four phase detectors as the
reference signal. The outputs of these phase detectors are a d.c. component depending upon the phase
differential between input signal and reference
signal. A demodulated output corresponding to
the input of three channels can be obtained by

Table 8.3 Multi-phase modulation conditions for 2- or 3-circuit multiplex

Channel polarity					Advance phase shift with reference to the preceding element (degrees)
8-phase			4-phase		
I	II	III	I	II	
Z	Z	Z	Z	Z	0
Z	Z	A	—		45
Z	A	A	Z	A	90
Z	A	Z	—		135
A	A	Z	A	A	180
A	A	A	—		225
A	Z	A	A	Z	270
A	Z	Z	—		315

combination of these four detector outputs. This phase shift is interpreted as a combination of the significant conditions of three (8-phase) or two (4-phase) multiplexed channels; the appropriate conditions are then delivered as output signals on the two or three channels. For example, if the phase shift is detected as 135° (compared to the reference phase), the output signals applied to channels I, II and III will be Z, A and Z respectively.

To produce the reference carrier on the receiving side, the demodulator has a local oscillator equipped with an automatic phase-control circuit which feeds back unbalanced components of the phase-detector outputs to the oscillator, with a large time constant. The carrier from the oscillator is differentially phase modulated by the demodulated output signals which supply the information contained in the preceding element. The reference carrier with the same phase as that of the preceding element and with no noise component is then obtained. This unique method of producing the reference carrier without noise component results in good error-rate characteristics in the presence of noise.

Channel filtering. Integrators in the demodulators perform channel filtering. When a sinusoidal wave is applied to an integrator operating for a period of $1/120$ s ($= 8\frac{1}{3}$ ms), if the frequency of the input wave is an integral multiple of 120 Hz the result of integration will be zero unless the wave has any phase shifts in the integration period. This fact indicates that in an FDM system with 120-Hz spacing, inter-channel interference can be eliminated at the output of such an integrator connected to the phase

detector in the demodulator. It is necessary that no phase shift appears in the integrating period for all carriers. Due to this restriction the duration of an element of modulation should be greater than that of the integrating period. Rectiplex is designed to form an FDM system with a modulation rate of 96 bauds with element duration $10\frac{5}{12}$ ms; this modulation rate is considerably higher than that of a conventional system with 120-Hz spacing, where channel filters limit it to something like 80 bauds.

From the restriction mentioned above it follows that the integrators have to operate in synchronism with the modulation and that all carriers received by the demodulators must have the same phase-shifting point. The former is accomplished by the synchronizing carrier, which is a phase-reversing signal with a modulation rate of 96 bauds; for the latter, all carriers are modulated in synchronism. Provision is made at the receiving end for equalizing delay characteristics of the line.

Synchronous multiplexing. On the sending side, the incoming 7·5-unit 50-baud start—stop signals are converted to 7·0-unit signals, with the character period of $145\frac{5}{6}$ ms standardized for synchronous systems. This results in a channel modulation rate of $50 \times 7/7.5 \times 150/145\frac{5}{6} = 48$ bauds. Two such channels, A and B, are element-interleaved by TDM, producing an aggregate signal with a modulation rate of 96 bauds. The start and stop signals are transmitted (unnecessarily) within the 7-unit code over the aggregate path; although this increases the modulation rate, and the signal capacity remains that of the 5-unit alphabet, circuit arrangements are simplified and the high circuit utilization is not impaired.

Difference between clearance rates of the synchronous path (411 char/min) and start—stop input (400 char/min) is resolved by inserting one extra Z element at the end of a character of the output signal from time to time; with continuous sending this is added every five characters in most cases. On the receiving side, duration of the stop signal becomes $(145\frac{5}{6} - 20 \times 6) = 25\frac{5}{6}$ ms; when the extra Z element is added (as above) the stop signal is increased by $20\frac{5}{6}$ ms ($1/48$ s).

Phasing. For maintaining correct phase in the TDM system, one $\frac{1}{4}$-rate channel is allocated for the regular transmission of a phasing signal of specified pattern. If the system falls out of phase, rephasing is initiated automatically and accomplished in about one second.

Subdivided circuits. The 96-baud aggregate circuits can be subdivided to provide lower-rate circuits. A 96-baud 2-circuit aggregate path may be divided to form four $\frac{1}{2}$-rate circuits; or eight $\frac{1}{4}$-rate circuits; or two $\frac{1}{2}$-rate plus four $\frac{1}{4}$-rate circuits.

Leased circuits. Whether full-, $\frac{1}{2}$- or $\frac{1}{4}$-rate, leased circuits terminate on a sequential-to-simultaneous converter which includes element-storage facilities as a buffer against random arrival of character signals. The simultaneous signals are sampled 48 times/s and the output from two channels is combined by element interleaving to form a 96-baud aggregate signal.

Telex circuits. Any derived circuit (other than a reduced-rate one) can be used for telex. Type A and type B supervisory signals can be transmitted (in multiples of 1/96 s). To reduce delay in transmitting a call signal, a bypass circuit is provided for telex signals. Dial pulses are unavoidably distorted by the multiplex timing; at the receiving end these are re-generated in a dial-pulse regenerator automatically inserted in circuit only during the period of dial-pulse transmission. As with leased circuits, a buffer of element storage is provided to cope with charac-ter signals arriving at random, by storing elements until called in at appropriate instants by the multiplex combiner.

Data Communication. For data communication, an equipment for modulation-rate conversion takes the place of the TDM equipment. Fig 8.40 illustrates a typical application for data at 1200 bauds by appro-priation of bandwidth proper to a block of five MCVF channels (600 Hz).

Performance. Compared with systems using con-ventional band-pass channel filters and post-discrimination low-pass filters wherein some of the side-band signal energy is lost, the integrator forms an optimum detection filter which results in improvement in error-rate characteristics. Applica-tion of the noise-free reference carrier improves the

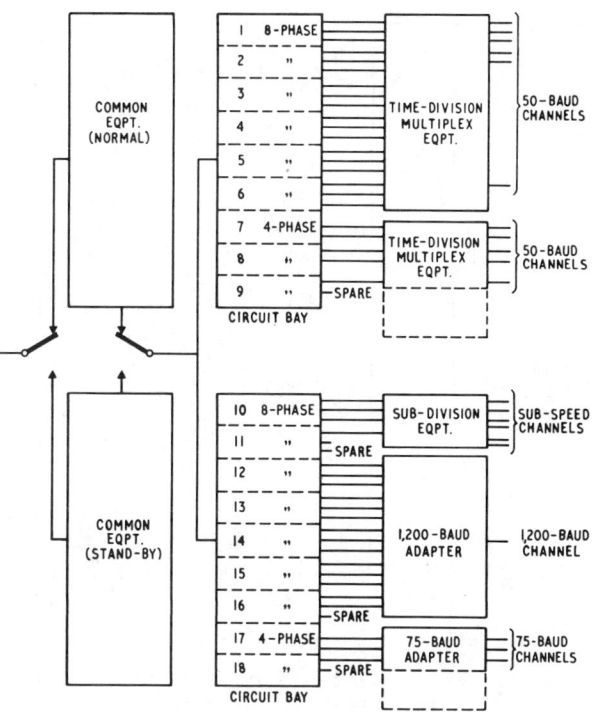

Fig 8.40 Example of utilization of Rectiplex system

performance against the presence of line noise. The adoption of phase modulation gives a theoretical improvement of 3 dB over other systems of modula-tion. The aggregate of these advantages results in a lower element error rate compared with more con-ventional systems. Fig 8.41 shows a graph of the element error rate plotted against signal/noise ratio.

Information capacities, in a 3-kHz band, of various systems are shown in Table 8.4.

Table 8.4 Information capacity (bits/s/channel)

22-circuit MCVF	22 × 50 = 1100
ARQ, 2-circuit*	22 × 2 × 50 = 2200
Cable multiplex	18 × 3 × 50 + 4 × 50 = 2900†
Kineplex	40 × 75 = 3000
Rectiplex	108 × 50 = 5400†

* With error-detecting code.
† Includes synchronizing channel.

Improved utilization. Basic transmission systems have been considered above on the basis of possible communication capacity. To obtain the most effic-ient utilization it is necessary to smooth the traffic; this is done by message storage in message-relay systems where unidirectional communication alone

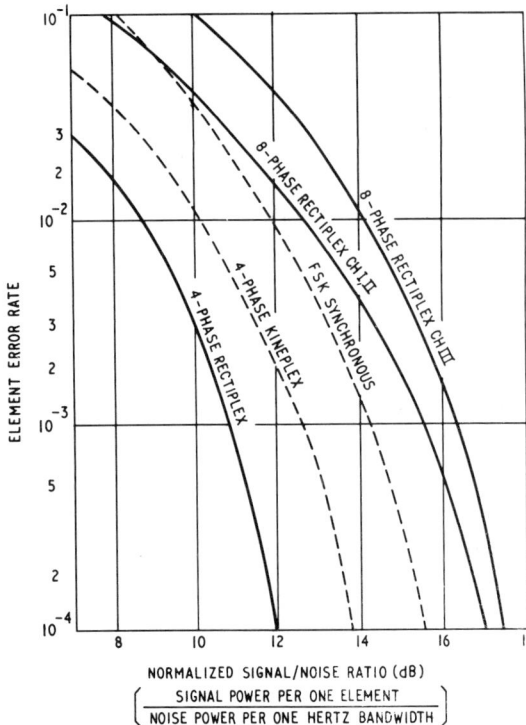

Fig 8.41 Rectiplex error-rate characteristics

is required. In systems of line switching, particularly for telex or other forms of direct customer service where 2-way communication is desired, the total utilization of channels approaches only 50%, i.e. in a half-duplex circuit only one direction of transmission is used at a time. The idle-return channel could, with appropriate switching elements, be used for other communications. Such a scheme is similar in principle to the Varioplex system for time-sharing circuits or to TASI* for time-sharing telephone circuits.

Varioplex circuits are generally arranged to run at a lower speed than any of the tributary circuits and, when no traffic is offered from a particular one, its time allocation is shared amongst the busy tributaries, permitting them to approach a disposal speed nearer their input speed. In this type of service 'semi-infinite' buffer stores are necessary. In TASI no storage is feasible and a certain degree of clipping at the commencement of a speech burst and an occasional 'freeze-out' when no circuit is available must be endured. In telegraphy, freeze-out resulting

*Time-assignment speech interpolation — see 'Data Tele-communication' (R.N. Renton), Pitman.

in loss of information cannot be tolerated so a certain amount of storage is inevitable, yet in 2-way communication the delay in transmission must be as short as possible. The traffic loading would have to be planned to permit only a reasonable delay in finding a free channel. For telex operation, even if coordination existed between the various messages, the necessary supervisory signals would absorb some of the capacity and in practice it is unlikely that the activity of channels could be significantly increased.

References

1 Wheeler, L. K., 'An Electronic Error-Correcting Multiplex Telegraph System', *POEE Journal* **50**, p. 44 (1957).
2 Croisdale, A. C., 'Teleprinting over Long-distance Radio Links' ibid. **51**, pp. 88 and 219 (1958).
3 Keller, P. R., 'Automatic Error-correction on HF Telegraph Circuits', *Point-to-Point Telecommunications* **2**, p. 15 (1958); and **3**, p. 21 (1959).
4 Van Duuren, H. C. A., 'Error Probability and Transmission Speed on Circuits Using Error-detection and Automatic Repetition of Signals', *IRE Transactions on Communications Systems*, **CS9**, p. 38 (1961).
5 Croisdale, A. C. and Harrison, A. F., 'Testing Radio-telegraph Automatic Error-correcting Equipment', *POEE Journal*, **54**, p. 245 (1962).
6 Keller, P. R. and Wheeldon, A. J., 'Recent Improvements in Automatic Error-correcting Systems for HF Telegraph Networks', *Point-to-Point Telecommunications*, **6**, p. 38 (1962).
7 Croisdale, A. C., 'Improvements in Automatic Error-correcting Equipment Used on HF Radio-telegraph Services', *POEE Journal*, **55**, p. 253 (1963).
8 Chesterman, D. A., 'An Automatic Error-correcting Radio-telegraph Multiplex System', ibid. **60**, p. 187 (1967).
9 Froom, R. P. *et al.*, 'A Monitor for Seven-unit Synchronous Error-Correcting Systems for use on Radio-telegraph Circuits', ibid. **53**, p. 1 (1960).
10 Chesterman, D. A., 'A Transistor-type Self-synchronising Seven-Unit Monitor for Automatic Error-correcting Radio-telegraph Systems', ibid. **57**, p. 37 (1964).
11 Robins, J. M. and Croisdale, A. C., 'An Electronic Telegraph Buffer Store Using a Magnetic Drum', ibid. **56**, p. 262 (1964).
12 Clark, T. H., 'A Telegraph Signal Buffer Store', *ATE Journal*, **12**, p. 107 (1956).
13 Hunter, N. C., 'The Telegraph Buffer Store Type DPS1', ibid. **20**, p. 95 (1964).
14 Hunt, C. S., 'Telegraph Character-storage Devices for Telex Services Routed over Radio-telegraph Circuits', *POEE Journal*, **61**, p. 119 (1968).
15 —, 'Stored-Program Character Storage for Automatic Error-correcting Telegraph Systems', ibid. **63**, p. 33 (1970).
16 Croisdale, A. C. and Hunt, C. S., 'Synchronous Multiplex Telegraphy on Intercontinental Submarine Telephone Cables', ibid. **60**, p. 52 (1967).

9 Telegram Services

This chapter deals with inland and overseas services, but the message-relay system for handling overseas telegrams is described in Chapter 11.

9.1 Inland telegrams

Over the years the inland-telegram service has suffered greatly from competition, firstly by increased penetration of the telephone in residential areas and, more recently, by telex service in the commercial field. The rising cost of labour has also had a profound effect on a service which inherently includes a large manual-handling component. As a result, traffic has fallen heavily and provided a spur for mechanizing operation as far as possible. These effects have not as yet been felt to the same extent in the overseas-telegram service, though the influence of international-telex service on commercial overseas telegrams is increasing; considerable mechanization of the overseas-telegram service has been introduced.

Handing in Telegrams. This term covers the methods by which a telegram is initiated by a member of the public or a business organization; a choice is available. Telegrams may be handed in over the counter at a Post Office. If this is a main Post Office it will have its own telegraph office, equipped with teleprinters in the same or an adjacent building and telegrams will usually begin their journey by pneumatic tube from the public counter to the telegraph-instrument room. At smaller Post Offices which accept telegrams, the volume of traffic does not justify the cost of providing, maintaining and staffing a teleprinter; telegrams are telephoned by Post Office staff into the nearest 'appointed' office where a telegraph-instrument room is available. These 'telephone—telegrams' pass between Post Offices either over the public-

telephone system or over a direct circuit, according to the volume of traffic. The proportion of this class of traffic tends to increase, owing to the fall in traffic at smaller offices, which can then no longer support the cost of teleprinter working. Facsimile transmission has been used to supersede telephone-telegrams but, although the chance of phonetic error is excluded, the economics of facsimile telegraph transmission for this purpose have not proved favourable, nor is operating time necessarily reduced in view of the need to obtain an acknowledgement of receipt on completion of transmission.

Telegrams can be telephoned into the nearest appointed telegraph office by any telephone subscriber, dialling the code (e.g. 190) for the *phonogram* service. Outside normal working hours, phonogram circuits are switched to a smaller number of 'night-appointed' offices which provide 24-hour service. Telegrams can be telephoned at any hour from a public telephone-call office, the charge for the telegram being paid by use of the coin box. Phonograms and telephone-telegrams together account for about 80% of all telegram acceptances.

Telex subscribers wishing to send a telegram to an addressee who is not a telex subscriber can dial the *printergrams*-service position in the nearest telegraph-instrument room and 'hand in' the telegram by transmission from the telex teleprinter.

Commercial organizations which originate or receive a large number of telegrams may have an exclusive direct teleprinter circuit connecting their premises with the nearest telegraph-instrument room for handing in (and receiving).

Forwarding Telegrams. This expression covers the transmission of a telegram from the office of origin to the destination office. In the United Kingdom this process is entirely carried out over a teleprinter automatic-switching (TAS) system which connects all main telegraph offices by a direct dialling system so that a telegram can be rapidly transmitted to the instrument room in the destination or delivery office nearest to the addressee's premises, without need for intermediate handling.

Delivery of Telegrams. As far as the method is reversible, a telegram is delivered to the recipient over the same medium used for handing in, e.g. via phonograms, printergrams or leased teleprinter circuit. Where none of these methods is applicable; the telegram is delivered by hand by a junior postman. Delivery costs represent a very high proportion of the total of dealing with a telegram.

9.2 Combined phonogram/teleprinter positions

The fall in telegram traffic, the flexibility of the TAS system and the increased proportion of acceptances by telephone (both phonogram and telephone-telegram) have led to the development of this rationalized system of 'combined working' in which the one operator handles both phonogram (including telephone-telegram) and teleprinter transmission and reception at the same operating position.

Fig 9.1 Combined phonogram/teleprinter position (*Courtesy of the Post Office*)

The equipment of the operating position for combined working is shown in Fig 9.1. In addition to the teleprinter, the TAS dial unit, and a file of routing codes, the position equipment includes a telegraph typewriter (i.e. one with the same key-board layout as the teleprinter, and printing in one case only) and a phonogram-key panel which accommodates 20–40 phonogram and telephone-telegram (TT) lines. The equipment is mounted on an L-shaped table to which the phonogram panel forms an extension. The telegraphist is seated on a swivel-chair to gain equal access to the phonogram panel, typewriter, dial unit and teleprinter. This method of working abolishes a good deal of movement of both staff and message forms, compared with the use of separate, specialized phonogram and teleprinter positions. At the combined position, the telegraphist has all the equipment to hand for the complete processing operation. This method dispenses with conveyor belts and for separate staff to deal with routing and distribution. Greater efficiency results from the entire transaction being conducted by the one operator.

The telegraphist accepts a telephone-telegram from the caller, typing the message as it is dictated; the TAS routing is then determined from the file, the teleprinter connection dialled and the telegram transmitted by teleprinter to the distant TAS destination office, from which it will be delivered to the addressee. If further phonogram calls are waiting, further telegrams can be accepted, the forwarding being left in abeyance for the time being. At normal traffic levels this allows occasional traffic peaks to be smoothed out. At times of extreme pressure, two telegraphists can be employed at the combined position, one to deal with phonogram acceptances and the other looking after routing and forwarding. The combined method enables a telegraphist also to deliver, by telephone, telegrams which are received from the TAS network. This method of working has been standardized for all but the largest telegraph offices (where phonogram automatic-distribution equipment is used) and the very small offices (up to four teleprinters) where the savings inherent in combined working are not readily realizable. Efforts to rationalize the system still further by taking down the telegram on perforated tape, concurrently with typing it on the acceptance form, to eliminate one keyboard operation, have not been successful owing to the high percentage of cases needing alteration of words due to amendment of text by the caller during or after dictation, or possibly operating errors in the taking-down stage.

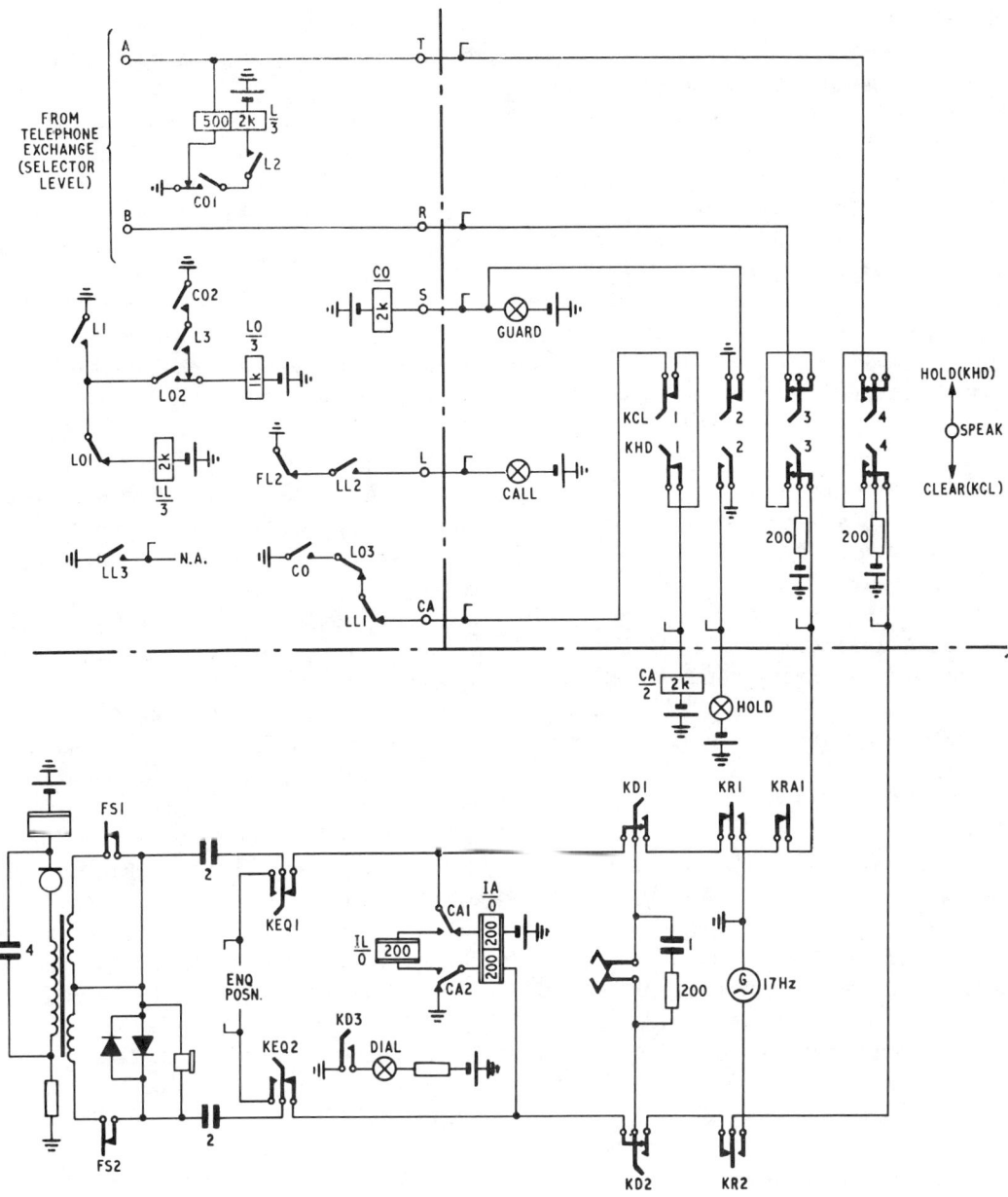

Fig 9.2 Phonogram and telephone/telegram panel

Phonogram Panel. This panel is fitted with a number of three-position keys, one key per telephone line. That illustrated is designed for a maximum of 30 lines, multipled over all phonogram panels in the instrument room. Thrown towards the operator the line keys are in the CLEAR position; the central position is labelled SPEAK, the upward position gives the HOLD facility. Associated with each

key are a CALL lamp and a GUARD lamp. In addition to the telephone dial there are two position keys, seen in the lower right-hand of the panel, labelled RING-DIAL and ENQUIRY TRANSFER – COIN BOX RING ANSWER. The operator wears a lightweight headset, which is plugged into the position jack.

The circuit shown in Fig 9.2 comprises line

relays, line keys and operator's-position circuit. The line-circuit equipment can be modified, by strapping appropriate U-points, to cater for different types of telephone-exchange line — automatic (from final-selector or group-selector levels), manual (CB or LB), or direct telephone-telegram line; as shown, the circuit is for lines from group-selector levels.

The call signal is -50 V on the A wire, which operates the L relay. **L1*** operates the LL relay. **LL2** lights the CALL lamps (on all positions); **LL3** closes the night-alarm circuit. If the call has remained unanswered after 30 s, the steady glow on the CALL lamp is changed to a flashing signal by FL2 in order to give priority to calls which arrived first.

Any telegraphist can answer a call by throwing the key of the calling line to the SPEAK position. At KCL2 the cut-off relay CO is operated and the GUARD lamp lights (on all positions) so that other telegraphists will not use this line. **CO1** holds the L relay (via **L2** and the second winding), disconnecting the operating winding from the line circuit; **CO2** operates the LO relay which holds at **LO2**. Contact **LO1** releases the LL relay to disconnect the CALL lamps at LL2. The T and R wires are extended to the telegraphist's telephone circuit via KCL3—4 and KHD3—4.

The telephone circuit incorporates a foot-operated switch FS1—2 to prevent room noise from being injected through the microphone into the receiver during dictation; the receiver is protected against excessive line-noise peaks by two diodes. High-impedance coils IL and IA (via CA1—2) provide a low-resistance holding loop to the exchange.

The telegraphist can ascertain whether the call is from a coin box by reception of a discriminating tone, and if so can open the coin chutes and check the coin insertion by operating the COIN BOX RING ANSWER key KRA.

On outgoing telephone calls, the telegraphist can dial a connection over the public-telephone system, using the DIAL key KD; contact **KD3** lights the DIAL lamp as a reminder to restore the DIAL key after dialling. The telegraphist can ring over direct telephone-telegram lines by use of the RING key KR.

The telegraphist can speak to the enquiry posi-

---

* A relay contact printed in heavy type, e.g. **L1**, is to be read as in the operated state.

tion, using the ENQUIRY-TRANSFER key KEQ, the calling line being held by operating the line key to HOLD to apply the 400-Ω loop with battery and earth potentials to the exchange; the HOLD lamp·serves as a reminder to restore the key later. The operator at the enquiry position can take up the connection with the caller by throwing the appropriate corresponding line key on a similar, multipled panel.

9.3 Phonogram and telephone/telegram automatic distribution and call-queueing system

The 'delayed-answer' flashing arrangement used in the 'combined' system is suitable for small installations to ensure that no call is kept long awaiting an answer. For use at offices which carry heavy phonogram and TT traffic, many lamps would be glowing and flashing at any instant and the occasional call would lose its priority.

An automatic distribution and call queueing system ensures that incoming telephone calls are answered in cyclic order and that the 'unfortunate call' — subjected to a time-of-answer well above average — is eliminated. The system also provides facilities for smoothing out sudden peaks of incoming traffic by automatically increasing the number of staffed positions to which incoming calls have access; should the traffic peak be sustained, the availability of staffed positions can be still further increased under control of the supervisor. Teleprinter transmission and reception are handled at a separate suite of positions.

The equipment provides automatic distribution of incoming calls to free, staffed positions on a cyclic basis. Calls are connected direct to a free operator's headset; when a free position is not available, calls are queued up in the order of arrival and are distributed, in that order, to free positions as they arise. Ringing tone is connected to the calling subscriber's line when the call is received by the phonogram equipment, whether it is connected to a disengaged position or stored; the tone ceases when the call is connected to the operator's headset. Busy tone is returned on all calls reaching the phonogram installation over and above the agreed maximum number of stored calls.

The block diagram of Fig 9.3 shows equipment for call queueing, automatic distribution, initiating

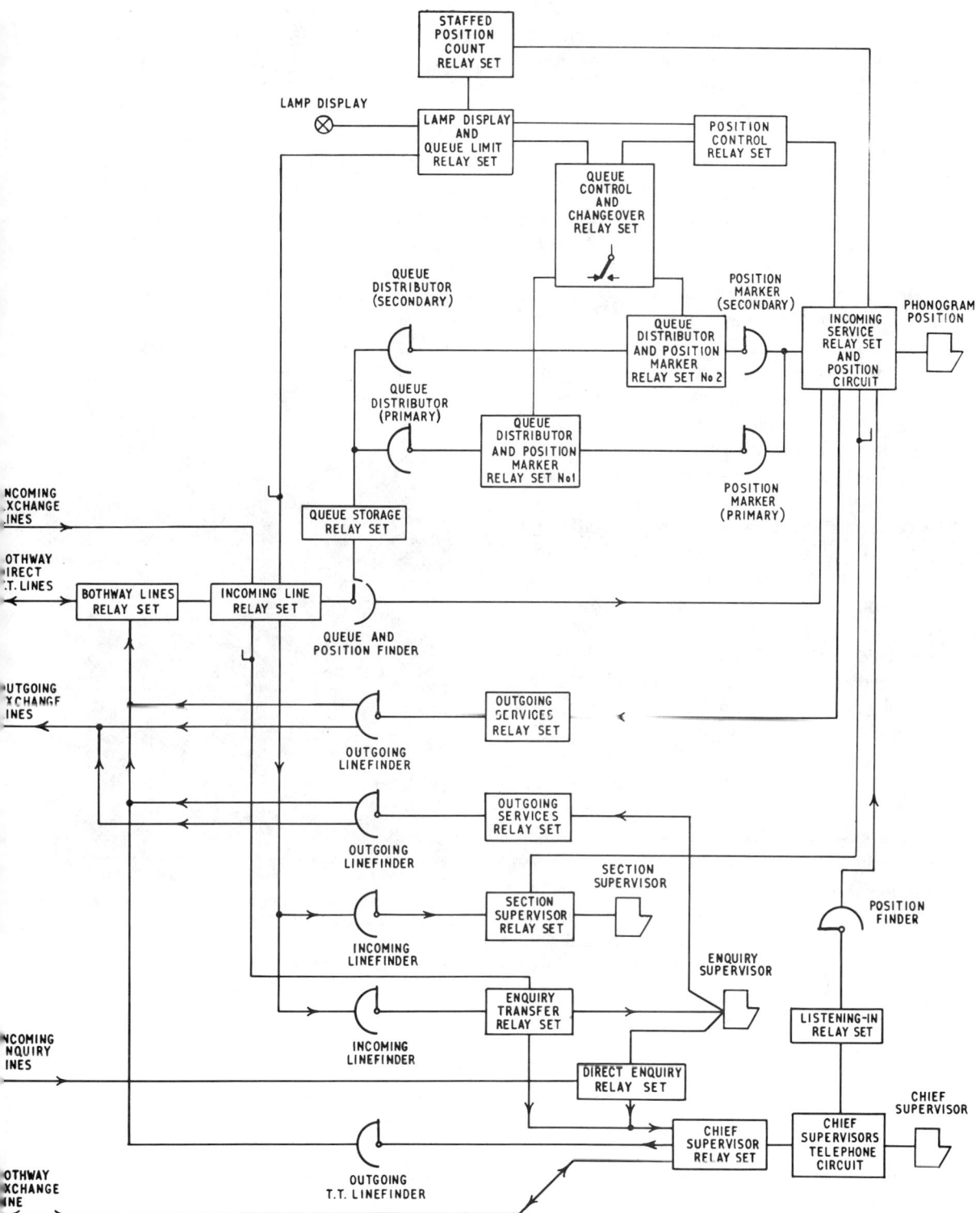

Fig 9.3 Phonogram automatic distribution — block diagram

outgoing calls and transferring calls. Operator's positions are divided into three groups — incoming, outgoing and bothway. Incoming calls are presented cyclically, in order of arrival, to staffed, disengaged, incoming positions. Bothway positions normally function as outgoing positions, but should all staffed incoming positions be engaged, a waiting call is automatically routed to a free bothway position. To meet seasonal traffic conditions, bothway positions may be converted to incoming positions by means of a master key controlled by the chief supervisor. When this key is operated, bothway positions are included in cyclic distribution of incoming calls, i.e. calls are presented to disengaged bothway positions in sequential order even though free incoming positions are available. Operation of a master key also permits outgoing positions to be converted to incoming positions, at the discretion of the chief supervisor, by means of individual keys.

The number of queue positions at any time may be varied in relation to the number of staffed positions available to incoming traffic. This variation may be performed manually or automatically. In the manual case, the setting of a rotary switch determines the maximum size of the queue; in the automatic case, removal or insertion of an operator's headset plug into the position jack causes a counting device to function, and the size of queue is automatically limited in accordance with the number of staffed positions available. The ratio of queue size to staffed positions is preset, but could be altered by means of soldered cross-connections.

A queue-lamp display is provided which functions only when queue positions are occupied and indicates the size of the queue at any instant. A QUEUE FULL lamp glows when the limit of the queue is reached. The supervisor is provided with a lamp display giving individual indications for each queue position and an audible alarm whenever the QUEUE FULL lamp glows. When automatic control of the queue is in operation, the number of staffed incoming positions is also indicated to the supervisor.

Should common-control equipment fail to function correctly, automatic change-over to standby equipment is provided; the standby equipment is positioned to agree with that of the regular equipment upon taking over. Manual change-over is also provided for test and maintenance purposes.

Outgoing calls may be originated over the telephone-exchange network, or to direct TT circuits by operation of the outgoing key on the position. Operation of the EXCHANGE outgoing key causes a uniselector to hunt over a common group of exchange lines for a free outlet; busy tone is returned should all outlets be engaged. Direct TT circuits are obtained by operation of the DIRECT TT outgoing key and dialling a pre-arranged code; automatic ringing is provided. Standard tones indicate the condition of the called line.

Fig 9.4 Phonogram automatic distribtuion — operator's position (*Courtesy of the Post Office*)

Position Equipment. The operator's position (Fig 9.4) is equipped with a telephone dial, three keys and three lamps. A white glow indicates that the position is staffed but no call connected; a green glow shows that the position is staffed and an incoming call connected. The green glow is maintained for a 15-s guard period after a caller has cleared, at the end of which the green lamp flashes until the CLEAR key is operated. The red glow shows that an outgoing call is in progress. Operation of the non-locking CLEAR key releases the position and removes the guard condition on an incoming call or enables a further call to be made at an out-going position. A non-locking TRANSFER key has two operated positions to enable any incoming call to be transferred to either the enquiry position or the section supervisor. A third key with two locking positions enables an operator to originate calls over the public-telephone system and over any direct TT circuit. The position is guarded against the connection of incoming calls when an outgoing call is being or has been set up. Withdrawal of the operator's headset plug from the jack guards the position against receiving incoming calls.

TELEPRINTER AUTOMATIC SWITCHING

The teleprinter automatic-switching (TAS) system connects all teleprinter telegraph offices in the United Kingdom; the teleprinters used (No. 11) operate at 50 bauds and the keyboard is designed to alphabet No. 2. It is an entirely separate net-work from that of the telex service, apart from the printergrams access. The design allows for subse-quent integration with the telex service, a method of joint operation used by some European adminis-trations.

9.4 TAS numbering scheme

Call charging is not a factor in considering the net-work plan for the telegram service, but the ready availability of alternative routes at times of traffic congestion or trunk-circuit interruptions assumes importance. To avoid complexities of routing translators and other discriminating and storage devices, an area-numbering scheme was adopted for the TAS system. As a result, the digits dialled to reach a given office cannot be uniform over the network and a system of dialling codes, different for each exchange area, is employed. A national directory is used in which the entry for each tele-graph office shows the name or letter code (well known to telegraphists) of the exchange which serves it, together with the exchange number, usually of three digits, e.g. BM531 designates a particular office connected to the Birmingham exchange. A dialling-code list, showing routing digits to be dialled to reach every exchange is displayed on the front of the dial unit at each operating position. To call another office the

telegraphist dials the routing digits for the exchange or switching centre serving it, followed by the digits corresponding to the exchange-line number of the required office. Supposing the routing code for reaching the Birmingham switching centre from London to be 31, the telegraphist would dial 31–531 to gain the office referred to above. The identifying code BM531 would be inserted on the telegram form at the circulating position in the originating office before the telegram is passed to the telegraphist for transmission; alternative routing codes are also quoted on the dialling list.

The TAS network was based upon a number of fully-interconnected zone-switching centres, supple-mented by a number of area centres each connec-ted to the appropriate parent-zone switching centre for access to the network. This is the plan of the telex network. Due to the reduction in telegram traffic, area centres were closed and the system served by a smaller aggregate number of zone centres, fully interconnected by circuits in direct VFT systems. A section of the network is shown in Fig 9.5 which indicates typical routing digits

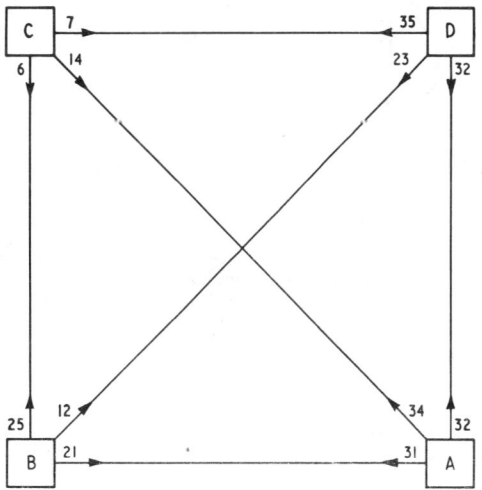

Fig 9.5 TAS trunk-network plan

to gain access from one switching centre to another. Using the triangulation method in which, for example, switching centre B – normally reached over a direct trunk route such as A–B – can also be reached via an intermediate switching centre C, a system of approved alternative routings is available by dialling the alternative routing code

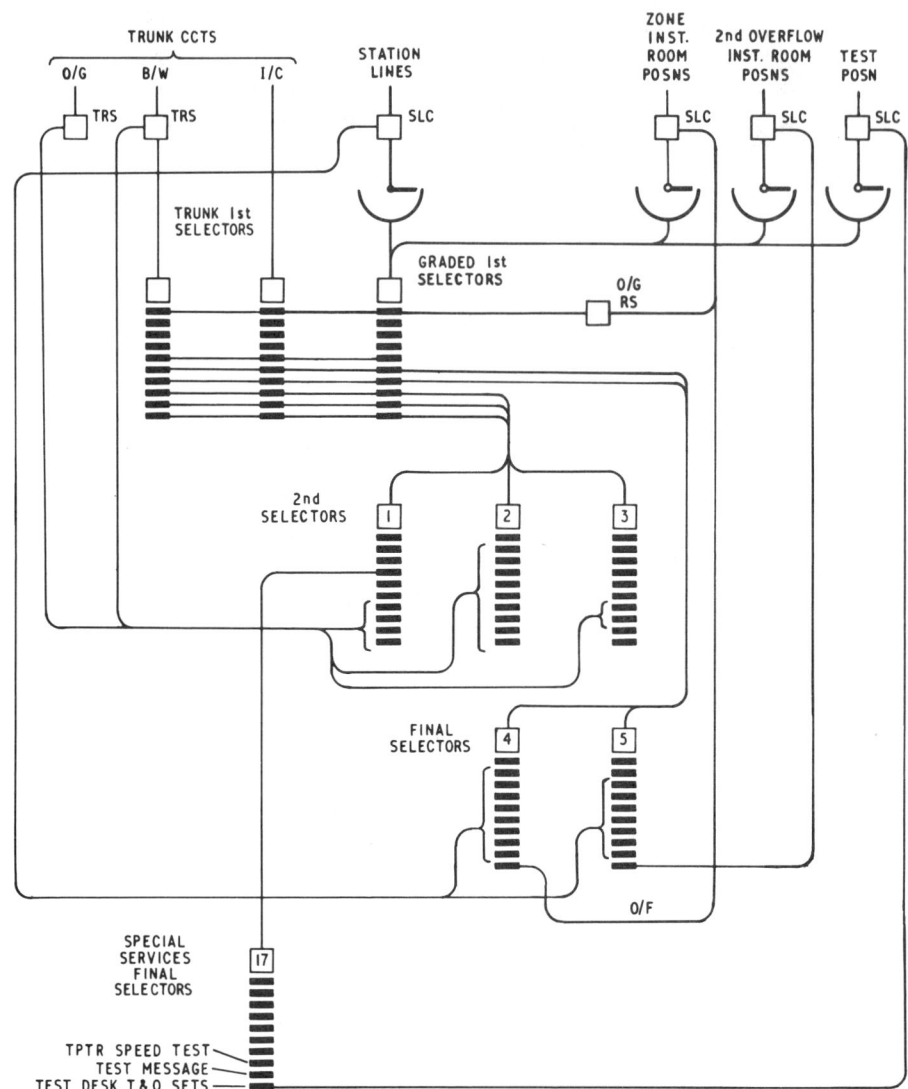

Fig 9.6 TAS switching centre — trunking diagram

(but the same exchange-line number) if difficulty is experienced on the primary route. For example, instead of dialling the normal routing 31–531 quoted above, an alternative routing to the same office would be by dialling 346–531. Alternative routings take account of transmission considerations and also the desirability for the primary routing code to consist of fewer digits than the alternative one.

A typical trunking diagram is shown in Fig 9.6. Lines from the operating positions terminate on uniselectors giving access to a common pool of 1st selectors. Certain levels of 1st selectors are trunked out to 2nd selectors giving access to trunk circuits to other switching centres; other levels from 1st selectors lead to final selectors giving access to the telegraph offices served by this zone-switching centre, and in this case bearing 3-digit exchange numbers. All group and final selectors are of the 200-outlet type.

In allocating selector levels consideration is given to traffic loading to keep dialling time to a minimum. Level 0 of 1st selectors gives direct dialling access to teleprinter positions in the telegraph office associated with the switching centre; this saves dialling time and equipment on circuits which carry heavy traffic. The 'overflow' level — level 1 of final selectors — is teed to level −0 outlets of 1st selectors to route overflow traffic to zone-office teleprinters. Level 9 of 1st selectors is reserved for use in conjunction with telex service. Level 1 of 1st selectors gives access (via 2nd selectors) to a special rank of final selectors providing test facilities. Incoming trunk circuits terminate directly upon incoming 1st selectors; outgoing trunk circuits are routed through bothway trunk-relay sets.

With the use of bothway trunk-relay sets there is always the possibility that on seizure a circuit may suffer intrusion from the distant end before the arrival of the call signal can guard the distant outgoing multiple. A propagation time of 40 ms is imposed by the VFT system and for this reason it is usual to use a high-speed relay, operating in less than 1 ms, to detect arrival of the call signal. Possibility of double-seizure is greatly reduced by reversing the order of connection of circuits to selector-bank outlets at the two ends — circuits are tested in reverse order at the two ends and the same circuit does not appear among the early choices at both ends. If the route also includes unidirectional outgoing circuits these are arranged to be the early-choice outlets.

Station lines work on a bothway basis, but at larger centres, separate forwarding (only) and receiving (only) positions are established. At offices where circuits are used bothway, those used primarily for forwarding traffic are arranged to be later choices in order of testing on selector banks.

A valuable feature of the TAS system is provision of a 'suspense' or waiting facility on final selectors Small groups of circuits are inefficient when handling automatically-switched traffic at the usual grade of service. The proportion of incoming calls which meet the engaged condition on groups of station-line circuits from final selectors may be greatly reduced by allowing a proportion of such calls to form a small queue and wait for a predetermined time for a line in the required group to become free. The 'queue' allows for only one or two calls to wait, and waiting time is limited to 30—60 s — a figure related to the average holding time of a connection (1 min) to avoid unnecessary congestion of preceding selectors. A line becoming free during

this period is automatically seized by the first waiting call and a second moves up to the head of the queue. If on expiration of the waiting period a line to the required station has not become free, the waiting call is automatically rerouted to an 'overflow' level of the final selector; this leads to teleprinter positions in the instrument room associated with the switching centre which serves the destination office. The overflowed telegram is subsequently retransmitted to its destination by staff at the overflow centre. If the group of lines *and* waiting positions in the queue are engaged, any further call for this office is rerouted to the overflow level without delay. This overflow feature restricts the number of repeat calls which would be necessitated by encountering engaged condition on station lines, with a consequent reduction in ineffective holding time on trunk circuits. Experience has shown that the TAS network can handle considerable traffic peaks without difficulty as the suspense facility ensures that a high proportion of incoming calls are effective.

The suspense feature enables traffic for offices closed at night and weekends to be immediately rerouted to the overflow office, which serves as a night-concentration office, by using the OUT-OF-SERVICE feature at the teleprinter position.

The service area of a zone-switching centre can be very extensive and may include more than one night-appointed office. Auxiliary overflow centres have been established in certain large instrument rooms not directly associated with the switching centre but well situated to serve as the night-appointed office for smaller offices in the vicinity. Access to these night-appointed offices can be from different final-selector units or from final-selector units in which two separate overflow levels are provided.

9.5 TAS signalling principles
These are basically those of the CCITT Recommendation U1 but not always identical to those of the telex system.

Disengaged Line. A space condition (+80 V) applied to both channels of the circuit.

Call Signal. A transition from space to mark (−80 V) condition on the sending channel.

Proceed-to-select Signal. This signal is returned to a station line when an outgoing call is made to indicate that a 1st selector is connected to the calling line and that dialling may commence. The signal is a 25-ms mark pulse on the receive channel.

Call-confirmation Signal. This signal is used only on trunk circuits and is sent back over the receive channel immediately following seizure of the send channel for a call; its purpose is to prove the continuity of the trunk circuit and the response of the switching equipment at the called switching centre. The signal is a 25-ms mark pulse sent from the called end of the circuit back to the calling end.

Dialling. Dial pulses have a nominal ratio of 67% space to 33% mark at a nominal speed of 10 pulse/s. The break of the dial-pulse contacts corresponds to the space signal.

Call-connected Signal. This signal is used to indicate to the calling station that connection has been made either to a teleprinter or to a source of start–stop service signals. It consists of a transition on the receive channel from space − to which condition the channel is held throughout the selection stage (apart from the proceed-to-select and call-confirmation pulses) − to mark condition and is followed either by the answer-back code of the called teleprinter or by service signal.

Clear Signal. This signal is used to release a connection and consists of a change from mark to space maintained for at least 300 ms. Circuit arrangements permit the clear signal to be initiated either by the calling or the called party, providing an 'either-party-release' system. Equipment is designed so that a clear signal initiated on either channel is repeated on the other channel of the circuit, ensuring that both channels and equipment revert to the disengaged condition.

Immediate-release (IR) Signal. This signal, consisting of 250-ms mark and 1250-ms space, continuously repeated, is applied to spare levels and spare numbers. It results in application of immediate forced-released condition to any call switched to a spare level or number.

Service Signals. These are provided so that, by means of characters printed on the teleprinter tape at the calling position, a calling telegraphist can be informed of congestion or fault conditions encountered during dialling. The signals are:

OCC	All lines to the called office are engaged.
NC	Trunk circuits engaged.
DER S	The selected station line is faulty.
DER T	The selected trunk line is faulty.
MOM	Wait.

9.6 Station equipment

The teleprinter No. 11B, printing on gummed-paper tape is used, in conjunction with a dial unit; the latter contains switching relays (operating at +80 and −80 V). At offices with not more than eight TAS positions, equipment is mounted on L-shaped tables which allow for flexibility in office layout and accommodate additional equipment. In larger offices the teleprinter and dial unit are fitted on long bench-like tables (Fig 9.7). Power supply for the motor (160 V d.c.) and relay operation is taken from the main ± 80-V battery so that service may be uninterrupted by any electric-mains failure.

Fig 9.7 TAS forwarding positions, bench-mounted (*Courtesy of the Post Office*)

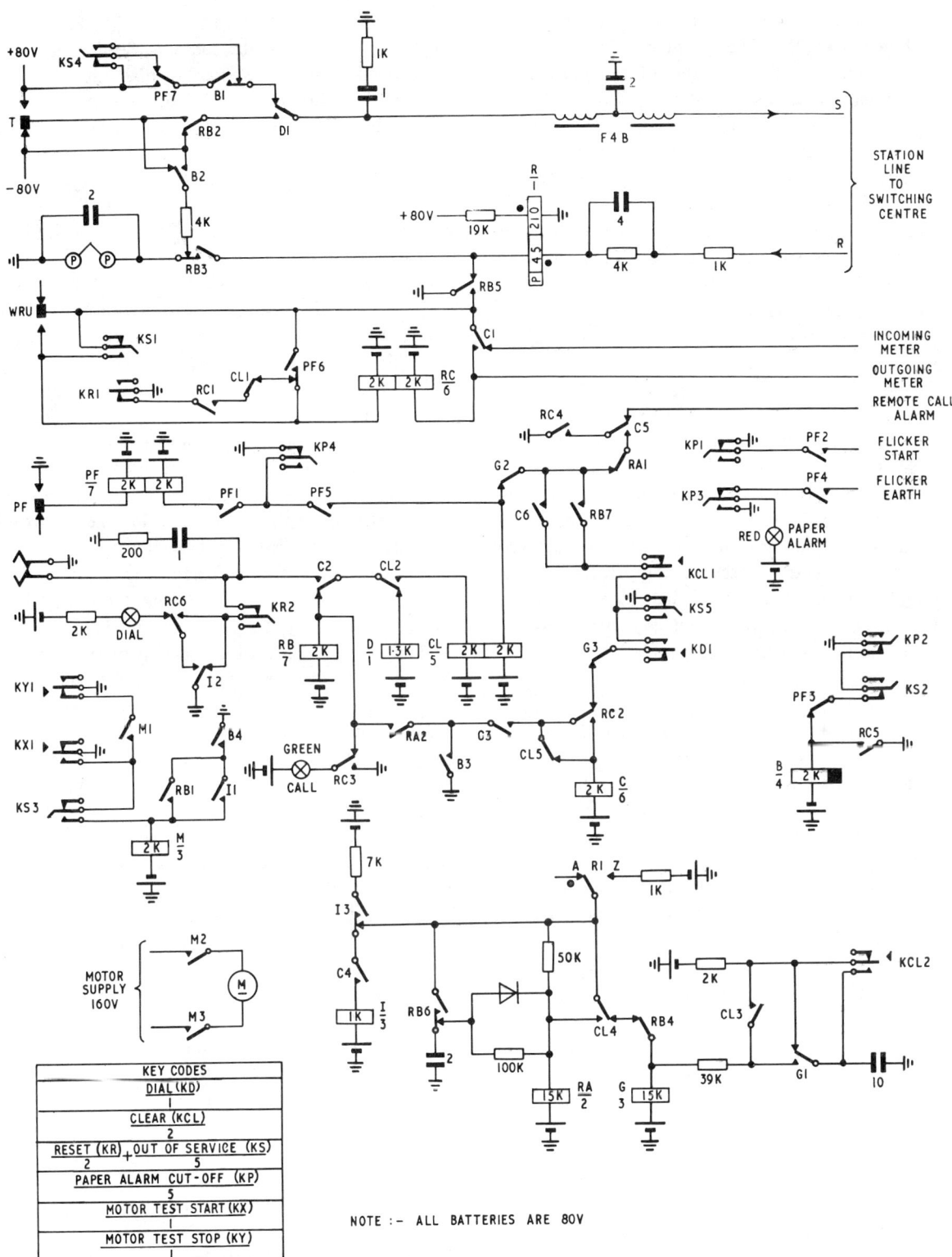

Fig 9.8 TAS station — circuit diagram

The circuit diagram for the position equipment is shown in Fig 9.8. The circuit is arranged for 2-way simplex operation, without local record, so that duplex transmission could be used. A filter No. 4B and a signal-shaping network are included. The return channel is normally used only for reception of the answer-back code — once (automatically) on completion of dialling and again (following depression of the WRU key) at the end of transmission to serve as an acknowledgement of receipt of the telegram. Service signals are also received over this channel.

Two supervisory lamps are provided on top of the dial unit — a green CALL lamp which glows during any connection and remains glowing after an incoming call is cleared until extinguished by the telegraphist throwing the RESET key; and a red lamp which glows in the event of paper-feed failure. A DIAL lamp, mounted near the dial, glows on receipt of the proceed-to-select signal, until the call matures. The dial (type 14) has make-break contacts; dial pulses are converted to double-current signals by the dial relay D. The B relay is normally operated, but its contacts are shown in the unoperated condition.

Outgoing Call. In the free condition, +80 V is applied to the S wire to the switching centre via KS4, PF7, **B1** and D1. The teleprinter magnet is held to the stop condition from −80 V at the transmitter contacts, via **B2**, the 4 kΩ resistor and RB3.

To make an outgoing call the telegraphist presses the non locking DIAL key **KD1**, and the C relay operates. C2 via CL2 operates relay D to the dial contacts; C3 holds the C relay. **D1** applies the call signal, −80 V, via RB2 on the S wire to the switching centre. When a 1st selector has been seized, a proceed-to-select signal (25-ms mark) received over the R wire operates the polarized relay R (normally held by spacing bias current in the second winding). **R1** closes a circuit for relay I via **C4** and I3 which holds the I relay. The RA relay, although connected in parallel with relay I, cannot operate to a 25-ms signal on account of the time constant of the RC combination which shunts it. **I1** via **B4** closes the circuit for the motor relay M which at **M2−3** starts up the teleprinter

motor. **I2** lights the DIAL lamp and also clears the short-circuit from the dial contacts. Seeing the DIAL lamp glow, the telegraphist dials the routing digits of the wanted office. **D1** transmits double-current dial pulses over the S wire to selectors in the switching centres. When the required station is found to be disengaged, a call-connected signal (inversion to −80 V) is returned over the R wire and operates the polarized relay R in series with the R wire. **R1** operates the RA relay which has an operating delay of about 60 ms. **RA2** lights the CALL lamp and also operates the RB relay. **RB2−3** switch the teleprinter transmitter and receiver to the S and R wires respectively; **RB5**, via **C1**, operates the RC relay. **RC6** extinguishes the DIAL lamp and restores the short-circuit over the dial contacts to prevent any false signals from the dial being misoperated. Receipt of the call-connected signal is followed almost immediately by the answer-back code of the called station, after which the circuit is available for transmission of the telegram. During reception of the answer-back and other signals, relay R responds to signals on the R wire; although **R1** pulses the RA relay, the latter remains held due to its delayed-release feature of 300 ms provided by the RC circuit which shunts it. During the call, relays B, C, D, M, RA, RB and RC remain operated. Before clearing the connection, the telegraphist presses the WRU key; receipt of the answer-back code is taken as acknowledgement of receipt of the telegram.

The telegraphist clears the connection by pressing the CLEAR key KCL until the CALL lamp is extinguished. **KCL1** operates the clear relay CL via **C6** and G2. Contact **CL2** holds the CL relay over its second winding via **C2** and the dial contacts and also releases the D relay. D1 applies +80 V clear condition to the S wire to release selectors in the switching centres. As a result, the clear-confirmation signal is applied over the R wire by the first selector to release. Relay R responds to this and R1 releases the RA relay after 300 ms. RA2 releases the RB relay; RB5 releases the RC relay; RC2 releases the C relay; RC3 disconnects the CALL lamp; C2 releases the CL relay; C4 releases the I relay; I1 releases the M relay. The circuit is restored to normal with the teleprinter motor stopped.

Alternatively, the connection could be released by the distant station, or by receipt of a service signal, or as a result of a line interruption exceeding 300 ms. Relay R responds to the +80-V clear signal. R1 releases the RA relay. RA1 operates the CL relay via **RC4**, **C5** and G2; release of the connection follows as described above.

Incoming Call. The incoming call signal (−80 V) operates relay R. Contact **R1** operates the RA relay. RA2 operates the RB and D relays and lights the CALL lamp. **D1** via **RB2** applies −80 V to the S wire to send the call-connected signal to the calling station. **RB1** operates the M relay to start the teleprinter motor; **RB2–3** connect the teleprinter transmitter and receiver to the S and R wires. After a brief delay to ensure that the teleprinter motor has reached full speed, the switching centre sends the WRU signal which causes the teleprinter to send the answer-back code. The WRU contacts close and operate the RC relay which holds at **RC1**.

After the telegram has been received (there is no need for a telegraphist to be in constant attendance at receiving positions) the clear signal is normally sent by the calling station. The +80-V signal causes the RA relay to release via R1 after 300 ms. RA2 releases relays RB and D. Contact RB1 releases relay M to stop the teleprinter motor. Relay RC remains held at **RC1**; contact RC3 maintains the circuit for the CALL lamp. The RESET key, depressed only when the gummed-up telegram has been collected from the teleprinter, releases the RC relay at **KR1** and extinguishes the CALL lamp at RC3. Any delay in pressing the RESET key does not prevent the arrival of a further incoming call.

An incoming call could be cleared by depression of the CLEAR key at the incoming end. KCL1 operates relay CL; contact **CL1** releases the RC relay; **CL2** releases the D relay. D1 sends the +80-V clear signal to the S wire. The clear-confirmation signal releases relay RA via R1 as described above.

Position out of Service. For various reasons, such as closing the office, giving maintenance attention or renewing the paper roll, it is sometimes necessary to close a position to incoming traffic. For this purpose the OUT-OF-SERVICE key KS is operated. **KS2** releases the B relay; **KS4** disconnects the S wire. (If the station is served by a VFT circuit, the out-of-

service condition is −80 V on the S wire, not followed by dial pulses.) The equipment at the switching centre responds to this signal and busies the line against incoming calls until the OUT-OF-SERVICE key is restored. For maintenance tests the teleprinter is connected in-local, with the teleprinter transmitter connected to the receiver via B2 and the 4-kΩ resistor. The KX and KY keys allow the teleprinter motor to be switched on and off.

Paper Failure. To avoid lost telegrams at unattended receiving positions it is necessary that any failure of the paper tape to feed, whether due to paper exhaustion or to mechanical failure, should be detected immediately. The teleprinter is fitted with paper-feed contacts PF which close if the paper fails to feed, and the PF relay operates. **PF1** holds the PF relay; **PF5** operates the CL relay which clears the connection, stops the teleprinter motor and extinguishes the CALL lamps at both stations. The sending telegraphist is made immediately aware that the transmission has failed. **PF3** releases the B relay. After a brief delay to ensure that the connection has cleared, *B1* disconnects the S wire to apply the out-of-service condition. **PF4** applies a flicker signal to cause the PAPER-ALARM lamp to flash.

Miscellaneous Features. The guard relay G is included in the circuit design to cover contingencies which arise under abnormal conditions. A remote-alarm extension is provided for use when the instrument room is unattended. The flicker-earth signal is provided by a self-pulsing relay, when PF2 operates on paper failure. The self-pulsing relay also operates the remote alarm. Incoming and outgoing meters provide traffic statistics.

9.7 The station-line circuit

The primary purpose of this uniselector equipment is to enable a large number of station lines to share a smaller quantity of 1st selectors, provided on a basis of traffic. This equipment also permits the station line to be used for either inward or outward calls, and it prevents intrusion from incoming calls when a line is in use or has been put temporarily out of service.

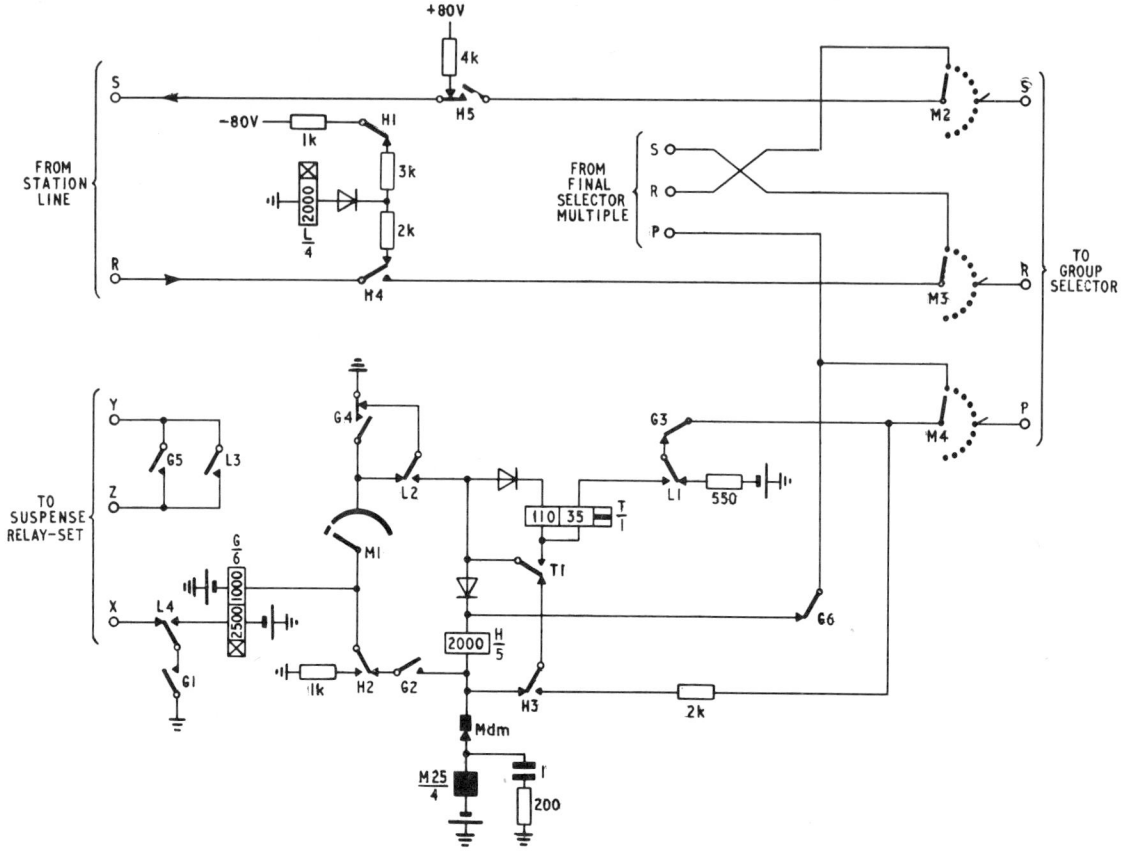

Fig 9.9 TAS station-line circuit

The circuit diagram is shown in Fig 9.9. The 24 sets of uniselector-bank contacts are multipled (i.e. connected in parallel) to other uniselectors with which the outlets to 1st selectors are shared. The incoming S, R and P wires from the multiple of final selector-bank contacts (centre of diagram) are used on inward calls. The third conductor, the P wire, is used on all selector and line relay-set circuits within the exchange; its purpose is to guard any engaged circuit against intrusion from searching selectors and also to maintain the holding condition for all selectors and relay sets which are in use on a switched connection. The X, Y and Z terminals provide the free/engaged/out-of-service discrimination conditions to the relay set which controls the suspense facility at the final selector.

The high-speed relay T is used in the standard battery-testing type of circuit when the uniselector searches for a free outlet. Relays L, H and G are used for the call-signal detection, line switching and guarding respectively.

Outgoing Call. The line relay L is connected via the diode to a point on the potentiometer which includes on the one side a resistor connected (via H1) to −80 V, and on the other side a resistor plus the line resistance. In the disengaged condition, +80 V is applied to the line at the teleprinter-station circuit; provided that the single-wire resistance does not exceed the limit of 1060 Ω, the junction of the resistors is at positive potential and the diode prevents the L relay from operating.

The −80-V call signal from the station equipment operates the L relay. **L1** disconnects the 550-Ω testing battery used by inward calls and applies the testing relay T, with its windings in series, to the P wire via G3 and M4. **L2** provides earth potential to the T relay-testing circuit and it also closes the self-drive circuit for the M magnet (via T1, H3 and Mdm). Due to its interrupter contacts, Mdm, the drive magnet steps the wipers around the bank outlets until a 550-Ω battery potential is found on the P wire. When found, this battery condition completes the circuit for relay T to operate. **T1** immediately breaks the M magnet circuit and short-circuits its 100-Ω winding to apply earth potential via its 35-Ω winding and prevent any other T relay from operating to the same 550-Ω battery; **T1** also opens the short-circuit across the H relay, which now operates in series with the M magnet (the latter cannot operate to the reduced current in series with the H relay). When the group selector is seized, the 550-Ω battery previously on its P wire is replaced by earth potential whose purpose is to release the T relay by short-circuiting it and to hold the H relay (via M4, **H3**, T1, and Mdm). **H4** and **H5** extend the R and S wires from the line through to the group selector, H4 disconnecting the L relay from the line circuit. Relay L has a delayed release to ensure that the H relay has an overlapping holding circuit provided by **L2** until earth potential from the 1st selector is established. **H1** breaks the alternative circuit for the L relay; the L relay has also a delayed operation to make sure that the relay is sufficiently fluxed to give its full release delay if the first-choice outlet is seized. **H2** operates the guard relay G; relay G has a delayed operation to ensure that the holding circuit for relay H is provided from the P wire before the operating circuit via G4 and **L2** is broken.

Incoming Call. For an incoming call the SLC is seized over the S, R and P wires from the final-selector multiple. While the circuit is free, and in service, the 550-Ω test battery is applied to the P wire of the final-selector multiple via L1, G3 and the M4 wiper standing on the home contact. The final selector applies earth potential through its testing relay on the P wire to operate the H relay, via G6 in series with the M magnet. (The diode presents a high backward resistance to prevent earth potential from the final-selector P wire from stepping the M magnet via G6, T1 and H3; it also raises the value of the T1—H3 shunt path across the H relay.) **H2** operates the G relay; **H3** holds the H relay to earth potential from the final-selector P wire (via home contact and

wiper M4, T1 and Mdm). Contacts **H4—5** extend the final-selector R and S wires to the station line via the home-position contacts of the uniselector.

Releasing. During an inward or outward call the H relay is held operated by earth potential on the P wire (from either a final selector or an outgoing trunk-relay set). The G relay is held operated by **H2**. Receipt of a clear signal results in removal of earth potential from the P wire and relay H releases. H4—. 5 restore the R and S wires to the normal disengaged condition; H2 disconnects the operating circuit of the G relay which has a delayed release.

Following an outgoing call, H2 also closes a self-drive circuit for the M magnet which causes the uniselector wipers to be restored to the home contact (via **G4**, the continuous M1 arc, **G2** and Mdm); relay G is held at **G4** via the homing arc and its circuit is not disconnected until the M1 wiper reaches the home contact.

In either case, **G3** delays restoration of the 550-Ω battery potential on the P wire for sufficient time to ensure that station equipment has fully restored before an incoming call can be offered. For a brief interval while relay H is releasing, after earth potential has been removed from the P wire, battery potential through the M magnet, Mdm, H relay, T1 and **H3** is present on the P wire. This might be interpreted, by the T relay of another SLC searching over the multiple, as a free-circuit condition if it were not for the inclusion of the resistor which limits the current to a value insufficient to operate the T relay.

Out-of-service condition. If the telegraphist throws the OUT-OF-SERVICE key on the dial unit, the circuit to the R wire of the station-line circuit is disconnected. Relay L operates from the local circuit via −80 V and H1, and causes the uniselector to hunt for a free 1st selector; relay H operates but the absence of the −80-V call condition on the R wire results in the release of the selector. Earth potential is not maintained on the P wire and relay H releases. The uniselector restores to the home position. Relay L was released at **H1** and **H4** so that the original operating circuit for relay H at L2 is also broken. When the H relay releases, relay L is re-operated at

H1. Relay G, which was operated at **H2**, is now held at **G1** and **L4**. Contacts **G3** and **G6** prevent the circuit from being seized over the P wire from the final-selector multiple.

If the station is connected to the SLC over a VFT circuit, the sequence of operation is slightly different. Operation of the OUT-OF-SERVICE key applies −80 V to the R wire, operating relay L and causing the seizure of a free 1st selector in the normal manner. Relay H operates and **H2** operates relay G. As the call signal is not followed by dial pulses, the 1st selector frees itself after a delay of 12–24 s, removing earth potential from the P wire. Relay H releases, the uniselector homes and relay L is re-operated from −80 V on the R wire; relay G remains operated as described above.

When the OUT-OF-SERVICE key is restored, relay L is now at a point on the potentiometer which is at positive potential and so releases. L4 releases relay G and the circuit is again open to receive calls.

Suspense Feature. The relay contacts connected to X, Y and Z terminals enable the testing circuit for inward calls to discriminate between the conditions shown in Table 9.1.

Table 9.1 Discriminating conditions for station-line circuit

Circuit condition	Relays operated	X, Y and Z terminals
Free	None	All disconnected
Engaged	H and G	X at earth potential (via L4 and **G1**) Y and Z looped at **G5**
Out of service	L and G	X disconnected at **L4** Y and Z looped at **G5**

9.8 The group selector

The function of a group selector is to step the wipers vertically to the required level on receipt of one digit; and then (during the inter-digit pause) drive the wipers automatically over the pairs of outlets in that level to search for the first free outlet. A single design of group selector is used throughout the system; the simple variations required between 1st, 2nd or trunk selectors are provided by selection of appropriate U-points and wiring on the shelf jacks.

The group selector (Fig 9.10) comprises A, B, BA and C relays for the vertical-stepping circuit; AH,

HA, BH and HB relays for testing the outlets; and the time-pulse relay TP for the forced-release condition.

Proceed-to-select and Call-confirmation Signals. Relay A is polarized by a diode so that it operates to −80-V call condition and releases to +80-V clear condition; the R and C elements improve the pulse waveform. The disengaged group selector has the usual 550-Ω test-battery potential on the P wire, via N3 and BA2. When the group selector is seized from the station-line circuit, from another group selector, or via an incoming-trunk circuit, relay A is operated by the −80-V call signal. **A1** (via N1) operates the B relay in series with the vertical magnet (which cannot operate to the reduced current in series with relay B). Contact **B1** holds the B relay; **B3** operates the C relay over its 700-Ω winding; **B4** applies earth potential to the P wire to hold the preceding equipment; **B7** operates the BA relay (via S2 and TP5).

When the selector is first seized, +80 V is applied to the S wire via B6, HB3 and HA3; when **B6** operates it changes this potential to −80 V at BA5. Relay BA has a delayed operation and, when **BA5** operates, it changes the S wire potential back to +80 V. This brief period (nominally 25 ms) while BA5 delays its operation constitutes the proceed-to-dial signal if the selector is a 1st selector, or the call-confirmation signal if the selector serves an incoming-trunk circuit. On other selectors which do not need to generate this pulse, the HB3 contact is connected direct to +80 V, the B6 and BA5 contacts being ineffective.

Vertical Stepping. On receipt of the pulse train, relay A releases to the +80 V ('break') A periods. Each time it releases, A1 short-circuits the B relay and the vertical magnet is fully energized (via **B1**, A1, **C1** and the 5-Ω winding of relay C). The vertical-magnet armature steps the wipers to the level corresponding to the digit dialled. Relay B does not release during the nominal 67-ms periods while it is short-circuited by A1. When the wipers first step off normal, N1 short-circuits the 700-Ω winding of relay C; this relay, however, is energized over its 5-Ω winding by the current which flows in the vertical-magnet circuit during the 66-ms periods, and it remains operated due to its 700-Ω winding being

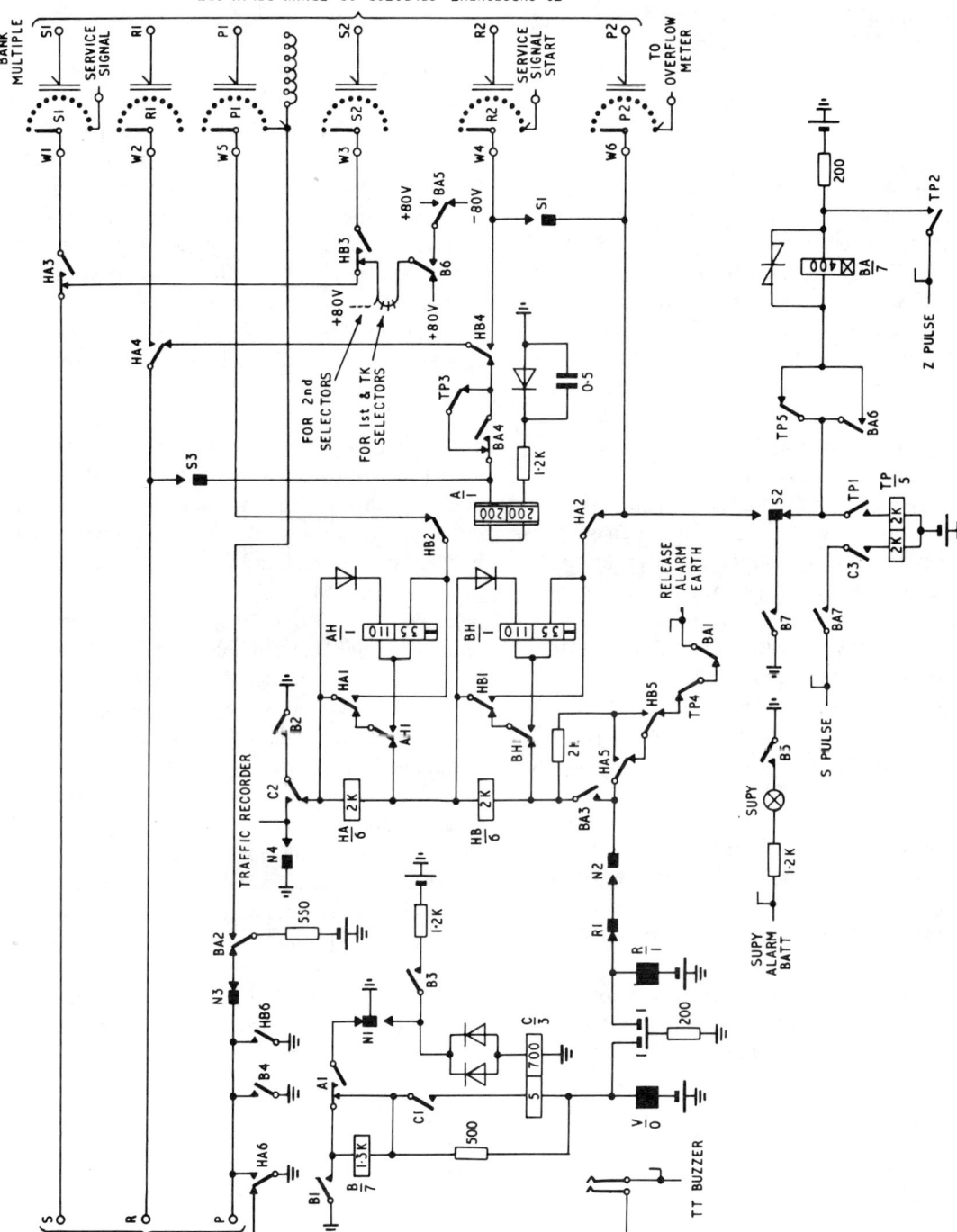

Fig 9.10 T AS 200-outlet group selector

short-circuited during pulsing. On completion of the pulse train, relay A remains operated. There is now no current in the 5-Ω winding of relay C, and its 700-Ω winding is short-circuited. Relay C releases 120 ms after the end of the final pulse into the vertical magnet.

Rotary Hunting. Release of relay C after the wiper shaft has been raised to the required level closes a self-drive circuit for the rotary magnet (via **B2**, C2, HA1, AH1, HB1, BH1, **BA3**, N2 and R1); the wipers are driven over the contacts of the level dialled. Two testing relays, AH and BH, using the standard testing circuits, are applied to the P1 and P2 wipers respectively. Immediately either wiper encounters the 550-Ω test-battery potential on the bank contact, relay AH (BH) operates and **AH1 (BH1)** breaks the circuit of the rotary magnet to arrest further movement of the wipers. **AH1 (BH1)** also removes the short-circuit from the HA (HB) relay which operates in series with the rotary magnet, the latter being unable to operate in series with the relay.

Should relays AH and BH simultaneously encounter a 550-Ω test-battery condition on both the P1 and P2 bank contacts at the same outlet position, both will start to operate together; priority will be given for relay AH to operate instead of BH because the operation of **AH1** breaks the circuit for relay BH.

When relay HA (HB) operates, HA3 and **HA4 (HB3** and **HB4)** disconnect the A relay and the +80-V condition from the S and R wires respectively and extends these wires through to the next equipment. **HA1 (HB1)** provides a path to the P1 (P2) wiper via HB2 (HA2) to hold the HA (HB) relay from earth potential shortly to be applied by the selector or relay set which has been seized. Until then earth potential is maintained by **B2** to hold the HA (HB) relay, guard the P1 (P2) wire and short-circuit the AH (BH) relay to release it. **HA5 (HB5)** completes the holding circuit for the HA (HB) relay. Contact **HA6 (HB6)** applies earth potential to the incoming P wire to hold the preceding selector.

With the release of A1, relays B and BA are also released. The circuit is extended to the next selector, only relay HA (HB) remaining operated throughout the call. If a connection needs to be 'traced' by maintenance staff, insertion of a test link between T8−9

will operate the rack buzzer if an 'even' (P2) outlet has been seized.

All Outlets Engaged. If all 20 outlets on the required level are found to be engaged, the wipers reach the eleventh step and relay AH operates to the 550-Ω test battery connected to the P1 contact via **BA2**. Contact **AH1** disconnects the rotary-magnet drive circuit and operates relay HA. The S contacts operate on reaching the eleventh step. **S3** maintains a circuit to hold the A relay, normally broken by **HA4**; contact **S2**, via **B7**, operates the traffic-overflow meter; **S1** closes the start circuit for the service signal; **HA3** applies the OCC signal from the S bank eleventh step.

Forced Release. If a 1st, or incoming, selector, after having been seized, fails to receive any dial pulses — as for example for the out-of-service condition — the selector automatically releases itself from the connection. On seizure, **BA7** and **C3** connect the TP relay to the S pulse lead which applies an earth-potential pulse every 12 s. Relay TP operates to the first pulse to arrive on this lead and holds at **TP1** to S2 and B7. Twelve seconds later an earth pulse receive on the Z pulse lead short-circuits relay BA, causing it to release. BA4 (with **TP3**) opens the circuit of relay A, which releases, followed by relays B and C. If a 1st selector is concerned the release of BA5, followed by B6 about 300 ms later, sends a −80-V pulse back over the S wire to reset the position equipment if the seizure of the selector was due to operation of the DIAL key. In any case, B6 re-applies +80 V to the S wire to clear the distant equipment. B7 releases the TP relay. TP4 closes a circuit for the rotary magnet (via release-alarm earth, BA1, HB5, HA5, N2 and R1) which steps the wipers to the twelfth position from where the wiper carriage restores to normal.

The TP relay is not used on 2nd selectors. If such a selector is held by a fault condition, the supervisory lamp remains glowing at **B5**, drawing its current through a rack-supervisory relay which raises an alarm after 6−12 min.

Selector Release. After a normal connection the selector is released by the removal of earth potential from the P wire by a succeeding selector or relay set.

This releases the HA (HB) relay and completes the circuit for the rotary magnet (via release-alarm earth, BA1, TP4, HB5, HA5, N2 and R1). The wiper carriage is driven to the twelfth step where, being unsupported, the carriage drops and restores to normal. All N contacts then restore, N2 breaking the circuit of the rotary magnet. N3 guards the selector from being seized during the releasing stage by disconnecting the 550-Ω test battery until the selector is fully restored. Any mechanical failure to restore prevents the release of N2 and maintains the circuit for the delayed release-alarm circuit to draw attention to the fault.

9.9 The final selector

The function of the final selector is to select the required station line on receipt of the tens digit (by stepping vertically) and the units digit (rotary stepping). All line circuits for a given office are arranged as a group of consecutive outlets in the final-selector multiple; a maximum of 20 station lines can be accommodated on one level of the final-selector bank multiple.

If the first line of a group serving an office is engaged, the final selector will automatically hunt for the first free line in that group; alternatively, if all lines in the group are engaged, hunting will be delayed for a brief period since, with the short holding times normal to this service, a line will probably become free within 60 s. Either one (small groups of lines) or two (larger groups) searching selectors can be held for delayed hunting to start as soon as a line becomes free. If a free line fails to become available within a period of 30–60 s, a waiting call is automatically re-routed to a different level of the final selector to search for an 'overflow' position in that level. To make more effective use of the available multiple, two overflow levels may be provided, each giving access to a different overflow office serving the telegraph offices nearer to it. For this purpose levels of the final selector are divided into lower and upper groups; each of these groups has one of its levels allocated as an overflow level. For any given call the overflow level to be used is determined by the level to which the wipers were originally stepped. An overflow level gives access to positions in a large instrument room nearest to the office which was dialled but unobtainable. If the entire group of lines to the office dialled is out of service (e.g. office closed) incoming calls are re-routed to the overflow level without delay.

The final selector automatically applies the WHO-ARE-YOU signal when the called station has been connected. Provision is made for dealing with calls to faulty or spare lines.

The circuit diagram for the 200-line final selector is shown in Fig 9.11. The A relay is permanently connected to the R wire since it has to respond not only to the dial pulses but also to detect the clear signal. Its circuit must be of high impedance and a sensitive polarized relay is used. Its second winding gives the relay a spacing (A) bias to ensure the release of a connection should the line become disconnected, and also to hold the relay to the space (A) contact while the selector is disengaged. The B and CD relays form the usual pulsing-circuit elements, together with relay E which changes the pulsing circuit from the vertical to the rotary magnet during the final inter-digit pause. High-speed relays AHA and AHB are testing relays in associated with switching relays HA and HB. Relays WA and WT operate somewhat similarly to relays AHA and AHB for the waiting or suspense facility. Relays OF, AOF, AO and VS function for the overflow-level stepping; relay F detects receipt of the answer-back code.

Seizure. The final selector is seized from a group selector which switches to the 550-Ω test-battery potential on the incoming P wire (via N3 and BA3). Relay A operates to the −80-V call signal and A1Z operates the B relay. Contact **B1** holds the B relay; **B3** operates the BA relay. BA2 applies earth potential to the incoming P wire to hold all preceding equipment in the same switching centre throughout the call. **BA4** operates the CD relay via OF1.

Vertical Stepping. On receipt of the tens digit, relay A responds to the pulses, closing A1A during 'break' periods and A1Z during 'make' periods. Relay B, energized during make periods, remains held during pulsing due to the e.m.f. induced while it is short-circuited during break periods. The 5-Ω winding of relay CD is energized in series with the vertical magnet (via **B1**, A1A, VS2, **CD1**, E2 and NR1) during

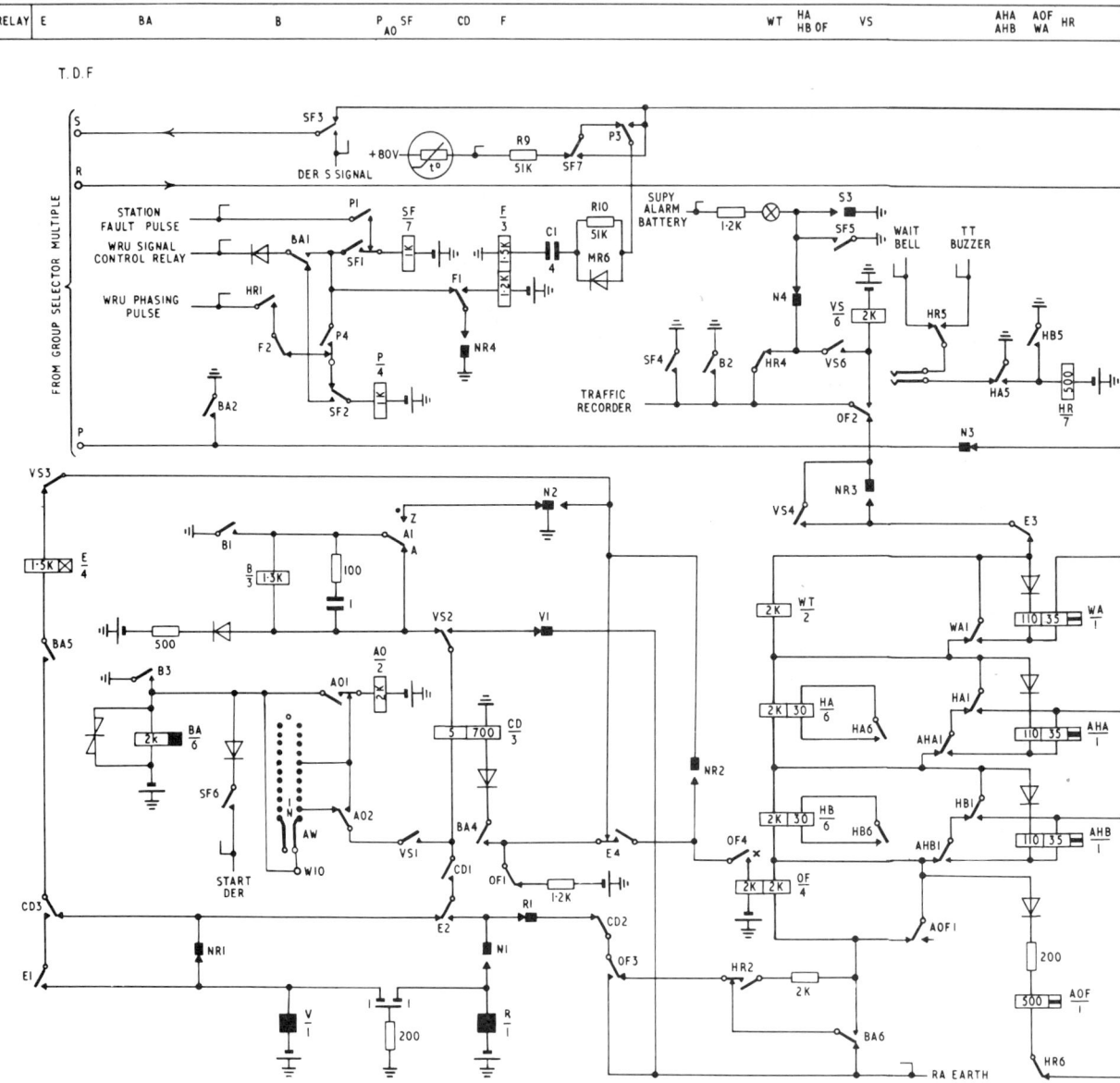

Fig 9.11 TAS 200-outlet final selector

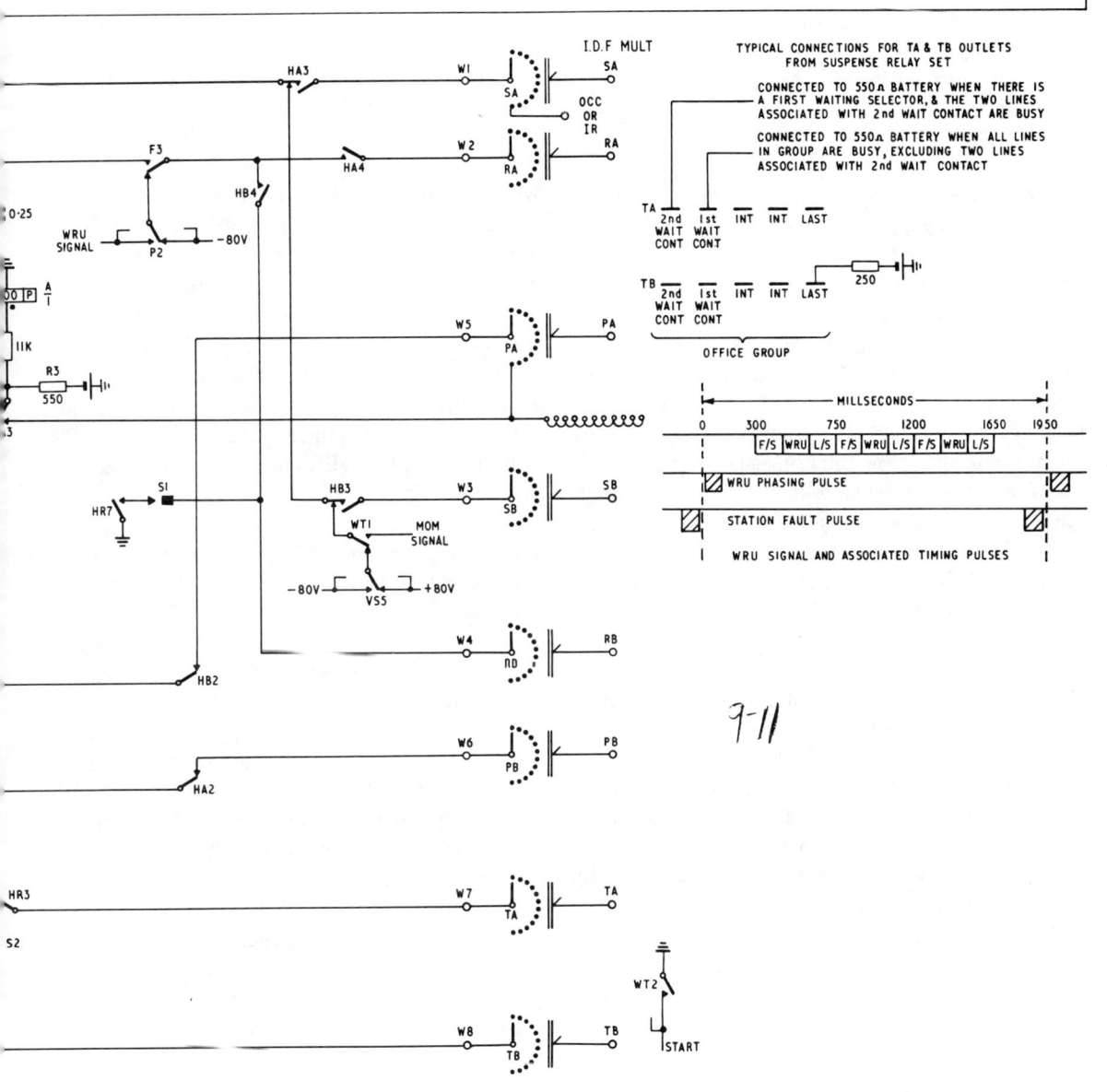

I.D.F MULT

TYPICAL CONNECTIONS FOR TA & TB OUTLETS
FROM SUSPENSE RELAY SET

CONNECTED TO 550Ω BATTERY WHEN THERE IS
A FIRST WAITING SELECTOR, & THE TWO LINES
ASSOCIATED WITH 2nd WAIT CONTACT ARE BUSY

CONNECTED TO 550Ω BATTERY WHEN ALL LINES
IN GROUP ARE BUSY, EXCLUDING TWO LINES
ASSOCIATED WITH 2nd WAIT CONTACT

TA
2nd 1st INT INT LAST
WAIT WAIT
CONT CONT

TB
2nd 1st INT INT LAST   250
WAIT WAIT
CONT CONT

OFFICE GROUP

MILLSECONDS

| 0 | 300 | 750 | 1200 | 1650 | 1950 |

F/S WRU L/S F/S WRU L/S F/S WRU L/S

WRU PHASING PULSE

STATION FAULT PULSE

WRU SIGNAL AND ASSOCIATED TIMING PULSES

9-11

START

break pulses. The vertical magnet steps the wipers to the level dialled. Relay CD remains held due to break pulses in the 5-Ω winding and to the e.m.f. induced in the 700-Ω winding which is short-circuited via **N2** and E4 as soon as the wiper carriage steps off normal; the CD relay releases 120 ms after completion of vertical stepping. **A1Z** remains operated during the inter-digit pause and enables the B relay to remain held at **B1**.

Rotary Stepping. When relay CD releases at the end of vertical stepping, CD3 operates the E relay (via N2, VS3, **BA5** and NR1) in series with the vertical magnet, the latter being unable to operate to the reduced current. E4 breaks the short-circuit path from the 700-Ω winding of relay CD, allowing it to re-operate; **CD3** with **E1** then holds the E relay. The units-pulse train is repeated by A1 which pulses the B and CD relays as described above. The break pulses are now directed to the rotary magnet via **E2** and **N1**. The rotary magnet steps the wipers to the outlet which corresponds to the units digit. On the first rotary step the NR contacts operate and **NR2** short-circuits the CD relay via **N2** and **E4** during rotary stepping, at the end of which it again releases.

Rotary Hunting. Assuming that a free line is available, the wipers are caused to drive if necessary in search of it. The hunting circuit is similar to that used in the 200—outlet group selector. During rotary stepping, relay E holds via **E1, CD3** and **BA5**. On completion of pulsing the units digit, relay CD releases and CD3 releases the E relay. A self-drive circuit is provided for the rotary magnet (via **B2**, OF2, **NR3**, E3, WA1, HA1, AHA1, HB1, AHB1, AOF1, **BA6**, HR2, OF3, CD2, R1 and **N1**) to drive the wipers around the bank. At the same time, the AHA and AHB relays are connected (via **B2**, OF2, **NR3** and E3) to the PA and PB bank contacts respectively to test for the free-outlet condition provided by the 550-Ω test-battery potential in the SLC. When the free line is found, relay AHA or AHB operates; if both outlets are free simultaneously, on the corresponding PA and PB outlet, **AHA1** disconnects the circuit for relay AHB. **AHA1 (AHB1)** breaks the rotary-magnet circuit and removes the short-circuit from relay HA (HB) which operates. **HA1 (HB1)** extends earth potential (from **B2**, OF2,

NR3, E3 and WA1) to guard the P wire seized, releasing the AHA (AHB) relay; contact **HA2 (HB2)** disconnects the PB (PA) circuit of the outlet, which has not been selected, in case it becomes free. **HA3 (HB3)** extends the S wire through to the SLC; **HA4 (HB4)** applies the −80-V potential call signal via P2 and F3 to the station equipment to start the teleprinter motor; **HA5 (HB5)** operates the relief-switching relay HR.

Who-are-you Signal. When the selected line has been seized, **HR1** applies the phasing relay P to the WRU phasing pulse circuit. At the commencement of the WRU signal cycle, a 50-ms pulse appears on this lead and operates the P relay which holds at **P4** to F1 and **NR4** for the duration of the cycle. **P2** replaces the −80-V call signal by the WRU signal cycle, which is applied to the RA (RB) wire to the called station. This cycle comprises three sequences of FIGURES-SHIFT, WRU, LETTERS-SHIFT, starting and ending with 300 ms at −80-V potential.

P3 applies the F relay to the S wire to detect the return of the answer-back code from the called station. In the idle state, capacitor C1 has been previously charged to +80 V through R9−10. When **P3** operates, capacitor C1 is charged negatively from the −80-V call-connected signal applied to the S wire by the called station. Diode MR6 being non-conducting to −80 V, the capacitor-charging current is limited by resistor R10 and is too small to operate the F relay; the time constant is approximately 200 ms. The start element (+80 V) of the first character of the answer-back code causes a heavy surge of current through diode MR6 in the conducting direction, and relay F operates. **F3** prevents further application of the WRU signals and extends the RA (RB) wire through to the SLC. Three separate WRU signals are provided in case the teleprinter motor at the called station has not reached full speed. **F1** disconnects the holding circuit of relay P; **F2** disconnects the operating circuit of relay P. When P3 releases it disconnects the F relay from the line; the connection is now through. Relays A, B, BA, F, HA (HB) and HR are held during the call.

Delayed Hunting ('Suspense'). The final selector is fitted with an additional 200-contact bank with associated wipers designated TA and TB. Each group of station lines is provided with a small queue facility comprising one (small groups of lines) or two bank outlets (large groups of lines) which constitute the 'wait' contacts. The arrangement of the TA and TB contacts is indicated in Fig 9.11. The first wait contact is the second TA contact of the group; the second wait contact (if provided) is the first TA contact of the group. The directory number which is dialled corresponds to the first wait contact – or to the second wait contact if two are provided. For each office group, the first (and second) TA contacts and the last TB contact of the group are connected to a suspense-relay set which applies the appropriate potentials; other TA and TB contacts are unused.

If all lines in the group are simultaneously engaged, the next call to arrive must wait. The suspense-relay set applies a 550-Ω test-battery potential to the TA contact at the beginning of the office group. Before hunting over the group of lines, the waiting final selector tests the TA contact with the WA relay, which operates if the 550-Ω test-battery potential is present. **WA1** breaks the self-drive circuit of the rotary magnet, applies a guarding (35-Ω) earth potential to the TA contact against seizure by another selector, and removes the short-circuit from relay WT which operates. It will be seen that **WA1** prevents any possibility of the AHA or AHB relays operating. **WT2** operates the MOM start circuit and **WT1** applies the MOM signal at 2-s intervals to the S wire to the calling station. The selector waits on this outlet, relay WA remaining held over its 35-Ω winding to the 550-Ω battery potential on the TA contact.

When a line in the group becomes free, the suspense-relay set disconnects the 550-Ω test battery, releasing relay WA. Contact WA1 restores the rotary self-drive circuit and testing circuits for AHA and AHB relays to search for the free line in the normal manner. WA1 also short-circuits the WT relay which releases.

For an office with one wait contact, the waiting test-battery potential is applied to the TA contact corresponding to the directory number. When two wait contacts are provided, these must be consecutive TA contacts having two separate test-battery potentials supplied from the suspense-relay set. The first final selector needing to wait arrives at the position corresponding to the directory number and, since

there is no battery potential on the PA, PB or TA bank contact, takes one further rotary step and is then held by the 550-Ω battery potential connected to the second TA (first-wait) bank contact. The suspense-relay set then applies the waiting battery potential to the preceding TA bank contact so that the next selector testing this number will be held on the contact corresponding to the number dialled. When a line in the group becomes free, the 550-Ω battery is removed from the first wait contact, enabling the selector standing on that contact to test, step if necessary, and switch to the free line. The waiting battery potential is then re-applied to the first wait contact but removed from the second, with the result that the second waiting call steps on to the first waiting contact. The waiting battery potential is then re-applied to the second wait contact to hold a subsequent final selector which calls this number.

Overflowed Calls. If there is no free line in the group, a selector is released from a wait contact after a period of 30–60 s by disconnection of the test-battery potential from the suspense-relay set. In these circumstances – and also if a further selector is stepped to this group of lines at a time when all lines and wait contacts are engaged – the wipers drive to the last outlet of the group. Here a 250-Ω test-battery potential is applied to the TB bank contact. This potential is permanently applied to prevent selectors from testing into a following group of station lines; its value of 250Ω allows two searching selectors to switch to it simultaneously if need be. Relay AOF operates to this potential (via **B2**, OF2, **NR3**, E3, WA1, HA1, AHA1, HB1 and AHB1). Contact **AOF1** breaks the rotary-magnet circuit and removes the short-circuit from relay OF which operates in series with the rotary magnet. **OF4** holds the OF relay via **NR2** and **N2**. Contact **OF2** operates the vertical-stepping relay VS and prevents the operation of relays WA, AHA and AHB when the wipers subsequently drive over the remaining outlets (of another office) on the same level. **VS5** returns −80 V for the S wire to hold the calling-station equipment; **VS6** holds the VS relay. **OF3** closes a circuit for the rotary magnet (via the release-alarm earth circuit) to drive to the twelfth step, restoring the NR contacts. NR2 releases the OF relay. The wiper carriage drops and restores to the normal position. The N contacts are now restored, N1 breaking the circuit of the rotary magnet. N2 and OF1 allow the CD relay to re-operate. **CD1** completes the circuit of the vertical magnet from the release-alarm earth, V1, **VS2**, 5-Ω winding of the CD relay, **CD1**, E2 and NR1. The selector self-drives vertically until the overflow level is reached, where the CD relay 5-Ω winding and the V1 contacts are short-circuited by earth potential from **B3** on the vertical-marking bank and wiper AW, AO2 and VS1. This holds the vertical magnet operated, but relay CD shortly releases, the 700-Ω winding being also short-circuited at N2. Contact CD1 disconnects the vertical magnet; CD2 closes the normal self-drive circuit for the rotary magnet which drives the wipers over the contacts of the overflow level until relay AHA (AHB) finds a free circuit. Relays HA (HB) and HR operated; **HR4** releases the VS relay. The call is switched through to a position in the appropriate overflow office and the WRU signal is applied.

Relay AO will be operated if the wipers, *when in their original dialled level*, were positioned on or above the pre-determined second overflow level. This is achieved by appropriate strapping on the vertical-marking bank contacts. **AO1** locks the AO relay (if operated); AO2, normal or operated, determines at which of the two overflow levels the short-circuit will be applied to V1 and the CD relay.

All Overflow Circuits Engaged. If all outlets on the overflow level are engaged, the wipers drive to the eleventh step and the S contacts operate. Relay AHA operates to the 550-Ω test-battery potential at R3 (via **BA3**, the PA wiper and HB2), operating relays HA and HR. The OCC signal is applied to the S wire via the eleventh SA outlet.

If it is required to omit the overflow facility from a group of station lines, the 250-Ω battery potential is omitted from the last TB contact. If the selector fails to find a free outlet the wipers are driven to the eleventh step where the OCC signal is applied. Such a group of circuits needs to be accommodated on outlets towards the end of the level or, if all outlets were busy, the selector could test the group of lines of another station accommodated later in the level.

Spare Levels and Lines Located Between Working Lines and 11th Step. A selector, stepped to a contact in a spare level or group of spare lines following the last working line on a level, hunts under self-drive conditions until it reaches the eleventh step where the circuit for relay AHA is completed via PA bank to 550-Ω battery. The conditions are then the same as reaching the eleventh step on the overflow level except that on spare levels the IR signal takes the place of the OCC signal. The immediate release signal (IR) consists of 1250 ms at +80-V potential followed by 250 ms of −80-V potential, repeated continuously. It is not accompanied by a printed service signal, but it causes the extinction of the DIAL lamp and stops the teleprinter motor.

Faulty Line. If a station line is faulty the answer-back code is not received in response to the WRU signals and relay F does not operate. A 50-ms station-fault pulse is fed via P1 to the station-fault relay SF towards the end of the WRU signal sequence. If **F1** has not released the P relay, the SF relay operates to the pulse via **P1** and holds at **SF1**, F1 and NR4; contact **SF2** releases the P relay; **SF3** applies the DER S signal to the calling station; **SF4** holds the HA (HB) relay so that it shall remain operated when the calling station clears, releasing relays A, B and BA. Contact BA2 removes holding earth potential from the incoming P wire and allows the preceding equipment to release. The selector remains standing on the outlet of the faulty line, busying it against further calls and drawing attention to the fault by closing the supervisory lamp and audible-alarm circuits at **SF5**.

Selector Release. During a call, **BA2** holds all preceding equipment connected in the same switching centre and **B2** holds the HA (HB) relay and the SLC of the called station. On receipt of the +80-V clear signal, relay A responds and A1A releases the B relay by short-circuiting it. B2 releases the HA (HB) relay; HA5 (HB5) releases the HR relay. HA3−4 (HB3−4) disconnect the S and R wires, HA3 (HB3) re-applying +80 V to the incoming S wire. B3 releases the BA relay. BA2 allows all the preceding equipment to release. The self-drive circuit for the rotary magnet is completed by HR2 (via **N1**, R1, CD2, OF3, BA6 to release-alarm earth). The selector steps to the twelfth position and then restores to normal NR4 releases the F relay. During release, **N3** disconnects the 550-Ω test battery to prevent the selector from being seized before it is back to normal. N1 disconnects the circuit of the rotary magnet.

9.10 The suspense-relay set

A suspense-relay set is provided for each group of station lines serving an office. It receives from each station line in the group an indication whether the line is free, engaged or out of service, and in response to this information the suspense-relay set applies appropriate marking conditions to the final-selector multiple contacts to determine whether calls shall delay their hunting operation.

The suspense-relay set provides the following functions:

(1) Applies battery potential to the first wait contact of an office group when all working lines are busy.

(2) Applies battery potential to the second wait contact when the first is already occupied by a waiting final selector (one wait contact only is provided for small groups of lines).

(3) Releases a final selector from the first wait contact when a line in the group becomes free.

(4) Releases a final selector from the second wait contact to step on to the first when the latter becomes free.

(5) Releases a final selector from a wait contact after it has waited for 30–60 s.

(6) Releases a call immediately to the overflow level if all lines are out of service, or if the wait contacts are already occupied.

The circuit diagram is shown in Fig 9.12. It will be seen in Table 9.1 that a free station line is indicated by the L and G relays both being released; the out-of-service condition by the L and G relays both being operated and the engaged condition by relay L released and relay G operated. The chain circuit of the group-control wire is completed to operate relay PT only when at least one station is engaged on a call and the remaining station lines are either busy or out of service.

Taking first the case of a small group with only one wait contact, relay PT is released and the TA bank wait contact is disconnected when all circuits are free or out of service. When there is no free line but some lines are engaged (and therefore in service), relay PT is operated via the chain of contacts. **PT1** applies the 550-Ω test-battery potential, with relay S in parallel, to the TA bank wait contact. The next calling final selector will wait on this contact, operating its WA relay in series with the S relay. When a line becomes free, relay PT releases; PT1 removes the 550-Ω test-battery potential from the TA bank contact, releasing the waiting selector which drives and switches to the free line.

If a line fails to become free within 30–60 s, a waiting selector is released to drive to the overflow level. Relay S is operated by the waiting selector. **S4** connects the time-pulse relay TP to the S-pulse wire. On arrival of the first 30-s pulse, relay TP operates and holds at **TP1** to **S5**. After 30 s, an earth-potential pulse present on the Z-pulse lead operates the M relay via **TP2**. Contact **M1** holds relay M to this Z pulse; **M2** disconnects the TP relay; **M3** disconnects the battery (and S relay) from the TA bank contact to release the final selector from the wait contact. The final-selector wipers are driven round the bank and, finding no free outlet, reach the battery potential on the last TB contact of the group. This operates the OF relay in the selector, causing it to be driven to the overflow level. Relays S, TP and M have now released.

When an office is closed down there is no battery potential on either of the wait contacts. A searching final selector drives immediately over the group of circuits until the battery potential on the last TB contact is encountered; the OF relay then operates and the selector is driven to the overflow level without delay.

For an office group with two wait contacts the sequence of numbering the wait contacts is as shown in Fig 9.11. As far as the first wait contact is concerned the operation is as described above. When the first wait contact is occupied, relay S being operated, S5 operates the SA relay. Relay PTA being operated by the second contact chain, **S3** via **PTA1** applies a 550-Ω test-battery potential to the second wait contact. The next final selector calling this group will operate its WA relay to this battery and wait on this contact. As soon as a line becomes free relay PT releases. PT1 discharges the final selector standing on the first wait contact to search for the free line, and

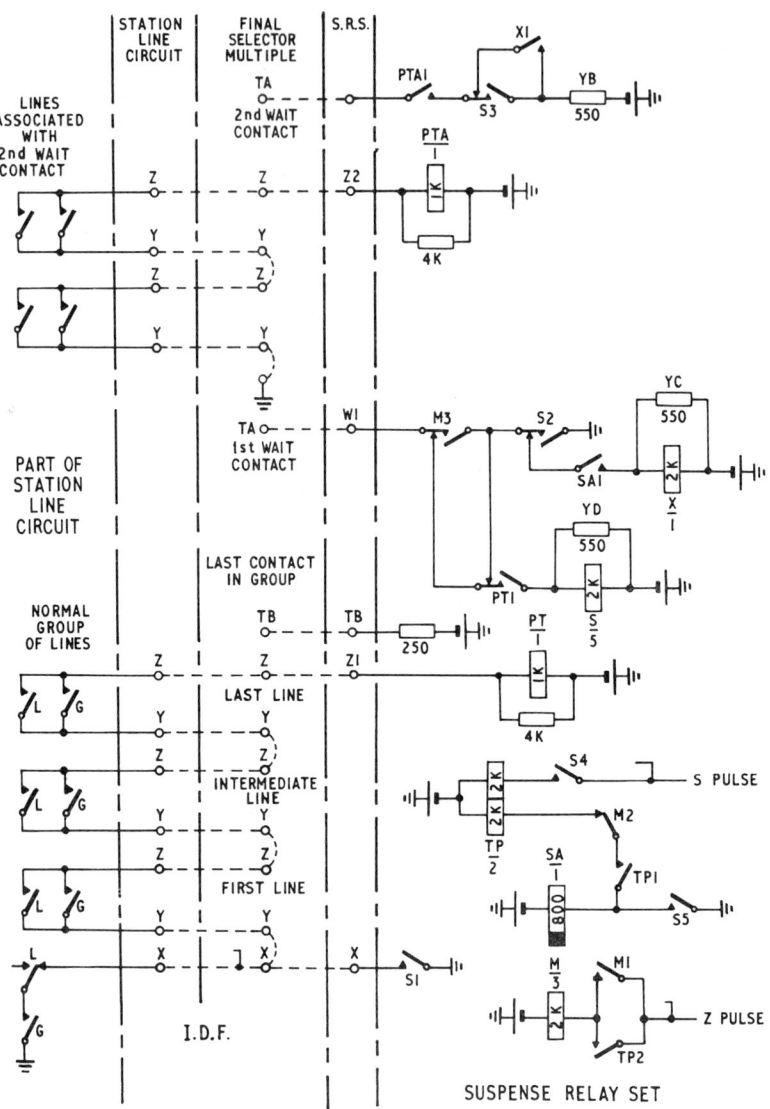

Fig 9.12 TAS suspense-relay set

relay S releases. S3 removes the battery potential from the second wait contact and the final selector which had been waiting on the second wait contact now steps on to the first where it is held by the battery potential applied to the TA contact via YC (with relay X in parallel), **SA1**, S2, PT1 and M3. This circuit is maintained during the delayed release of **SA1** which is long enough to cover the time required for the selector discharged from the first wait contact to drive, switch to the free line and re-operate the PT relay. This arrangement is necessary to prevent incorrect release of two waiting selectors when only one line becomes free. During delayed

release of **SA1**, relay X is operated by the selector standing on the first wait contact. **X1** applies battery potential to the second wait contact to arrest any further selector which may arrive there. If the selector standing on the first wait contact is released by the time-pulse operation, the selector on the second wait contact is stepped on to the first wait contact as already described.

The final selector has 200 outlets but only 100 *sets* of contacts (SA, RA, PA, SB, RB, PB, TA, TB). If the two sets of S, R and P contacts corresponding to every wait contact were not used for station lines, the effective size of the bank multiple would be seriously reduced. The two sets of S, R and P contacts associated with each wait contact are put to use. A selector waiting on a first wait contact with its WA relay operated will release its WA relay and instead operate its AHA (AHB) relay, without further stepping if a line on the contact set associated with that wait contact becomes free. When working lines are associated with the second wait contact, a complication arises because a final selector cannot step backwards from the first contact to seize circuits 1 or 2 associated with the second wait contact. To overcome this difficulty, such lines are connected so that they do not influence release from suspense of the first waiting selector. A free line associated with a second wait contact removes the 550-Ω test-battery potential from the TA bank of the second wait contact and allows a selector to switch to it even

though this may already be held on the first wait contact. Lines are not usually associated with the second wait contact until they have been already allotted to all other available contacts in the group. The circuit is arranged so that the group-control wire in the suspense-relay set does not include the SLCs of the lines associated with the second wait contact; these are instead connected to relay PTA. Contact PTA1 controls the battery potential applied to the second wait contact so that, if one of the associated lines becomes free, the selector standing on the second wait contact is allowed to switch to this line, whilst that standing on the first wait contact is unaffected.

For some services such as enquiry positions it is required to provide delayed hunting without overflow. This is achieved by omitting the connection to

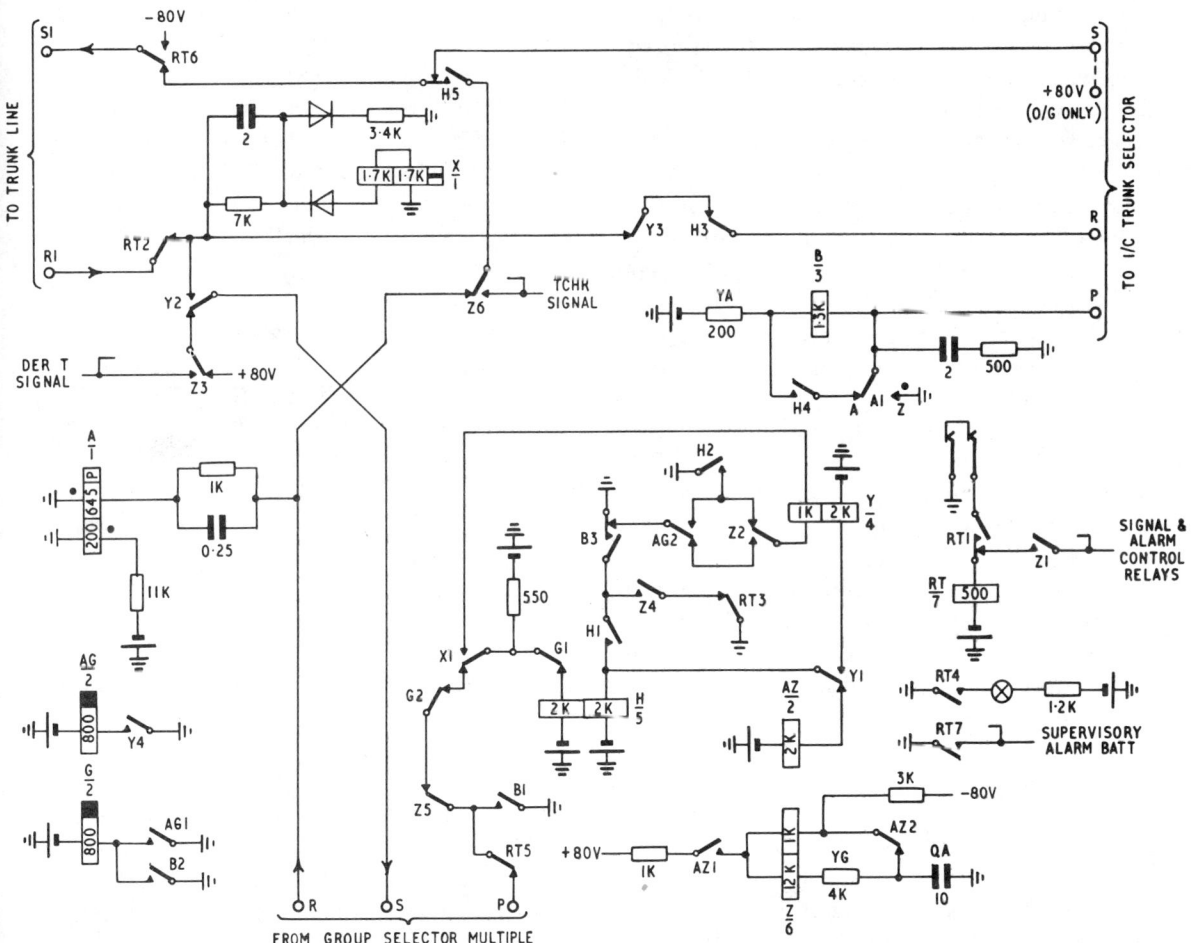

Fig 9.13 TAS bothway trunk relay set

the TB tag and also by arranging that there are no other working lines after the last line of the enquiry group in the final-selector multiple. The OF relay cannot operate and, if all enquiry circuits are engaged, the selector drives to the eleventh step and returns the OCC signal to the caller.

9.11 Bothway trunk-relay set (Fig 9.13)

This relay set meets the two basic needs for a bothway-trunk circuit — it switches the trunk circuit to the incoming or outgoing path according to the direction from which it is seized and it holds and guards the selectors in the switching centre since the P wire is not extended over the trunk circuit.

For a unidirectional outgoing-trunk circuit only the latter need arises, yet it is economic to use the bothway-relay set for this purpose since most of the relay-set components operate for the outgoing call. In this use it is necessary to apply +80 V to the lead which would normally be connected to the S wire of an incoming-trunk selector.

On release from an outgoing call it is important to guard the outgoing-trunk multiple long enough to ensure that the equipment in the distant switching centre will have fully released. Means are also included in the relay set to check the correct functioning of the circuit.

A unidirectional incoming-trunk circuit terminates directly upon an incoming 1st selector and has no need of a relay set.

Incoming Call. In the disengaged state the trunk-circuit S and R wires are connected direct to the incoming selector via H5 and H3; the selector applies +80 V to the S wire. The high-speed relay X responds to the −80-V call signal; its purpose is to disconnect the 550-Ω test-battery potential at **X1** on the P wire of the outgoing multiple as quickly as possible so that the trunk circuit cannot be seized simultaneously from both ends. The −80-V call signal also operates the A relay in the incoming group selector which applies earth potential to the selector P wire and operates relay B in the relay set. **B1** applies earth potential to guard the outgoing P wire multiple (since relay X will respond to all signals present on the R wire); **B2** operates the G relay. The relay set remains in this condition until the call is cleared,

when removal of earth potential from the incoming-selector P wire releases the B relay. B1 removes earth potential from the outgoing P wire multiple, but the 550-Ω test-battery potential is not re-applied until after the delayed release of G2, relay G being released by B2 and relay X releasing to the clear signal. This additional guard is necessary to cover any differences in the timing of the B relays in the selectors in the two switching centres involved, as well as differences in the release times of the selectors.

Outgoing Call. The outgoing-trunk circuit is seized from a group-selector level by switching to the 550-Ω test-battery potential on the P wire (via Z5, G2 and X1). When the group selector operates its HA (HB) relay, earth potential which it temporarily applies to the P wire operates relay H which holds at **H1** and **B3**. The call signal operates relay A, which is permanently connected to the R wire to detect the call and clear signals. **A1** operates the B relay; **B1** holds the preceding selectors and guards the circuit against seizure by another selector; **B2** operates the G relay; **H5** extends the call signal over the trunk circuit to seize the incoming selector in the distant switching centre; **H3** (with **H5**) disconnects the trunk circuit from the incoming selector.

The distant incoming selector returns the 25-ms −80-V call-confirmation signal over the R1 wire, operating relay X briefly. **X1** closes a circuit for relay Y to operate (via, Z2 and **H2**); **Y1** holds the Y relay over its second winding to **H1** and **B3**. Contact **Y2** completes the R1−S path of the trunk circuit. **Y4** operates the AG relay; **AG1** holds the G relay. The trunk circuit is now connected through to the group-selector level.

Trunk-Circuit Release. Following seizure in the outgoing direction, relay A responds to the receipt of the clear signal and A1 short-circuits the B relay via **H4**. Relay B releases after 300 ms. B3 releases the H relay and disconnects one winding of relay Y, but this relay holds over the other winding (via B3, **AG2**, Z2 and **X1**) until relay X responds to the complementary clear signal. X1 releases relay Y; contact Y4 releases the AG relay; AG1 releases the G relay. The joint release delays of relays AG and G totals 600−900 ms — the period which elapses before the outgoing appearance of the trunk circuit is again marked free by G2 on the P wire. This allows ample time to ensure release of the connection in the distant switching centre.

Trunk Circuit Faulty. A trunk circuit is regarded as faulty if the call-confirmation signal is not received within 200 ms of applying the call signal; this delay is measured by the Z relay. When the outgoing circuit is seized, relays B and H are operated as already described and relay AZ is operated via **B3**, **H1** and Y1. Contacts **AZ1−2** close the operating circuit for the Z relay, which has two windings connected differentially. The flux in the winding due to the current which charges capacitor QA opposes the flux building up in the second winding, hence the relay will not have sufficient flux to operate until the charging current approaches zero.

If relay Y has not operated (to denote receipt of the call-confirmation signal) before relay Z operates, contact **Z2** disconnects the circuit of relay Y and the trunk circuit is regarded as faulty. The DER T signal is returned to the calling station over the S wire via **Z3** and Y2.

In case the trunk-circuit fault is of a transient nature, an automatic re-test of the circuit is applied in the following manner.

Z1 connects the RT relay to the signal and alarm-control relays, one of which operates to start the DER T and trunk-circuit-hold-and-retest (TCHR) signal-distribution circuits, and to energize a group of timing relays which control timing for automatic re-testing of the trunk circuit. At this stage the resistance in the lead to the signal and alarm-control relays is too great to permit the operation of relay RT.

Contact **Z6** connects the TCHR signal to the trunk circuit. This signal consists of −80 V interrupted by 500-ms pulses of +80 V at intervals of 29·5 s. The effect of this is to cause the release and re-seizure of the distant trunk selector at 30-s intervals, this being accompanied by the return of the call-confirmation signal if the trunk circuit is no longer faulty.

Meanwhile, **Z4** provides a holding circuit for relays H and AZ, independently of B3 which restores when the caller clears on receipt of the DER T signal. **Z5** disconnects the battery potential

from the P wire, busying the trunk-relay set to further callers.

If on re-test the call-confirmation signal is received, relays X and Y operate if B3 is normal (i.e. if the caller has cleared) and **Y1** disconnects the circuit for relay AZ which releases. AZ1 releases relay Z. Contact Z4 disconnects the holding circuit for relays H and Y and the trunk-relay set is restored to normal and able to accept further calls.

If, on the other hand, a satisfactory response is not received to the automatic re-testing, after 2·5 min the signal and alarm-control relays apply earth potential to the signal and alarm-control lead; this allows relay RT to operate and lock via **RT1**. Contacts **RT4** and **RT7** energize supervisory alarms. **RT6** applies −80 V to the trunk circuit to busy the trunk-relay set at the distant end. **RT2** disconnects the incoming R wire: **RT5** disconnects the P wire to force-release the connection if the caller has not already cleared, and to busy the trunk-relay set to outgoing traffic. **RT3** disconnects relays H and AZ; contact AZ1 releases relay Z.

This method of automatically testing the trunk circuit ensures that short-duration faults do not produce an alarm; this avoids wasting the time of maintenance staff in attending to transient fault conditions. At the same time, by temporarily busying the circuit, it cannot be taken for service while the fault condition persists.

9.12 Relay set outgoing from selector levels (Fig 9.14)

Large groups of station lines — such as those to the instrument room associated with the switching centre — are served direct from group-selector levels. It is necessary to connect these circuits via relay sets which provide holding and guarding conditions on the P wire and also enable the answer-back codes to be taken automatically. In these respects, the relay set provides certain functions corresponding to those already described for the final selector.

The relay set is seized over the P wire, the 550-Ω test-battery potential being found in the SLC circuit. Polarized relay A is permanently connected to the R wire to detect call and clear signals. On receipt of the call signal, relay A responds and **A1Z** operates the B relay. **B1** holds the preceding equipment; **B2** operates the H relay; **B3** applies the −80-V call signal

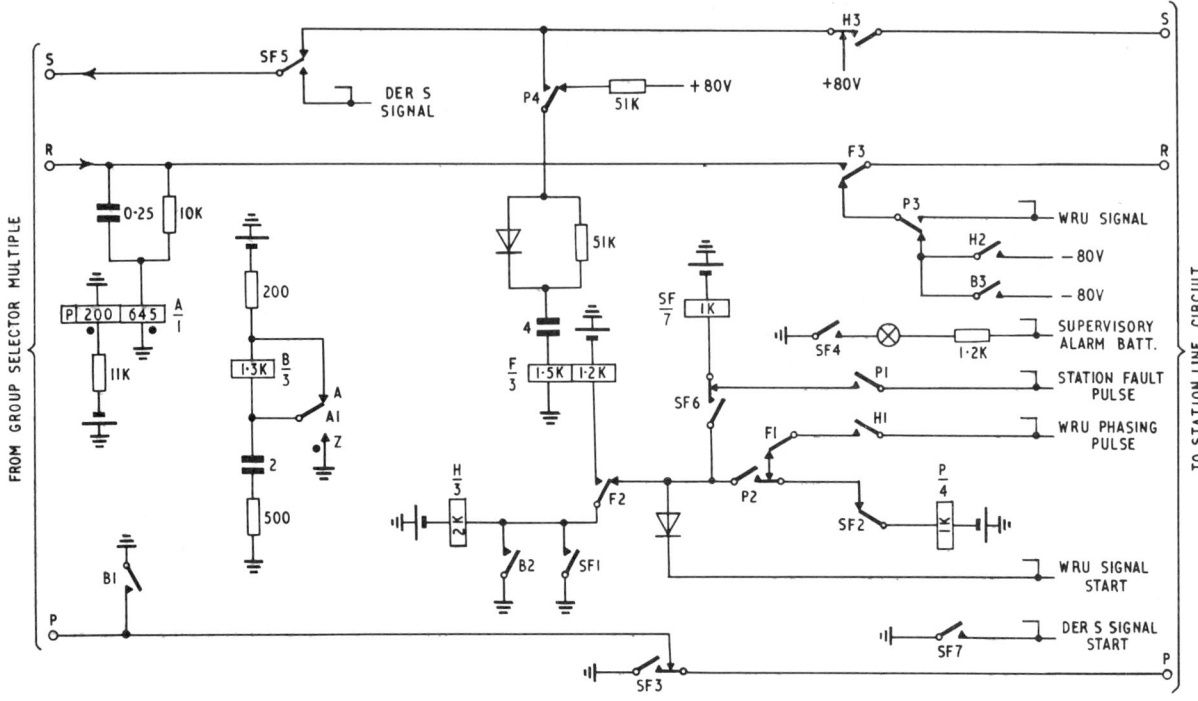

Fig 9.14 TAS outgoing relay set from selector levels

to the station line R wire; **H1** applies the phasing relay P to the WRU phasing pulse lead. Relay P operates to an earth pulse present at the commencement of the WRU cycle. **P2** holds the P relay to **B2**; contact **P3** applies the WRU signal sequence, via F3, to the R wire to the called station. **H3** extends the S wire through. **P4** connects relay F to the S wire to detect receipt of the answer-back code signals. When relay F operates, **F1** disconnects the circuit to the WRU phasing pulse; **F2** holds the F relay to **B2** and releases the P relay; **F3** completes the circuit for the R wire; P4 disconnects the F relay from the S wire. The circuit is now through, relays A, B, F and H remaining operated.

Failure to receive an answer-back code results in operation of relay SF. **SF1** holds the H relay; **SF2** disconnects the P relay; **SF4** provides alarm conditions; **SF5** returns the DER S signal to the calling station; **SF6** holds the SF relay. **SF3** provides earth potential to the final-selector multiple to prevent the faulty circuit from being seized there, since this group-selector level may be teed to the overflow level of the final selector. At the same time

SF3 permits the group selector to be released when B1 restores.

Relay A responds to the clear signal; A1A short-circuits the B relay which releases after a delay; B1 releases the preceding equipment; B2 releases the F and H relays.

9.13 The Special-services final selector
This selector differs somewhat from the regular final selector, its purpose being to give access to various testing facilities and not to normal station lines. The circuit diagram (Fig 9.15) is simpler than that of the regular final selector as the provision of waiting, overflow, automatic WRU signal transmission and station-fault alarm are unnecessary. The facilities provided are:

(1) Returning holding conditions to preceding equipment.

(2) Vertical and rotary stepping under the control of dialled pulses.

(3) Hunting for a free outlet in the required group.

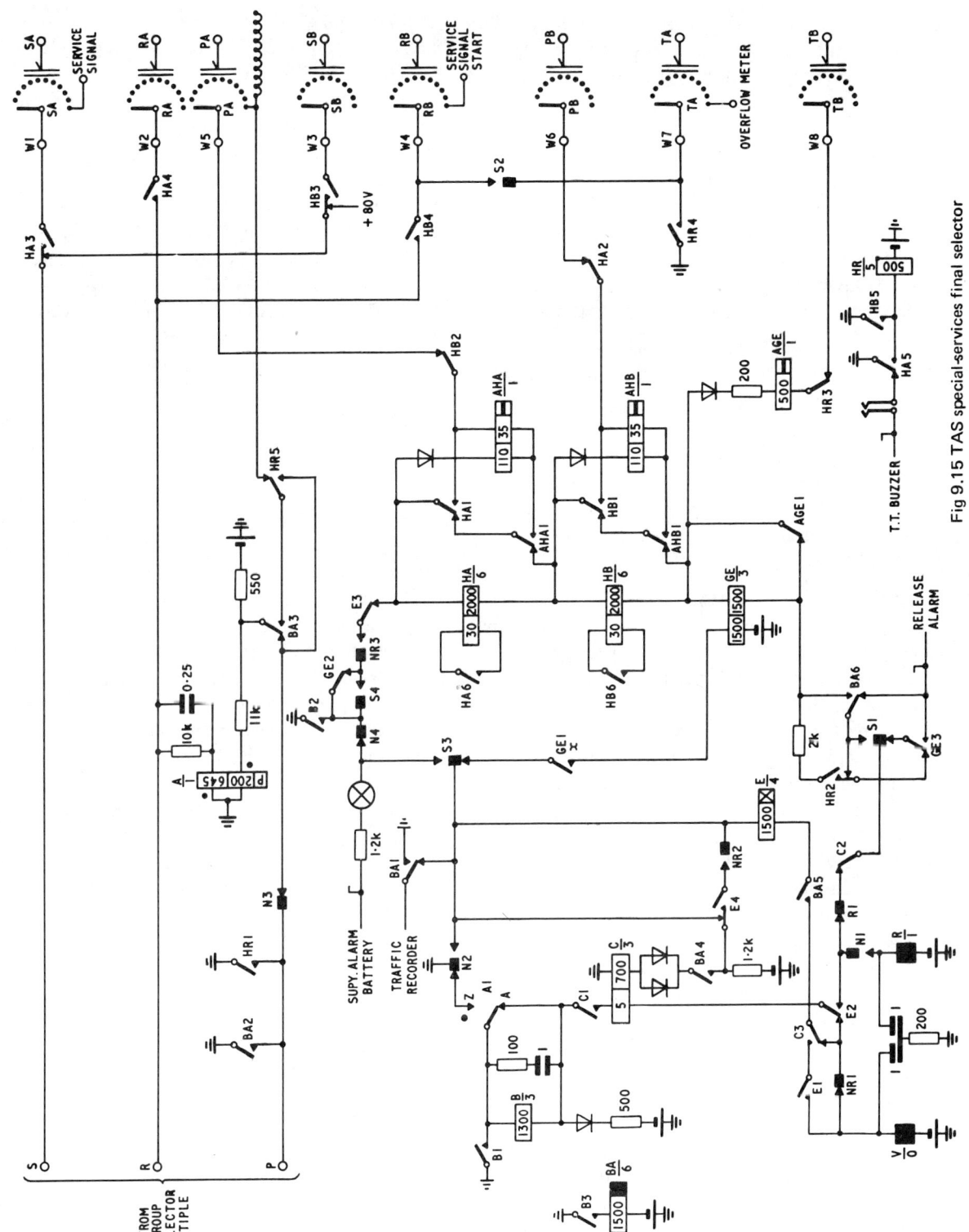

Fig 9.15 TAS special-services final selector

(4) Returning the OCC signal to the calling station if all circuits in the required group are engaged.

(5) Releasing the selector and the preceding equipment when a clear signal is received.

(6) Guarding the selector against seizure during release.

(7) Giving audible and visual alarms if a selector is seized and not stepped; or if it is held on the eleventh step; or if it fails to release due to a mechanical defect.

(8) Applying the IR signal to release the connection forcibly if a spare number is dialled.

The circuit operation is very similar to that of the regular final selector, as far as it applies, with the absence of the P, F, SF, WA, WT and VS relays. One difference is that when this final selector encounters battery potential on the TB contact to indicate the end of the group, relays AGE and GE operate to drive the selector to the eleventh step and return the OCC signal to the caller.

9.14 Switching-centre design

Though not identical, arrangements for service-signal generation and distribution, and rack design for selectors and relay sets are similar to those for a telex exchange.

Three types of motor-driven signal generator for service signals, pulses and test signals are provided in duplicate. The standby set comes into operation automatically in the event of a fault, otherwise the two sets are changed over at weekly intervals. Generator speed is controlled by a comparator circuit which continuously checks the nominal 50-Hz signal from the generator against the 50-Hz output from a valve-maintained tuning fork. The 5-unit start—stop signals are generated from commutator segments. The printed service signals do not terminate with an automatic clear signal, reliance being placed upon trained telegraphists to clear a connection on receipt of a service signal on an ineffective call.

Signal distribution is via high-impedance valves to polarized relays, to avoid wear on the commutator segments by arcing under heavy loads. Distribution is duplicated on an odd/even basis with either manual or automatic change-over to prevent interruption of service signals on failure of distribution equipment. All service signals are monitored continuously to detect immediately any failure in supply.

Trunking and Grading* Grading is used wherever the number of trunks exceeds the availability — 24 for uniselectors and 20 for group selectors — and is carried out on separate centralized trunk-distribution frames (TDF). Bank multiples from selector-rack shelves are cabled to the TDF and terminated on horizontal strips of grading tags, corresponding to the 20 (24) outlets of a bank multiple; grading is carried out by soldering vertical strips of bare tinned copper wire to connect those multiple tags which have to be commoned according to the grading design. During periods of light traffic it is possible for a faulty early-choice selector, repeatedly seized, to disrupt service. To minimize such an effect, station lines are spread over two separate series of early-choice first selectors, a procedure used in the grading design at all stages.

Provision of Plant. Little or no provision needs to be made for growth in this service. Although telegraph-traffic density varies during the day, traffic carried by the automatic switching network remains reasonably constant during working hours. Traffic can accumulate in an instrument room to some extent during peak periods before reaching the TAS network, but peaks tend to be flattened by this storage process and traffic level in switching centres is rendered more uniform. Public-telegraph traffic accurately reflects national events and has a seasonal variation which shows peaks at times immediately prior to public holidays.

Traffic records are taken between 10.30 a.m. and 12.00 a.m. to ensure that busy-hour traffic is recorded. The average value of the busy-hour/day traffic ratio is 1 : 6·8, rising to 1 : 8 at seasonal traffic peaks. An average call duration of 60 s is used for traffic design. Circuit provision is based on average busy-hour traffic during the busiest three consecutive months in the year.

Groups of station lines are provided as bothway groups up to 16 circuits if with suspense facilities; or 20 circuits if without. Requirements in excess of this are met by division into unidirectional and bothway components. For office groups up to four lines, one 'wait' contact is provided on the final selectors; for groups of five lines and over, two wait contacts

---

* Described in *The International Telex Service* (R. N. Renton), Pitman.

are provided. Table 9.2 shows examples of traffic loadings used in determining the scale of circuit provision to teleprinter offices with and without suspense facilities. Although certain assumptions and estimates are made for some of the factors in this traffic — which can be extremely complex on account of variation in holding times, the extent of batching

an excessive number of repeat calls due to insufficiency of plant.

The proportion of the total cost of handling inland telegraph traffic represented by the switching equipment is relatively small. Traffic design assumes 'pure-chance' characteristics which requires the number of originating sources to be very large. With the

Table 9.2 Provision of station line and trunk circuits

No. of circuits	Capacity of circuits at the approved grade of service					
	Overflow centres		Other offices		Trunk circuits	
	Messages	Erlangs	Messages	Erlangs	Messages	Erlangs
1	—	0·002	18	0·30	1	0·02
2	4	0·065	45	0·75	13	0·22
3	15	0·25	69	1·15	35	0·59
4	32	0·53	99	1·65	64	1·07
5	54	0·90	140	2·33	98	1·63
6	79	1·32	176	2·94	134	2·23
7	108	1·80	216	3·60	172	2·87
8	139	2·31	258	4·30	214	3·56
9	171	2·85	288	4·80	256	4·26
10	206	3·43	324	5·40	299	4·98
11	241	4·02	360	6·00	343	5·72
12	278	4·63	396	6·60	389	6·48
13	316	5·27	438	7·30	435	7·25
14	355	5·92	480	8·00	482	8·04
15	395	6·58	528	8·80	530	8·83
16	436	7·26	576	9·60	578	9·63
17	477	7·95	—	—	627	10·44
18	518	8·64	—	—	675	11·25
19	561	9·35	—	—	724	12·07
20	604	10·07	—	—	775	12·91

(the transmission of more than one telegram per call), and the effect of offering overflowed traffic to the destination office — the table is found to offer a satisfactory basis for circuit provision.

The suspense facility is not provided on station lines trunked from group-selector levels to a main instrument room. Since the instrument room and associated switching centre are generally in close proximity, normally in the same building, the cost of line provision is not an important factor. These circuits are provided to a grade of service of 0·002 (availability 20) from reference to the standard traffic-capacity tables; this group of circuits also carries overflow traffic from final-selector units at the same grade of service.

Trunk circuits are provided to a design grade of service of 0·02. This figure takes into account the relatively low cost of providing circuits by MCVF systems and the inefficiency which would result from

TAS system this is far from being the case; on the other hand, traffic originated per line is relatively high. The effect of this limited number of comparatively heavily-loaded sources is to introduce an element of smoothing into the traffic, a tendency which is increased at each switching stage. With smoothed traffic, fewer trunks are required to carry a given volume of traffic at the same grade of service.

Trunks from SLCs to first selectors are provided to a grade of service of 0·005 (availability 24). Between ranks of group selectors the same grade of service (0·005) is used, the availability being 20. Trunks from group-selector levels to final selectors, and also trunk circuits which interconnect switching centres, are provided at a grade of service of 0·02 (availability 20).

All selectors and relay sets are cabled via the intermediate distribution frame (IDF) to permit an orderly cabling system and to enable rearrangements of plant to be carried out. External circuits pass to the main distribution frame (MDF) where fuses are included; the MDF is also a convenient point for disconnection and access to the line for testing and localizing faults between internal and external plants.

With the relatively small number of switching centres serving the entire TAS network, damage or destruction of a switching centre — by fire, for example — could have a profound effect on the service. Mobile automatic switching units have been provided so that in such circumstances the switching centre affected could be quickly replaced.

9.15 Instrument-room layout

The layout of operational equipment in telegraph instrument rooms varies according to services provided, volume of traffic handled and shape of available accommodation.

In the interests of efficiency, with particular regard to reduction in movement of staff and telegram forms within the instrument room, certain basic design principles are followed. If phonogram and telephone-telegram (TT) positions are in a separate room, tables used for circulation, segregation and addressing duties would be located between outgoing and incoming teleprinter positions. Early-choice positions of each group of teleprinter circuits would be adjacent to these tables as would a position provided for the night concentration of phonogram lines: this enables night-staffed positions to be closely concentrated. Telegram forms from incoming positions to the circulation, segregation and addressing tables would be carried by conveyor bands. If volume of traffic and distance involved were sufficient, telegrams forms would also be carried by conveyor bands to convenient distribution points near to outgoing positions. Overflow positions may be located away from circulating positions and they would be provided with circulating information in respect of offices for which the instrument room acts as overflow centre. If phonogram positions share the main instrument room with teleprinter positions, circulation, segregation and addressing tables would be located between outgoing teleprinter positions and the phonogram and telephone-telegram suite for ease of distributing outgoing telegram forms. A small number of positions, sufficient to handle overflow and incoming traffic under night concentration, would

take precedence of location over late-choice outgoing circuits; the remainder of incoming positions can be located away from circulation positions, which are reached by band conveyors.

Circulation of telegram forms within the small telegraph office presents no problems. In medium-sized and large offices, telegrams may arrive from the public counter, incoming TAS positions, phonogram and telephone-telegram positions and printergram positions (the latter mostly at zone centres); these telegrams have to be forwarded to the delivery room, or over outgoing TAS positions, phonogram and TT positions or printergram positions. Handling a large volume of telegram forms within the office presents a problem of some magnitude, bearing in mind the need to guard against loss or damage to a telegram form, minimize delay, avoid excessive staff movement and keep down costs. Use of conveyor bands in instrument room, and pneumatic tubes in the building, may be of great assistance; on the other hand, at medium-sized offices where combined phonogram/TAS positions are used, the need for conveyor belts is avoided.

All inward telegrams arrive at the circulation table — the focal point where the next stage in their routing is decided. In larger offices, telegrams would be placed on a multiple-band conveyor for transfer to various distribution points placed as near as possible to circuits from which telegrams will be forwarded. Telegrams for local delivery pass to the segregation-table position where it is determined whether it should be delivered by telephone, telex or messenger and whether any special delivery instructions are applicable. This position is placed immediately adjacent to incoming positions to eliminate need for separate distribution of telegram forms and to allow segregation work to be combined with incoming 'gumming' work at quiet periods. From here, all telegrams for local delivery pass over a slow-moving band to addressing tables where, after addressing and enveloping, they travel by pneumatic tube to the delivery room. For large organizations having registered abbreviated addresses these must be 'unpacked', i.e. decoded: stencils would be used for addressing these envelopes. Addresses of all telegrams for delivery over telephone-telegram circuits are examined at the segregation table to ascertain whether addressees are telephone subscribers in order that such telegrams may be telephoned direct

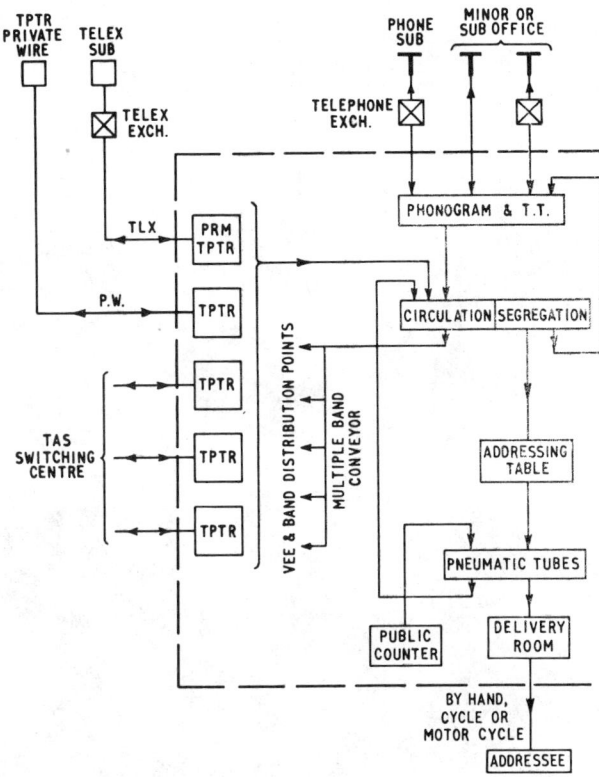

Fig 9.16 Telegram circulation in instrument room

to them, with saving of time and cost. The circulation procedure is summed up in the diagram of Fig 9.16.

Teleprinters with gummed-paper-tape rolls are used in inland telegraph office instrument rooms for all purposes; any extraneous signals, such as corrections, can be omitted when gumming the tape to the 'C' telegram form for despatch to the addressee. Times of handing in and arrival at a receiving position are entered on telegram forms using a 'Blick' stamping machine which prints date and time on the form.

At TAS receiving-only positions it is unnecessary for a telegraphist to be present while the teleprinter is receiving the telegram. To avoid staff movement over a group of receiving positions, consoles are provided with three teleprinters mounted in a 'stack' (Fig 9.17) in such a way that the tape feeds are accessible with minimum movement to a telegraphist seated at the adjacent table. The machine nearest to hand for a seated telegraphist receives most work, being connected to an earlier-choice outlet of selector level than the other two; the dial and keyboard are used for testing purposes.

For smaller offices having up to eight TAS positions, the L-shaped single-position teleprinter table is used, lending itself to a variety of layouts to suit available accommodation. The teleprinter positions would be used for both forwarding and receiving. Pneumatic tubes would probably be provided for reception of telegrams from the public counter and for the despatch of received telegrams to the delivery room, unless these two points were adjacent to the instrument room; band conveyors would be unnecessary. A desk would be provided for circulation, addressing and enveloping. At times when it is required to concentrate traffic at a few positions, surplus positions can be closed by using the OUT-OF-SERVICE key.

For offices having eight or more TAS positions, a double-table is the standard method adopted for mounting teleprinters and dial units. Two tables are placed back-to-back with sufficient space between them to install a conveyor belt to carry telegram forms. Cable chases are provided underneath the table top at the rear to run line and power cables

Fig 9.17 Three-tier stack of receiving teleprinters (*Courtesy of the Post Office*)

and signalling leads to instrument positions in each suite of tables. Wiring for each position is brought out through traps mounted over the cable chase. Only apparatus which is essential from an operating point of view is mounted on the table — position relay sets are mounted on racks leaving only the dial together with the DIAL and CLEAR keys on the table surface; the lesser-used keys (out-of-service and paper-fail, alarm reset) and the supervisory lamps are mounted on the sloping fascia panel behind which runs the vee-band conveyor belt carrying telegram forms to or from the circulating position.

At medium-sized offices a common group of station lines may be provided for both outgoing and incoming traffic and there will be no overflow positions. The last few circuits in the group of station lines, up to the number required to handle busy-hour outgoing traffic, would be located nearest to circulation positions. Early-choice circuits from selector levels, which carry heavier loads of incoming traffic, would be allocated the most favour-

able remaining positions. The rest of the group of teleprinter station lines, which will carry a reduced load, may then be allocated to tables which are more remote from circulation positions. Enquiry positions would preferably be located adjacent to outgoing and early-choice incoming positions.

At 'appointed' offices the instrument room contains phonogram and telephone-telegram positions and also possibly printergram positions. Phonogram and TT positions are the 'combined' phonogram/TAS type except at very large centres where automatic distribution equipment is used; in the latter case, phonogram positions may be installed in a separate room to reduce effect of room noise. The layout of such an office might be as in Fig 9.18. Receiving teleprinters are in a three-tier stack where the incoming-message tape is gummed on to 'C' forms and passed to the addressing point, after dating and timing with the Blick machine. Telegrams to be telephoned to the addressee pass on to the outward phonogram operator, who may also handle forward-

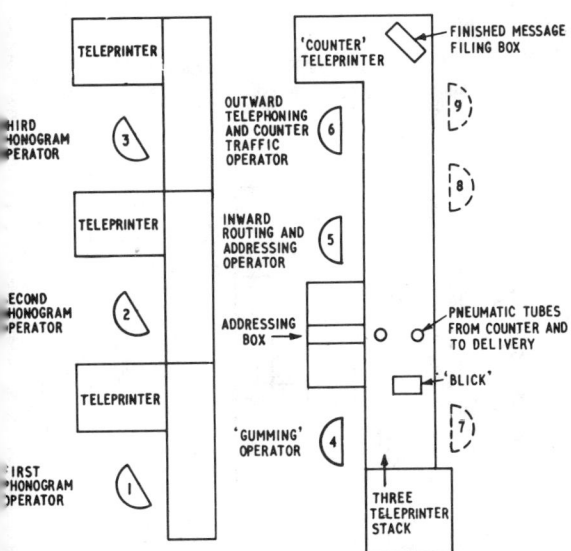

Fig 9.18 Layout of medium-sized telegraph office

ing, by teleprinter, of telegrams received over the public counter.

At a very large modern central telegraph office, a new layout has been designed following traffic studies using work-flow diagrams and charts. The plan for this office, which has something like 180 operating positions handling every type of telegraph service in one large room, is shown in Fig 9.19. Lay out of the instrument room falls broadly into two main divisions. The general plan is that phonogram traffic is accepted and forwarded at one end of the room, while teleprinter-received traffic, including printergrams, is handled at the other. The two main streams of work converge during treatment towards the middle of the room where finished traffic accumulates and certain miscellaneous operations are carried out.

In view of the large volume of phonogram and telephone-telegram traffic, phonogram automatic-distribution equipment is used, with separate TAS forwarding positions. For lighter traffic loads at night and weekends, a suite of 14 combined phonogram/TAS positions is installed. By day these positions are used as teleprinter forwarding-only positions; at other times they are used for combined forwarding and acceptance by being paralleled with 14 normal phonogram positions so that one or other, but not both, can receive calls from the phonogram distribution queue. At peak periods, these positions —

seen in Fig 9.20 with TAS forwarding positions in the background — can be used as additional phonogram-acceptance positions by provision of additional telephone circuits (outside the queue).

On the incoming side, teleprinters for both TAS and printergram circuits are mounted in three-tier stacks. Provision is made, by key-switching, for concentrating these circuits on a small number of positions at nights and weekends. Incoming teleprinter stacks are at one end of a 70-ft table and received traffic is passed, mainly by conveyor bands, through positions for 'unpacking' abbreviated addresses, enveloping and addressing, tracing difficult addresses, telephone delivery or teleprinter transmission, ending up in finished traffic racks.

Printergrams traffic received from telex subscribers, as well as traffic received from other printergram centres, for delivery within the area of this office is received at adjacent teleprinter stacks and disposed of on a common group of TAS/telex forwarding positions. These positions have the facility for operating into both the TAS and telex networks; the latter is reached by dialling an initial access digit which routes a connection from a TAS 1st group-selector level to a telex 1st group selector, giving direct dialling access to all telex subscribers. The Model 11 teleprinters at these positions have end-of-line lamps which glow shortly before the nominal 69 characters have been sent to remind the telegraphist to press the CARRIAGE RETURN and LINE FEED keys when sending to a telex subscriber (who uses a page machine).

Positions for documentary facsimile telegraph machines for receiving telegrams from branch offices are shown on the plan, as also are private-wire teleprinter terminations.

'A' printergrams are telegrams received from telex subscribers, via the telex exchange 01 level, for onward transmission via the general telegraph service.

'C' printergrams are inland telegrams of all types for transmission to telex subscribers. They are received from all TAS offices, via level 791 of the TAS switching centre, for forwarding at the combined TAS/telex forwarding positions. The 'A' and 'C' nomenclature is used generally for telegrams accepted and for delivery respectively by any other means — phonograms, TAS, etc.

'RQ' is a request for correction of a telegram and 'BQ' is the reply to this request. These queries are

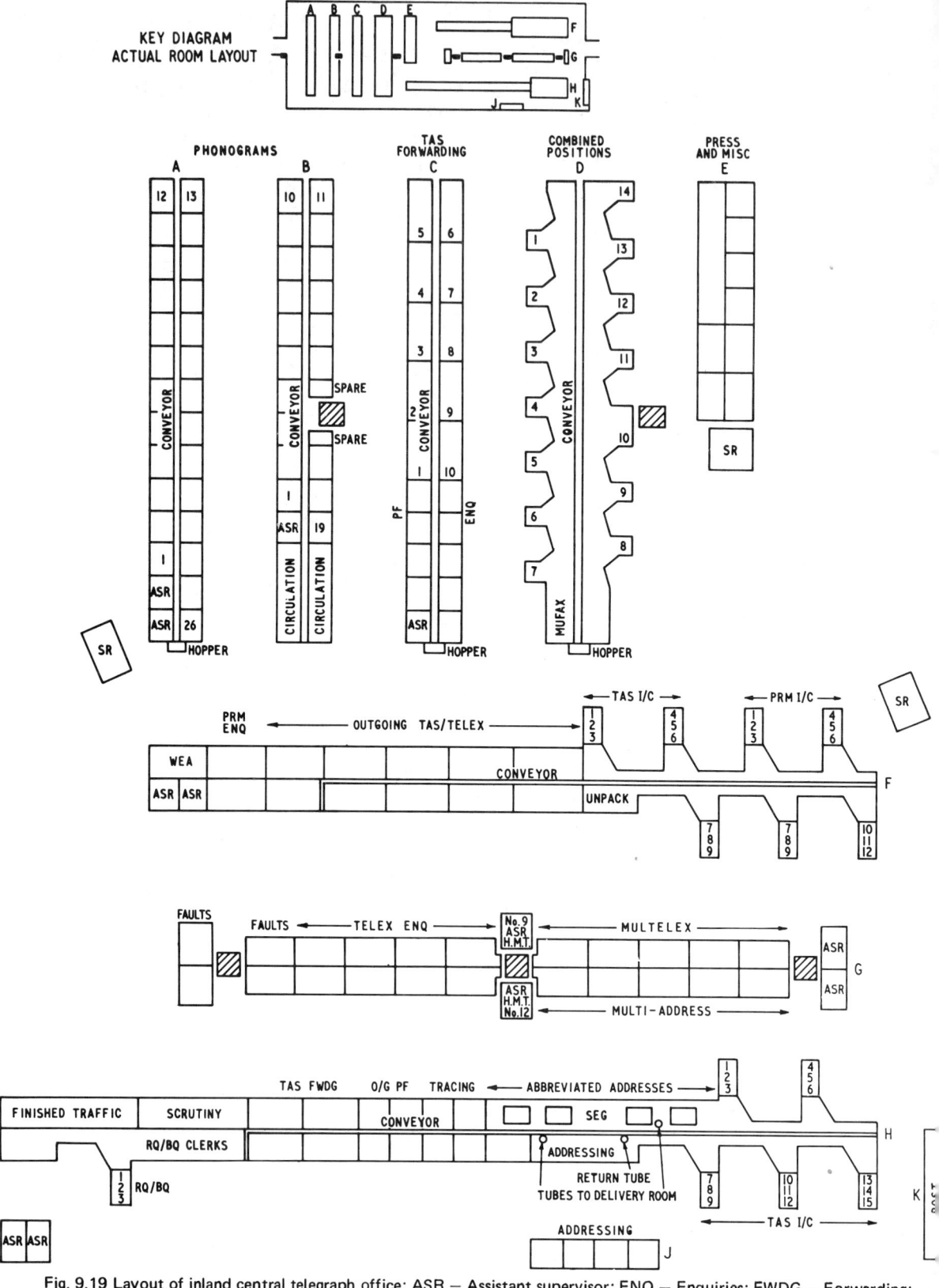

Fig. 9.19 Layout of inland central telegraph office; ASR — Assistant supervisor; ENQ — Enquiries; FWDG — Forwarding; I/C — Incoming; O/G — Outgoing; PF — Phonograms; PRM — Printergrams; TAS — Teleprinter automatic switching; TLX — Telex; WEA — Weather.

Fig 9.20 Combined phonogram-acceptance/teleprinter- forwarding positions in central telegraph office
(Courtesy of the Post Office)

handled at the special RQ/BQ positions accessible over level 777 in the TAS switching centre.

Multiple-address telegrams — i.e. the same text to be sent to a number of different addresses according to a list furnished — are forwarded from the special positions designated over the TAS or telex networks.

Multelex is a similar multi-address service available to telex subscribers; calls arriving over the 2009 telex level are received on a reperforator and transmitted to each required address by an automatic transmitter.

Enquiry positions for telex and printergrams, arriving over telex levels 03 and 02 respectively, handle all enquiries for these inland services.

Power Supplies and Cabling. Power supplies for small offices are taken from the electric-mains supply; the ±80-V supplies are obtained from a rectifier unit,

individual to each position, mounted inside the L shaped table. In case of a prolonged failure of the electric-mains supply, telegrams would be temporarily handled by telephone. Individual line circuits are separately cabled directly to a connection block on the operator position. Similarly, an individual mains-power feed and signalling earth wire are connected to each position and terminated on a power-fuse switchbox or circuit-breaker.

At large offices, continuity of power supply is essential; power for the teleprinter motor, line signalling and switching are taken from a ±80-V battery floated from a rectifier installation.

Cabling for telegraph lines and power are run in underfloor chases, either to each L-shaped table or to each long-table suite; in the latter case, ±80-V supplies are fed via distribution boxes to a distribution fuseboard at the end of each suite of teleprinter tables. A telegraph-signalling earth wire — always separate from a mains-protective earth wire —

is taken to each teleprinter position. From the distribution point on the table suite, larger-capacity cables carry the line and miscellaneous signalling connections to the IDF and other equipment. The power-distribution box is cabled back to the main power-distribution board of the power plant.

9.16 Maintenance

Telegraph equipment requiring maintenance may be broadly classified as station equipment, MCVF equipment and exchange equipment.

Maintenance of station equipment is described below, of MCVF equipment is dealt with in Chapter 7. Maintenance of TAS exchange equipment closely follows the procedure for telex exchanges.*

TAS Dialling Units. TAS stations are concentrated at relatively few large telegraph offices, at each of which a portable tester is held for fault finding on dialling units.

Facilities provided by the tester include line-current measurements, checks of functioning of CALL and CLEAR operations, and timing checks on relays. The teleprinter and an 8 pulse/s dial are used as sources of pulse signals for checking relay performance. For example, the holding of the supervisory relay (RA) is checked by sending plugged LETTERS-SHIFT signals, inverted in the tester to give 20-ms negative and 130-ms positive signals. Similarly, the 20-ms negative pulse from a single inverted LETTERS-SHIFT character checks the maximum operating lag of the I relay, and a 40-ms pulse from a single inverted character V is used to check the minimum operating lag of the RA relay. The 8 pulse/s dial in conjunction with a relay provides nominal pulses of 250 and 350 ms for checking minimum and maximum release lags of the RA relay.

Dial speed and ratio testing require co-operation from the exchange test desk.

Teleprinter-speed check. The speed of the motor on the No. 11 teleprinter can be checked using an electrical stroboscope. This unit comprises a neon

*See *The International Telex. Service* (R. N. Renton), Pitman.

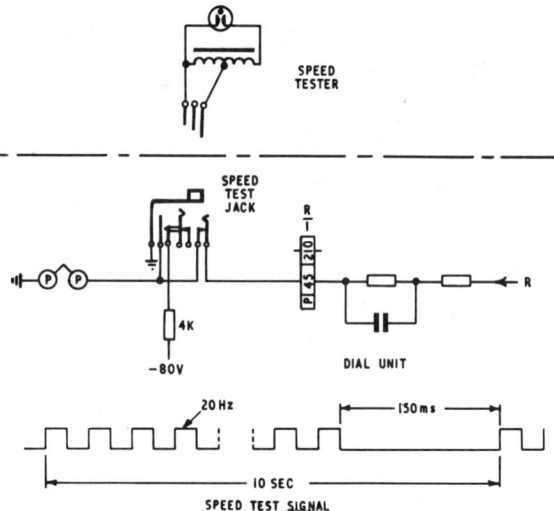

Fig 9.21 Using speed-test signal at a TAS station

lamp and an auto-transformer fitted in a small cylindrical case. On dialling the code for the speed-test signal, 20-Hz reversals are received over the station line from the exchange-signal generator (controlled from a 50-Hz tuning fork in the exchange). The plug of the tester is inserted in the SPEED TEST jack (Fig 9.21) of the dialling unit; this disconnects the teleprinter electromagnet from the line, holding it at −80 V. The ±80-V reversals at 20 Hz are fed to the step-up transformer of the speed tester to strike the lamp; this lamp illuminates a stroboscopic disc mounted on the teleprinter-transmitter shaft running at 400 rev/min. The 20-Hz reversals are interrupted at 10-s intervals for 150 ms to time rotation of the stroboscopic display for determining the speed error; not more than two segments of the stroboscope pattern should drift during the 10-s period. Alternatively, the speed may be checked locally or at the exchange test desk using a TDMS.

Teleprinter-transmitter Distortion and Receiver Margin. These may be measured using one of the portable TDM sets described in Chapter 2. Receiving margin may alternatively be measured by dialling into the TAS network to obtain access to the test-signal generator. This machine emits the standard test sentence THE QUICK BROWN FOX JUMPS . . . , firstly without distortion, followed by test sentences

with separate early and late distortion graded at
±20, ±25 and ±30%. Each of eight transmissions of
the test message is preceded by LETTERS SHIFT,
a CARRIAGE RETURN and a LINE FEED signal,
and an announcement in the form TEST DISTOR-
TION $X\%$, the whole of this preamble being sent free
of distortion since it is important that the receiver
should correctly record these signals. At the exchange, the speed-test circuit-relay set operates to a
phasing pulse which ensures that a call is switched
to the test message only immediately before the
beginning of an announcement. On the No. 11 teleprinter, receive margin may be measured by adjustment of the orientation device while receiving undistorted signals.

9.17 Overseas Telegrams
At the London overseas central telegraph office,
which carries traffic for the overseas telegram service, something like 60 000 international telegrams
are handled daily over approximately 120 overseas
circuits. Roughly one-third are transit telegrams,
routed via London between addresses in two foreign
countries; of the remainder, about one-half are for
Central London and one-half for Outer London and
provincial addresses. Methods of acceptance and
delivery follow conventional inland practice, except
for the gentex offices. The equipment of this central
office includes a phonogram automatic-distribution
system, TAS incoming and forwarding positions and
'overseas' printergrams positions reached via the telex
04 level.

The traffic pattern shows two peaks, one in the
morning, the other in the afternoon. These differ
basically in that the morning peak shows a main
traffic flow from overseas to inland circuits, with
the reverse flow during the afternoon. The afternoon
peak between 17.00 and 18.00 h is the greater and
represents the station busy hour.

Mechanical handling and processes involved in
cross-office transmission on such a vast scale are very
costly. Only in recent years, now that older forms of
overseas telegraph systems have been largely sup-
planted by teleprinters working on highly-stable MC
VF systems in submarine telephone cable, error-
correcting radio-teleprinter or satellite systems,
has it become possible to introduce mechanized

processing and reduce manual handling and cross-
office movement of telegram forms.

Gentex operation is suitable on highly stable
routes where telegraph channels are plentiful and
relatively cheap and on which automatic switching
is possible. There are many overseas routes which
do not satisfy these criteria and which have hitherto
necessarily been operated on a point-to-point retransmission basis. Message-relay operation offers a
preferred alternative which does not place the burden
of re-equipment on overseas centres which do not
choose to mechanize their service.

An automatic message-relay system (see Chapter
11) has been introduced to carry overseas (extra-
European) traffic. With this system, telegrams are
briefly stored at the central office and offered in the
form of a queue to the outgoing channels. In this
way, it is possible to achieve much higher channel
loadings than with the pure-chance-access gentex
systems. By using serial numbers to safeguard against
loss of telegrams, answer-back working is unnecessary
and the return channel can be used as a separate
traffic channel.

Gentex Service. This is the service by which international telegrams are handled over an automatic
exchange system to enable a telegraphist at an originating office to forward a telegram direct to a
delivery office in another country. This system
replaced the less efficient method of retransmitting
telegrams on a link-by-link basis between telegraph
offices and the international central telegraph offices
in each country.

CCITT Recommendations allow participating
countries to use existing (or planned) telegraph networks for this service. At the present time, gentex
service operates mainly between European administrations and is being steadily extended. Some administrations have separate gentex and telex networks;
others have integrated networks in which telex subscribers are barred from access to the gentex component of the joint network; on the other hand, gentex
offices have access to the telex network for
delivering telegrams. Some administrations, as yet
without their own gentex network, have their gentex
offices connected as station lines on a gentex
exchange in a neighbouring country.

Use of type-A or B signalling is permitted, using

dial or keyboard selection; as for telex, outgoing international trunk circuits must conform to the type of signalling used by the incoming country. The type-C signalling system makes provision for discrimination between gentex and telex on integrated switching systems. Use of overflow centres is included in the scheme, for acceptance of traffic when a calling office fails to obtain a free line to the delivery office. Receipt of the answer-back code at the end of transmitting the telegram is regarded as acknowledgement of receipt.

In the United Kingdom, to avoid establishing a third telegraph network, gentex service is operated over the telex network. Gentex offices, known as *overseas telegraph area offices*, are sited at a number of key points in London and certain provincial cities. Telegrams from the United Kingdom for overseas gentex offices reach (UK) gentex offices over the national telex network, from final selectors in the local telex exchange. Gentex offices also handle overseas printergram traffic for telex subscribers in their area, who dial the special printergrams code quoted in the telex directory. Gentex receiving positions use tape-printing teleprinters, without local record. These positions (Fig 9.22) use a modified form of the TAS dial unit which conforms to CCITT gentex Recommendations*; in this unit, the equivalent relay to RA in Fig 9.8 has a release delay of not less than 400 ms.

Gentex forwarding positions (Fig 9.23) use page-printing machines to conform to the page layout of machines customarily used in European gentex

Fig 9.23 Gentex forwarding position (*Courtesy of the Post Office*)

offices. A modified form of TAS dial unit is used which provides a local record path for the page teleprinter (unless countermanded by a switch), a call-in lamp operated by the J-contact, and the extended time delay for detecting the clear signal. Facilities are also provided for using a printing reperforator or automatic transmitter, with full operational safeguards.

For outgoing traffic the basic trunking diagram for the gentex exchange, situated alongside the London international telex exchange, is shown in Fig 9.24; equipment is of the same type as that for the telex service. Apart from ARQ radio circuits, most international trunk circuits carrying gentex traffic operate on a unidirectional basis. For outgoing traffic, high telegram loading on circuits carrying gentex traffic justifies provision of direct circuits from each gentex office in the United Kingdom to gentex selectors in the London international telex exchange. Station lines from gentex offices terminate directly on gentex 1st selectors which give access from level 0 to the inland telex network.

The trunking diagram shows the distinction between outgoing international routes using exclusive gentex circuits and those using joint gentex/telex circuits. Each group is further divided according to whether type-A or B signalling is used. Outgoing international trunk-relay sets are arranged to permit or suppress operation of the international accounting meters according to the point of input to the relay set. On all circuits used for gentex traffic, operation

Fig 9.22 Gentex incoming position (*Courtesy of the Post Office*)

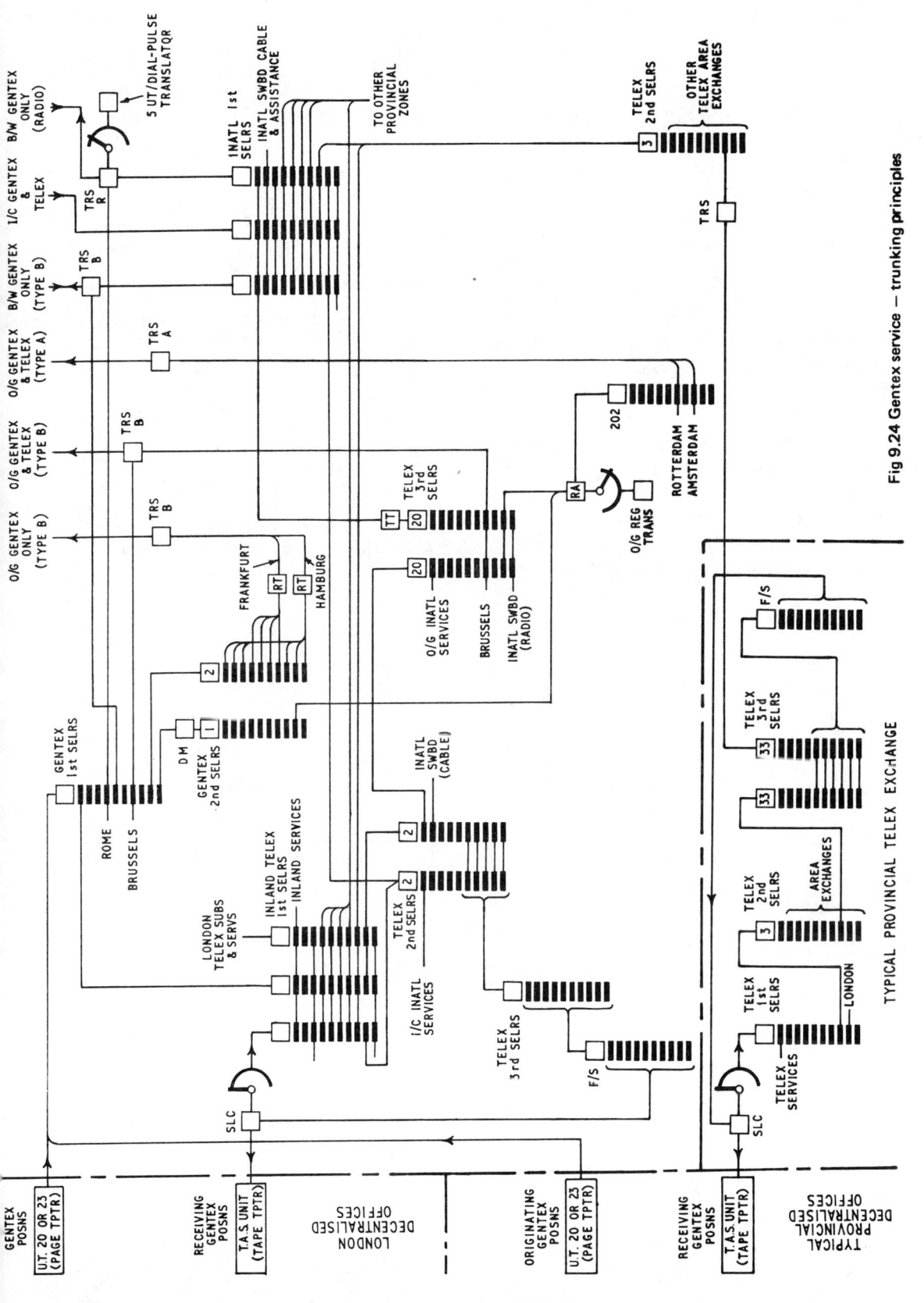

Fig 9.24 Gentex service — trunking principles

of accounting meters is prevented since other arrangements are made.

For shared gentex/telex routes, use of de-metering relay sets DM in the trunks from gentex 1st selectors to 2nd selectors is necessary. A common group of register translators is shared with telex traffic. Entry to a national gentex network at more than one point is possible by using a discriminating digit. From the digits dialled, the register translator is able to determine the correct route for a multi-entry and also to provide dial-routing pulses followed by 5-unit start–stop selection signals to operate the distant register system. For calls over outgoing ARQ radio routes, the gentex telegraphist dials the routing code, e.g. 7, and then sends from the teleprinter key-board 5-unit start–stop selection signals corresponding to the number of the required gentex office.

Incoming international trunk circuits, which may be either exclusive gentex or joint telex/gentex circuits, terminate directly on incoming 1st gentex selectors; exceptionally, if the circuits are provided on ARQ radio systems, the 5-unit/dial-pulse translation equipment shown is necessary. Level 9 gives access to the cordless telex switchboard; levels 1–8 give direct access via 2nd selectors to the national telex network for calls to reach the gentex offices on final-selector multiples.

References

1 Neate, R. A., 'Combined Working in Telegraph Instrument Rooms', *P. O. Telecommunication Journal*, **13**, p. 112 (1961).
2 Wilcockson, H. E. and Walker, H., 'Phonogram Automatic Distribution', *POEE Journal*, **42**, p. 149 (1949).
3 Wilcockson, H. E. and Mitchell, C. W. A., 'The Introduction of Automatic Switching to the Inland Teleprinter Network' *IPOEE Printed Paper No. 195* (1949).
4 Coulman, A. H. and Goss, C. E. G., 'Emergency Mobile TA Units', *POEE Journal*, **53**, p. 268 (1961).
5 Stripp, W. A. 'The New CTO at Fleet', *P. O. Telecommunication Journal*, **15**, p. 42 (1963).
6 Mitchell, C. W. A. and Gray, A. T., 'Speeding the Overseas Telegraph Service', ibid, **21**, p. 9. (1969).

10 Teleprinter Private Services

Most private teleprinter circuits operate at 50 bauds using page teleprinters, either the Model 7 or the more recent Model 15, both operating at 6·6 char/s. With page teleprinters, a local printed record of the message being sent is almost always required, not only to indicate to the operator when the end of a printed line is reached but also to serve as a file copy.

Some renters require teleprinter circuits which operate at 75 bauds; in this case the Model 75 page teleprinter may be used or the newer Model 15, both operating at 10 char/s.

Apart from short lines (up to about 20 miles) which are more economically provided by cable pairs with d.c. transmission, circuits in standard VFT systems are used for leased teleprinter circuits. According to the location of the stations to be connected, either a direct VFT single-link circuit would be available or a permanent multi-link circuit would be provided which could, if necessary, include up to five VFT circuits in tandem; in practice, the need for leased inland circuits comprising more than three links in tandem rarely arises. At a modulation rate of 75 bauds, if a circuit can be provided using not more than two VFT channels in tandem, standard 120-Hz-spaced MCVF systems give an adequate performance; otherwise, MCVF channels with 240-Hz-spacing are used.

10.1 Leased point-to-point teleprinter circuits

The standard private teleprinter installation is supplied with a steel office table which supports the teleprinter and the signalling unit. On installation it is necessary only to connect the a.c. mains power supply, the telegraph line circuit and the signalling earth lead. The teleprinter plugs into the signalling unit for all its supplies.

Leased Circuit with Teleprinter No. 7. Earlier installations use the Model 7 teleprinter with a 160-V d.c. motor. The circuit (Fig 10.1) follows normal simplex principles. A local record is provided via the teleprinter send–receive switch SR. While operating in local, the distant teleprinter is held to the stop condition by −80 V from the LOCAL key. If connected to a switchboard, or on a broadcast network, the clear (or acknowledge) signal is sent by applying +80 V to the S wire for about 5s from the CLEAR key. A switch is provided to select either a visual or audible alarm to operate from the J or answer-back unit contacts. The diagram shows the arrangement for the power supply from a.c. mains using a rectifier unit (No. 66C) to provide 160 V to drive the motor, and ±80 V via protective resistor bulbs for line and local signalling currents. Fuses are included in the rectifier unit and a circuit breaker is used for protection against fire risk should the teleprinter motor stall or become overloaded. The mains earth lead is used for protective purposes only; a separate earth lead is used for telegraph-signalling purposes.

An automatic transmitter may be associated with the teleprinter station and one arrangement is shown in Fig 10.2. The teleprinter circuit is as described above: when a perforated tape is placed in the transmitter head, tape-out contacts TO close, operating the OC relay. OC1 operates the transmitter-clutch solenoid and tape commences to feed; OC2 changes the S wire over from keyboard transmitter to tape transmitter; OC3 connects an interrupt-detector relay to the R wire; OC4 provides a local record from the tape transmitter, unless countermanded by operation of the DISCONNECT MONITOR key KDM.

When receiving a message from a keyboard, a distant station can attract the attention of the sending operator, if necessary, by sending signals over the return channel. The effect of this is to mutilate the local record due to operations of the send–receive switch while signals are being sent. This form of interruption is not possible if a local record is

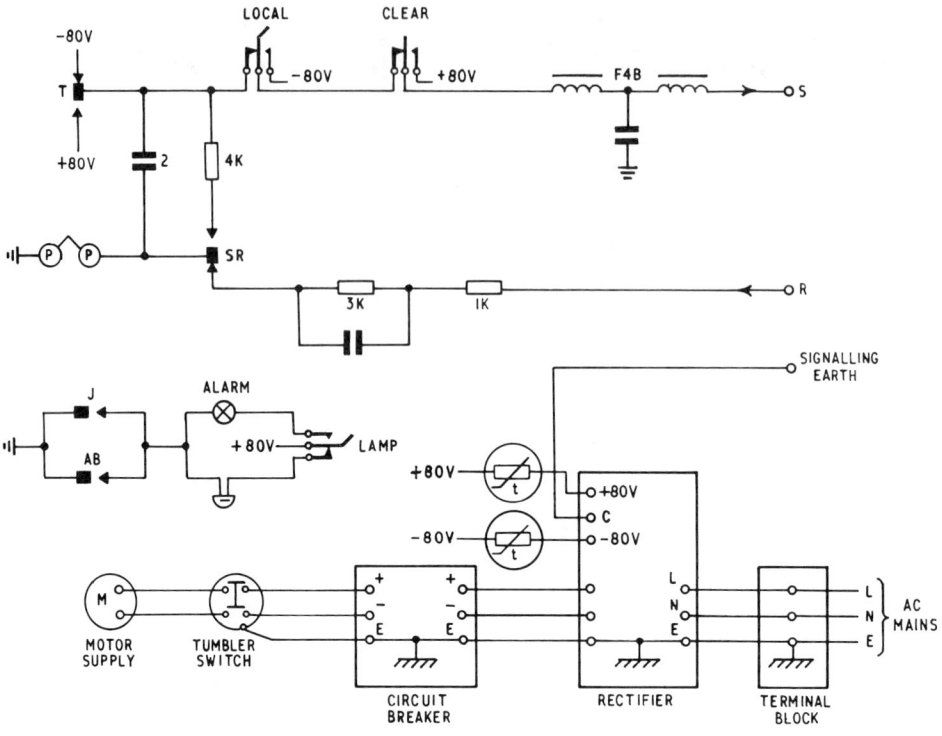

Fig 10.1 Private-station equipment for teleprinter No. 7

being taken from tape transmission, and it is usual to terminate the R wire with an interruption-alarm relay such as AL which responds to signals received over the R wire. The AL relay operates to positive signal elements, and having done so it holds via **AL1** to the −80-V condition; **AL2−3** provide alarm conditions; **AL4** disconnects the clutch to stop tape feed. The clutch will also be disconnected by tight-tape contact TT if for any reason the tape feed is impeded. Otherwise, when the tape has run through the transmitter head, the tape-out TO contact restores, releasing the OC relay.

Leased Circuits with Teleprinter No. 15. The Model 15 teleprinter is normally equipped with an a.c. (synchronous) motor; at large installations where continuity of service is essential during electric-mains failure, 160-V d.c. motors are fitted and supplied from a ±80-V battery installation. Optional tape transmitter and reperforator devices are integral units of this teleprinter.

The signalling unit used with the Model 15 tele-

printer is illustrated in Fig 10.3 with an interior view shown in Fig 10.4; essential details of the circuit arrangement are shown in Fig 10.5. CCITT signalling is not used for private circuits — the standing condition on the S and R wires is −80 V. The circuit operates normally in the simplex mode; duplex operation (without local record) can be used if required. Keyboard and tape transmitters are wired in series and a local record is obtained from either, via send−receive switch SR. Other mechanically-operated contacts are M (motor on-speed), TR (tape-transmitter off-normal), AL (alarm), TE (reperforator tape exhausted), and J (operated from combination 10 of alphabet No. 2). Solenoids RM and TM are used to trip the reperforator and tape-transmitter mechanisms respectively from the signalling unit.

The power pack supplies ±80 V (smoothed) for line signalling and −50 V for relay operation. Circuit breakers CBR are integral with the mains transformer and the teleprinter motor. An alarm lamp and buzzer are fitted to operate from the J signal or on receipt of incoming signals during outgoing tape transmission or while the LOCAL key is operated; a second alarm

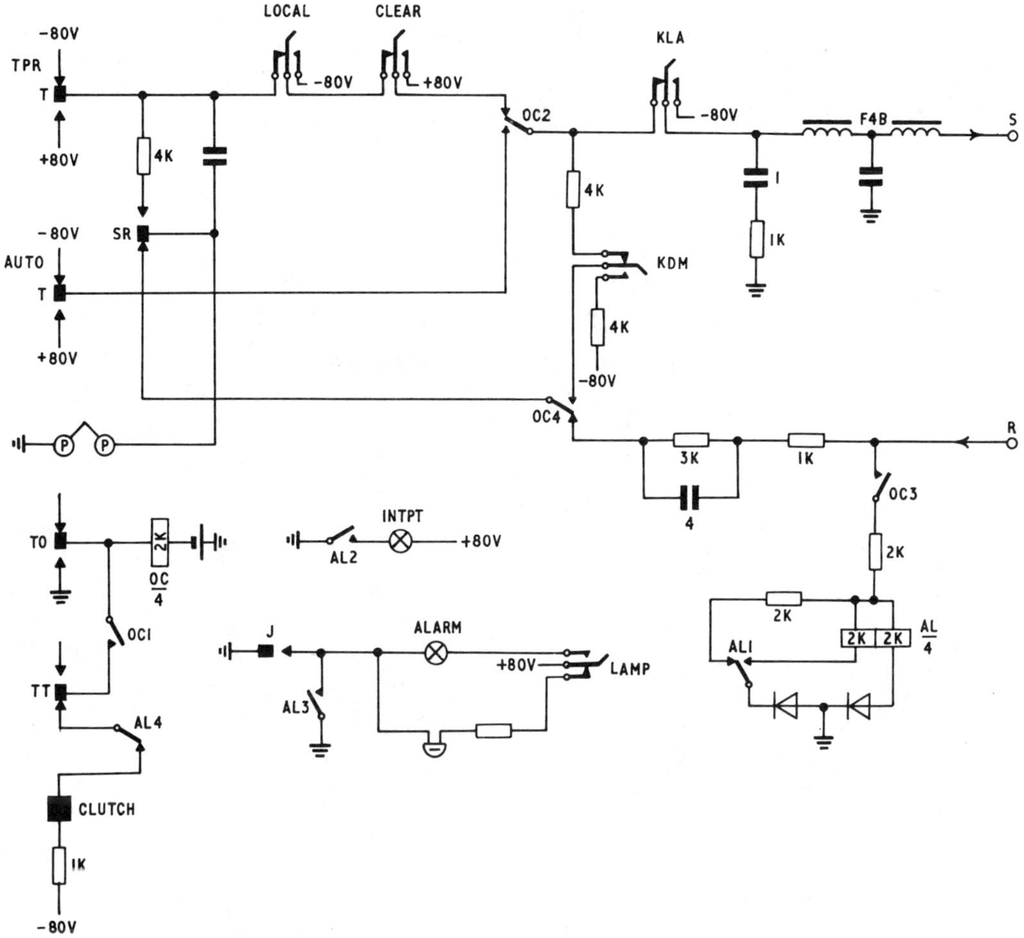

Fig 10.2 Private-station equipment for teleprinter No. 7 with automatic transmitter

Fig 10.3 Signalling unit for private station using teleprinter No. 15 (*Courtesy of the Post Office*)

Fig 10.4 Signalling unit — interior view (*Courtesy of the Post Office*)

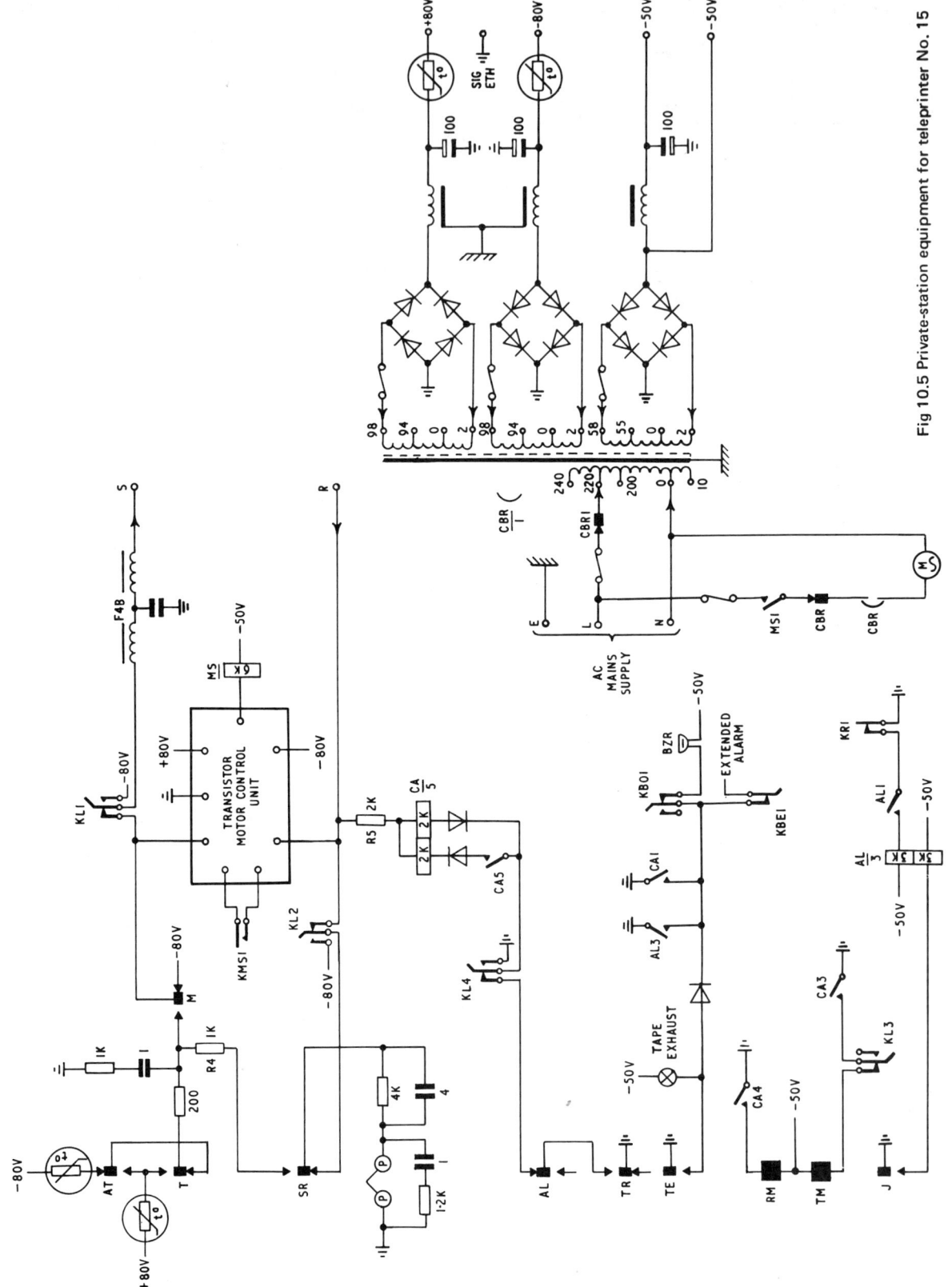

Fig 10.5 Private-station equipment for teleprinter No. 15

indicates the approach of reperforator-tape exhaustion.

When the START button KMS1 is pressed, the MS relay operates. **MS1** applies power to the teleprinter motor. When correct running speed is reached the on-speed contact M switches the keyboard and tape transmitters in series to the S wire, via the filter No. 4B. A local record is provided via R4 and SR; in the normal position, SR connects the teleprinter magnet to the R wire, via the signal-shaping net-work.

Tape transmission may take place by loading the perforated tape and pressing the tape-transmitter ON button: the keyboard is meanwhile mechanically held inoperative. Tape transmission ceases when the tape has run through the transmitter or the tape-transmitter OFF button is pressed, or when the transmitter-release magnet TM is operated. A perforated tape is obtained in addition to the page copy if the reperforator ON button is pressed or if the reperforator release magnet RM is operated.

Tape transmission may be interrupted by the distant station sending keyboard signals; the first start element received on the R wire operates the CA relay (via R5, KL4, AL and TR). **CA1** operates the buzzer; **CA2** lights the alarm lamp; **CA3–4** operate the transmitter and reperforator-release magnets respectively; **CA5** holds the CA relay to the negative line signals. The CA relay can be released by pressing the tape OFF button to release TR. Receipt of the characters FIGURE SHIFT J operates the AL relay which holds at **AL1** via the reset key KR1; contacts **AL2–3** operate the alarm lamp and buzzer.

The teleprinter can be used in local for preparing perforated tape by throwing the LOCAL key KL. Contact **KL1** applies −80 V to the S wire; **KL2** disconnects the R wire and provides the −80-V holding condition for the teleprinter magnet at SR. Contact **KL3** prevents interruption to the tape transmitter by CA3; contact **KL4** connects the CA relay to the R wire to give alarm conditions if the distant station calls. When the roll of reperforator tape reaches a certain low level, **TE** lights the TAPE EXHAUSTED lamp and operates the buzzer; TE is restored by fitting a new roll of tape.

Provision is made for repeating incoming signals to an external printing reperforator, if required; and for connecting a polarized relay in the R wire circuit if relayed-line conditions apply.

Motor On/Off Electronic Control Unit. To conserve electrical energy and reduce wear and noise it is arranged for the motor to switch off automatically if teleprinter signals do not occur for a period of 45–90 s.* As a result it is necessary for the motor to start up on receipt of the next start signal, or on operation of the START button KMS. For the teleprinter No. 15 these features are provided by the electronic control unit shown in Fig 10.6.

This unit controls operation and release of the motor-start relay MS in the output circuit. A positive potential transition will cause the relay to operate; it will release 45–90 s after the last positive transition. Two inputs are provided and a positive transition on either will operate the circuit.

In the disengaged condition, negative potential exists on both inputs (via keys KL1 and KL2), back-biasing diodes D1 and D2. Transistor TR1 is cut off, and TR2 and TR3 are conducting; the base of TR4 is at negative potential; TR4 is conducting, holding TR5 and TR6 cut off.

Positive potential applied to resistor R2 will forward-bias diode D2; a pulse of positive potential will be applied to the base of TR1 via C2; transistor TR1 will conduct, and C9 will be charged to negative potential via R4 and TR1. The base of TR2 goes more negative as C9 charges, and the emitter of TR2 follows. Transistor TR3 cuts off when the emitter of TR2 is at about earth potential, and C10 charges to +12-V potential via R15 and D5. The base of TR4 goes positive as C10 charges, and TR4 is cut off. The base of TR5 goes negative, and both TR5–6 saturate rapidly; relay MS then operates.

When the positive transitions cease at the input (the nominal minimum being one element − 20 ms) capacitor C9 is fully charged. With absence of positive transitions from the input, TR1 is cut off and C9 commences to discharge via R16 and R5. As TR2 is connected as an emitter follower, the base circuit presents a very high impedance. After approximately 5 s the emitter of TR2 will have risen to about earth potential and TR3 conducts. The collector potential of TR3 drops to near earth potential and back-biases D5. Capacitor C10 is charged to +12 V and will start to discharge to −12 V via R13–14. In approximately 90 s, capacitor C10 will have discharged to earth potential and D5 will be forward-biased to prevent

*Recommendation S7.

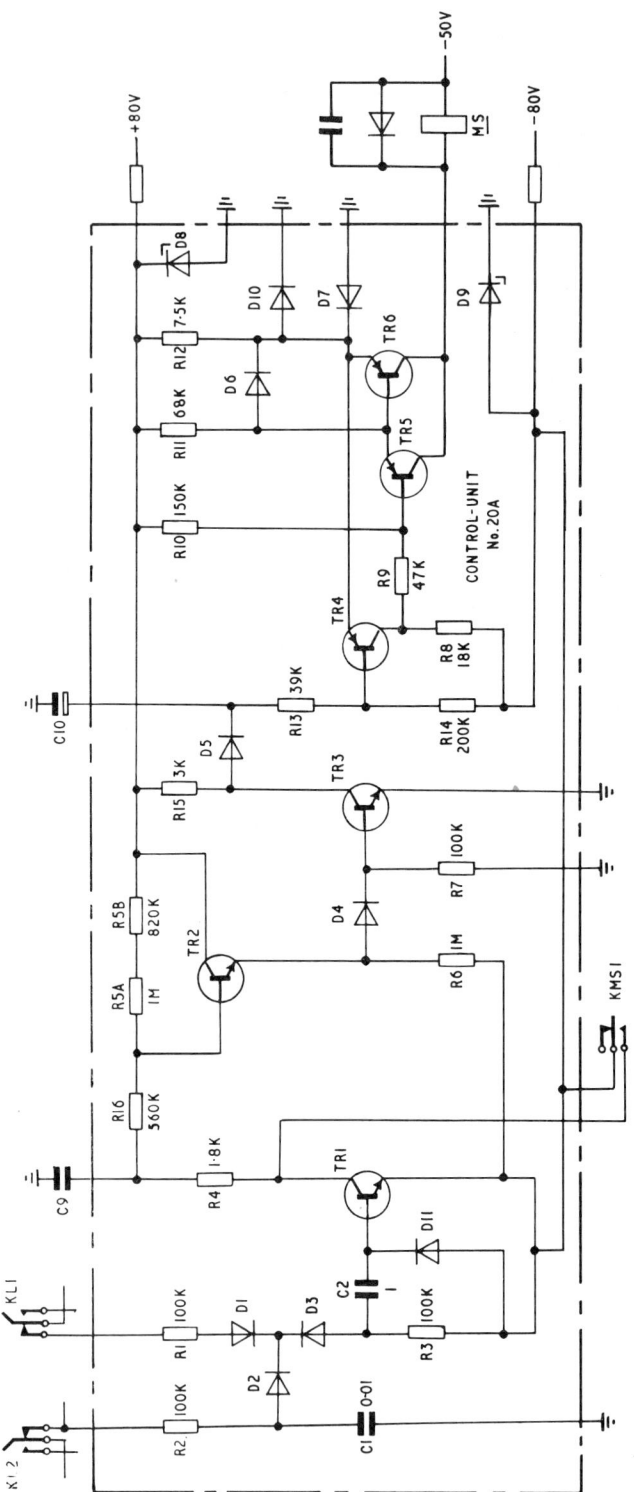

Fig 10.6 Electronic ON/OFF switch for motor

any further discharge. The base of TR4 is now held negatively by the potential divider R13–14, and TR4 conducts. With TR4 conducting, TR5 and TR6 are cut off, and relay MS releases.

10.2 Switchboards for private teleprinter networks

For inter-communication between a number of teleprinter stations served by private teleprinter circuits, a switchboard on which all the leased circuits are terminated is installed at a central point. Switchboards of various capacities are available to meet standard requirements, installations are also specially designed to meet particular needs.

The following description refers to a typical 30-line teleprinter-switchboard installation. The lines terminate upon three rows each of ten switchboard jacks in the left-hand vertical panel; these jacks are used for both answering and calling. Associated with each jack is a CALL lamp and an ENGAGED lamp. Provision is made for simultaneous transmission, or 'broadcasting' to selected lines up to the full 30 lines. For this purpose, the lines to the jacks are intercepted by 30 BROADCAST keys, in the right-hand panel; with each key is associated an ACKNOWLEDGE lamp and an ENGAGED lamp.

The keyshelf accommodates ten cord circuits with ANSWER and CALL cords and separate ANSWER and CALL clearing lamps. A PRINT & MONITOR key for each cord circuit enables the switchboard operator's teleprinter (on an adjacent table) to communicate with any calling or called station, or to monitor an established connection. On the keyshelf is a 3-position key associated with the operator's teleprinter; this key can split a cord circuit to enable the operator to communicate separately with either the calling or called station. In the normal position of the key, monitoring on both channels of the communication can be carried out. Night-alarm, line-testing equipment and other miscellaneous facilities are provided. Apart from the operator's teleprinter, all equipment associated with line circuits, cord circuits and position circuit is housed within the switchboard.

Line Circuits. The circuit arrangement of the teleprinter stations and for the operator's teleprinter circuit is as shown in Fig 10.1.

The line-circuit terminations and also the broadcasting equipment are shown in Fig 10.7. The standing condition on the line is −80 V on both the S and R wire, the station teleprinters being held to the stop condition by −80 V applied to the S wire from the switchboard jack. A station calls by tapping the teleprinter space bar. The call relay C, shunted by a diode to prevent operation to −80 V, flashes to the +80-V elements of the calling signals. C3 flashes the CALL lamp of the calling line; C2 operates the PILOT lamp and rings the night-alarm bell if this is switched on at **KNA1**.

When the operator answers by inserting the ANSWER plug of a free cord circuit into the line jack, the C relay is disconnected from the R wire, as is also the −80 V from the S wire. The S and R wires are extended to the tip and ring wires of the cord circuit. The auxiliary jack springs close a circuit to light the ENGAGED lamps associated with both the line jack and the BROADCAST key to guard the circuit.

Cord Circuit and Position Circuit (Fig 10.8). The switchboard operator plugs into the jack of the calling line with the position key operated to the ANSWER side (KCA) and the PRINT & MONITOR key KPM of the cord circuit operated. KPM1–2 close a circuit for relay CC to operate via resistor YO in the position circuit. Relay CC holds via **KPM4** and **CC2x**. Contact **CC1** applies earth potential to the common point where YO feeds the operating windings of the other cord-circuit CC relays. Consequently, if a KPM key of any other cord circuit is subsequently thrown inadvertently, the associated CC relay cannot operate; this prevents more than one cord circuit being connected at a time to the position circuit and avoids putting station circuits into contact unintentionally. **CC3–4** and **CC5–6** divert the tip and ring circuits of the ANSWER and CALL cords through the operator's position circuit.

The position circuit contains two polarized relays MA and MC, each with three windings; these relays respond to signals from the ANSWER and CALL sides respectively. In the quiescent condition, the central winding of each relay is energized with a current of 8 mA to give the relay a *mark* bias, from −80 V at RBC; the other two windings are disconnected. MA1 and MC1, in series, hold the

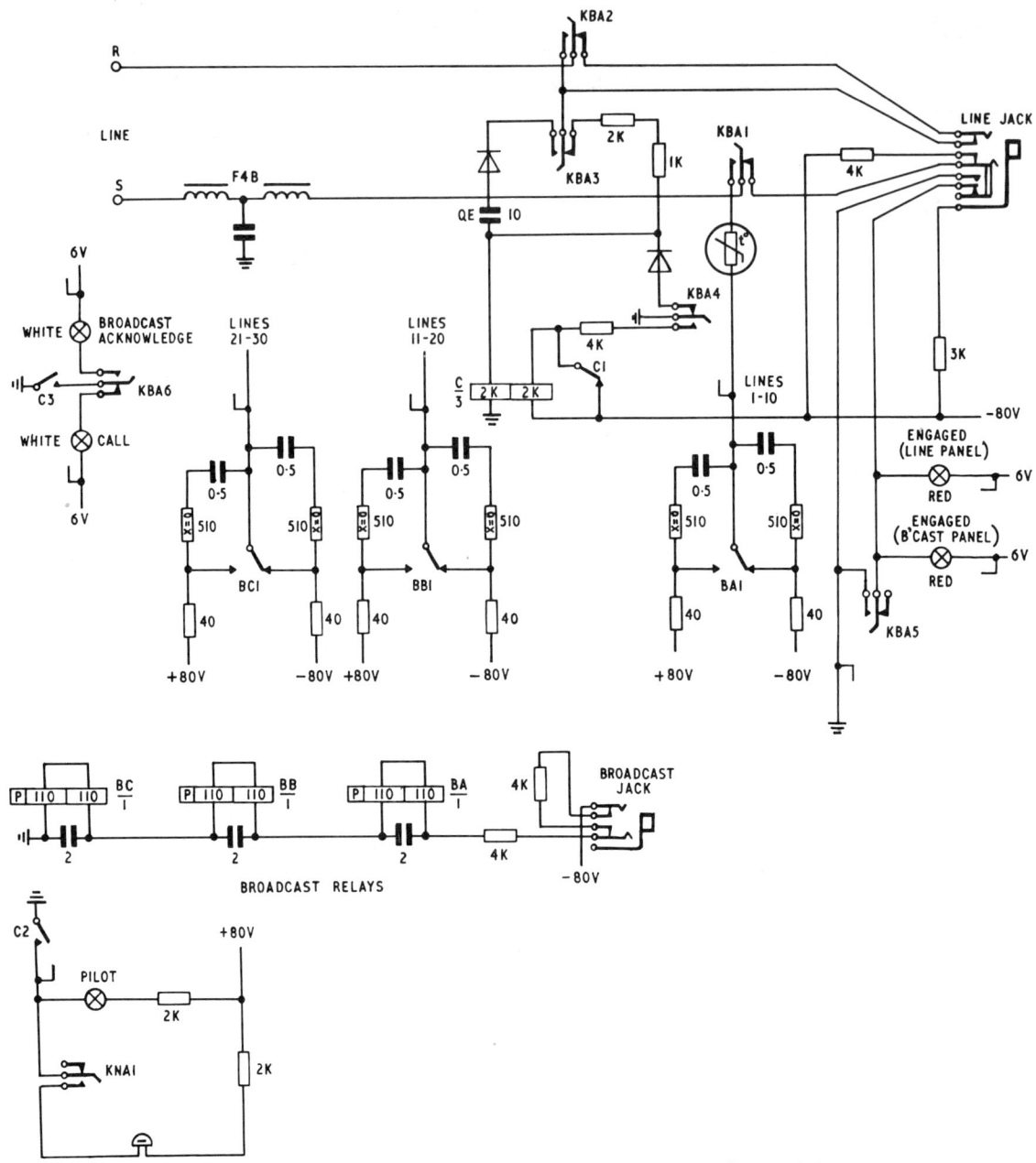

Fig 10.7 Private teleprinter switchboard – line circuit

position-teleprinter electromagnet to the stop position from −80 V at RBA; operations of MA1 and MC1 will record signals on the operator's teleprinter. While **KCA2** or **KCC2** are operated, signals from the operator's keyboard are fed via MC1 and MA1 to provide a local record to this teleprinter.

When a cord circuit is connected to the position circuit, the operation of **KPM3** gives both relays,

MA and MC, an 8-mA *space* bias from the +80 V via R3 and R4. Under this condition the two bias windings of each of the two polarized relays are neutralized: operation of the relays will now depend upon the magnitude and direction of current in their line windings. When **CC3** and **CC6** are operated the tips of the ANSWER and CALL plugs are connected respectively to the third windings of the **MA**

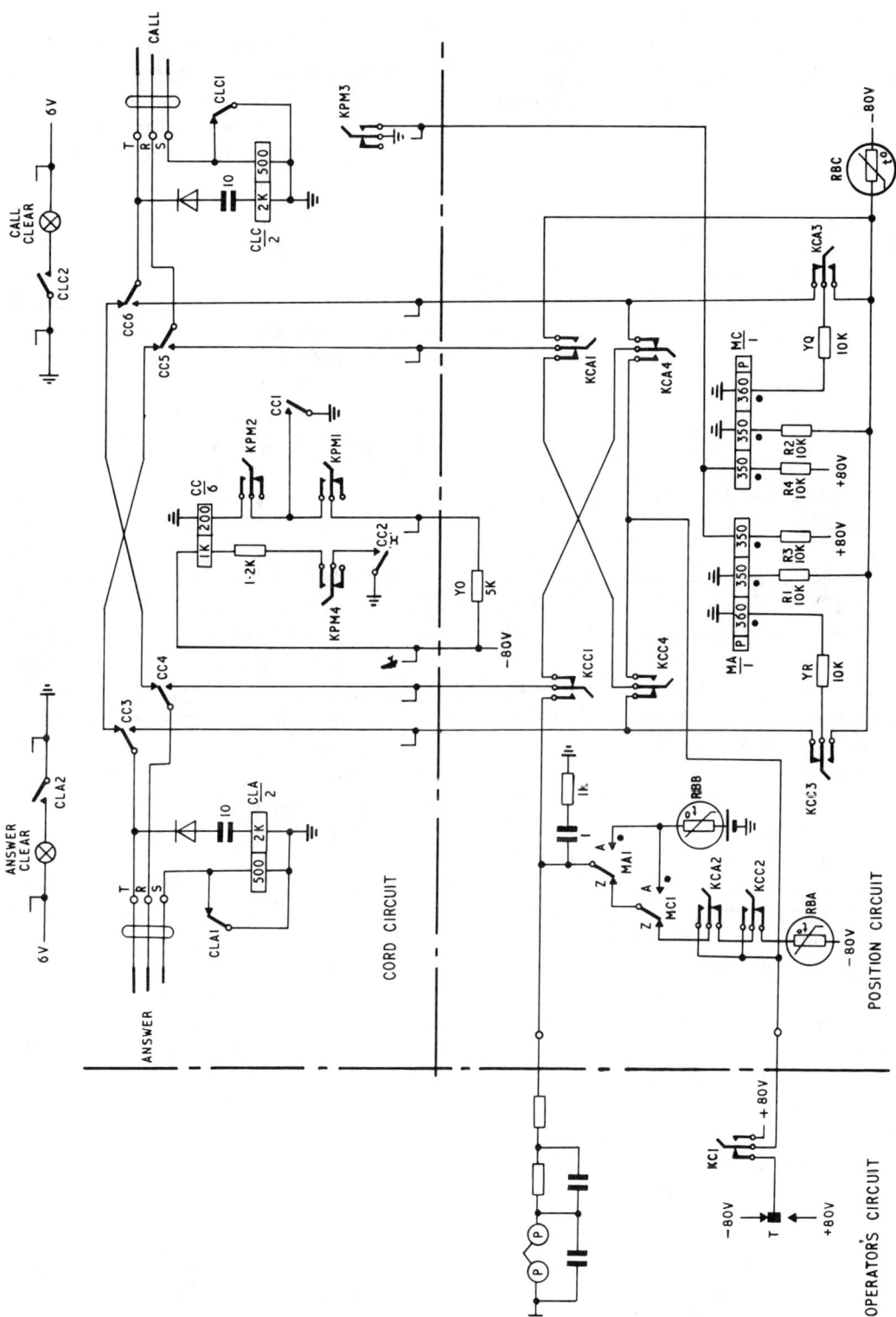

Fig 10.8 Private teleprinter switchboard — cord and position circuits

and MC relays via KCC3 (on the answer side) or
KCA3 (on the call side). The line currents (about
8 mA, due to resistors YR and YQ) from the calling
and called stations control the operations of the MA
and CA relays to operate the switchboard teleprinter.
While setting up a connection, the third winding of
the called-side relay MC will be disconnected from
the tip of the plug and given a mark bias via RBC
by operation of the ANSWER key **KCA3**; or the
calling-side relay MA will be similarly disconnected
and biased by **KCC3**.

The transmitter of the switchboard teleprinter, as
well as providing the local record, is connected to the
ring of the ANSWER plug (via **KCA4**, KCC1 and
CC4); or, alternatively, to the ring of the CALL plug
(via **KCC4**, KCA1 and **CC5**).

With the position ANSWER key KCA and the
cord circuit PRINT & MONITOR key KPM operated,
the operator is able to exchange teleprinter communication in either direction with the calling station. To
call the wanted station, the switchboard operator
inserts the CALL plug, of the cord circuit in use,
into the line jack of the wanted station; the position
key is moved to the CALL position KCC. Signals
from the space bar of the switchboard teleprinter
start up the teleprinter motor of the called station;
the operator can send the WRU signal to obtain the
answer-back code of the called station. Under these
conditions, the tip of the plug – the R wire of the
calling station – is disconnected at **KCC3–4** and the
caller cannot interrupt while the call is being set up.
On the other hand, the answer-back code signals
from the called station which are fed into the MC
relay via **CC6** and KCA3 to operate the switchboard
teleprinter, are repeated also from MC1 via **KCC1**
and **CC4** to the calling station. The calling station
receives a record of all signals exchanged between the
switchboard teleprinter and the called station. After
obtaining the called station, the operator advises the
calling station to proceed with the communication.
If the operator should need further communication
with the calling station, key KCA is operated instead
of KCC; the called station is meanwhile on the line,
but any transmission from the latter is prevented by
KCA3 from operating the MC relay which would
otherwise interfere with signals between the switchboard and calling-station teleprinters. At the same
time, the teleprinter of the called station is temporarily held to the stop position by −80 V applied from

RBC via **KCA1** and **CC5**.

When the called station has been obtained the
switchboard operator retires from the connection by
restoring the KPM key, releasing the CC relay and
connecting the calling and called stations together
via CC3, CC4, CC5 and CC6. The two stations can
now exchange communications.

Monitoring. If the switchboard operator needs to
monitor the connection, operation of the cord circuit PRINT & MONITOR key KPM operates the CC
relay as before, diverting the connection, but without interruption, through the position circuit (via
CC3, KCC4, KCA1 and **CC5** – also **CC6**, KCA4,
KCC1 and **CC4**); the position key is in the MONITOR
(unoperated) position. The two monitor relays MA
and MC are connected over high-impedance leak
circuits through YR and YQ, one to each channel of
communication. According to the direction of transmission at any instant, either **MA1** or **MC1** will
repeat signals to the switchboard-position teleprinter,
the other relay remaining side-stable. If both stations
transmit simultaneously, a mutilated monitor copy
would result. During monitoring, the transmitter of
the switchboard teleprinter is disconnected (at KCA2,
KCC2, KCC4 and KCA4); consequently, no interference to the communication results should the
monitor teleprinter receive a WRU signal or should
the operator accidentally touch the keyboard.

Clearing. When the connection is to be terminated,
both stations send a +80-V clear signal by operating
their CLEAR key for 5 s. The clear relays CLA and
CLC are in leak on the tip circuits of the ANSWER
and CALL plugs respectively.

During normal teleprinter signalling, these relays
receive only the small, brief charging currents of the
$10\mu F$ capacitors respectively, and cannot operate.
While the call is in progress the normal negative
potential on the line charges QF (QC) negatively.
Owing to the high backward resistance of the diode
and the short duration of the positive elements, the
negative charge on the capacitors is substantially
unaltered. When a 5-s +80-V clear signal is received,
they receive a positive charge but, due to the

diodes, the current magnitude is again too small for the relay to operate. At the end of the 5-s clear signal, the line potential reverts to −80 V. The positive charge on the capacitor is now discharged in the forward direction of the diodes and this current, together with −80-V charging current *in the same direction*, is sufficient to operate relay CLA (or CLC). The relay holds due to **CLA1** (**CLC1**) inserting the second winding in series with the plug-sleeve circuit to the −80 V on the bush of the jack. **CLA2** (**CLC2**) lights the CLEAR lamp associated with the ANSWER (CALL) plug. The switchboard operator, seeing both of the CLEAR lamps lit, clears the connection by removing both plugs from the line jacks; the hold circuits for the CLA and CLC relays are disconnected and the relays release, extinguishing the CLEAR lamps.

Broadcast Transmission Connections. Any number of selected lines up to the maximum of 30 can be connected to receive simultaneously a message transmitted from the switchboard-operator's position or from any station requesting this facility. The station requiring to broadcast a message calls the switchboard in the normal manner and passes details of the required connections. The operator throws the BROADCAST key KBA (Fig 10.7) of each line which is to receive the broadcast message, provided that it is not already engaged. At **KBA1** the S wires of selected stations are connected in parallel to contact BA1 (or BB1 or BC1) of the broadcast relays; key **KBA2** intercepts the R wire circuit and together with **KBA3−4** rearranges the circuit of the line-calling relay C to convert it to a clearing or 'acknowledge' relay, by a circuit arrangement similar to that of CLA and CLC in the cord circuit; **KBA5** lights the ENGAGED lamps on the line jack and broadcast-key panels; **KBA6** changes over the circuit of C3 from the CALL lamp to the ACKNOWLEDGE lamp.

The station which is to transmit and has called the switchboard is already connected via an ANSWER plug to a cord circuit; the CALL plug is now inserted in the BROADCAST jack. This extends the transmitter and R wire of the calling station to the group of three polarized relays, BA, BB and BC in series;

the contact of each relay is commoned to a group of a maximum of ten S wires. Having connected all the selected receiving stations, the switchboard operator then informs the calling station. The switchboard operator restores the PRINT & MONITOR key. The calling station now has control of the broadcast relays and can proceed to transmit the broadcast message. The BROADCAST jack applies a −80-V holding condition to the S wire to the calling station, which will receive a local record from the station transmitter. All the connected stations receive the message repeated from **BA1**, **BB1** or **BC1**.

A protective resistor RBD is connected in the broadcast-transmission feed to each line which gives visual indication of any line suffering from an earth fault; without this protection other lines switched in parallel would also suffer effects of this fault.

The switchboard operator can monitor the broadcast message in the normal way − from the cord circuit but not from the BA1 (BB1, BC1) contacts in the line circuit.

A broadcast message can also be transmitted from the switchboard teleprinter by inserting a CALL plug into the BROADCAST jack and operating the KCC key and the appropriate KPM key.

On satisfactory receipt of the broadcast message each station presses the CLEAR key for 5 s. On restoration of this key, relay C in each line operates as described for the CLA and CLC relays. C1 enables the C relay to hold to **KBA4**; contact **C2** operates the pilot lamp or night-alarm bell; **C3** lights the broadcast ACKNOWLEDGE lamp. When the calling station sends the clear signal, and a broadcast-acknowledge signal has been received from each connected station, the operator restores the BROADCAST keys. After advising the calling station that all receiving stations have acknowledged receipt of the broadcast message the operator withdraws the plug from the line jack and broadcast jack, restoring all circuits to normal. Any station which fails to acknowledge the message can be communicated with separately.

The 40-Ω resistors, each capable of dissipating 160 W, are provided to limit current under fault conditions (e.g. bunching of the BA1 . . . contacts). The resistance value is a compromise between ensuring sufficient protection under fault conditions and avoiding excessive voltage drop under working conditions when ten lines are being fed in parallel.

10.3 Automatic switching for private teleprinter networks

According to the renter's requirements, for a large private teleprinter network either the message-relay system is available or, alternatively, automatic switching equipment provides full intercommunication by direct dialling between all stations connected to the network.

For the latter service, the private automatic switching centres form an integral yet distinct unit of the TAS centres of the inland telegraph service, with which they share only power supplies, signal generators and maintenance service. Because of the relatively small size of most private teleprinter networks, the number of switching centres on the network tends to be small and this leads to small groups of long, expensive station lines. The suspense facility of the final selector, with overflow traffic routed to a main controlling office, is essential to such systems to achieve a high circuit efficiency for a given grade of service. With this feature, when an office is closed, messages are routed to the overflow centre without delay. Other facilities required include a precedence service to give priority treatment to urgent messages under congestion conditions and a multi-address service for transmitting the same message to selected stations. On private networks the use of page teleprinters is universal; automatic transmitters may be required in conjunction with the teleprinter. Printing reperforators are associated with the overflow positions so that overflowed messages can be retransmitted by automatic transmitter to the required station with a minimum of time and effort. These overflow positions can also be reached by direct dialling; reception on a printing reperforator provides a useful facility for transferring messages either to the telex system or to any private circuits not associated with the switching network.

Station Equipment. This equipment is similar to that provided for a telegraph office of the inland telegraph service, except for the use of page teleprinters and a modified design of dial unit. An alarm for the J-bell signal is provided and a switch enables the local-record circuit to be cut off so that duplex transmission can take place over a dialled connection; alternatively, an additional teleprinter can be brought into use for duplex operation to give local-record printing also.

At busy offices, one or more 'off-line' teleprinters with perforating facilities are provided to prepare perforated tapes for automatic transmission. For multi-address traffic the required number of tapes is prepared in this way for consecutive automatic transmission to the appropriate stations. An automatic transmitter may be shared between two adjacent positions by key-switching.

The circuit for the dial unit is basically that shown in Fig 9.8; differences in design are shown in Fig 10.9. The local record from the teleprinter keyboard is provided by the 4-kΩ leak path, (via R11, the send–receive switch, SR, AT2 and KLC1) which connects the teleprinter electromagnet to the tongue of the teleprinter transmitter; in the rest position, SR connects the electromagnet to the R wire for reception. When an automatic transmitter is in use, relay AT is operated; **AT1** connects the automatic transmitter to the S wire in place of the teleprinter transmitter; **AT2** provides a local-record path via R6 and KLC1. The local record may be cut off from both transmitters by throwing the LOCAL RECORD CUT key, which at **KLC1** disconnects the electromagnet from the send–receive switch and connects it direct to the R wire. Duplex operation can take place if the keys KLC are thrown at the two connected stations.

Reception of the FIGURE-SHIFT and J signals closes the J contact, lights the J lamp, and operates the J relay; **J4** flashes the alarm lamp.

Automatic-transmitter Positions. After a connection has been established by dialling, the automatic transmitter may be started by operating the AUTOMATIC TRANSMITTER START key KATA (Fig 10.10). This is a double-throw key, non-locking when operated upwards and locking when operated down. If the key is operated in the non-locking position, the automatic transmitter starts and when it has reached governed speed it is connected to the S wire. Tape starts to feed and the AUTOMATIC TRANSMITTER START lamp glows; the key may now be released. When the end of the tape passes through the head of the automatic transmitter, the connection automatically clears.

If the key is operated in the locking position, the

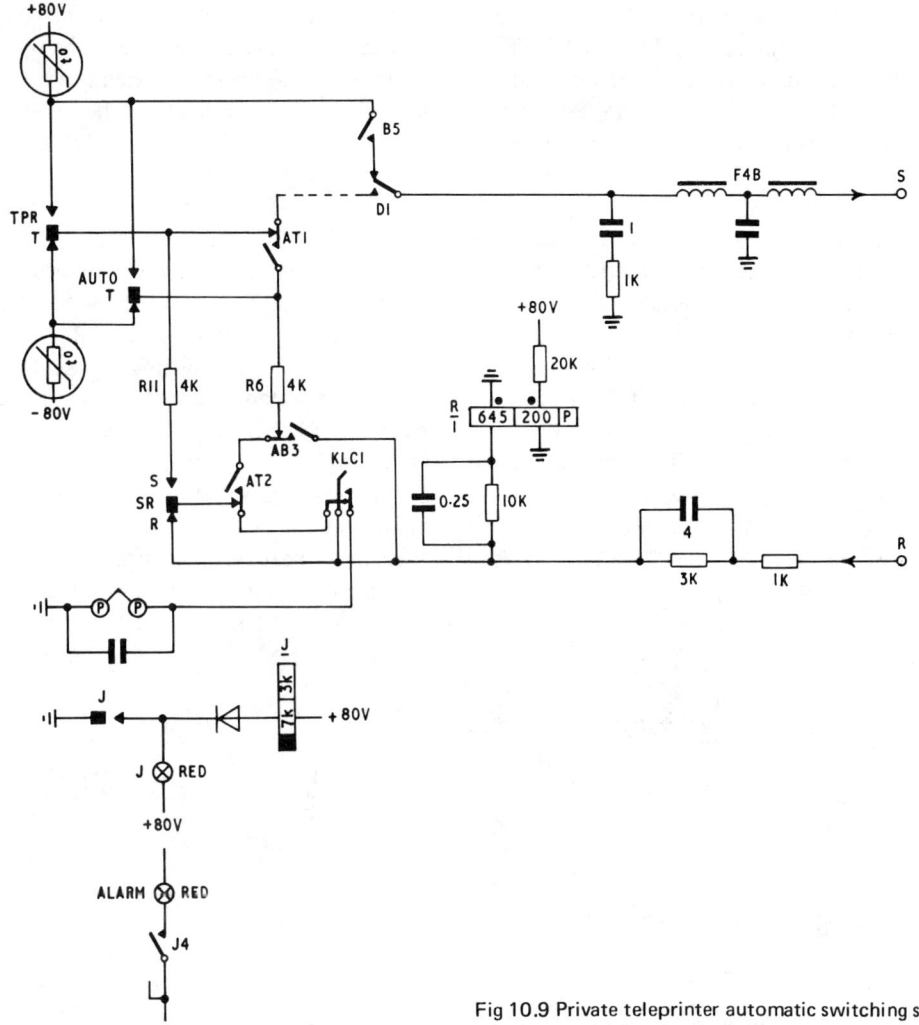

Fig 10.9 Private teleprinter automatic switching station — line circuit with local record

automatic transmitter starts as described, except that the key is not restored. The end of the tape passing through the transmitter head causes the transmitter to stop and the teleprinter is reconnected to the S wire: the ALARM lamp flashes and the audible alarm is operated to indicate that transmission is completed. The alarms are reset by restoring the AUTOMATIC TRANSMITTER START key and the connection can be cleared by using the CLEAR button.

During automatic transmission a local record is obtained on the teleprinter. A WRU signal on the tape would operate the answer-back units of the local and the distant machines; transmission of the local answer-back code is suppressed and a relay,

operated by the local answer-back off-normal contact, disengages the clutch from driving the automatic transmitter. While the automatic transmitter is stopped, the distant answer-back code is printed on the local machine. When the local answer-back unit is restored to normal, the relay operated by the answer-back off-normal contact is released and the clutch of the automatic transmitter is re-engaged.

Should a clear signal be received from the distant station while automatic transmission is in progress, the +80-V clear signal on the R wire of the local station releases the relay set to cause the return of +80 V on the S wire in the normal manner. In addition, because a perforated tape is in the transmitting head of the automatic transmitter, a relay is operated which lights the ALARM lamp and the CIRCUIT RELEASE lamp and extends out-of-service conditions to the switching centre. The alarms are reset and the position restored for service by operating the RESET key.

Referring to Fig 10.10, insertion of a tape into the automatic transmitter head operates the tape-out contact TO. If the non-locking key KATA is operated, **KATA**2 operates the automatic-transmitter motor-start relay MA. When the transmitter motor has reached governed speed, the centrifugal switch M closes and relay AT operates (via **TO**, KR3, CL2, AT3, **KATA**1 and **M**). Contacts **AT**1 and **AT**2 function as described for Fig 10.9; **AT**3 holds the AT relay to the **TO** contacts; **AT**7 provides a hold circuit for the motor relay MA (against the release of KATA2), lights the AUTO TRANSMITTER START lamp, and operates the clutch of the automatic transmitter (via the tight-tape/send contacts). The tape

commences to feed through the transmitter head and key KATA can be released. The signals from the perforated tape are transmitted over the S wire, a local record being printed on the page teleprinter. When transmission is complete the end of the tape passes through the transmitter head and the tape-out contacts TO restore. The CL relay now operates (via TO, KS4, KATB3 and **AT**6) to indicate the clear signal, automatically, but otherwise as described for Fig 9.8. The TO contact also disconnects the AT relay (which has a delayed release).

Alternatively, the locking key KATB is operated after a perforated tape is inserted in the automatic-transmitter head; relay MA operates, the motor starts, relay AT operates, the AUTO TRANSMITTER START lamp glows, the transmitter clutch is operated and the tape commences to feed through the transmitter head, giving a local record, all as described above (but via KATB contacts). When the end of the tape passes through the transmitter head,

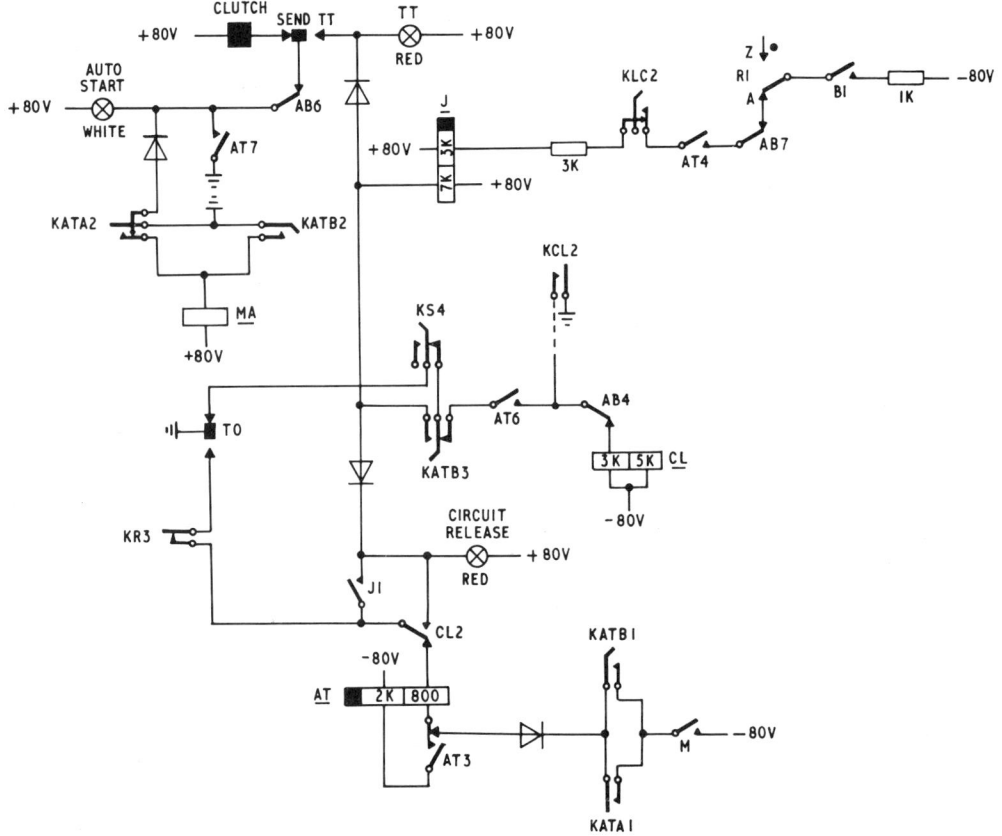

Fig 10.10 Private teleprinter automatic switching — control of automatic transmitter

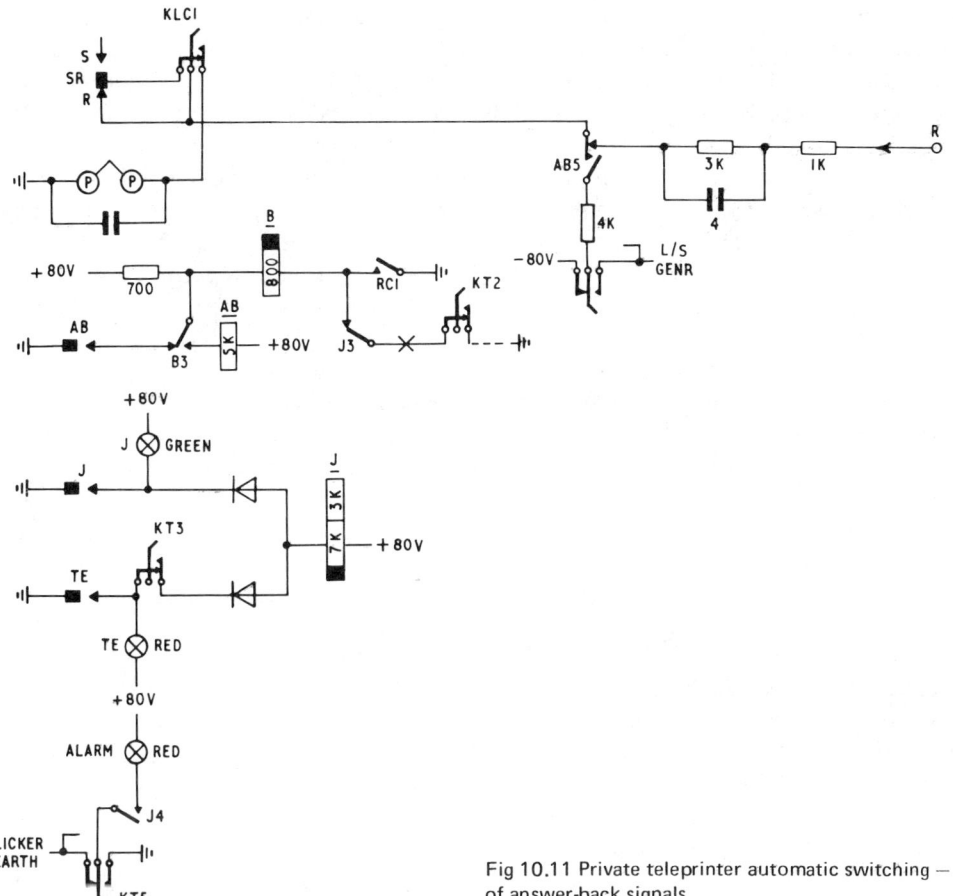

Fig 10.11 Private teleprinter automatic switching — control of answer-back signals

contact TO restores; in this case, TO releases relay AT but operates relay J (instead of relay CL); contact **J4** flashes the ALARM lamp. The operator can then restore key KATB to release the J and MA relays and dim the ALARM lamp. The operator may either send from a further perforated tape or clear the position in the normal manner, using the CLEAR button **KCL2** to operate the CL relay.

If a WRU signal is present in the perforated tape it will be transmitted by the automatic transmitter over the S wire to the distance station and also to the local teleprinter which is printing a local record. Both machines will recognize the signal and cause their answer-back off-normal contact to close, followed by the answer-back units rotating to transmit their respective answer-back codes. The distant station will transmit its answer-back code; at the local station the answer-back off-normal contact will operate relay AB (shown in Fig 10.11). **AB3** disconnects the

local-record circuit and connects the R wire to the teleprinter receiver; **AB6** disconnects the clutch of the automatic transmitter; **AB7** prevents relay J from operating when relay R responds to the answer-back signals being received. The distant answer-back code received on the R wire is printed by the local teleprinter; the local answer-back code is ineffective, being disconnected from the S wire by **AT1** and from the local teleprinter receiver by **AB3**. The automatic transmitter ceases transmitting until the answer-back code has been transmitted and the answer back off-normal contact opens. When the AB contact opens, relay AB releases, reconnecting the local-record circuit at AB3 and operating the clutch of the automatic transmitter at AB6; the automatic transmitter then resumes transmission.

Should the distant-station operator wish to interrupt an automatic transmission (the local receiver being connected from the R wire for a local record, without a send–receive switch), this can be done by sending signals from any key of the keyboard. Contact **R1** responds to these signals and operates the J relay; **J4** flashes the position ALARM lamp.

Should the distant station clear the connection while automatic transmission is actually in progress, the connection is broken down and the teleprinter positions (at both stations) are restored in the normal manner. At the local station, however, the operation of **CL2**, due to the clear signal, releases the AT relay and operates relay J (via TO); **AT7** releases the clutch of the automatic transmitter and also releases the motor-start relay MA. Contact **J1** (via, KR3 and **TO**) lights the circuit RELEASE lamp and locks the J relay; **J4** flashes the ALARM lamp; **J3** disconnects the B relay (see Fig 9.8) to apply the out-of-service condition to the S wire at B5. The alarm can be cleared and the station restored to service by operating the RESET key KR, to release relay J at KR3 and re-operate relay B at J3.

If the perforated tape should become tight during a transmission (due for example, to the occurrence of a pause when the tape is being fed from reperforator at the same time) the tight-tape contacts TT operate, disconnecting the clutch to stop the tape feed and lighting the TIGHT TAPE lamp (via AT7 and AB6); relay J is also operated, flashing the ALARM lamp at **J4**. When the tape again becomes slack, the TT contacts restore, cancelling the foregoing conditions.

Printing-reperforator Position (Fig 10.11). The motor is started up in the normal way when the DIAL key is pressed for an outgoing call or on receipt of an incoming call. The arrival of an incoming call is followed by receipt of the WRU signal from the final selector. The answer-back unit operates and the answer-back off-normal contacts AB close, operating the AB relay via **B3**. Contact **AB5** disconnects the electromagnet of the printing reperforator from the R wire and applies −80 V to ensure that the answer-back code will not be perforated in the tape over the local-record circuit (SR does not operate with the answer-back unit). Relay AB restores at the end of the answer-back code transmission.

If the LOCAL RECORD CUT key KLC is

operated **KLC1** disconnects the electromagnet from the send–receive switch and connects it instead to the R wire. On receipt of the WRU signal on an incoming call, relay AB operates and **AB5** via **KLC3** connects the electromagnet to the letter-shift generator so that LETTER-SHIFT characters are punched in the tape while the printing reperforator sends out the answer-back code. This occurs on operation of the answer-back unit at the beginning and end of a message and provides a useful message-separation signal which is indicated by the Maltese cross printed eight spaces later than the corresponding punched character.

Tape-exhaustion contacts TE are provided so that as the supply of unperforated tape in the roll nears exhaustion the TE contacts close to light the TAPE EXHAUSTED lamp and operate the J relay; **J4** flashes the ALARM lamp; **J3** releases relay B. Contact B5 applies the out-of-service conditions, clearing the connection. Operation of the TAPE EXHAUST ALARM CUT OFF key (KT) releases the J relay to stop the alarm. When the paper supply is renewed, the TE contacts are restored; restoration of the KT key allows the B relay to re-operate and the position is re-opened for service.

Exchange Equipment. The station-line circuit, group selector and final selector are those of the TAS system; in addition, each final-selector multiple is

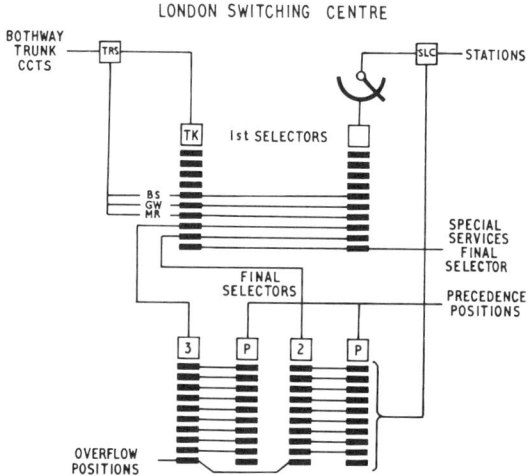

Fig 10.12 Private teleprinter automatic switching – trunking diagram

equipped with one precedence final selector for disposal of priority traffic under congestion conditions; access to the precedence final selector is restricted to overflow positions. The period for which a precedence final selector can wait to be switched to a free line is unrestricted, but should the wait contact be already engaged by a waiting call, this regular final selector is automatically released (to the overflow level) by the precedence final selector; the wait contact is then occupied by the precedence final selector.

Certain offices require 'order' traffic to be routed to positions where the teleprinter is equipped with pre-printed order-form paper. Lines serving such offices are divided over two separate final-selector groups, one for message traffic, the other for order traffic.

A typical trunking diagram for one of the switching centres is shown in Fig 10.12. Such a network caters for over 100 station lines serving 40–50 offices. Bothway-trunk circuits are provided to an agreed grade of service, normally 0·02.

10.4 Private automatic telegraph exchange (P.A.T.X.)

Each station is equipped with teleprinter No. 15 and dialling unit.

Standard facilities permit dialling between stations via the P.A.T.X. to which they are connected or via a maximum of three other P.A.T.X.s which may form an integrated private network fully interconnected by trunk circuits. Answer-back signals are received automatically after dialling. Printed-service signals are provided — OCC for engaged conditions or unallocated trunk codes, ABS for a station line shut down or unallocated. Either party to a connection may clear it down.

At the exchange, up to five groups, each of two or more consecutively-numbered auxiliary lines, can be accommodated on a final-selector level, provided that the aggregate number of these lines per level does not exceed 10.

Optional facilities enable messages to be broadcast among any stations connected to a P.A.T.X., or to any stations connected to a second P.A.T.X. which forms part of the same private network. A broadcast

is initiated by dialling the code 01. For convenience of setting up a broadcast network, stations on the P.A.T.X. may be arranged in four groups of different patterns, any group being selected by dialling appropriate digits. For example, a broadcast network is set up by dialling the codes of the stations wanted; or by dialling code 55 if a broadcast to all stations on the same P.A.T.X. is required; or dialling code 51, 52, 53 or 54 to select one of the patterns of stations in the pre-selected groups. The station initiating the broadcast dials a final digit 0 (i.e. a total of five digits — 01–51 (etc.)–0); this final digit causes the selected stations to be connected, if disengaged, to the output of the broadcast unit. Selected stations return their answer-back code within 10 s of transmission of the WRU signal from the broadcast unit; an engaged station is indicated by return of the OCC signal. If a distant P.A.T.X. is to be included in the broadcast its routing code is dialled prior to the five digits quoted above. The broadcast network is cleared down by operation of the CLEAR key at the initiating station.

10.5 Teleprinter broadcast system

An important application of teleprinter operation is the service which provides a simultaneous, or broadcast, transmission of messages from a central office to a number of out stations; such a system requires the provision of direct permanently-available circuits so that immediate response of the out-station teleprinters can be gained without delays due to switching or the risk of finding one or more of the stations engaged on other traffic when required to receive a broadcast message. There is no practical limit to the length of correctly-planned transmission circuits for this purpose, and transmission can take place by d.c. or VFT modes.

With modulation at 50 bauds a broadcast telegraph-relay contact may have to perform one million operations into an inductive line and filter in a period of a day or so; despite the use of spark quenches and radio interference-suppression devices, the wear on these contacts and the consequent maintenance attention which is needed becomes significant. For this reason mercury-wetted relays are used for broadcast systems.

Fig 10.13 shows a broadcast unit suitable for simultaneous teleprinter transmission over a group of 3–5 lines. Incoming signals are applied at terminal R1, tag R2 being normally strapped to tag E to complete the circuit. At BR1, signals are repeated over a group of 2–5 lines via protective resistors each serving two line circuits.

The line capacity may be extended by connecting a number of BR relays in series under control of the broadcasting teleprinter. Tags R and R2 are then used to effect the series connections, using R1 and E on the final 5-line unit to include the 10-K current-limiting resistor and the earth potential.

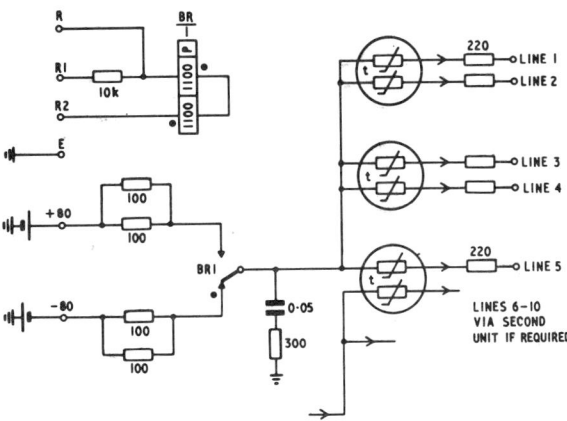

Fig 10.13 Teleprinter broadcast unit

The limitation of five lines to each BR1 relay contact is imposed by surge currents into the filters No. 4B and the line; these may reach 200 mA. The mercury-wetted contacts are limited to a load of 100 VA.

For a more extended broadcast network any line at an *out station* can be connected to one or more similar broadcast units to repeat the signals to a further five or more lines in cascade. Simplex communication between the central-office teleprinters and any selected out station may be provided by a switchboard using keys, or plugs and jacks as already described (Fig 10.7). Lamps for calling and broadcast-acknowledgement signals from the out stations are provided on the switchboard.

It is not usual to provide intercommunication between out stations on a system provided primarily for broadcasting since such connections might have to be frequently interrupted for broadcast messages.

10.6 Teleprinter-conference system

A teleprinter-conference system enables a small number of permanently-interconnected teleprinter stations at remote points to have 2-way communication between all stations. One station only may transmit at a time: the signals are received simultaneously by all stations, including the sending station.

The system depends upon a polarized relay with four equal windings; three of these carry the line currents, the fourth acting as a space-bias winding. To cater for five lines, two relays RA and RB are used (Fig 10.14), the sixth winding being given a mark bias. The current in each receive line is adjusted by RV1–5 to be equal to half the value of current in the bias winding. In this way the magnetic effect of the bias winding is neutralized by the joint effect of any two of the line windings so that the line current in a third line winding – positive or negative – has control of the relay operation.

Contacts RA1 and RB1 are connected in cascade to control relay BR – a mercury-wetted relay. The send line into each teleprinter circuit is connected via a resistor bulb to the BR1 contact so that all stations receive the signals. The transmitting station also receives its own signals back via the relays to provide a local record of the signals sent. The usual send–receive switch method of providing a local record at an out station must be avoided, otherwise mutilation of the local record would occur from two sources due to the operations of the send–receive switch.

A fault on any line would throw the relay out of electromagnetic balance; incoming wires are connected through break jacks so that a faulty line can be temporarily plugged up and a negative potential applied to the relative winding of the relay RA or RB. If less than the full number of lines are connected, spare windings must be held to negative potential, using the break jacks, to preserve the balance. The 50-Ω resistors (carrying the surge currents for all lines) in the ±80-V feeds to mercury-wetted relay contacts limit the value of current which can arise under fault conditions.

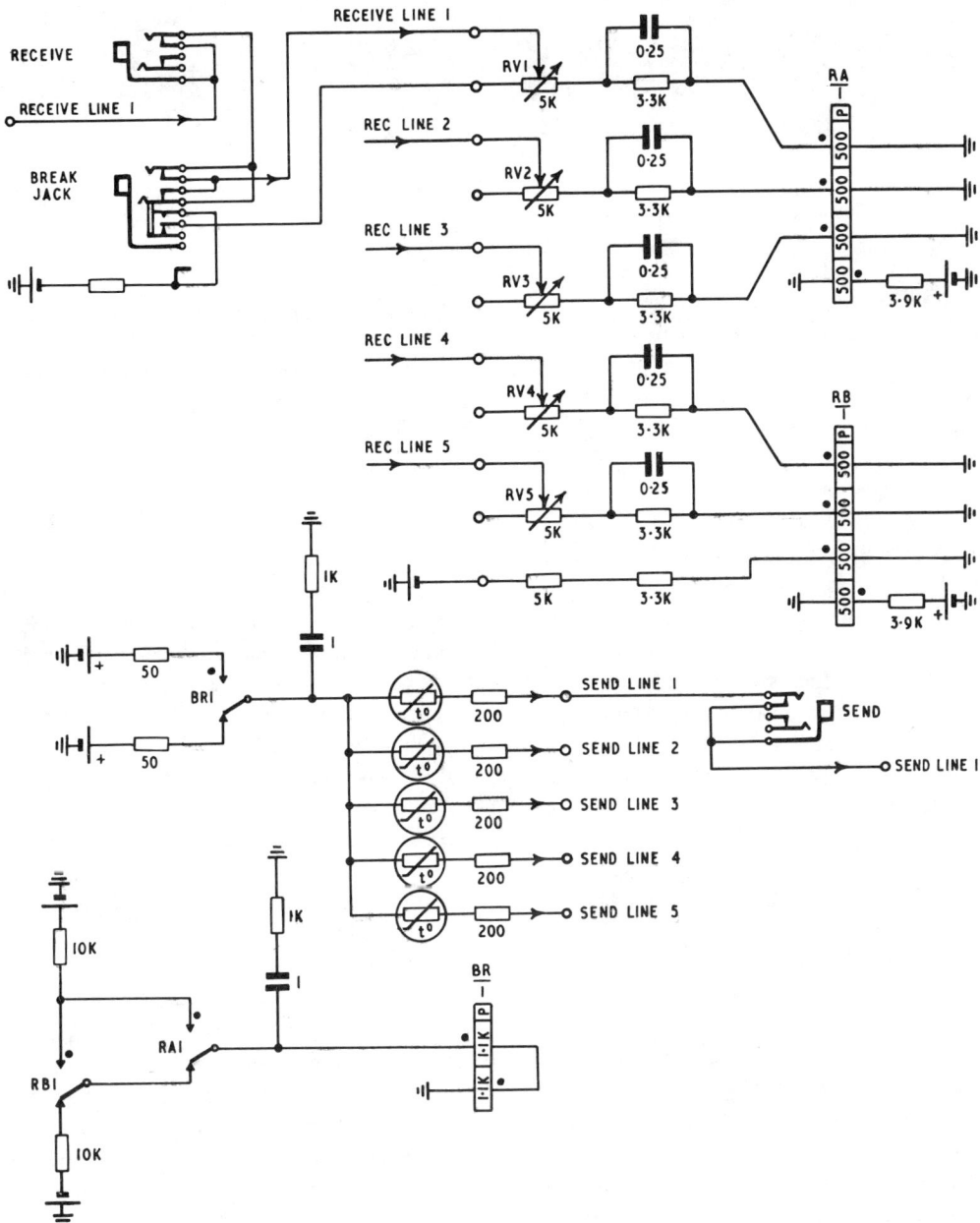

Fig 10.14 Teleprinter-conference circuit — 5 lines

References

1 Walker, H., Jeffery, D. A. and Pollock, D. R., 'Teleprinter Private-Wire Automatic Switching', *POEE Journal*, **50**, p. 245 (1958).

2 Sheeky, B. and Turbin. C. A. R., 'Private Automatic Telegraph Exchanges – P.A.T.X. No. 1A', ibid. **64**, p. 250. (1972).

11 Message-Relay Systems

11.1 Characteristics

In network systems, circuits are switched as links in a chain to set up 2-way communication paths between any two stations; calling and called parties are simultaneously on the circuit and a teleprinter 'conversation' or an exchange of data and acknowledgements can take place.

In network systems described in this chapter, at no time is there any direct connection between originator and addressee of a message. A message from the calling station is retransmitted or relayed at one or more points, link by link, with a brief delay at each intermediate station until the objective station is reached. The message-relay system, sometimes called 'message switching', is the older, traditional method of telegraphy in which telegrams were, until recent times, retransmitted at a number of stations in the process of reaching the objective station. New techniques have transformed older methods with a view to saving costs and delays and improving efficiency.

The advantage of circuit switching is firstly that the two parties to a communication are put into contact and confidence is gained in the receipt of messages exchanged between them; queries can be resolved at the time. This service can only be achieved at the cost of liberal provision of circuits between switching centres and to some extent it fails if the desired station line is frequently found to be engaged when required by a caller, calls having to be repeated and occupying trunk circuits in the process. The return channel is used ineffectively, being occupied only by the return of supervisory signals, answer-back codes etc. though duplex working could be provided.

In message-relay systems the originator entrusts his message to an administration, an agency or a carrier* and has no specific acknowledgement of

* Carrier — a private agency which undertakes to forward communications to their destination.

arrival of his message at its destination; responsibility for safe delivery rests with the administration or carrier. The main advantage of this system is that so long as traffic exists, circuits can be fully loaded and used at their utmost efficiency; since machine transmission is always used (with the possible exception of the initial link from the originating station which may be sending from a keyboard) a circuit is always operated at its maximum traffic rate. Since information is passing only in a forward direction the two channels forming a circuit are used in the duplex mode independently and the circuit is effectively loaded. The advantage of this efficiency is particularly marked on long and expensive circuits; it is also important to relatively small but long-distance private networks where the provision of groups of circuits for circuit switching is not economically justified and it is important to use circuits at their highest traffic loading. Other advantages given by the message-relay system are that messages of high priority can be more readily accorded preferential treatment to evade circuit-congestion delays: and that multiple-address traffic can be more easily handled by this means. An incidental advantage may be the easing of electrical transmission problems since signals are freshly generated at every relay centre.

Message-relay systems may be divided into three classes according to the method of handling messages: (1) manual systems in which at any relay centre the arriving messages are handled by operators who direct them on the next stage of their journey; (2) semi-automatic methods (e.g. 'push-button') in which message routing is effected by mechanized means under the control of operators; (3) automatic systems in which routing information contained in the message heading is read by equipment, usually electronic, and the message is directed without human aid.

All these methods necessitate temporary storage

of at least a part of the message. The traditional medium for receiving messages for temporary storage and subsequent retransmission has been perforated paper tape. This has given rise to an alternative name — 'tape-relay' — for these systems. The first, and possibly the second class of system, are sometimes referred to as 'torn-tape' systems since it is characteristic of the manual system that the message must be fully received at a relay centre before it can be processed; the completed message, represented by a length of perforated tape, is torn off at a receiving position and manually transported to a retransmitting position. In the alternative 'continuous-tape' method, perforated tape issuing from a receiving machine is fed into an adjacent transmitting machine, the tape running continuously from one to the other without being cut or torn. In automatic systems the storage medium is magnetic, by tape, core, disc or drum. From the nature of processing, message-relay systems are also known as 'store and forward systems'.

Among requirements to be met, a particularly useful one is the disposal of the multi-address message which has to be directed to a number of selected stations. In private networks a high proportion of traffic consists of multi-address messages used as a means of disseminating information quickly throughout an organization. Such a facility is more readily provided by a message-relay system than by a circuit-switching system.

The successful operation of a message-relay system, particularly those which include international circuits, depends upon rigid operational procedure and two essential features are: (1) serial numbering every message at every stage of its journey, to guard against lost messages; and as a means of identification to enable any mutilated message to be retrieved and retransmitted; (2) the use of an agreed message format to enable the address and any special forwarding instructions to be immediately and unequivocally recognized.

11.2 Message routing

A message-relay system may have a single relay centre but more usually comprises several interconnected relay stations. The form which a network takes depends upon a study of traffic and

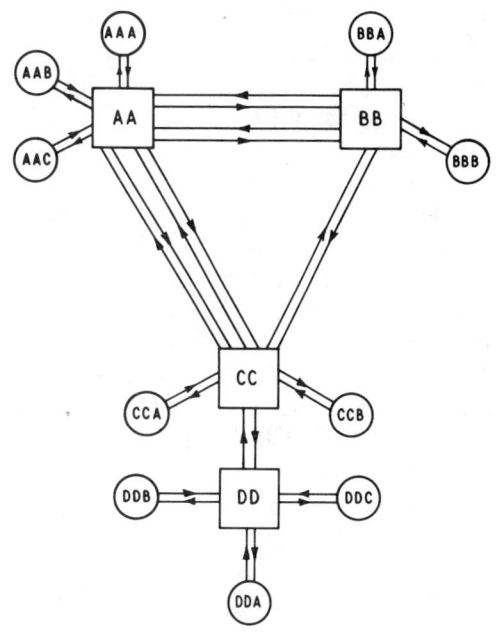

Fig 11.1 Message-relay routing

economics. The relay centres may be national or situated in different countries or continents.

A typical network based upon four message-relay centres is illustrated in Fig 11.1. Here relay centres have been given two-letter codes AA to DD and terminal stations are designated by adding letters A, B, C . . . to relay-centre codes. A message from station AAB to station DDA is sent firstly from the originating station AAB into the home or 'entry' relay centre AA; here the message is redirected to an intermediate station CC from whence it is again retransmitted into the 'exit' relay centre DD which repeats the message into the objective station DDA. This necessitates the message being retransmitted three times.

If the route between centres AA and CC had been congested or faulty the message from AA would have been sent over an alternative route via centre BB, from where it would have been routed into centre CC for onward retransmission to the exit relay centre DD. Alternative routing can always be provided by a system of 'triangulation' of routes such as the connections between centres AA, BB and CC. In the network shown there is no alternative route between centres CC and DD. It is presumed that centre DD is of lower importance or carries less traffic; in the event of emergencies, messages between CC and DD would depend upon other means such as availability of reserve circuits or perhaps temporary use of telex service.

The example given refers to a station-to-station or single-address message. Should station AAB wish the same message to be repeated to station AAC, BBB and DDA, for example, the message would bear the multiple address 'AAC BBB DDA'. The following action would then take place. Station AAB makes a single transmission to relay centre AA. On receiving the message, centre AA makes three copies of it: one is sent to AAC, one to centre BB for onward transmission to station BBB and one to centre CC for onward transmission to centre DD and thence to station DDA. Relay centres are equipped with facilities for producing multiple copies of a message. When manually operated, the copying process entails some further delay to messages; in automatic relay centres this processing can be carried out at very high speed.

At a relay centre, message routing is determined by the address code or 'routing indicator'. In the case quoted the presence of the address DDA at the heading of the message would leave no doubt at intermediate stations of the required routing. The foregoing principles would apply whatever the handling system adopted — manual or automatic.

In a small manually-operated network the relay-centre codes may be abbreviations of place names. In extensive automated systems a 'logical' routing code must be used which signifies the territorial division into routing regions and areas.

If the network includes triangulation and alternative routes a system of responsibilities must be specified for sending on multiple-address traffic, otherwise either duplicated or omitted messages could result. In one method 'diversion indicator' is inserted in the message heading; a receiving relay station may have to delete the diversion indicator and the redundant routing indicators (i.e. those already acted upon) in multi-address messages before retransmitting the message.

In telegraph service, a lost message from whatever cause is not permissible. In a message-relay system, security is provided by a rigid system of serial numbering applied to all messages at all stages of their journey; in the case quoted, station AAB

would apply a serial number to each outward message (a new series being started daily at midnight). At the relay centre, a new serial number would be applied unique to the route AA to CC, with a separate numbering series on the route AA to BB. At each centre, a log or similar record would be kept relating the serial number of every incoming message to its outgoing serial number. In this way any message can be traced through the system and the fact of a missing one would be quickly revealed by a careful check of sequential serial numbers on the incoming-message record at every station.

11.3 Message format

In addition to routing-indicator codes and serial numbers, other information which must be included is an indication of priority and the address or code of the sending station; it is also important to know precisely at what point a message is concluded. For the effective operation of the system a number of other general instructions have to be inserted in the message; the whole of these instructions — apart from certain specific exceptions — are retransmitted with the message from centre to centre.

Every message is made up from two main parts, (1) the text and (2) the heading and the ending, sometimes referred to jointly as the 'envelope' for the text. To avoid lost or mutilated messages, reduce delay and promote efficient operation it is essential for the originating station to include routing and other instructions in correct sequence to the agreed format. If relay centres are spread over a number of countries this discipline is particularly important. Again, if the system is automatic, correct routing depends entirely upon information contained in the heading, which then becomes vital.

Different message formats have been introduced by various organizations although an internationally-agreed one has been published by the CCITT as Recommendation F31. The essence of this agreed format, primarily designed for automatic message-relay systems, is illustrated in Table 11.1. Such a unique format is invaluable for compatibility between different designs of automatic systems and also between private networks which may have a common meeting point, e.g. at an airport installation.

Table 11.1 CCITT message format (F31)

Item	Start of message (SOM) signal (Note 1)	Channel sequence numbers (Note 3)				Telegram identification group (Note 4)	End of line
		Channel indictor	Sequence number	Channel indicator	Sequence number		
Number line (Note 2)	ZCZC →	GEB ↑	099	→↑ AFA ↑	135	→↑	LX↑74	<≡
	Destination indicator — Country / Office	Priority and tariff (Note 5)	Origin indicator — Country / Office	Number of chargeable words	End of line		
Pilot line (Note 2)	→ GB LO →	HL →	UR WA →↑	013	<≡		
Preamble line	→ WASHINGTON→ ↑ 13/12→13→1205 <≡≡≡						
Paid service indications	→ LT<≡						
Address line (Note 6)	→↑MIDBANK→LONDON↑→<≡≡≡						
Text	→FORWARD→SDONEST→PRESENT→ACCOUNT<≡ ↓BALANCE→JCNES→NUMBER→↑78↓A↑765<≡						
Signature	→↑→↑→↑JOHNSON<≡						
Collation	↓COL→LT→↑78↓A↑765<≡≡						
End of message (EOM) signal (Note 1)	↓NNNN						

Key to symbols representing functional combinations:

→ = Space
↑ = Figure Shift
↓ = Letter Shift
< = Carriage Return
≡ = Line Feed

Notes

1. The SOM and EOM signals are the same for every message; the SOM signal may be preceded by the page-alignment function < ≡ ↓ and the EOM signal may be followed by a message-separation signal of 10 × ↓.
2. Items in the number and pilot lines are identified because they constitute the message heading presented to the automatic-routing equipment.
3. Channel sequence numbers are included for every channel over which the message has already been transmitted.
4. The telegram identification group may contain up to 12 printed characters plus shift combinations.
5. The priority indicator may be A, B, C or H; the tariff indicator may be N, O, P, L, M, U, D, I or Q.
6. The hyphen signs in the address line are provisional and may be omitted.
7. If several channels in tandem are concerned, the new channel sequence number is added first, followed by any previous ones.
8. This group is optional and is inserted by the 'linked entry office' (i.e. the home relay centre). The group comprises 1 SP FIGS (if necessary) and not more than 12 printable characters (letters or figures, including shifts, but no spaces).

A hypothetical printed message to this format might appear as follows:

EXAMPLE OF MESSAGE LAYOUT TO RECOM-
MENDATION F31

ZCZC GEBO99 WY79 (not more than 12 printed characters)
GBLD HL URWA 013 57825 (customer-identification group not sent beyond first retransmission centre)
WASHINGTON 13/12 13 1205

(Three line-feeds)

LT
—MIDBANK LONDON—

(Three line-feeds)

FORWARD SOONEST PRESENT ACCOUNT
BALANCE JONES NUMBER 78A765
JOHNSON

(Three line-feeds)

COL LT 78A765

(Ten line-feeds)

NNNN (plus ten letters-shifts)

A LETTERS-SHIFT at the commencement, followed by a delay of about 2 s, ensures that a dormant telegraph machine will start up and reach operating speed before the first character is received. The alignment functions (CARRIAGE-RETURN and LINE-FEED) are required to present a suitable format on a teleprinter page; they also have a particular significance in automatic message-relay systems for identification of various stages in the format at which certain actions must commence or cease. The group of line feeds at the end is used to separate one message from another on a teleprinter page; the final group of letters-shifts separate one message from another on a paper tape and so produce a place at which the message tape can be torn and separated.

Alphabet No. 5 makes provision for special codes such as SOH (START OF HEADING) for message-relay instructions; use of these codes reduces transmission time and storage requirements.

11.4 Manual system

From the equipment point of view (see Fig 11.2) this is the simplest form of message-relay system. Each incoming channel terminates upon a printing reperforator. Each outgoing channel is fed from an automatic transmitter with which is associated a serial-numbering transmitter; a monitoring teleprinter, located in an adjacent room, is connected in parallel to take a file copy of the message. The automatic transmitter is preferably of the multi-gang type, incorporating two or three heads. With this provision, while transmission is in progress from head No. 1 further tapes can be inserted in

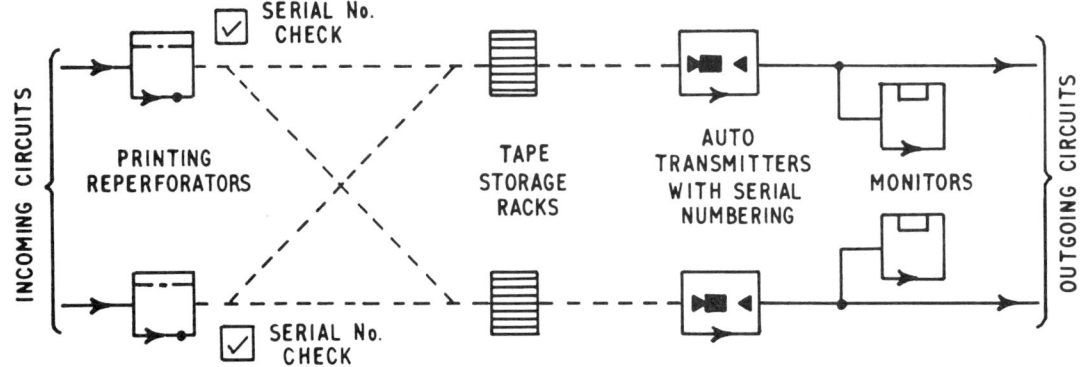

Fig 11.2 Manual tape-relay system

heads Nos 2 and 3; as soon as transmission from head No. 1 is completed, head No. 2 automatically takes over the outgoing channel followed in due course by No. 3, and again by No. 1. In this way, no channel time is lost between transmission of successive message tapes.

Operation at the relay centre is as follows. On receipt of a message the receiving operator checks that the sequence number is correct for that circuit and records this on a log together with the address for which the message is destined; the operator also checks the message generally for apparent freedom from mutilation and looks for the 'end of message' indicator. The message tape is then torn at this point from the machine and placed in a rack or clip to await attention.

The tape is collected by a routing operator who reads the destination address on it and carries it to the appropriate transmitting position. Here message tapes awaiting transmission are put into racks or clips in order of arrival for that route.

A message which bears more than one address is taken from the printing reperforator to a tape-copying section, which is equipped with a teleprinter, an automatic transmitter, six printing reperforators and a bank of keys to enable any of these machines to be switched into a local circuit. The message tape to be copied is put into the automatic-transmitter head, and the appropriate number of printing reperforators are switched into circuit. The automatic transmitter is then run ('off-line') and the required copies are produced on the printing reperforators. If any additional instructions need to be interpolated they can be inserted by using the teleprinter keyboard in place of the automatic transmitter. The copied tapes are then taken to the appropriate transmit positions as though they were single-address messages. The message details are recorded on a log at the copying position.

The transmitting operator takes up the waiting message tapes in sequence and, after checking that address and route correspond, inserts them in the automatic-transmitter heads. Before transmission of a message, a new serial number is transmitted from a serial-numbering transmitter. After transmission of the message, the operator enters the sent and received serial numbers in a log. At frequent intervals incoming and outgoing logs and tape-duplication logs are inspected to guard against delays or lost messages.

Since an operator's attention is not required during transmission or reception of a tape, transmitting and receiving machines are conveniently grouped in consoles so that one operator can control tapes of several machines. Fig 11.3 shows consoles each housing three printing reperforators and also 3-gang automatic transmitters arranged in two levels. The suite in the centre background is of teleprinters and single-headed automatic transmitters for acceptance and despatch of messages by telephone and pneumatic tube.

Serial-numbering transmitters are housed separately. Earlier installations use multi-gang automatic transmitters (No. 4A), each head supplied with a pre-

Fig 11.3 Printing reperforators and automatic tape-transmitter positions (*Courtesy of The Post Office*)

cut tape giving sequential numbers 001 to 999 followed by a LETTERS-SHIFT signal as an indicator for switching the line over to the message head; the tapes are reset to 001 daily. Electronic serial-numbering transmitters have been recently introduced.

A jackfield in the monitor room provides facilities for redistributing traffic within the centre in exceptional circumstances, or for diverting lines from faulty machines.

Each tape-transmit position serves three channels and is wired for two 3-gang automatic transmitters. With two-tier mounting, each head of the upper machine operates sequentially with that immediately below it to give double-headed transmission; for lightly-loaded channels a single-headed transmitter would be used. In either case message transmitters operate sequentially with remotely-housed automatic serial-numbering transmitters. Whenever a message tape is loaded into a transmitter head the associated numbering transmitter starts up, sends a serial number to line and then switches the line to the loaded message head which proceeds to transmit. A second tape in a double-headed message transmitter would set off a similar sequence of operations. An electromagnetic digital counter on the transmit position operates in sequence with the numbering transmitter to advise the operator of the serial number for the 'tick'-sheet check log.

The lower part of each console houses power supply, fuses and miscellaneous apparatus. Rear access is provided for machine maintenance and replenishment of paper-tape supplies.

The relay centre illustrated contains 45 printing reperforators, 20 automatic message transmitters, 70 teleprinters and 12 automatic numbering transmitters. Printing reperforators, in sound-reducing and dust-proof consoles, occupy suites on two opposite sides of a square, suites of automatic transmitters forming the other two sides. This arrangement gives a minimum of staff movement between incoming and outgoing positions. Message queries are dealt with by staff of the monitor room.

Alarm lamps with reset keys on the face of printing-reperforator consoles indicate line failures, the failure of a motor to start up, receipt of a 'J-bell' signal and warning of near exhaustion of paper-tape supply; a key provides tape run-out if the end of a message tape has not been ejected from the cabinet.

Lightly-loaded station lines are worked simplex. As the send and receive terminations at the relay centre are remote from one another, out stations are equipped with supervisory-signal units; when the out station sends a CALL signal to the relay centre this locks the associated transmitter at the relay centre and lights an ENGAGED lamp at that position until a CLEAR signal is later received.

A number of lightly-loaded station lines may be connected to a line concentrator at the relay centre. Typically, 25 simplex circuits could be concentrated upon 6 printing reperforators, 6 automatic transmitters and 6 monitor teleprinters – a considerable saving in machines and accommodation; each circuit would retain its individual numbering transmitter. Incoming calls automatically seize a free printing reperforator; for outgoing messages a push-button panel enables any automatic transmitter (with its associated monitoring teleprinter) to be connected to any marked channel; supervisory-lamp indication is provided. This feature is, in fact, an application of the push-button semi-automatic system.

Fig 11.4 shows the general arrangement of the concentrator. A load-distributor uniselector LD allocates printing reperforators to accept calls in cyclic order via line-finder uniselector LF. For message identification a time-injection unit prints the time of day immediately after the serial number. Routine-testing features are built into the system.

A manual message-relay system requires a number of operators and involves high labour cost. In a medium-sized manual office, staff movement may be considerable; in a large office, conveyor bands would be provided to despatch message tapes from receiving positions to forwarding and duplicating suites. Successful operation of the system depends on careful supervision to ensure rapid and secure message forwarding. The large number of machines entails maintenance costs, but the amount of ancillary equipment needed is very slight.

11.5 Semi-automatic system

A semi-automatic system reduces the amount of cross-office movement of staff and message tapes, resulting in staff savings; one operator can route several hundred messages an hour. The types of machine used are similar to those in the manual system.

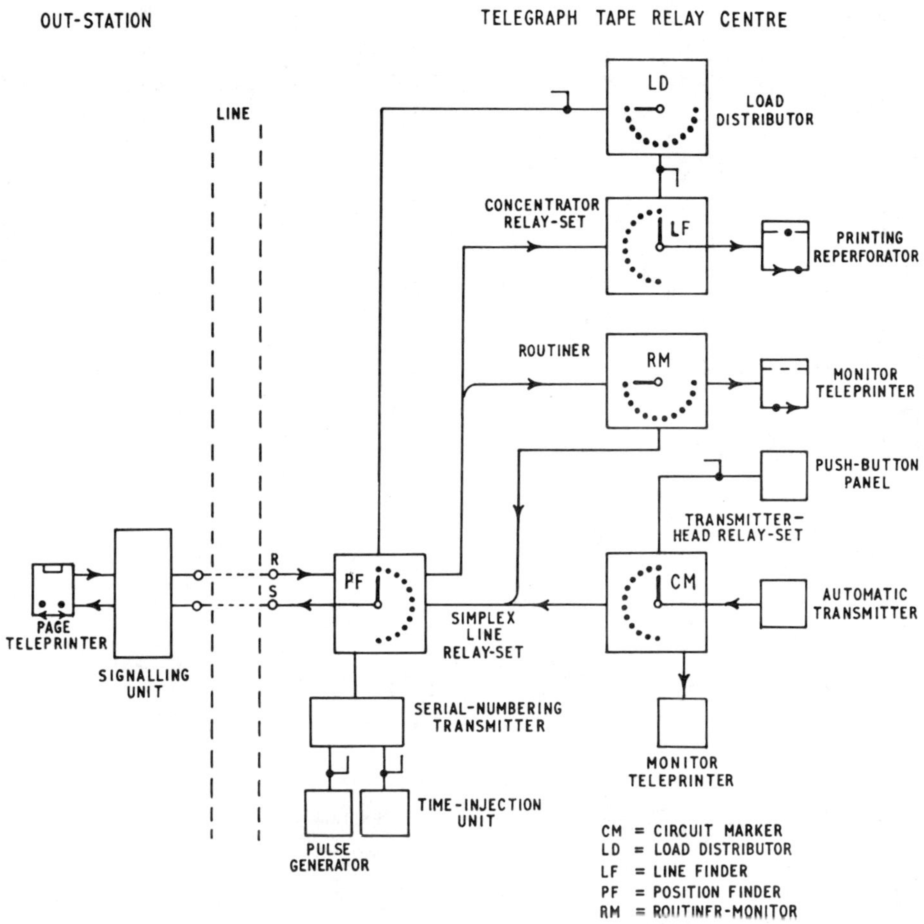

OUT-STATION

TELEGRAPH TAPE RELAY CENTRE

LD — LOAD DISTRIBUTOR

CONCENTRATOR RELAY-SET

LF

PRINTING REPERFORATOR

ROUTINER

RM

MONITOR TELEPRINTER

PUSH-BUTTON PANEL

TRANSMITTER-HEAD RELAY-SET

PF

CM

AUTOMATIC TRANSMITTER

PAGE TELEPRINTER

SIGNALLING UNIT

LINE

R

S

SIMPLEX LINE RELAY-SET

SERIAL-NUMBERING TRANSMITTER

TIME-INJECTION UNIT

PULSE GENERATOR

MONITOR TELEPRINTER

CM = CIRCUIT MARKER
LD = LOAD DISTRIBUTOR
LF = LINE FINDER
PF = POSITION FINDER
RM = ROUTINER-MONITOR

Fig 11.4 Concentrator for lightly-loaded lines

Several printing reperforators in consoles are closely stacked forming a suite, together with a number of automatic transmitters and a push-button panel situated at the centre of the combined suite so that all are within easy reach of the operator (Fig 11.5). Printing reperforators terminate incoming channels as in the manual system, but automatic transmitters are associated with automatic switching equipment controlled from push-buttons. The operator reads the destination on the incoming-message tape and tears off the completed tape from the printing reperforator. In some systems the message heading may be displayed on a cathode-ray screen. The tape is inserted into any disengaged automatic transmitter and the push-button associated with the required outgoing channel is pressed. The operator may then deal with another incoming-message tape. Meanwhile,

Fig 11.5 Semi-automatic system — operator's position
(*Courtesy of The Post Office*)

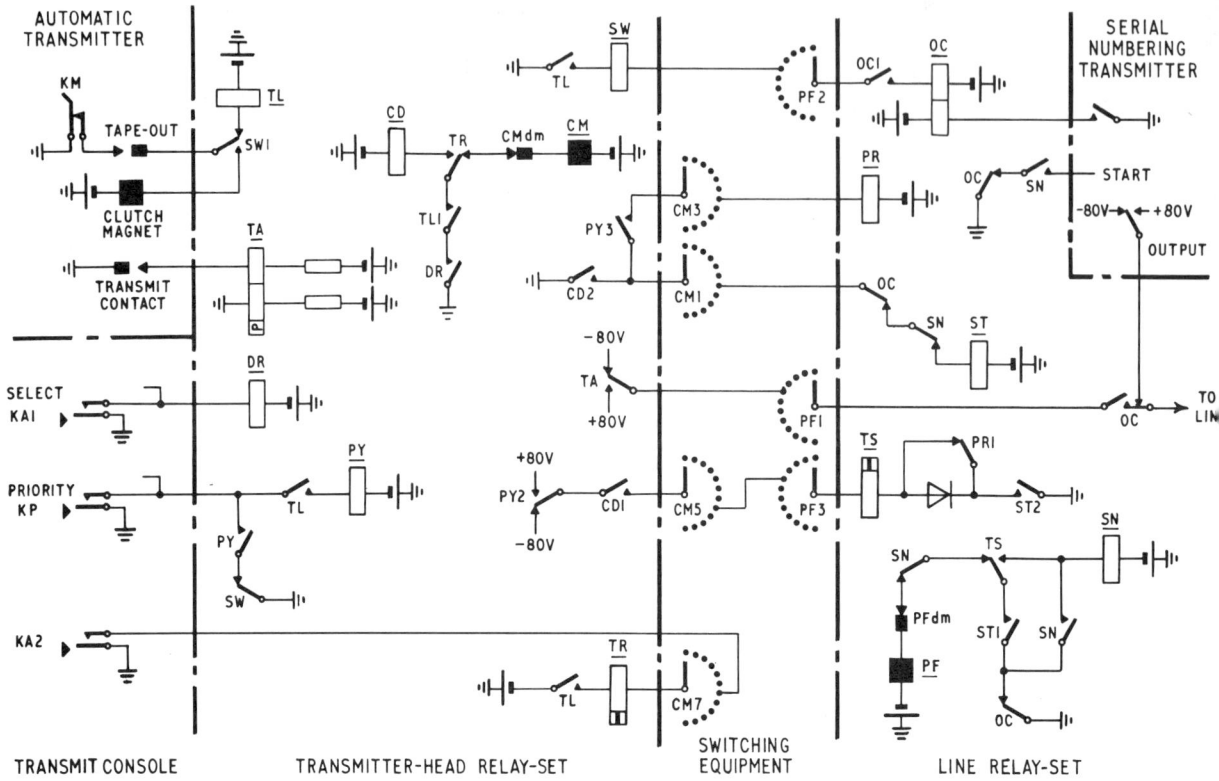

Fig 11.6 Semi-automatic system — outgoing line-selection circuit

the transmitter is automatically connected to the required outgoing channel and the message tape transmitted after waiting, if need be, for the circuit to become disengaged, and following automatic transmission of the serial number.

The method of outgoing-channel selection uses a circuit-marking uniselector CM and a position-finder uniselector PF (Fig 11.6). When a tape is loaded into the transmitter message-head, relay TL in the transmitter head-relay set operates; **TL1** prepares a circuit for the associated uniselector CM. When the required route-selection key KA is depressed, relay DR operates from **KA1** and completes the drive circuit for uniselector CM; **KA2** applies earth potential to mark the appropriate contact on arc CM7. When the wiper CM7 reaches the marked contact the high-speed relay TR operates, cutting the CM magnet-drive circuit and operating relay CD. Contact **CD2** applies earth potential to the start lead of the line-relay set. **CD1** applies −80 V to mark the bank of the uniselector arc PF3 which is associated with the required channel. When the

channel is free (i.e. neither the OC nor the SN relay is operated) the start relay ST in the line-relay set operates; **ST1** causes uniselector PF to drive until the required transmitter position is located; **ST2** prepares for the high-speed relay TS to operate as soon as the −80 V marked position is located. A contact of relay **TS** causes the operation of relay SN which starts up the serial-number transmitter to send the serial number to line; on conclusion, the serial-numbering transmitter operates relay OC and **OC1** operates the switching relay SW in the transmitter-head relay set. Contact **SW1** energizes the clutch magnet of the automatic transmitter and the message is sent to line. Another OC contact has meanwhile switched the line from the serial-numbering transmitter to the message transmitter. The polarized relay TA is used to convert single-current signals generated by this type of transmitter head into double-current signals.

While a line is engaged, either relay SN or relay OC is operated and relay ST cannot operate for a fresh message. To avoid the operator waiting for the

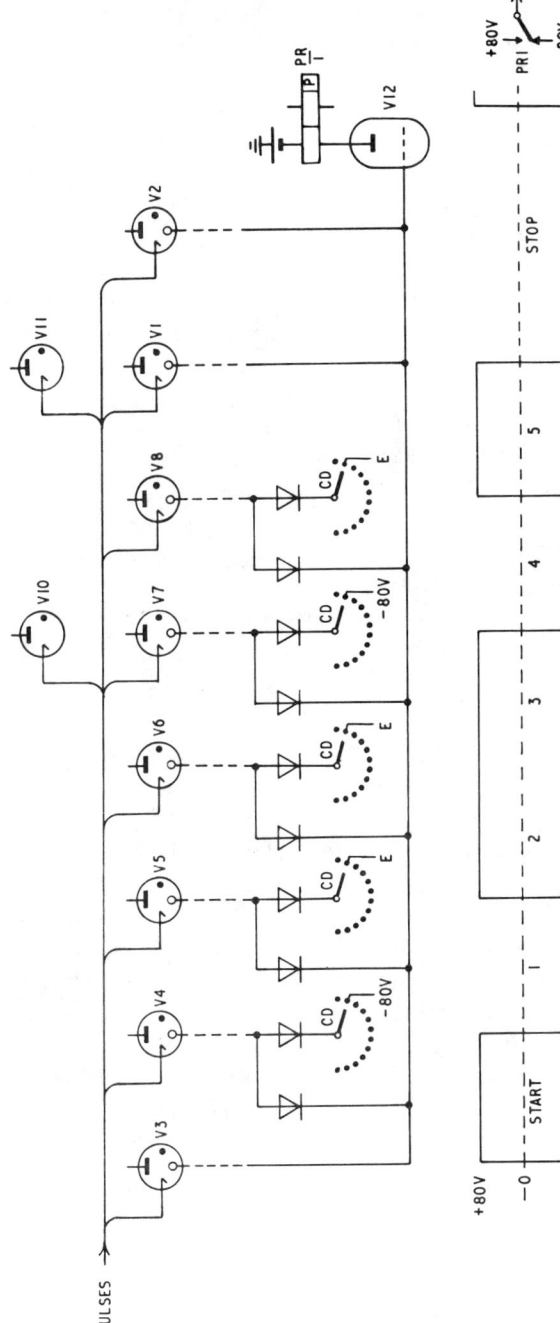

Fig 11.7 Serial-number transmitter — signal generation

line to become free, a lamp on the push-button panel lights to indicate that the call has been accepted as soon as the marked circuit has been found and relay CD has operated. The operator can then release the push-button and proceed with the next message.

On completion of transmission of a message the TAPE OUT contacts open, releasing relay TL and restoring the circuits to normal.

The equipment provides precedence for priority messages. If the priority key KP is also depressed, relay PY operates. Contact PY3 operates relay PR in the line-relay set while PY2 changes the polarity applied to CM5 from −80 to +80 V. At the same time PR1 has polarized relay TS by inserting the diode in circuit. The position finder PF will drive past any other (−80 V) marking potentials until the +80 V of the priority call is located and relay TS is able to operate. Relays PY and PR release when relay SW is later operated.

In large installations, if there is more than one channel in a given route, the first free channel of that route will be selected when the circuit button is pressed. Alarm conditions are provided to guard against undesirable conditions such as a failure to connect to a required line within a specified time.

11.6 Serial-numbering transmitter

The triple-headed automatic transmitters used as serial-numbering transmitters are costly to maintain and operate; in new installations they have been superseded by electronic equivalents designed to reduce running costs and give more reliable service.

An electronic serial-numbering transmitter comprises a ring counter (see Appendix B) using cold-cathode valves, four uniselectors and a number of relays. One uniselector acts as a code distributor, the other three act as a 4-digit number register. Provision has been made in the design to generate up to 46 character signals; these include four numerals providing serial numbers from 0001 to 1999 together with other signals such as CARRIAGE RETURN, LINE FEED, LETTERS SHIFT, the station code and the START OF HEADING code.

The ring counter is fed at 20-ms intervals from a pulse generator and is brought into operation each time a new message tape is inserted in a message transmitter. Fig 11.7 illustrates the method of signal

generation. When the numbering transmitter is seized the cold-cathode valve V2 in the ring counter is caused to strike; the other valves V3 to V8 and V1 in the ring counter strike sequentially to succeeding pulses and as each strikes the preceding valve is extinguished. Five of the valves in the ring counter (V4 to V8) are each connected to a separate CD wiper of the code distributor and also to the control grid of a pentode valve which operates as a switch; a polarized relay in the pentode anode circuit provides double-current output signals. These five paths determine the polarities of the five code elements; of the remaining valves, V3 provides the start element, V1 and V2 provides a 40-ms stop element. V10 and V11, striking in conjunction with V7 and V1 respectively, control the stepping of the code distributor CD during each stop signal, the uniselector magnet being energized with the striking of V10 and de-energized when V11 strikes.

The potential of the V12 control grid is determined by the combination of the voltage change from a cold-cathode valve each time it strikes to a pulse, together with the bias potential from the associated uniselector wiper. The diagram also shows the circuit bias potentials (−80 V and earth) appropriate for generating the code elements for character 'D' in alphabet No. 2.

The potentials applied to the pentode control grid are shown in greater detail in Fig 11.8. The polarized relay PR is normally biased to the mark contact by its 200-Ω winding and it moves to the space contact only when anode current flows in V12. With the transmitter at rest the cathode of V12 is at earth potential and the control grid is at −80 V so that the valve will not conduct. When the transmitter is started the V12 cathode potential is lowered to −42 V due to the 900 + 1000Ω potential divider in the cathode circuit. With the CD4 wiper standing on a contact at earth potential, this has no effect upon the control grid due to the high backward resistance of MR8 in this path. When V4 strikes to a pulse, its cathode potential rises from −80 V to about −10 V; this pulse is applied through the low forward resistance of MR7. With −10 V on the V12 control grid and −42 V on the cathode, the grid is driven relatively positive to the cathode and the valve conducts for the ensuing 20 ms until the arrival of another, different, pulse. Earth potential applied from CD5 wiper to the cathode of V5

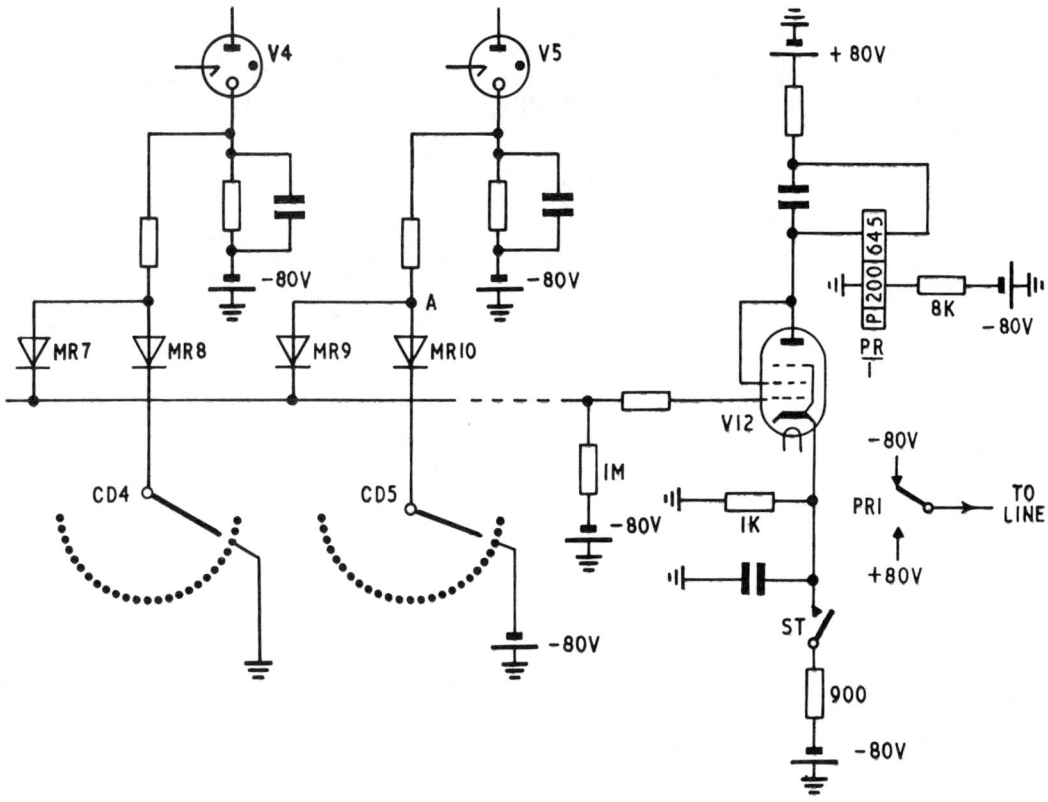

Fig 11.8 Serial-number transmitter — detail of signal generation

while it strikes would maintain V12 anode current for a further 20 ms. But with the condition shown (−80 V on CD5), when V5 strikes it cannot raise the potential at point A above −80 V due to the low forward resistance of MR10 to −80 V at CD5. The control grid of V12 remains at −80 V, negative to its cathode at −42 V, and so V12 does not conduct. In this way the polarity of signals at PR1 is determined by the potentials at CD4 to CD8 wipers; the uniselector banks are strapped according to the signals to be provided.

Serial numbers are generated and advanced by one digit each time the number generator is used (Fig 11.9). Four consecutive contacts on uniselector arcs CD4 to CD8 are allotted to the 4-digit serial number. The number range is from 0001–1999; the thousands digit changes only from 0 to 1 and in alphabet No. 2 this involves a change only to the first code element. The first of the four allotted contacts ('thousands') is permanently strapped at CD5, CD6, CD7 and CD8, according to the code, but at CD4 (element one) the potential is deter-

mined by leading this bank contact out to an arc of the hundreds uniselector (H3); the H3 bank contacts, strapped as shown, produce the required potential change after the hundreds wiper passes the 9th hundred. For the hundreds, tens and units, the three remaining sets of bank contacts are led out to wipers of the hundreds, tens and units uniselectors (H4, T4, U4 . . . H8, T8, U8); all these bank contacts are strapped as required.

The units uniselector is stepped once each time the numbering equipment is cleared down. On seizure, relays ST, STA and STB operate in sequence; they release in the same sequence on clear down. The units uniselector magnet is energized during the combined release lags of relays STA and STB, the release of relay STB being delayed by shunting its winding. The tens uniselector is similarly energized, but only each time the units wiper U2 has taken ten steps; the hundreds uniselector is similarly stepped depending upon ten steps by the tens-uniselector wipers. On clear down, CD uniselector is driven to the home position.

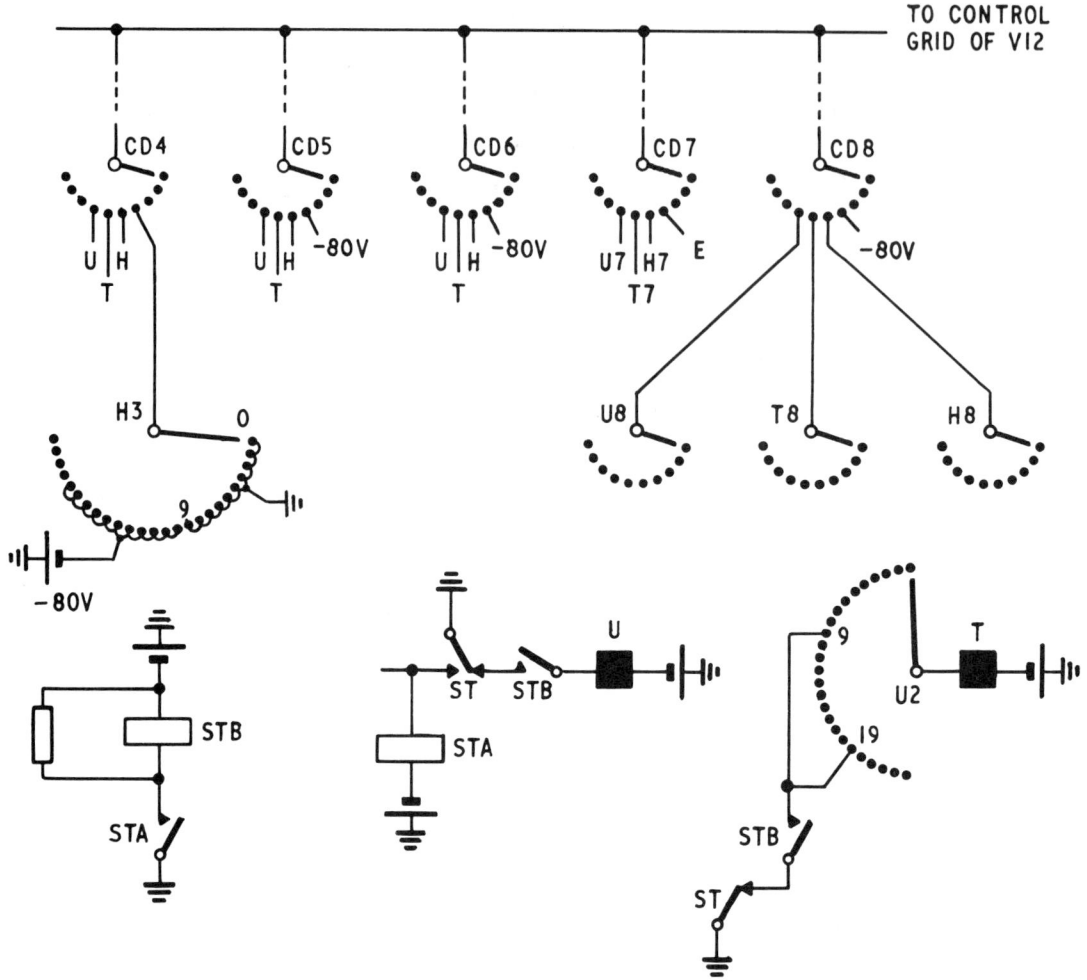

Fig 11.9 Serial-number transmitter — numbering circuit

The circuits illustrated show the method of generating 25 characters; if a greater number is required the CD uniselector steps round twice in conjunction with additional relays which prevent a second transmission of the serial number and change polarities on the CD uniselector banks.

11.7 Automatic systems

Automatic message-relay systems operate without the need for staff at the relay point, apart from those required to exercise general traffic supervision and equipment maintenance. It is essential that messages are prepared to a standard format. Unless the system is restricted to a limited list of address codes, considerable equipment complexity arises in achieving automatic operation.

Different approaches have been made to the design of automatic systems, but in all cases certain common features must be provided. Since the routing information is given within the message (usually very near the beginning) and the whole message has to be retransmitted, storage is required at least for that part received before the routing can be determined. If the required outgoing channel is not immediately available, the store must cover everything received until such time as the required channel becomes free. The capacity of the store has to be capable of storing the entire message, though if retransmission commences as soon as the route has been determined, only a part of it will be in store at any moment. Some systems deliberately store the whole message before starting to retransmit in order that detectable errors or cancellation instructions in any part can be used to divert the message for the supervisor's attention instead of allowing onward transmission. Again, if the selected outgoing channel is operating at a much higher character rate than the channel on which the message originates then, in order to avoid wasting outgoing channel time, complete reception is necessary before beginning retransmission. This may arise when a message received from a channel operating at 50 bauds ($6\frac{2}{3}$ char/s) is to be retransmitted over a channel capable of 75 bauds (10 char/s); or when receiving from an ARQ channel subject to delays from repetition; or receiving at hand speed; onward transmission will in all these cases be at the full cadence of automatic transmission.

Another common feature is that routing information contained in the message has to be read at least twice, once to the equipment which determines routing and a second time when outgoing transmission takes place. Received messages have to be marshalled into outgoing queues to ensure that they are retransmitted in accordance with priority and with time of arrival at the centre; finally, arrangements have to be provided for transferring messages from input to store and from store to output channel.

The equipment consists of a number of discrete sections concerned with message storage, translation of routing information and switching. Each function may be carried out using electromechanical (relays and selectors) or electronic (including magnetic) techniques and a complete system may employ either or both types of equipment. Use of magnetic storage eliminates maintenance of start—stop machines and also saves a significant annual expenditure on paper tape. Conventional relays and selectors with their circuit techniques are well-proven; configurations employed in equipment based on their use ensures that faults tend to cause gradual degradation rather than complete failure in service from a system. Wholly electronic systems may be achieved by the use of a large number of units wired together in such a manner as to perform the required operations: these are the so-called 'wired-logic' or 'wired-program' methods. On the other

hand, equipment may be based on the digital computer which is instructed, by a program stored in the machine, to carry out the required processing of input information; this type is referred to as the 'stored-program' method.

Systems in service have included: (1) perforated-tape storage (continuous tape rolls) with electronic wired-program translation and conventional relay and uniselector switching; (2) magnetic-drum storage, electronic wired-program translation and switching. The wired program tends to lack flexibility and, because in general each function is carried out by a separate equipment unit, such a system tends to be large and expensive; (3) magnetic-core and magnetic-tape storage, conventional relay and motor-uniselector switching with a mixture of electronic circuits and relays for translation; these systems are relatively inexpensive, but the equipment requires a fair amount of maintenance; (4) magnetic-disc or core storage with stored-program computer (including magnetic-core storage) for translation and switching; such systems operate at very high speed, reducing message delays and increasing operational efficiency, and provided the system is designed to handle the required number of lines and traffic load, delay between reception and transmission is reduced to a minimum. Another advantage of this class of system is that once the complex program for the computer has been designed, the time required for installation and commissioning ready for service can be very short. It is relatively economical in equipment in that a common set of units is made to serve each operational function in turn under control of the program. With a stored-program system, if major revisions of operating procedure are required they can be implemented with only a short break in service while the new program is 'read-in'; although the new program may have taken considerable time to design and perfect, this work can be completed off-line. Automatic systems, being autonomous, are not restricted to a given modulation rate, code or traffic rate. The same relay-centre equipment can handle telegraph or data traffic at modulation rates ranging from 45—2400 bauds, using various codes.

The stored-program type of message-relay system is expensive in initial expenditure but it can result in considerable annual saving in staff and maintenance costs. Saving in power consumption and in accommodation are also achieved since use of micro-miniature techniques permits considerable reduction in size of the computer and its power consumption. The high cost of stored-program message-relay systems may be justified only for larger centres; smaller centres can benefit by sharing a system designed to handle message traffic of two or more independent networks.

Stored-program, computer-based systems have great versatility. Because of their inherent data-processing ability they are capable of providing a number of ancillary features such as format generation and conversion, code conversion, accounting, traffic statistics, etc. Since each feature requires a section of program which may contain a large number of instructions and takes up storage and processing time, the cost of providing such extra functions may be significant.

Operational Features. Operations applied to signals in the automatic message-relay centre are basically fourfold: (1) reception, (2) analysis, (3) processing, and (4) retransmission. In the course of steering a message through the centre a number of other features apply.

Different storage characteristics, from considerations of bit capacity, access time and cost, are required for different stages of processing. A complete storage system usually includes three categories of store, all based upon magnetic devices: (1) fast access (e.g. magnetic cores with a typical cycle time of $1 \cdot 5 \ \mu s$) required for storing data while it is being analysed or processed, for storing the complete computer program and the routing information; (2) medium-speed access (e.g. the magnetic drum) for storing traffic while it is in transit through the system; (3) slow access (e.g. magnetic tape or discs) for storing a 'historical' or long-term file record of traffic which has passed through the centre over a specified period, e.g. 30 days, in case of a traffic or accounting query arising; also for storing transit traffic in the event of an unusual delay due to an outgoing route becoming faulty, seriously congested or subjected to a scheduled closure; and thirdly, for storing diagnostic programmes which may be used for rapidly locating equipment faults when these occur.

Messages for an outgoing route must always be transmitted in order of arrival from any one in the whole range of incoming channels; this is done by forming a queue when necessary on any outgoing route and marking a place in the queue for each message as it arrives for that route. This order of transmission can be overridden only by the need to give precedence to messages of higher priority; systems are designed to recognize three or four grades of priority. In searching for the presence of any message awaiting transmission over a given route, the equipment first examines the grade of priority of waiting messages; the message to be sent first is the one which has waited longest in the highest grade, followed by others in chronological order with due consideration to priority.

Checking incoming-channel numbers, inserting outgoing-channel serial numbers, and maintenance of the agreed message format are essential requirements. Automatic alternative routing may be provided.

While in transit through the centre all messages are entirely in electronic form, but it is required to display any specified message in printed form to a supervisor who can change or edit it if, for example, the format or address is found to be incorrect. As with any switching system, flexibility is provided to meet growth or changes in the system loading.

A wide range of optional features can be met by the computer-based system and the more important are: (1) a log or record, by serial numbers, of all messages which pass through the system; it is possible to retrieve any message for examination with little delay; (2) inserting date or time in a message; (3) converting code, speed or format; (4) error control, either detection or correction if the incoming signals use a protected code; (5) automatic dialling and answer-back techniques to enable the computer to set up its own call over the public-telephone network or telex system.

System Operation. Fig 11.10 shows the general appearance of a particular system, while Fig 11.11 illustrates the basic relationship of the main units of a system which would handle several thousand messages/h for about 100 lines.

A computer-based message-relay system operates to a program which is a sequence of simple instructions and tests executed in order to carry out the specified operation. The main, or executive, pro-

Fig 11.10 Computer message-relay installation (*Courtesy of the Marconi Co. Ltd*)

gram may be modified to some extent by the conditions encountered, e.g. priority.

The program governs the following series of operations on any incoming message:

(1) collection of signals from incoming channels and their transfer to a temporary store;

(2) detection of the START-OF-MESSAGE signal;

(3) build-up of characters from each channel into a message within the store;

(4) analysis, recognition and translation of the address information contained in the message;

(5) placing the message into the queue for the appropriate outgoing channel, depending upon the priority level and time of arrival of the message;

(6) transmission of each character of the message from the store to the marked outgoing channel;

(7) detection of the END-OF-MESSAGE signal and termination of transmission.

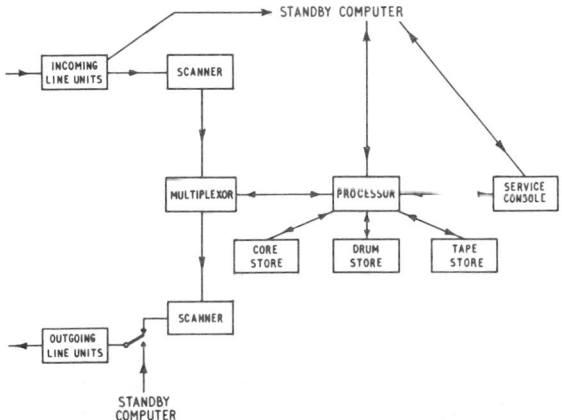

Fig 11.11 Principle of computer message-relay system

All the above operations are performed virtually over a single link within the computer system; a large number of incoming and outgoing channels are connected to this. Telegraph and medium-speed data signals are much slower than computer operating speeds (e.g. 1 μs per cycle) so that time-division multiplexing can be used between computer and line circuits. The computer operating speed must be adequate to handle the designed number of lines and traffic loading. Incoming and outgoing channels are subjected to a continuous scanning process to determine (reception) or apply (retransmission) the signal condition on any channel at any instant. For this purpose each channel has its line termination at which multiplexing can be arranged, as well as conversion from line voltage (e.g. ±80 V) to computer voltage (e.g. 3, 6 or 12 V), and any other conversions required, such as speed or code.

Incoming sequential signals are presented to the central processor in parallel or sequential form with start and stop signals suppressed and usually with a sixth parity element added to 5-unit code signals for checking throughout the processing stages: alternatively, a simple Hamming protected code may be used for transferring data within the computer. Typically a computer word may comprise 12, 18 or 24 bits (plus a word-parity bit) so that two, three or four telegraph characters can be packed into one machine word.

As soon as a START-OF-MESSAGE signal is received the computer gives the message a system reference number and allocates storage space; this may be in the form of core storage from which the message may be transferred to a magnetic drum as each core-storage block is filled. The 'storage address' of the information is recorded to enable storage blocks to be linked together on retransmission.

When the address information in the heading is received the processor will analyse this, check the incoming channel identity and serial number for correctness and decode the priority. The processor then inserts the address of the first drum-store section of the message into the queue, if any, awaiting retransmission on the appropriate channel. When the outgoing channel is disengaged, messages for that channel are transmitted from the drum, in the assembled queue order, the 'address' of the drum-message stores having been recorded in the queue. Before retransmission starts, the channel-identity code and serial number are interpolated in the heading. When transmission is completed the message is transferred from the drum to the magnetic tape or disc for the long-term record and erased from the drum. A multi-address message would be treated as a series of single-address messages. Any code conversion required is likely to take place in the processor.

In the event of an address being received which cannot be decoded, or of an error in the identity or serial number of the incoming channel, the processor will initiate an alarm and pass the message to the supervisor.

Control consoles are provided for general service requirements. A console would also be provided to enable a supervisor to monitor and control the overall operation of the system; for example, to obtain information regarding delays and queues, or to initiate alternative routing due to a line fault. These operations may be effected by means of control keys or by sending coded instructions to the processor from a keyboard. Digital displays and control keys are provided on the console. A printed record of any message can be presented to the supervisor by a high-speed output printer on a visual-display unit.

Separately, or as part of the supervisor's console, one or more service consoles would be provided for handling any traffic rejected by the system due to incorrect format or routing indicators. Such a message would be selectively displayed before the service operator, who is given the facility of amending or editing the heading from a keyboard; characters to be changed are first indicated by a movable underline marker on the display screen and then amended from the console keyboard, the edited message being checked on the display before being passed on to the processor for subsequent transmission. Messages on any specific channel, e.g. a radio channel, can be supervised if need be. An engineering console would also be provided to permit technical monitoring and control, testing and the application of diagnostic routines.

Reliability. With a computer-type message system, reliability in service assumes paramount importance because the operation of the whole centre depends on a single common 'highway' in the computer; a breakdown here could lead to complete stoppage. For public service, systems are required to operate daily for 24 h without major breakdown, without faults which would result in a lost message and without the possibility of taking the system out of service for general maintenance.

With the stored-program computer, the required reliability can be achieved only by duplication of those parts of the equipment which are in common service for all operations. Two entire equipments could operate in parallel, one as the main system, the other in a standby capacity ready to take over

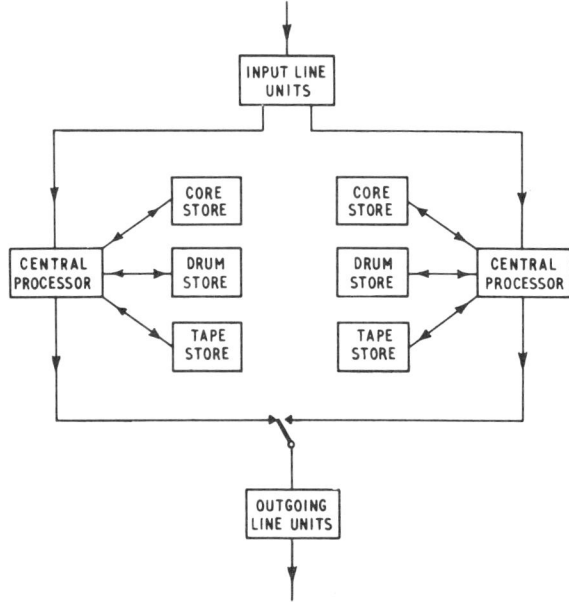

Fig 11.12 Automatic message-relay system with dual parallel equipment

at any instant; or duplicate equipment could be provided on a 'ladder' or series-link basis. In either case continuity of service is dependent on the possibility of diagnosing and locating the fault, replacing a faulty unit and rechecking the whole system by off-line diagnostic routines in a very short time, for until this work is completed the system is running without standby.

A system using duplicate, parallel computers is shown in Fig 11.12. With this arrangement each computer accepts all character signals from the incoming channels over independent connections. Processing decisions, made independently, updating the standby queue and the contents of outgoing queues in each central processor are automatically subjected to continuous comparison in fault-detection circuits.

Normally the outgoing-channel units are connected to only one system; if that system becomes faulty the output units are automatically switched to the other system and service is maintained.

When the computer is put out of service due to a fault the prime consideration is to protect the system against any loss of message. With parallel operation, the two computers, although in message synchronism, are not necessarily in character synchronism. With the standby computer brought into service, any messages which were being transmitted at that instant are artificially terminated and then completely transmitted again before transmitting later messages. Meanwhile, the faulty processor is then checked using diagnostic programs to locate the fault. Modular or unit construction using printed-circuit plug-in boards allows speedy replacement of a faulty unit and the program rechecks the computer before restoring it to service as the standby. Systems are available in which automatic changeover from main to standby has been achieved without loss of telegraph characters; this refinement requires bit synchronism at the output multiplexers.

An alternative method of utilizing the same equipment for main and standby is to allocate prime responsibility for approximately one-half of the line circuits to each of the two computers which, however, have access to all the stores; should either computer fail the other automatically takes responsibility for all the line circuits. This method, known as 'space diversity', is shown in Fig 11.13. With dual systems the total computer capacity must be twice

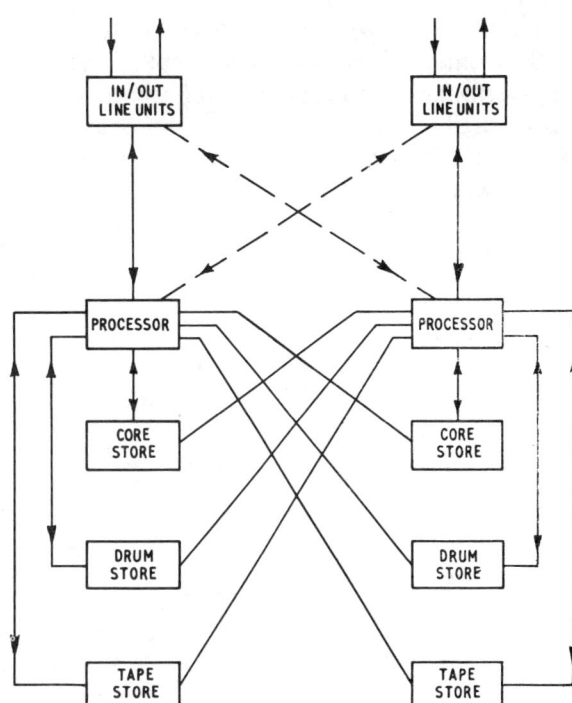

Fig 11.13 Automatic message-relay system with space-diversity dual equipment

that needed for the traffic, i.e. 100% redundancy. For large systems this may be uneconomical and an alternative is to use a larger number of smaller computers to reduce the redundancy while meeting the reliability requirement. The space-diversity method, in which each computer is responsible for a certain section of the line circuits, can then be applied to this multi-computer operation.

11.8 Automatic system for overseas telegrams

This automatic message-relay system, installed in the overseas central telegraph office, is based upon electromechanical switching using 100-outlet motor-driven uniselectors and relays. Magnetic storage is used — ferrite cores for temporary storage of routing and similar information, and magnetic tapes for temporary storage of overflowed telegrams. Electronic techniques using semiconductors are applied wherever possible. Uniselectors are positioned from a single control unit which sets up switched paths one at a time for message transmission.

Provided that telegrams are prepared in accordance with the approved page-printing format, routing is automatic. Telegrams which do not comply, or have significant variations from this format, are automatically routed to one of the 14 semi-automatic routing positions, where routing is completed by an operator at a push-button panel. From the put transmission are at 50 bauds, but within the system retransmission takes place at 75 bauds to reduce equipment-holding times. Each telegram entering the system must be temporarily stored while the destination indicator is examined and the transmission path determined and established.

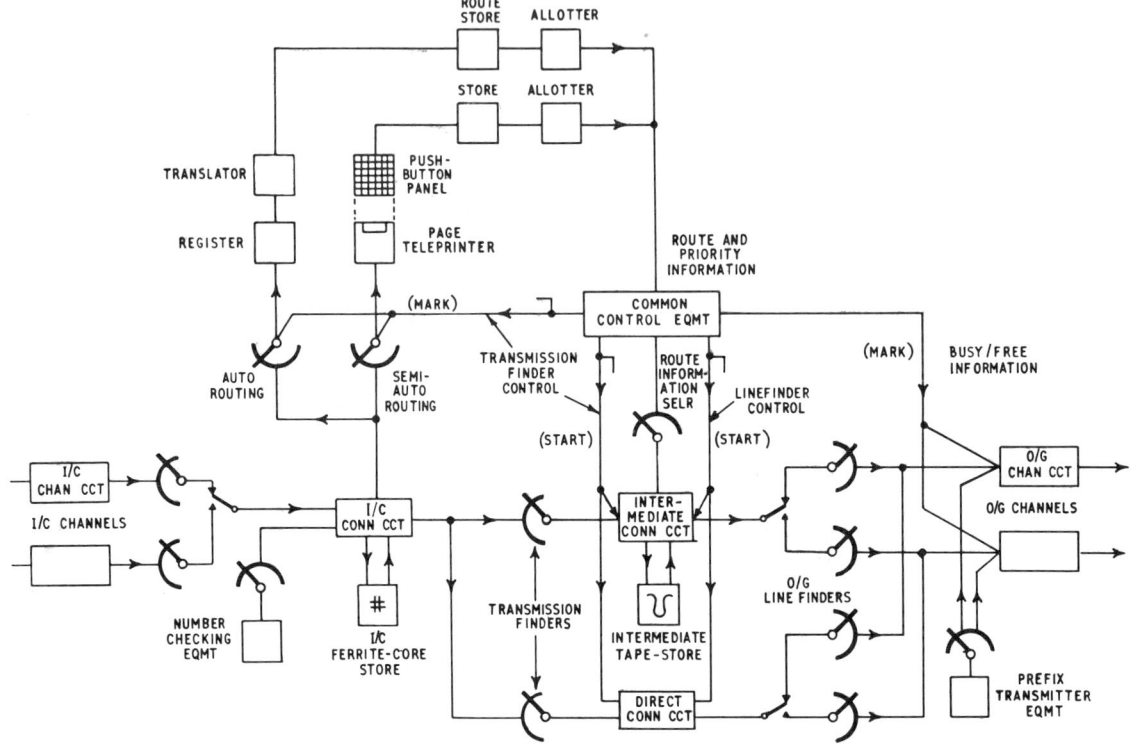

Fig 11.14 Automatic message-relay system for overseas telegrams

moment a telegram enters the system, retransmission can commence automatically within 10–20 s. With about 50 overseas channels (on submarine-cable or satellite VFT systems or on ARQ radio routes) and 50 inland channels, the system is designed to process about 40 000 telegrams/day with a busy-hour load of 3000 telegrams; the average message duration is slightly below 60 s. The installation is capable of extension to about twice the number of channels with a peak flow of 5000 telegrams/h. The system eliminates the need for manually handling cross-office traffic; at the same time it reduces the transit time for each telegram passing through the centre.

Incoming Circuits. A simplified block diagram of the system is shown in Fig 11.14. Input and out-

A group of 75 incoming stores – each with a 2000-character ferrite-core matrix – is provided to serve some 100 incoming channels with a grade of service of 1 in 10^4. When a telegram arrives, an incoming connecting circuit associated with a store is switched to the incoming channel by an incoming-line finder. Two groups of line finders with relay switching are employed to accommodate more than 100 lines, using 100-line selectors. Rapid line finding is essential to ensure that the incoming store receives all the telegram message following the SOM (start-of-message) sequence. Five non-homing motor-driven uniselectors hunt simultaneously for the calling channel as soon as the second code element of the first Z character in the SOM (ZCZC) signal is detected. This is done by relays in the incoming circuit which count signal transitions and start the line finders if two Z—A transitions are separated by one Z element. In addition to the Z, other characters such as CARRIAGE-RETURN and LINE-FEED will start the line finders, but character combinations which are most commonly thrown up as the result of line interruptions — T, O, M, V and letter-shift — will not operate the call detector. One of the five line finders will find the calling line in an average time of less than 100 ms and in no case will this exceed 330 ms, so that connection to the incoming connecting circuit will always be effected during reception of the SOM signal. Switching from the incoming-line circuit to the connecting circuit is effected during a stop signal so that a storage unit cannot receive signals out of phase.

The line terminations differ slightly according to whether the incoming channel is in a submarine-cable or satellite VFT system, or an ARQ radio system, and whether the originating country prepares the telegram format in accordance with F31 or not; these variations are taken into account by strapping appropriate connections in the line-termination. This discriminating information is passed forward to the connecting circuit for action. The channel remains connected to the incoming store until a detector recognizes the EOM signal NNNN and disconnects the circuit.

Automatic Routing. Twelve registers and a single translator are provided for decoding routing and priority information from the pilot line of the telegram and passing on appropriate signals to common control equipment. When 41 characters have been received by an incoming store, a register is connected to the store via the store finders. The incoming store retransmits the information at 75 bauds to the register until the second carriage return and line feed have been detected. During retransmission, each character is immediately rewritten back into the store after read-out so that a telegram is always available in store until the arrival of a new one. Writing into and reading out of the store are effected independently; this permits simultaneous reception and retransmission and also the use of different input and output speeds if required. While this telegram heading is being received, the register checks the format of it and, if correct, stores the 4-

character destination indicator and single-character priority indicator in 5-unit code form on a bank of capacitors. A teleprinter is associated with each register to monitor the telegram headings presented to the register. When the translator is free, the register is connected to it, and the route and priority information are transferred from the capacitor store to one of three route stores.

Routing may be effected either on the two first characters or on all four characters of the destination indicator, dependent on strapping in the translator. The destination indicator is translated into a 2 x 2-out-of-5 code on relays in the route store appropriate to one of 100 routes. There are four grades of priority — A (immediate), B (urgent), C (ordinary) and H (letter). The route store receives the priority information on one of four relays. As soon as the information has been passed to the route store, the translator is released to deal with calls from other registers. The register is partly released, enabling it to receive information from another incoming store. A connection is maintained from the route store to the incoming store in readiness to set up the outgoing connection for the incoming store. The information remains in the route store until the common-control equipment (CCE) is ready to route the telegram. The CCE, capable of processing 6600 telegrams/h, deals with all incoming calls, one at a time.

Outgoing Channel Free. Outgoing line terminations are associated with each outgoing channel and are adaptable to various types of line requirements by strapping — for example, to supply on selected channels a call signal followed by a delay of 1·5 s before transmission of a telegram, or the transmission of a series of letter-shift signals after the EOM signal.

If the route for which a telegram is destined has one or more channels free, the incoming store is connected to the outgoing channel via a direct connecting circuit. Each connecting circuit has two motor-driven uniselectors — a transmission finder for connection to the output of the incoming store and an outgoing line finder for connection to the outgoing channel. The incoming store holds a telegram while its routing is determined, after which the telegram is transmitted directly to the outgoing channel if it is free.

On accepting the routing information from the route store, the CCE tests the channels of the outgoing route, marks the appropriate bank contact of the outgoing line-finder multiple which is associated with the free channel, marks the bank contact of the transmission-finder multiple associated with the store, allocates a direct connecting circuit and finally causes the two uniselectors to hunt for the marked contacts. When connection is established, the CCE and the route store are disconnected and available to deal with another incoming telegram. A call signal is sent to start up the monitor teleprinter and also the distant receiving machine.

Before the transmission path from the incoming store is extended to the outgoing channel, a prefix-signal transmitter is connected via the line circuit to transmit a new SOM signal, followed by the outgoing-channel identity and the next sequence number for the telegram about to be transmitted. The read-out from the incoming store takes place in two stages. The first includes the original SOM signal up to and including the first stored letter space. As these characters are not required they are suppressed; to save time they are read out at 75 bauds while the prefix signal is being transmitted to the outgoing channel. The second stage, which follows the prefix, commences at the first character of the incoming-channel identity and is read out (at 50 bauds) until the EOM signal, when an EOM detector causes the connection to be cleared.

Outgoing ARQ Radio Channels. Telegrams for ARQ radio channels are routed via an intermediate magnetic-tape store. When retransmitting, read-out from the tape store is controlled by the ARQ pulses at $411\frac{3}{7}$ o.p.m.; the ARQ buffer store associated with each radio channel is replaced by this intermediate store.

Outgoing Route Busy or Out of Service. A group of 125 intermediate magnetic-tape stores, each capable of storing 80 telegrams, is provided to hold telegrams received from the incoming stores whenever they cannot be retransmitted immediately

because no free channel is available. The intermediate tape store is used also on telegrams for outgoing ARQ radio channels on account of 'cycling' delays. Each intermediate store is associated with an intermediate connecting circuit and two motor-driven uniselectors – a transmission-store finder and an outgoing-line finder.

An intermediate store is allocated by the CCE on the basis that each store should contain telegrams only for one route and all with the same priority classification. The number of stores allocated to each priority classification is pre-determined by strapping, and in general it is equal to the number of outgoing channels in the route. The CCE directs subsequent telegrams in cyclic order into the allocated stores while making the allocation with the first telegram by operating one of four relays in the intermediate connecting circuit associated with each store to mark the store with the appropriate priority. The CCE also marks the store with the route allocated by positioning the route-information selector.

When the CCE accepts routing information from the register-translator group and determines that the telegram must go into an intermediate store, it marks the appropriate contact on the bank multiple of the transmission finder and controls the finder to hunt for the marked contact. The CCE is disconnected and ready to deal with a fresh telegram. The telegram, with the original SOM suppressed, is read out at 75 bauds from the incoming to the intermediate store. On detection of the SOM signal, the connection between these stores is broken.

If a route is out of service for long periods, the allocated stores for one or more priority classifications can become full; in this event the CCE allocates further stores one at a time as the need arises.

Transmission from Intermediate Store. When a channel becomes free on the required route and a telegram is waiting in the intermediate store for that route, the CCE is again brought into operation, this time by a signal over the busy/free information path. The CCE selects the store, or stores, with telegrams for that route by examining the route-information selector; then examines the priority-marking relays in the intermediate connecting circuits, selects the intermediate store containing a telegram with the highest priority. The CCE marks the bank-contact multiple of the outgoing-channel finder and starts the finder. When the marked outlet is found, the CCE disconnects itself to deal with another telegram.

Before the transmission path is extended to the outgoing channel and the telegram read out from the intermediate store, a prefix transmitter is connected to the outgoing-line circuit to transmit a new SOM signal, followed by the outgoing-channel identity and the next sequence number. The connection is cleared when the EOM signal is detected by an NNNN detector in the store-output circuit.

Other telegrams held in store when a route becomes free are read out as described above, the group of stores with the highest priority classification being read out before a group with lower priority. Stores within a group are read out one telegram at a time in cyclic order, starting with the longest-waiting one.

Semi-automatic Operation. On routes from administrations which are unable or unwilling to comply with the prescribed message format, the strapping of the line terminals results in such telegrams being routed directly to one of 14 semi-automatic routing

Fig 11.15 Automatic message-relay system — semi-automatic routing positions (*Courtesy of Philips Tele-communication Industry*)

positions (Fig 11.15). Each position is equipped with two 75-baud teleprinters, which display the telegram heading in printed form, and a panel of 100 push-buttons, together with miscellaneous control buttons for routing the telegrams. Two teleprinter displays are provided on each routing position so that while the routing is being determined for one telegram, a second telegram heading may be recorded on the second teleprinter. Using the two teleprinters, one operator can route 300 telegrams/h through a position.

After 41 characters of the telegram have been received by the incoming store, the telegram is retransmitted at 75 bauds to a teleprinter on a semi-automatic position allocated by the store finder; further characters of the telegram continue to be printed as they arrive until either the operator routes the message or the EOM is detected, whichever occurs earlier.

When the operator has observed sufficient characters to determine the routing and priority he presses the button corresponding to the required routing and one of four priority buttons. Each button has a plastic cap marked with code symbols which are illuminated by an internal lamp when the button is pressed, enabling the operator to check his selection; a CANCEL button can be pressed if need be. If satisfied, the operator then presses a START button. This stops further recording on the teleprinter and the information regarding route and priority is transferred to a store associated with each push-button panel.

This store has a similar function to the route store for automatic routing — it retains the route and priority information until the CCE becomes available. The telegram is then routed by the CCE, direct from the incoming store, either to the outgoing channel or to the intermediate store. When either connection has been established the push-button store is cancelled and the push-button panel lamps are extinguished, enabling the operator to handle a fresh telegram on either teleprinter display.

Telegram Security. The system has three features which protect telegrams against loss.

Firstly, an incoming channel-sequence-number-checking equipment indicates whenever the number of a telegram received by the system from any channel is not one integer more than the last number received on that route. This enables telegrams not received by the system to be identified at the sending end and a request for a re-run. Sixteen channel-sequence-number-checking equipments are provided and normally a free equipment is associated with an incoming connecting circuit before it accepts an incoming telegram. The checking equipment is connected in parallel with the input of each incoming store so that whenever a store is recording the numbering line of a telegram the same sequence number is stored also in the checking equipment. The equipment identifies the channel from the position of the incoming-line finder connected to the channel, and it stores the 3-digit sequence number when it detects the figure-shift signal which precedes the number. To guard against a telegram being correctly checked by the channel-sequence-number equipment but failing, due to a fault, to be stored in an incoming store, the latter is designed to give an alarm should this occur. The number-checking equipment compares the currently-stored number with the previous number increased by one and, if the comparison shows the number to be out of sequence or incorrect, the numbers involved in the comparison together with the channel indicator are recorded on a teleprinter in the traffic-control area. Periodic inspection of this teleprinter record allows print-outs caused by telegrams which are received simply out of sequence to be ignored, and those indicated as missing can be picked out.

Secondly, measurement of the delay time in storage safeguards telegrams in transit through the system; transmission paths are also automatically checked periodically; alarm conditions are provided if either result is unsatisfactory. The telegram contained in a faulty store can be read out on a teleprinter in the traffic-control area. The transmission-path checking equipment tests the incoming and outgoing channels and certain internal paths, e.g. the input path to each intermediate store. Up to 100 transmission paths can be checked in a cycle lasting about 20 s, each being examined for at least 170 ms. The nature of the test depends on whether the path is shown to be free or engaged; if active, at least one signal transition will occur during the minimum 170-ms testing period. On radio channels, when suspension of transmission is shown by the absence of ARQ pulses, the testing is

similarly suspended.

Thirdly, on every outgoing channel a teleprinter with tape-perforating attachment and tape winder is connected to monitor all outgoing telegrams. The page copy is used in preparation of international accounts and provides a file copy of every telegram. Automatic accounting cannot be employed due to the presence of a significant number of telegrams which do not comply with the F31 format. The tape enables a transmission to be repeated if it has not been delivered; for this purpose an automatic transmitter is plugged into the outgoing channel on the monitor position (Fig 11.16). Alarm conditions are provided to cover the occurrence of any teleprinter faults.

Fig 11.16 Automatic message-relay system — part of monitor room (*Courtesy of Philips Telecommunication Industry*)

Traffic-control Facilities. To provide operating staff with current information on the state of traffic and occupancy of stores, so that early remedial action can be taken on the occurrence of a fault or exceptional traffic conditions, the traffic-control console (Fig 11.17) is provided with the following information and facilities:

(1) the number of magnetic-tape stores occupied at any time.

(2) the number of telegrams in each magnetic-tape

store. Operation of a button on the console for each store displays the number of telegrams in that store;

(3) the total number of telegrams waiting on each route; operation of a button for each route produces a digital display;

(4) by means of a print-out, the storage situation at any instant and for any route, including the number of telegrams in each priority and the stores involved; a button on the console initiates the print-out.

The supervisor's console is divided into five sections:

(1) Incoming section. Each incoming line has the following indications and controls: (*a*) Incoming-line engaged lamp – glows when the incoming line is seized and remains alight until the EOM is detected in the incoming store. (*b*) Incoming-line alarm lamp

Fig 11.17 Automatic message-relay system – traffic-control position (*Courtesy of Philips Telecommunication Industry*)

– glows if the transmission-check unit reveals a fault condition. (*c*) Incoming-number-comparison alarm. This alarm lamp glows if the number-checking equipment finds a discontinuity of the number series. The supervisor operates a read-out button to obtain the last correctly-received number and the number causing the alarm.

(2) Outgoing section. Each outgoing line has the following: (*a*) Outgoing-line engaged lamp. (*b*) Outgoing-line alarm lamp. (*c*) Outgoing blocking button. Operation of this will block the outgoing line against

further transmission; the engaged lamp will flash under these conditions. (*d*) Outgoing-number read button. Operation of this will cause the number of the last telegram transmitted over the corresponding line to be displayed.

(3) Meters and miscellaneous alarms. Meters for each storage group – both incoming and intermediate – give continuous indication of the traffic level in storage groups. Alarms show when none, five or 20 stores remain free. Other alarms relate to the prefix transmitters.

(4) Intermediate and routing section. Print-out facilities are provided for the store engaged for a particular route. Warning facilities are given when the number of telegrams in intermediate stores exceeds a pre-determined number.

(5) General alarm and display section.

If the queue of telegrams for a route becomes excessive, or if the route is out of service, traffic may be diverted automatically or manually to alternative routes. Automatic diversion is possible for up to 20 routes; operation of a push-button allows traffic to overflow from one route to an approved alternative route. Manual diversion is provided for all routes; operation of the appropriate button causes overflow traffic to be routed back to the system input so that the traffic enters for a second time to be directed at the semi-automatic routing positions to the alternative route; either high or low-priority traffic may be so directed.

Equipment Design. The equipment is mounted on racks (Fig 11.18) enclosed against dust and fitted with glass-panelled doors. Relay sets in the CCE are duplicated with automatic change-over in the event of failure. Duplicate control paths are also provided for the more vital relay sets.

The motor-driven 100-outlet uniselectors are operated from a common mechanical drive for each suite of racks. The drive is coupled to shafts running along the base of the suite and is transmitted to shafts along each uniselector shelf through a sprocket and chain drive (seen in Fig 11.19) and a torque-limiting clutch. Each uniselector is controlled by a coupling electromagnet and a detent electromagnet, the connections of which depend on whether step-by-step or free-running operation is being applied. Operation of the coupling electro-

Fig 11.18 Automatic message-relay system — part of equipment room (*Courtesy of Philips Telecommunication Industry*)

magnet connects the wiper assembly, via a gear-train and friction clutch, to the shelf drive-shaft. Operation of the detent electromagnet disengages the wiper detent lever. Testing to a marked outlet and stopping the uniselector is carried out within 2 ms. In free running, the wipers hunt at 300 outlets/s; for step-by-step operation the speed is 60 outlet/s.

Fault Localization. 'Time-out' and sequence-checking alarms are built into most of the units. An automatic routiner is provided to maintain a systematic functional check and check transmission standards of incoming and intermediate stores. Outside busy hours, the routiner continually circulates a test message, with adjustable, pre-set limiting distortion, to a special outgoing line-circuit termination. The lowest priority indication is used in the preamble to reduce interference with normal traffic. The test message is switched alternately via incoming and intermediate storage groups so that every store is checked. If a faulty store is found,

the test message is stopped and the alarm indicates the store and connecting circuits in use at the time. The check includes all message stores, translator, common-control equipment, prefix transmitters and the incoming number-check equipment. A test trolley is provided for checking stores; input and output of a store may be looped, permitting a message to be circulated and monitored.

Incoming Stores (Fig 11.20). The storage medium is a ferrite-core matrix with a capacity for storing 2000×5-unit telegraph characters, each element requiring one core. The matrix, of three-dimensional form, is constructed of $50 \times 25 \times 8$ cores. For explanatory purposes it is convenient to consider the matrix as consisting of $100 \times 20 \times 5$ cores in the x, y and z planes respectively. The 2000 vertical columns of the matrix, each comprising five cores, are used to store the five elements of a character. The relay selector steps once every character period and the uniselector steps once every time the relay selector completes a 20-step cycle. By applying coincident half-currents through the relay selector and the uniselector arc, each column of cores may be selected in turn.

When writing a character into the store, the writing relay selector and uniselector firstly apply coincident half-reset currents to reset the selected column or cores. The five elements of the incoming character are then stored in the selected column by the receiving distributor. This converts A elements of the character into parallel form as half-set currents on the appropriate wires of the 5-wire output, coincident with the application of half-set currents

Fig 11.19 Automatic message-relay system — incoming connecting circuit showing uniselector drive and core store (*Courtesy of Philips Telecommunication Industry*)

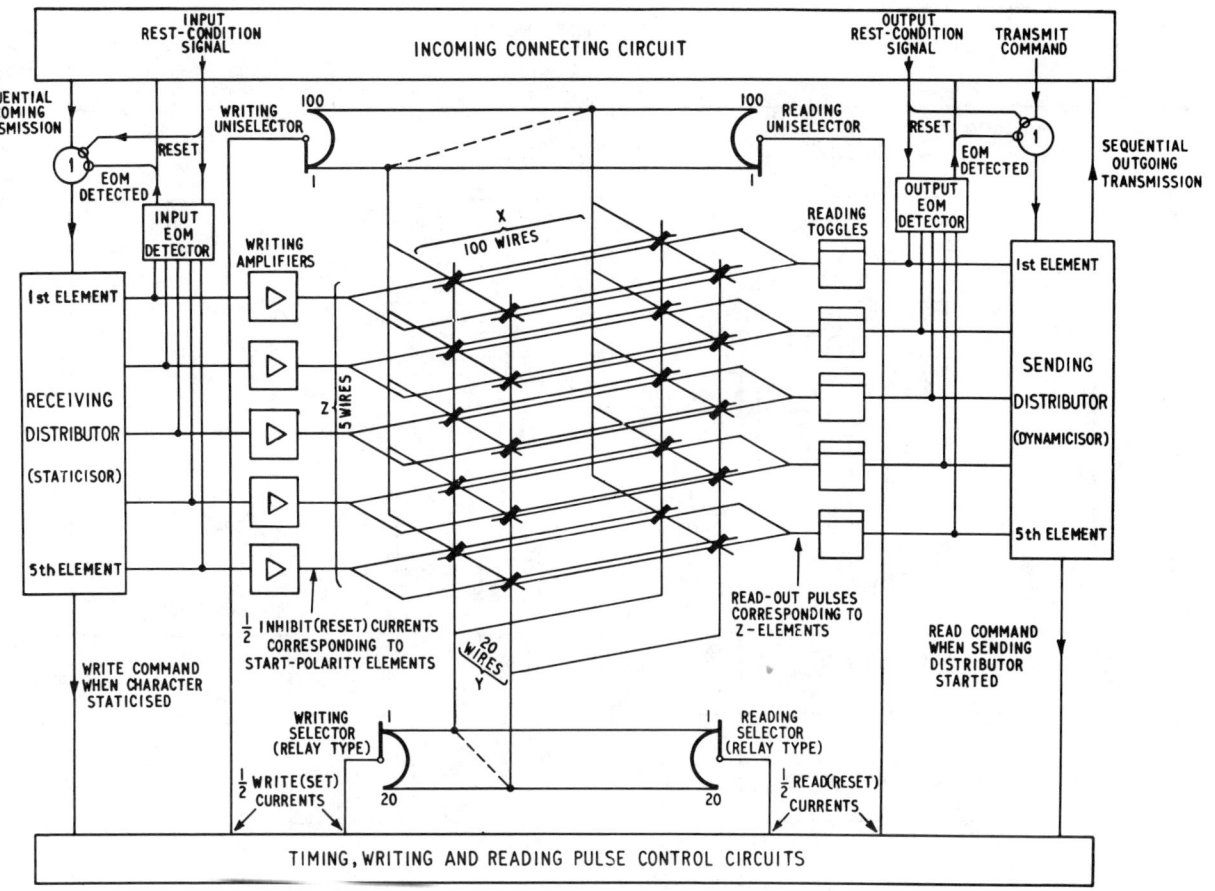

Fig 11.20 Block diagram of incoming stores

from the writing relay selector and uniselector; only cores corresponding to Z elements will be set.

Stored characters are read out by the reading relay selector and uniselector applying coincident half-reset currents to the appropriate column. Those cores in the column which had previously been set due to the storage of Z elements will be reset, causing read-out pulses on the respective wires in the z plane to operate the reading toggles; the sending distributor converts the character into a sequential output with start and stop elements added. As the stored information has now been destroyed, the same character is immediately rewritten into the same column in the matrix.

Each write or read cycle takes approximately 400 μs. Simultaneous write and read cycles are prevented by control pulses from the timing and pulse-control circuits, which permit only alternate write and read cycles.

Intermediate Store (Fig 11.21–2). This magnetic-tape store has a capacity of 40 000 characters on a closed loop of tape. Each store contains a mechanical and an electronic section, consisting of independent receiving (writing) and transmitting (reading) parts. The mechanical section is a panel containing a cassette holding the tape, above which are read and write heads and stepping motors. The tape (20 m x 12·5 mm) contains six tracks, five for telegraph code elements and the sixth a synchronizing track used to determine the position of the elements on the tape. The tape is advanced by friction drive from the stepping motors which are pulse-controlled from the electronic system.

In the electronic section, the receiving and sending distributors perform similar functions to those in the incoming store. Six write and six read heads are aligned across the width of the tape in positions corresponding to the six tracks. When a character

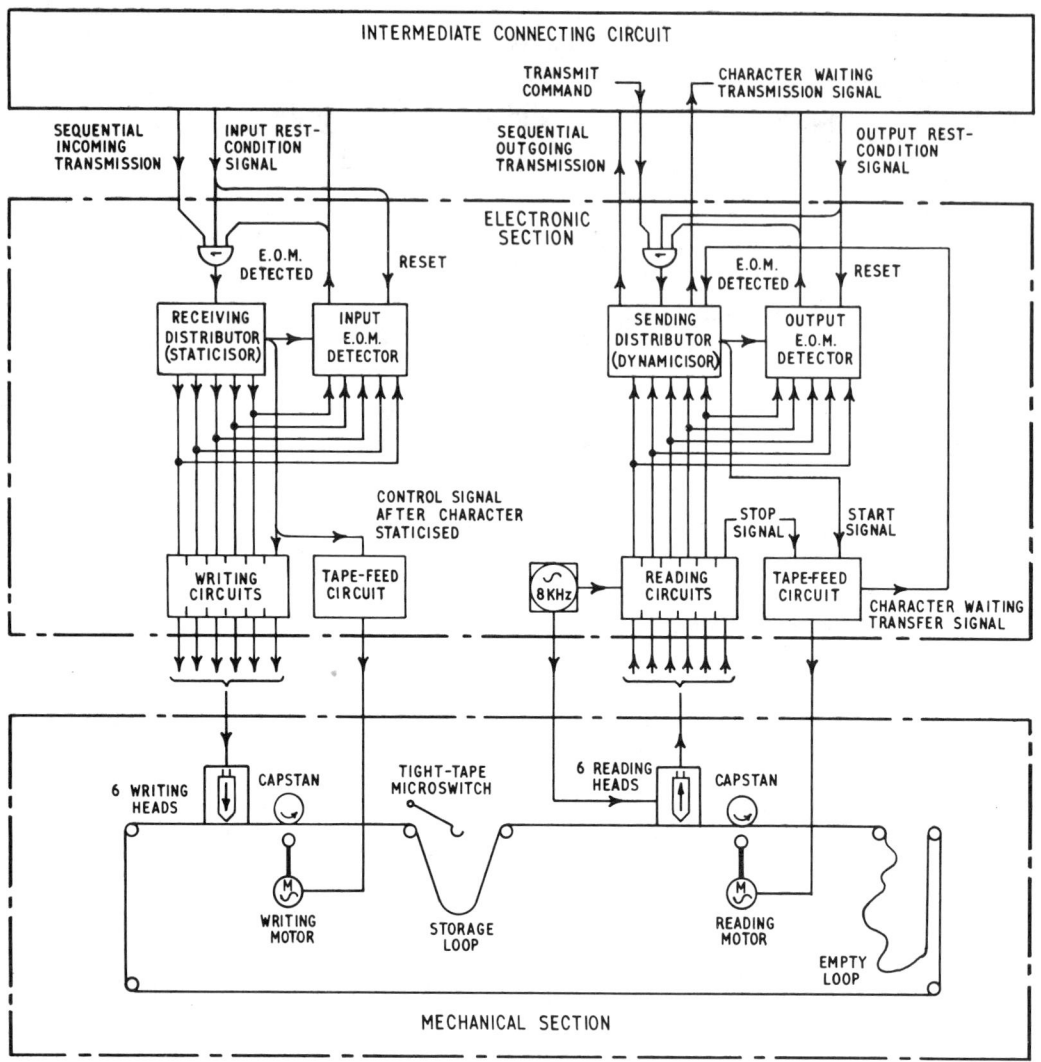

Fig 11.21 Block diagram of intermediate store

is fed to the receiving distributor, the polarity of its elements determines the direction of current, in five of the write heads; that through the sixth is reversed for each character recorded. The writing motor is stepped simultaneously with the application of these currents, which are sufficient to saturate magnetically 0·5 mm of the tape track.

As characters are received they are stored in the portion of the tape loop between the write and read heads until the EOM signal has been detected. The receiving distributor is then blocked until a command signal is given from the switching equipment to store further messages. Slackening tape

between the write and read heads causes the tight-tape microswitch to release, enabling the reading motor to step. As soon as the first character appears under the reading head, where it is detected by a reversal of polarity on the synchronizing track, the read motor is stopped; the transmit side of the store then waits until a command is received from the connected outgoing-line circuit to transmit the message.

On receipt of a command signal from the outgoing-line circuit, the character is transmitted in serial form with start and stop signals added. The transmitting distributor may be operated under

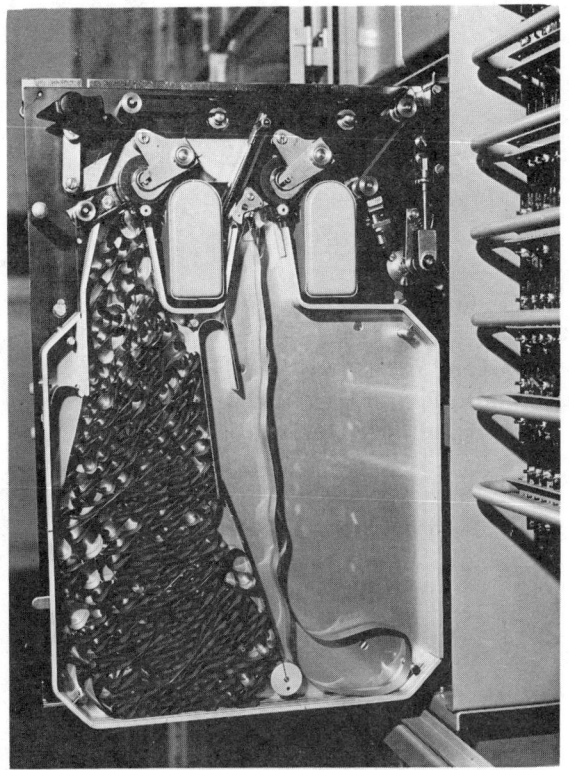

Fig 11.22 Mechanical section of intermediate store
(*Courtesy of Philips Telecommunication Industry*)

The daily traffic handled by this installation has risen to about 75 000 telegrams. To cope with growth of traffic, a larger, computer-type system is to be installed to handle up to 100 000 telegrams daily over 500 lines, at greater speed than possible with the present system.

The system will have three automatic processors: one will handle the traffic flow; a second will be on standby to take over instantly if required; the third will be used for operational control and program development.

The computer will be programmed to give a rapid recall of telegrams from magnetic discs or tape stores in case of enquiry, and for automatically providing information for international accounting. It will be designed to identify names of towns and cities, and also registered telegraphic addresses, to reduce aggregate transmission time. It will also be able to convert telegraphic addresses into telex subscribers' numbers. (Nearly half of the overseas telegrams arriving in the United Kingdom are for telex subscribers.)

Operators at visual-display units will deal with any telegrams which are not in the F31 format.

pulse control, for example when transmitting to ARQ radio channels.

When the EOM signal has been detected, the transmitting distributor is blocked until a command is received to transmit the next message, if any, in the store.

If the last available message is being transmitted, a point will be reached where the tight-tape condition occurs and stops the reading motor, but the last part of the message will still be stored between the reading and writing heads. To enable it to be transmitted, the writing motor will be started — it will have stopped because the EOM signal has been detected on the receive side — to allow sufficient slack in the tape for the remainder of the message to be transmitted. Because no polarity reversals are written on the synchronizing track during this tape-feeding process, then, with detection of the EOM signal on the transmit side, the reading motor will continue to step until the tight-tape microswitch again operates.

References

1 Roberts, M. E., 'Ministry of Transport & Civil Aviation Telegraph Centre, London Airport', *POEE Journal,* **49**, p. 118 (1956).

2 Meredith, L. A., 'A Concentrator for a Telegraph Tape-relay Network', ibid. **55**, p. 250 (1963).

3 Marsh, H. and Lee, F. W., 'A Push-button Torn-tape Relay System', ibid. **57**, p. 175 (1964).

4 Marsh, H., 'An Electronic Telegraph Serial-numbering Transmitter', ibid. **55**, p. 195 (1962).

5 Bubb, E. L., 'Gatwick Airport Electronic Relay System for Telegraphic Traffic', *P.O. Telecommunication Journal,* **14**, p.33, (1962).

6 Bell, H. V. and Pate, O. T., 'Introduction to Automatic Telegraph Message Switching', *Plessey Communication Journal,* **1**, p. 25 (1966).

7. Clark, L. G. S. and Smith, G. G., 'Message Switching by Computer', *Systems & Communications,* Sept. 1965, p. 42.

8 Laver, F. J. M., 'An Introduction to Electronic Computer Systems for Office Use', *POEE Journal,* **52**, p. 13 (1959).

9 Laver, F. J. M., 'On Programming Computers', ibid. **55**, (1962).

10 Austin, D. N. and Huck, J. F. J., 'A New Message Relay System for the Overseas Telegraph Service', ibid. **62**, p. 44 (1969).

12 Facsimile Telegraphy

Facsimile telegraphy is a system of *telegraphy which provides reproduction in the form of fixed images (photographic or otherwise) of the form, and possibly of the depth of tone or of the colours, of an original document, whether written, printed or pictorial.* The term facsimile telegraphy includes phototelegraphy, which pays special regard to tonegradation or 'half-tone' reproduction and uses photographic processing in reception; the term also includes systems which provide direct reception in black and white only, without regard for half-tones — known variously as 'black-and-white facsimile telegraphy', as 'documentary facsimile telegraphy', as 'direct-recording facsimile telegraphy' or as 'facsimile telegraphy' without any adjectival qualification.

The field of use of such systems lies in information such as press photographs, weather maps and line drawings which cannot be sent by coded or alphabetic telegraphy; printed documents are also conveniently sent by this means. Facsimile telegraphy requires the bandwidth of a telephone-type circuit.

12.1 Principle of facsimile telegraphy

In facsimile telegraphy every elemental area of the picture or document to be transmitted is explored or *scanned* by a beam of light. The intensity of reflected light carries information to a photocell which converts energy of the light signal into electrical energy suitable for modulating a carrier wave for transmission. At the receiver, a blank sheet of sensitive material is synchronously scanned and subjected to the information signal, e.g. a controlled beam of light or an electric current according to the method of reception.

Scanning is effected by clipping the document to be transmitted, and also the receiving blank, closely around a drum or cylinder. The drum is rotated at uniform speed while a relative traversing motion

parallel to the drum axis takes place between the scanning device and the drum. In this way the original picture is scanned in a close spiral around the drum and the received picture is built up as a series of parallel lines. Transmitting and receiving drums must rotate at precisely the same speed and be exactly in phase so that corresponding small areas are simultaneously scanned at transmitter and receiver. In some machines the original document is mounted upon a flat bed: scanning then takes place much as the eye reads a page — scanning a line at a time from beginning to end, followed by a rapid return of the scanning device from the end of a scanned line to the start of a fresh one. Most machines use a drum for transmission; for business documents and weather maps where one document may follow another in succession it is convenient to use a roll of the recording medium so that successive documents or maps can be received without need for reloading the machine each time a message is to be received.

12.2 The scanning aperture

A vital factor in fidelity of reproduction by facsimile telegraphy is the part played by the scanning aperture, particularly that at the transmitter. The size of the element of picture being examined at any instant must be comparable with the definition required in the received picture; in the limit this is a question of economics since it affects not only the cost of equipment but also the transmission time and the circuit-bandwidth requirement. For scanning lines to be invisible at normal viewing distance to the unaided eye, the minimum scanning density (pitch, or rate) is of the order of 100 lines/in., i.e. the scanning pitch or line width is 0·01 in., though some machines use much finer scanning.

For estimating the fundamental modulation frequency which results from the scanning process it is customary to assume that definition is required to be the same in horizontal and vertical directions: the height of the elemental area scanned is the same as the width and the finest details which can be scanned take the form of a black and white chequer-board in which the sides of the elemental square are equal to the width of a scanning line (Fig 12.1); any finer

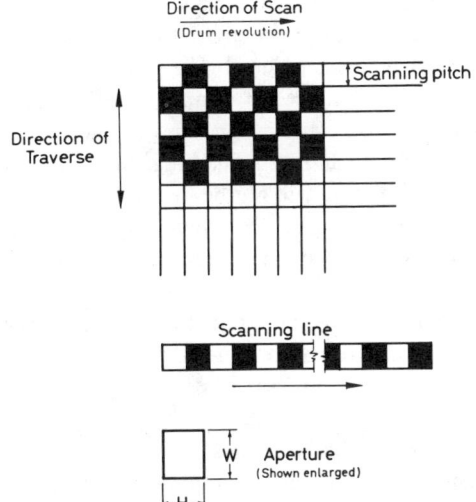

Fig 12.1 Scanning elements

detail is disregarded. It is possible for the definition to be greater in the direction of the scanning line length if the aperture height is less than the scanning pitch. The width (W) of the scanning aperture is that measured in the same direction as the width of the scanned line, i.e. parallel to the axis of the drum; while the height (H) is measured at right-angles to this — in the same direction as the length of a scanning line and parallel to the developed circumference of the drum.

At the transmitter the scanning aperture is inevitably a source of distortion. This is illustrated in Fig 12.2 which shows part of a scanning line made up of equal black and white squares, of sides equal to the scanning-line width; at (a), (b) and (c) are shown three apertures each with width equal to the width of the scanning line but with different heights. Scanning in this diagram takes place from left to right.

With the square aperture (a), maximum and minimum signals occur only when the aperture instantaneously coincides with a square which is completely white or black; at intermediate positions the field seen by the aperture comprises both black and white

areas in various proportions. This results in the triangular waveform shown and produces considerable waveform distortion. With aperture (b), whose height is reduced to 50% of the width, the field examined will be all-black and all-white for periods corresponding to 50% of the scanning element. At other times the instantaneous scanned area contains proportions of both black and white, resulting in the trapezoidal waveform shown. At (c) an aperture with very small height is able to scan areas of all-white and all-black for relatively long periods, resulting in a signal waveform much closer to the shape of the field being scanned. As the height of the aperture is reduced, distortion is also reduced;

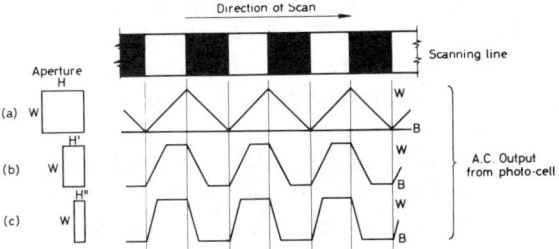

Fig 12.2 Aperture distortion

only with an aperture of infinitesimal height would a square-wave signal be produced when scanning the alternate black and white squares. A limit to the practical reduction in aperture height is set by the minimum amount of light which the aperture can usefully pass to the photoelectric cell. The usual compromise is to have an aperture height somewhere between 0·5 and 1·0 times the width.

Analysis of the waveforms shows them to be represented by a d.c. component (representing the average shade of the picture) together with a series of sine-wave odd-harmonic components of the fundamental frequency. If the amplitude of the fundamental is rated at 100%, the relative amplitude for the third and fifth harmonics are respectively 33 and 20% for the square wave but only 11 and 3% for the trapezoidal wave. Aperture distortion is equivalent to a frequency-discriminating network which reduces the amplitude of high-frequency components. Its effect could be partly compensated by the use of inverse equalizer networks at the transmitter.

At the receiving machine, the facsimile signal will

have already suffered distortion from the finite height of the transmitter aperture; the effect of the height of the receiver aperture is to increase the distortion. The optimum height of the receiver aperture is conditioned by the amount of light which in a given time will pass to the recording medium, and the desirable height will be influenced by the recording characteristics of the medium.

In a phototelegraph receiver (Fig 12.15) a solenoid is used to double the aperture height when the scanning speed is doubled so that the amount of light passed to the recording medium for a given tonal gradation remains independent of scanning speed. Width of the receiver aperture is set by the need for adjacent scanning lines to 'but' together exactly; increase in aperture width results in *overlap* at the edges of two adjacent scanning lines, while a reduction in aperture width causes *underlap*. In both cases the practical effect of incorrect aperture width is to make visible the line structure of the received picture. When initially setting up a receiver, aperture width is adjusted to give neither overlap nor underlap; aperture height is set for correct tonal values, determined by a series of trial pictures. At the transmitter the only effect of overlap and underlap is to vary the amount of reflected light seen by the photocell.

It is possible to calculate the fundamental modulation rate resulting from scanning the chequer-board pattern by a square scanning aperture of dimensions equal to the pitch. Denoting height of scanning aperture by a in., drum diameter by D in., and drum speed by n rev/s, the drum circumference is πD in. The scanning line will contain $\pi D/2a$ *pairs* of black and white squares, each pair corresponding to one cycle of the facsimile signal; in one second the number of cycles is $\pi nD/2a$ so that the maximum fundamental modulating frequency f is:

$$f = \pi nD/2a \text{ Hz}$$

Since $a = 1/F$ then $f = \pi DFN/2 = \pi Mn/2$, where $M = DF$ is the *index of co-operation*.

Considering a transmitter whose characteristics are as follows:

$n = 1$ rev/s, $D = 3.52$ in., $a = 0.01$ in.
(i.e. pitch $F = 100$ lines/in.)

then $f = \dfrac{\pi \times 1 \times 3.52}{2 \times 0.01} = 552$ Hz

This is the maximum fundamental frequency to be transmitted; the minimum frequency is zero, i.e. when the picture is of uniform shade. The facsimile signals would cover a range from 0–550 Hz. If the height of black and white elements of the chequerboard pattern is reduced below the value of the scanning pitch, higher frequencies are generated but the difference between maximum and minimum values of photocell current is rapidly reduced as height of scanning elements is reduced. The picture will then be resolved by the receiver as rapid variations of a medium-grey tone rather than black and white elements, and it can be assumed that fine detail in the received picture will not be markedly improved by transmitting frequencies appreciably higher than the fundamental modulating frequency according to the formula above.

If the motor runs the drum at double speed the bandwidth requirement will be doubled.

12.3 Apparatus characteristics

To ensure that facsimile telegraph-transmitting and receiving machines can interwork, whatever their origin, certain essential parameters must be established, the more important of which are summarized here.

Index of Co-operation. This index is in fact an aspect ratio which ensures that the ratio between height and width of the scanning line, and therefore of a picture also, will be preserved even though the area of the received picture may be greater or smaller than that at the transmitter.

For apparatus using drums with diameter D in. and scanning pitch P in. (or scanning density F lines/in.) the index of co-operation M is

$$M = D/P = DF$$

For apparatus using flat-bed scanning, if L in. is the length of the scanning line:

$$M = LF/\pi$$

(The factor π appears in the second expression since the diameter, not the circumference, was originally adopted for the drum index.) Index of co-operation and scanning pitch (rate, or density) are proportional to one another for a given drum.

Drum Factor. This coefficient is specified to ensure that the usable length of the receiving drum will be adequate to receive the full extent of the picture. For phototelegraphy, drum factor is the ratio:

$$\frac{\text{usable drum length}}{\text{drum diameter}}$$

Before a picture is transmitted it is necessary to verify that

$$\frac{\text{transmitter drum length used}}{\text{transmitter drum diameter}}$$

is not greater than the receiver drum factor.

The effect of discrepancies between certain characteristics of the transmitter and the receiver

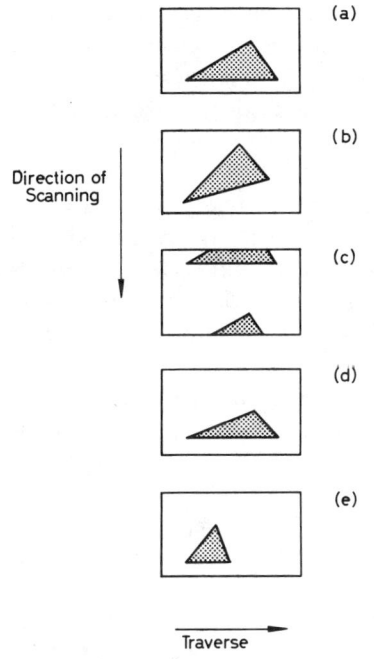

Fig 12.3 Effect of apparatus aberrations

are displayed in Fig 12.3. The direction for scanning an original picture (*a*) is shown. If transmitter and receiver drums do not run at exactly the same speed there will be *skew* in the received picture, i.e. the corners will not be right-angles. For a constant-speed error the base of the triangle becomes oblique as at (*b*), instead of horizontal, since each scanning line at the receiver is completed more quickly than at the transmitter (the illustration shows receiver running faster than transmitter). If a 10-in. picture is being

recorded at 100 lines/in. the drum makes 1000 revolutions during the transmission. If a 1 per cent skew were permissible the two drums must not be more than 0·01 of a revolution out of phase at the end of the transmission; this involves a tolerance not exceeding 10 parts in 10^6 in frequencies controlling drum speeds.

Incorrect phasing would cause displacement of the received picture – an extreme case being shown at (*c*) by a split picture.

If the indices of co-operation disagree, either width (*d*) or height (*e*) of the picture will be augmented.

12.4 Transmission characteristics

Compared with speech, transmission of pictures imposes more stringent requirements upon the channel. Provided that the described information is fully conveyed, the ear can tolerate considerable impairment of quality; with a picture the required information lies in accuracy of fine detail which is, moreover, recorded in permanent form.

Good-quality facsimile transmission requires the use of high-grade 4-wire telephone circuits. Two-wire circuits would be used only where a photograph of some highly important topical event is necessarily transmitted from a telephone exchange where 2-wire circuits only were available; they are inferior for facsimile transmission on account of the reflections which occur at the 2-wire/4-wire terminations.

Permanently-leased 4-wire circuits would be used for facsimile transmissions if the volume of traffic justified provision; characteristics of these circuits would be specially designed with regard to use for facsimile transmission. For international phototelegraph transmissions, the circuits used will frequently be those normally in use for speech but nominated for facsimile transmissions when required, for which purpose they are selected as having suitable characteristics*. When a telephone circuit is in use for facsimile transmission, the signalling terminations and any echo suppressors are temporarily disconnected from the circuit; the circuits also bypass the long-distance terminal exchanges to reduce the likelihood of disturbance. Phototelegraph stations are usually connected by permanent 4-wire 'local ends' to the centre at which these long-distance circuits

* See p. 81

are extended. Desirable characteristics for satisfactory facsimile telegraph transmission are described below.

Carrier Frequency. Amplitude modulation is the normal mode of phototelegraph transmission on landline circuits. On account of its relative simplicity and lower cost, double-sideband transmission is normally used for facsimile signals, vestigial-sideband transmission being rarely used at present, although capable of higher speed in a given bandwidth.

Fundamental facsimile modulating frequencies commence at zero and typically cover a range of 550 or 1100 Hz, dependent upon scanning speed; higher component frequencies generated in the scanning process may not be transmitted on account of bandwidth limitation. Unless the original facsimile signals are eliminated, the carrier frequency needs to be at least twice the fundamental modulating frequency, otherwise overlapping will occur between higher frequencies of the original facsimile signal and lower frequencies of the modulated lower sideband; intermodulation products would then arise (the Kendall effect) causing impairment to the received picture. If the carrier frequency is set too high, not only will higher frequencies of the upper sideband be liable to higher attenuation near the line cut-off frequency, but effects of excessive delay distortion are also likely to be felt. The Kendall effect may be eliminated by using a much higher carrier frequency (e.g. 7 kHz) and then remodulating the lower sideband to bring the modulated signal back to the telephone-speech band.

For use on (loaded) audio-frequency circuits, the frequency of picture-telegraph carrier current is fixed at about 1300 Hz, which gives least delay distortion on lightly-loaded cable circuits. For use on high-frequency carrier-type circuits, a picture-telegraph carrier-current frequency of about 1900 Hz is recommended; this is in the centre of the standard transmission band (300–3400 Hz) provided for telephone circuits. With a scanning density of about 100 lines/in. the minimum bandwidth at 1 rev/s is about $1900 \pm 550 = 1350$–2450 Hz; or at 2 rev/s the bandwidth requirement will be double this, $1900 \pm 1100 = 800$–3000 Hz. A third harmonic of these fundamental frequencies will be transmitted only for picture elements of height slightly greater than the scanning pitch, and then only at the lower

scanning speed. With amplitude-modulated photo-telegraph signals, the level of the output signal is greatest for white and least for black; for weather charts the opposite applies. It is desirable that the amplitude of the transmitted signal should vary linearly with the photocell voltage, and that no correction for tone scale should be made at the transmitting station.

SCFM (sub-carrier frequency modulation) is the normal method for transmission over radio circuits, including any landlines which are used to connect the radio station to the phototelegraph station. AM cannot be used because of frequency-selective fading which is characteristic of high-frequency radio propagation. SCFM is also used for photo-telegraph transmission over very long inter-continental submarine telephone-cable circuits on account of the great sensitivity of picture transmissions to the small level changes which occur on operation of the automatic level regulators.

Most phototelegraph equipments are designed for an AM input and output. For FM transmission, an AM/FM converter is interposed between equipment and line, a convenient arrangement when the same phototelegraph apparatus is required to operate alternatively over landline or radio circuits. Converters can be installed at either radio station or photo-telegraph station, according to convenience.

SCFM characteristics are, for phototelegraphy, a nominal centre frequency $f_0 = 1900$ Hz with 2300 Hz for black and 1500 Hz for white (for meteorological charts these limits are reversed). Taking the characteristics used above, the SCFM bandwidth requirement is ± 550 Hz about these limits, i.e. $(1500 - 550)$ to $(2300 + 550) = 950$–2850 Hz at 100 lines/in. and 1 rev/s. According to Kupfmüller and others the maximum facsimile modulating frequency should be multiplied by a factor of 1·6, giving approximately 700 in place of 550 and a bandwidth of 800–3000 Hz at 1 rev/s. With deviation of ± 400 Hz and facsimile modulating signal of maximum fundamental frequency 550 Hz, the modulation index is $400/552 = 0.74$. The bandwidth requirement for SCFM is greater than for AM(DSB).

Signal Power. Compared with speech signals, facsimile signals are liable to contain relatively more energy at higher frequencies; the possiblity of causing interference increases with frequency and for this reason the power injected into a line by a facsimile transmitter must be limited.

For AM transmission the phototelegraph signal corresponding to maximum amplitude (white) should be so adjusted that the absolute power level of facsimile signal at zero relative level point of the circuit is 0 dBm; the level of signal corresponding to black is usually about 30 dB lower than white-signal power.

For FM transmission the absolute power level of signal at zero level point of the circuit is −10 dBm0. The reason for the lower level with FM transmission is that average power in the system tends to be higher because the same power level is used for black, white and any intermediate tones. With increased loading of transmission systems by FM data signals, it is possible that individual power levels may have to be reduced to avoid overloading common amplifiers on wideband systems.

Attenuation. The overall line attenuation permissible for standardized telephone circuits is satisfactory for facsimile transmission. For systems using AM, attenuation/frequency distortion between facsimile stations should not exceed 8·7 dB (1·0 Neper) in the band of frequencies to be transmitted for photo-telegraphy. Unequal attenuation of upper or lower sideband components results in production of an unwanted quadrature component of the modulation envelope which gives waveform distortion. This can seriously impair the quality of a received picture: the practical effect is to produce an irregular black and white trail following a marked white-to-black transition, together with a general loss in sharpness of definition. For wideband circuits, e.g. used for newspaper-page transmissions, attenuation distortion should preferably not exceed 3 dB.

With AM, phototelegraphy is extremely sensitive to any level changes which occur during transmission, an abrupt change of as little as 0·25 dB being apparent as a tonal change, although greater variations can be tolerated if the level changes take place gradually. With SCFM transmission, phototelegraphy is immune to sudden level changes as great as 10 dB.

Delay Distortion. In a practical transmission system having a non-linear phase characteristic varying with frequency, the component frequencies of a complex wave will be delayed by different amounts and waveform distortion will arise; this form of distortion, of little importance in speech transmission, has serious effects on facsimile transmission. Considering the variation in the phase characteristic over the facsimile frequency band this usually has a minimum value (B_{min}) somewhere near to the middle of the frequency band with greater values B' at the edges; the differential delay distortion is then defined as the difference between these values, $(B' - B_{min})$ seconds. The choice of carrier frequency (AM) or mean frequency (FM) has been made so that it lies as near as possible to the frequency which suffers minimum group delay over the frequency band used.

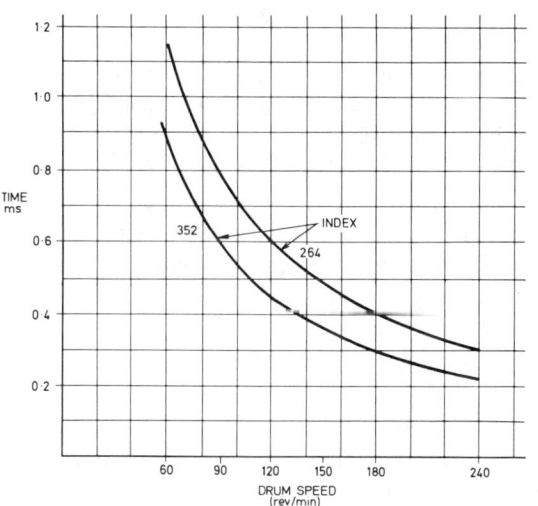

Fig 12.4 Permissible values of delay distortion

The CCITT recommends that for phototelegraphy this delay distortion should not exceed $1/2f_p$ seconds — where f_p is the maximum modulating frequency corresponding to definition and scanning speed; this permissible delay distortion for indices of co-operation 264 and 352 at various speeds is plotted in Fig 12.4, assuming a square aperture of sides equal to scanning pitch. For the examples of phototelegraph machines used above, permissible delay distortion at a speed of 1 rev/s would be approximately $1/(2 \times 550) = 0.9$ ms. For high-speed newspage transmission the problem becomes more acute and the formula given suggests that a maximum value of delay distortion would be around 25 μs.

The incidence of delay distortion is due to loading in audio-type circuits, and in modern HF circuits mainly from channel filters. Its value is cumulative when a circuit comprises a number of HF channels in tandem, otherwise it is independent of length of circuit. For this reason also, use of channel 1 or 12 of a 12-channel group (1 or 16 of a 16-channel group) is deprecated for facsimile telegraphy since the effect of group filters is felt in addition to delay distortion imposed by the channel filter.

For phototelegraph transmission over long intercontinental submarine-cable circuits, which may be extended over long landline circuits within the terminal continents, a maximum value of ± 300 μs has been suggested for the cable link. In quoting values for delay distortion it is important to specify at what frequency limits they are to apply.

The practical effect of excessive delay distortion is blurring or reduction in sharpness of definition in the received picture, and it is particularly evident as a characteristic fringe at sharp boundaries between black and white areas. On the received picture, outlines which should be well defined will be bordered with a fringe of alternately light and dark bands, the total width of the fringe depending upon the amount of delay distortion and speed of scanning. For a given bandwidth, delay distortion is the main factor which mitigates against use of higher scanning speeds, because sidebands then extend to extremes of frequency range where delay distortion becomes excessive. Delay distortion is a serious factor in black-and-white transmissions — for example in typescript the prolongation of letters in the direction of scanning reduces legibility. It may be considered that delay distortion equivalent to $\pm \frac{1}{4}$ the time of the elemental area will give very good definition; at $\pm \frac{1}{2}$ the elemental area the effect will be not very noticeable; but at ± 1 times the elemental area the result will be only tolerable.

The effects of attenuation/frequency distortion and phase/frequency distortion are similar. Circuits are usually readily equalized for attenuation over the relevant frequency band; the residual phase distortion is then a disadvantage. Phase equalizers can be constructed but they are complex and expensive bearing in mind that facsimile transmission represents a very small proportion of traffic carried by telephone plant; their use may not be justified unless required also for other purposes such as TASI or data transmission. In the limit, the extent of acceptable impairment will often be decided by a subjective assessment of suitability for commercial use of a given received picture and of the desirability for avoiding a high scanning speed; delay distortion is rarely a problem in pictures scanned at 1 rev/s.

Echo. Whenever a change of impedance along a line is encountered, some transmitted energy is reflected from this point. This occurs, for example, when the 2-wire line and the balance network of a 2-wire/4-wire hybrid termination are not well matched; an echo path for telephone speech is then produced around the 4-wire circuit which could result in double or multiple images on a received picture. Impedances in the 2-wire path, e.g. line and exchange, may also be mismatched. It is for this reason that 2-wire (repeatered) circuits are quite unsuitable for phototelegraphy and the preferred method is to use the 4-wire circuit as two disymmetrical channels, in which case there should be little likelihood of echoes due to mismatched impedances. If echo is present, the distance separating the double outlines on the picture would depend upon the propagation time of the circuit; for echoes due to imperfect balances at 2-wire/4-wire hybrid terminations, the time difference would be due to loop-propagation time around the circuit. Distortion produced by multiple low-level echoes may be difficult to distinguish from the effect of delay distortion.

Noise and Interference. Facsimile transmissions are particularly sensitive to single-frequency interference, the result of which is to produce patterns on the received picture. Phototelegraph transmissions are very sensitive to all forms of interference, which impairs the detail on a picture by changing the tonal values. The effect of interference from other circuits may be due to additive energy resulting from cross-talk or induction, or to a process of modulation if a non-linear device is involved. With additive noise, for a given level of interference the effect is most visible in black areas (with AM) of a picture; an interference pattern is not produced in white areas unless a sufficiently high level of it is applied to cause general break-up of the picture. A single interfering frequency produces in each scanning line a number of dots equal to the ratio of the interfering frequency (after demodulation) to the product of drum circumference and speed. The pattern produced

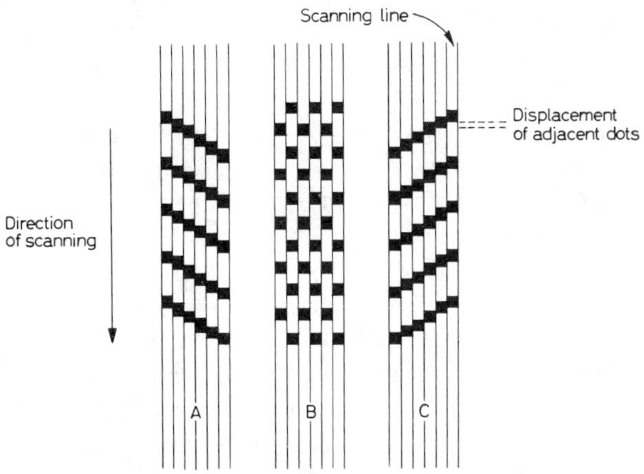

Fig 12.5 Formation of interference pattern

(see Fig 12.5) depends upon the value of interfering frequency and upon its ratio to receiver-drum speed. If this ratio remains constant, adjacent scanning lines form a pattern of sloping lines across the picture, slope depending upon whether the ratio is greater (A) or less (C) than a multiple of half the drum speed, becoming a horizontal line (B) when the ratio is a whole number. Spacing of the lines depends simply on the value of interfering frequency (i.e. number of dots per unit length) but visual effect depends critically upon slope of the lines produced. If the interfering frequency varies, the pattern produced consists of curved lines, and this may produce striking pattern effects.

A particular case of interference may arise from the presence of 100-Hz modulation from rectified power frequency in HF channel equipment; after demodulation an interference pattern corresponding to 100 Hz may be present in all parts of the picture, i.e. black or white areas, and not confined to black areas as with simple addition of noise.

For immunity against the effects of random-noise effects on the received picture, a minimum signal/noise ratio of 35 dB is regarded as essential for AM, with preferred values of 45–50 dB for high-quality pictures. At 30 dB the noise effect would be noticeable and at 25 dB a picture would be uncommercial. With FM transmission the signal/noise ratio is less stringent.

Multipath Distortion. On HF radio transmissions, multipath distortion is serious and is the limiting factor to speed of transmission. The difference in transmission times between various paths may be as much as 2 ms and it is possible during a very short time for signals arriving by different paths to become consecutively of maximum amplitude and so take control of the limiter. With an 88-mm drum at 1 rev/s, a time displacement of 1 ms is equivalent to 0·276 mm on the received picture; the effect on the received picture is to record detail randomly out of phase — straight lines transverse to the scanning direction will appear to be staggered.

Subscribers' Black-and-White Documentary Facsimile Service. In a subscribers' facsimile telegraph service over the telephone-exchange network, most of the foregoing considerations apply. Effects of echo and excessive delay distortion are particularly harmful since they prevent sharp black–white transitions and reduce legibility of small typescript or poorly-formed handwritten characters. Black-and-white facsimile systems are less sensitive to interference on account of the large white areas which are usually present, representing no-signal, and therefore immune to brief interruptions or drops in power level; the limited extent of black areas and virtual absence of grey tones in typescript or weather-charts reduce the probability of interference coinciding with useful (black) information. These systems are naturally sensitive to such interference as operator monitoring, time signals, signalling tones, impulsive noise etc.

For various reasons, FM has been agreed as the preferred method of transmission for this service with centre frequency at 1700 Hz and deviation equal to ± 400 Hz, i.e. 2100 Hz corresponding to white and 1300 Hz to black. The use of FM solves the problem of varied line attenuations experienced over a range of dialled calls to different destinations; it also coincides with frequencies chosen for one system of data transmission.

12.5 Phototelegraph transmitter

This description refers to the Muirhead Model K220 (Fig 12.6); being lightweight it is equally suitable for use as a portable transmitter or as a permanent

Fig 12.6 Phototelegraph transmitter (K220) (*Courtesy of Muirhead & Co. Ltd*)

installation. The block diagram (Fig 12.7) shows component units including push-buttons which control the complete operation; semiconductors are used throughout. The OFF button disconnects power from the equipment, a condition cancelled by operating any other button. The STANDBY button allows power to the electronic equipment so that it shall

have reached a stable temperature ready for operation; in this state neither the motor nor the scanning lamp is energized.

The Driving Motor. The picture drum is driven with power at 1020 Hz, derived from a crystal oscillator to ensure high precision and constancy of speed. The frequency of oscillation of the crystal is 4080 Hz, adjustable over a range of ± 10 parts in 10^6 by a small capacitor. To maintain constant frequency, the crystal is mounted in an oven whose temperature is sensed by a thermistor connected in the base-potential divider of one transistor of a 'long-tailed' pair. The long-tailed pair compares the voltage across the thermistor with the voltage across a zener diode and the difference is used to control the current through a power transistor mounted inside the oven which is heated by the power dissipation of the transistor. The oven reaches a steady temperature within 10 min.

The crystal-oscillator frequency is divided by four by means of a pulse shaper and two bistable elements giving a square wave signal at 1020 Hz

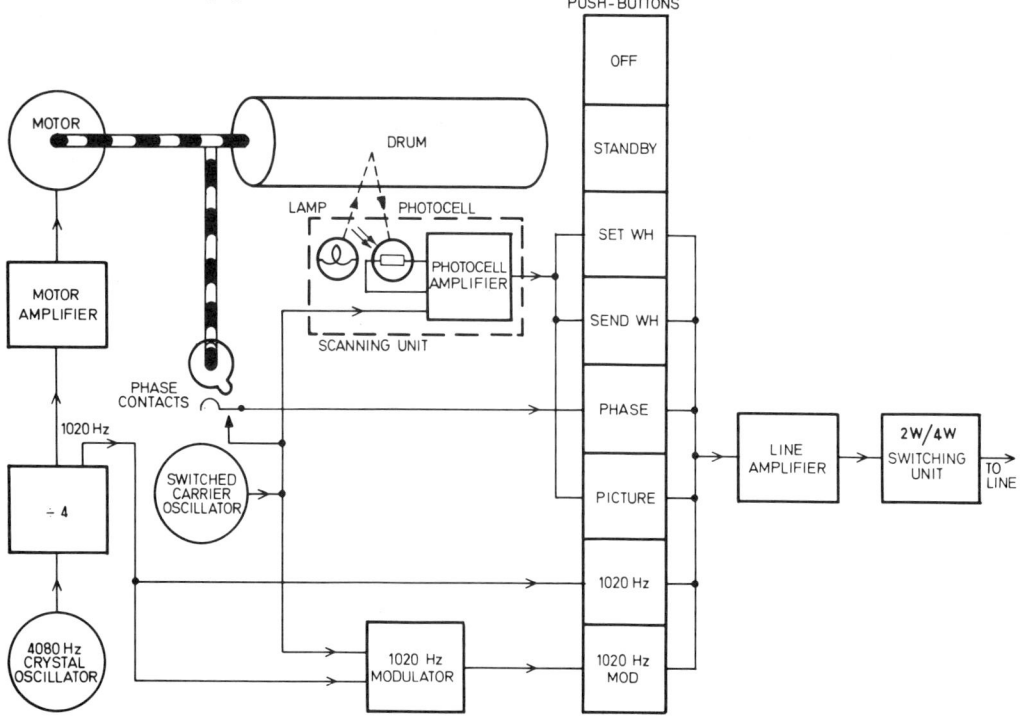

Fig 12.7 Phototelegraph transmitter — block diagram

which is filtered to remove third- and higher-order harmonic frequencies and produce a sinusoidal waveform. The output signal at 1020 Hz is fed via a potentiometer to the 1020-Hz modulator, and also to the motor amplifier which consists of a class B push–pull power amplifier with phase-shift capacitors providing a two-phase supply for the synchronous hysteresis motor.

The motor runs at the synchronous speed of 10 200 rev/min, driving the drum and a leadscrew, through reduction gearing with choice of gear ratio, at either 1 or 2 rev/s as selected by an operating lever. The drum shaft carries a cam to operate a pair of phasing contacts once per revolution and also a clutch sleeve which enables the drum to be coupled up to the motor from a mechanical linkage associated with the PICTURE button.

Scanning Unit. The optical unit is mounted on a carriage to traverse the length of the rotating drum. This unit comprises the scanning lamp (a 4-V filament lamp), a sealed optical unit, the photocell (a double-diode semiconductor) and photocell amplifier.

The arrangement of the optical system is shown in Fig 12.8. The lamp is fed with a.c. at carrier frequency to obviate 'hum'. Light from the lamp is collected by the condenser lens and focused by the front lens as a small spot (the scanning spot) on to the drum surface. Light reflected from the picture on the drum surface returns through the centre of the front lens and the objective lens, and is then turned through 90° on to an aperture screen by the small mirror mounted in front of the condenser lens. The photocell diode is positioned behind the aperture which determines the effective size of the scanned area of the light received from the drum surface. Maximum intensity of light is reflected from white areas of the picture, with minimum from black areas, and intermediate values coming from the range of grey tones. This concentric optical system is an improvement on an earlier method in which defocusing could occur due to slight changes in optical-path length caused by any eccentricity of the drum or by unevenness of the picture surface on the drum.

Resistance of the photocell diode varies with intensity of the light collected from the picture; this property is used in amplitude modulation of the carrier supply which is fed into the 3-stage photocell amplifier. The small internal capacitance of the photodiode produces an unmodulated component of the picture signal which is neutralized by feeding a similar signal to the emitter of the first stage of the

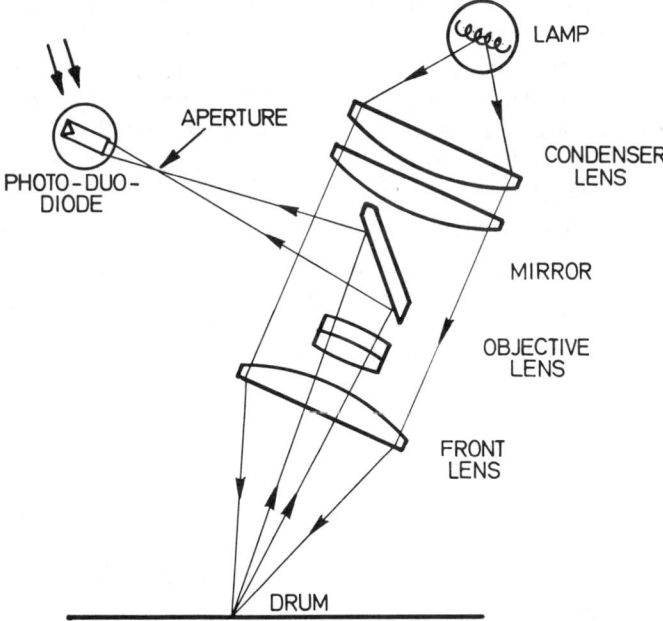

Fig 12.8 Optical system of phototelegraph transmitter

amplifier. A thermistor in the feedback loop of the amplifier compensates for thermal variations in the amplifier and photodiode.

The optics carriage is traversed along the length of the drum by means of a leadscrew which has a ratchet secured to one end, driven by a pawl from an eccentric mounted on the drum shaft. The leadscrew is coupled to the optics carriage by a half-nut which may be disengaged by a reset lever if it is necessary to position the carriage manually. The optics carriage travels along three guide rails on which it is spring-loaded to maintain close contact and preserve the correct vertical and horizontal position. Electrical connections are made to the optics carriage by phosphor-bronze pick-up contacts sliding along six rhodium-plated conductor rails which carry supplies to the scanning lamp and photocell amplifier and also the modulated picture-output signal.

For initially setting up the equipment, adjustments are provided for positions of the scanning lamp and photocell; a magnifying viewer enables the scanned area in the aperture to be seen when adjusting the scanning lamp, and the photocell is positioned for maximum output when a white surface is being scanned.

Carrier Oscillator. The carrier supply is provided from a multivibrator generating a squarewave signal followed by a frequency divider ($\div$ 2) and a low-pass filter which removes the odd-harmonic frequency components to give a sinusoidal output. A frequency of either 1300 or 1900 Hz can be selected by means of a switch which controls the base-bias current of two transistors forming the multivibrator. A thermistor in the multivibrator transistor-base circuit compensates for frequency drift due to temperature changes. Frequency accuracy ($\pm$ 1%) of the carrier supply, which at the receiver will be subjected to envelope detection of the modulated carrier, is less critical than that of the motor supply. In addition to the photocell modulator-amplifier, carrier supply is fed to the 1020-Hz modulator and to phasing contacts.

Synchronizing and Phasing. If both stations have available a standard of frequency which is better than $\pm$ 5 parts in 10^6 it can be safely assumed that

they will be independently in synchronism. Otherwise it is customary for the transmitting station to send 1020-Hz tone, which bears an unvarying relationship to the transmitter-drum speed, over the line, by pressing the '1020-Hz' button prior to transmitting a picture so that receiving equipment can be brought into synchronism. If the line circuit being used is one in a high-frequency carrier system it is always possible that a small change in frequency ($\leqq$ 2 Hz) of the 1020-Hz synchronizing tone may occur during transmission on account of the line modulating–demodulating processes; in this event the receiver oscillator and drum speed might be set fast or slow with relation to the transmitter drum. This difficulty is prevented by sending instead a *modulated* tone in which picture carrier frequency (e.g. 1900 Hz) is modulated by the synchronizing tone of 1020 Hz; at the receiving station, 1020-Hz synchronizing frequency is restored by detection and it is independent of any frequency change occurring in transmission. Synchronizing tone is often known as 'fork tone' since it was traditionally generated by a tuning-fork oscillator.

After synchronizing and before sending a picture, the transmitting station sends phasing signals for a short period; each consists of a short pulse of carrier frequency, the pulse being timed from a pair of contacts operated once per revolution by a cam on the drumshaft. This phasing signal enables the receiving drum to be started in phase with the transmitter drum. Phasing signals occur when the scanning beam is passing across the *dead sector* – the area corresponding to the position of the clip which holds the picture to the drum, in the position of two of the opposite margins of the picture.

Line Circuit. Picture, synchronizing phasing and speech signals are fed via an amplifier adjustable for power level into the line. Associated with this is a hybrid line transformer to enable transmission to take place over a 2-wire telephone circuit; alternative switching enables the equipment to be terminated on a 4-wire telephone circuit – the preferred method – with a 3-dB adjustment to compensate for omission of the 2-wire hybrid transformer. A telephone handset is provided for communication with the receiving station; additional contacts on the telephone set operate a relay which connects the telephone to the line when required since the variable impedance of the microphone must not be connected across the line during picture transmission. A small loudspeaker with its amplifier is connected to monitor all signals coming from the receiving station (except when the telephone is in use).

tight box or cassette which must be loaded in the dark room with either photographic film or paper and later opened in the dark room for photo-processing; the cassette fits the receiver to engage the drum-drive mechanism, and reception takes place in full daylight. A photograph of the receiver

Table 12.1 Muirhead picture-telegraph transmitter K220 (Technical characteristics)

Index of co-operation	352 (or 528)
Drum speeds	1 and 2 rev/s
Scanning density	100, 135·4 or 150 line/in.
Drum dimensions — diameter	2·6 in. (66 mm) or 3·52 in. (89 mm)
length	7·75 in. (197 mm) or 10 in. (254 mm)
Drum factor	2·98
Transmission time	17·5 min @ 1 rev/s
	8·7 min @ 2 rev/s

A power unit supplies stabilized voltages, with a voltmeter and multi-position switch for checking supplies at various points. Technical characteristics for this machine are listed in Table 12.1.

appears in Fig 12.9, a block diagram in Fig 12.10 and the control panel in Fig 12.11. The machine is transportable with a separate power-supply unit; it needs the facilities of a photographic dark room.

12.6 Phototelegraph receiver D700

This is a companion machine to the model K220 transmitter, though both machines are compatible with any other phototelegraph equipment designed to CCITT standards. The receiver is designed equally for *positive reception* on photographic paper or *negative reception* on photographic film.

Transmitter and receiver are basically similar as far as the drum driven at constant speed, the traversing optics and control panel are concerned. In the receiver, the drum is housed in a detachable light-

The Driving Motor. This is a self-starting synchronous hysteresis motor, driven from a valve-maintained tuning-fork 1020-Hz oscillator via an amplifier. The tuning fork vibrates in the fields of two coils which are transformer-coupled in the grid and anode circuits of a differential amplifier designed to have high frequency stability independent of changes in supply voltage or valve characteristics. Fine adjustment of the fork frequency is afforded within ± 50 parts in 10^6 by a potentiometer in the anode-coil circuit of the fork; the oscillator output is fed to a power amplifier and thence to the motor; a 1020-Hz supply is also fed to a fork-frequency comparison circuit.

The hysteresis motor runs at a speed of 3000 rev/min and drives the mainshaft for the drum through reduction gearing via a gearbox at which the drum speed can be selected at 1 or 2 rev/s. The drive to the drum is taken through a clutch comprising twin ratchets and pawls; the ratchet wheels are identical and drive as one unit about the mainshaft but they are displaced by half a tooth pitch relative to one another to improve phasing accuracy on pick-up. The clutch is engaged by operation of a solenoid at the appropriate instant, causing one of the two pawls to engage with one of the ratchet wheels.

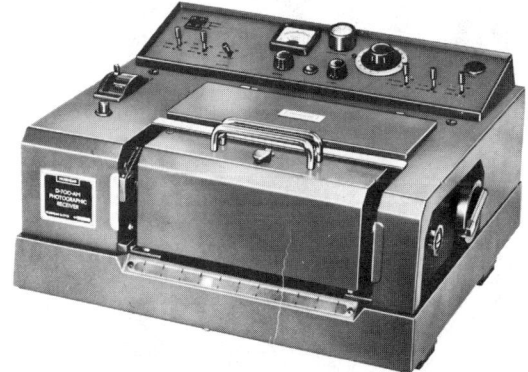

Fig 12.9 Phototelegraph receiver (D700) (*Courtesy of Muirhead & Co. Ltd*)

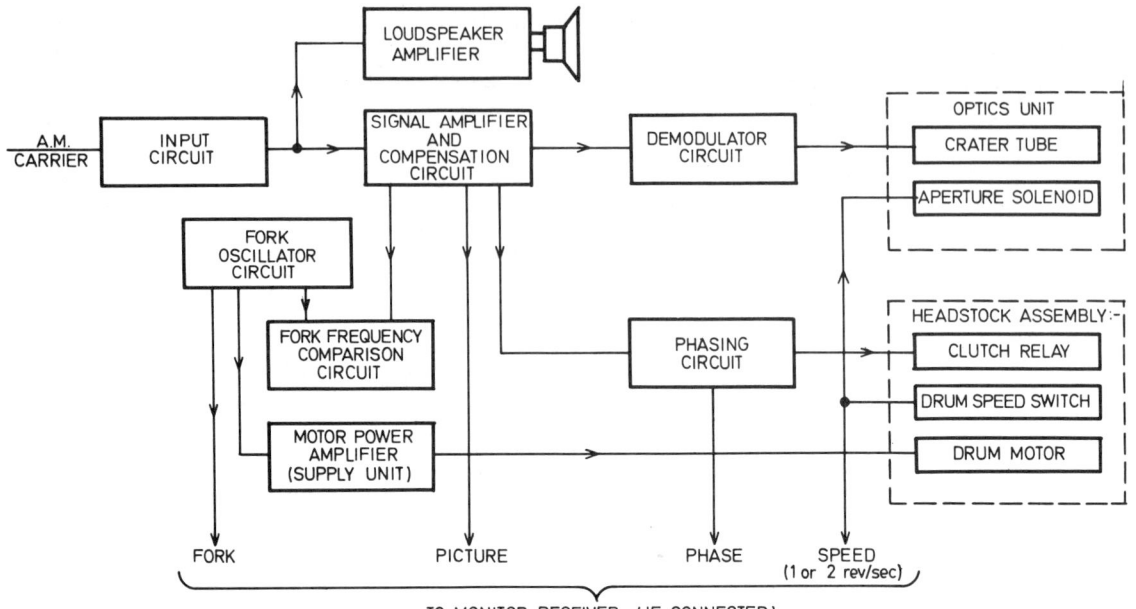

Fig 12.10 Phototelegraph receiver — block diagram

The Scanning Unit. The optical unit is mounted upon a carriage to enable it to traverse the length of the rotating drum. A drawing of the optical system is given in Fig 12.12.

The light source is a crater tube, a gas-filled lamp whose light output is proportional to the current in the lamp; the glow approximates to a point source. The picture signal received over the line is first compensated and then demodulated by a full-wave diode rectifier V4 (Fig 12.13). The demodulated signal is fed via a low-pass filter to remove any carrier-signal component and then to an amplifier valve V5 with

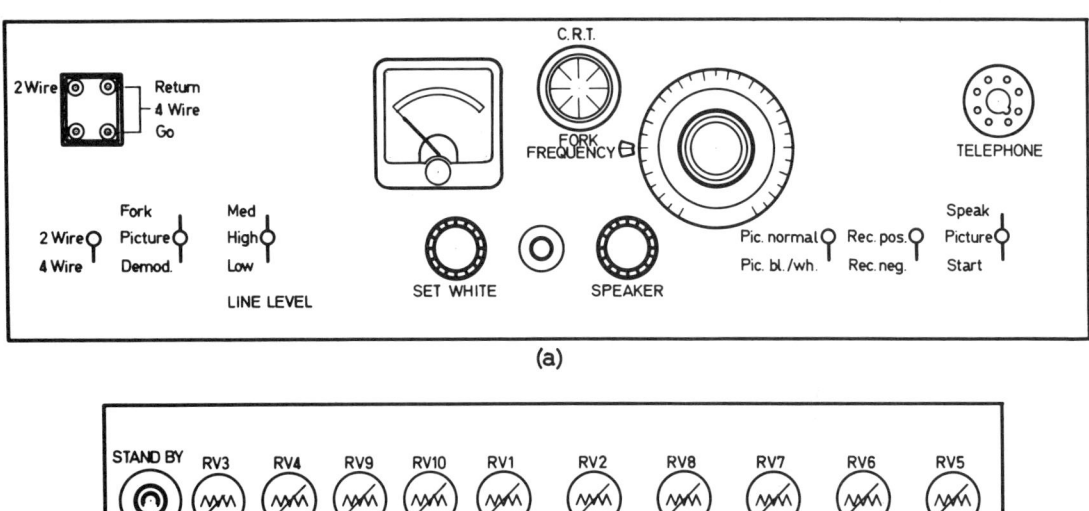

Fig 12.11 Control panel, phototelegraph receiver: (a) operating controls; (b) pre-set controls

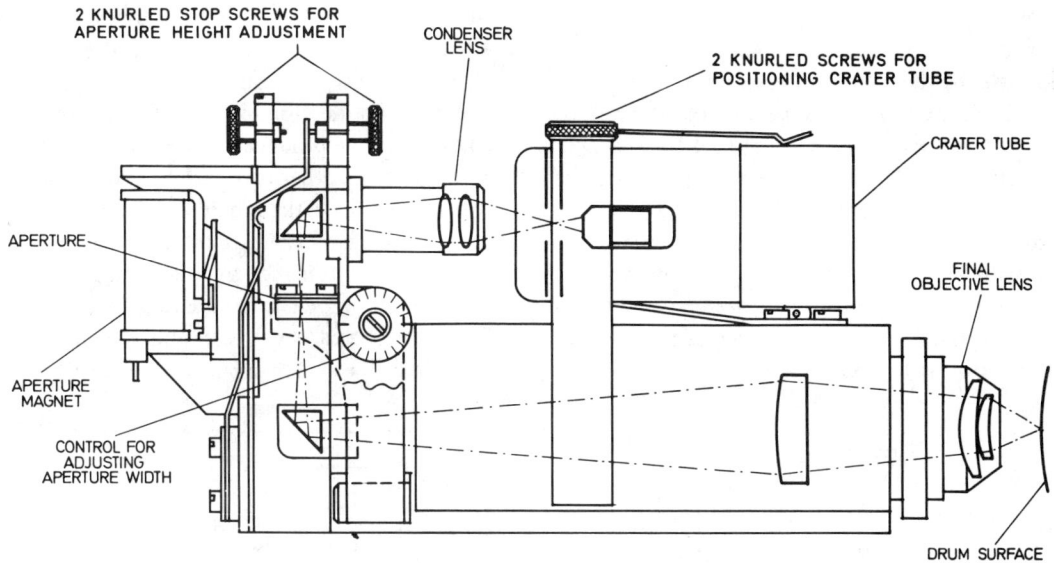

Fig 12.12 Optical system of phototelegraph receiver

the crater lamp in its anode circuit; apart from testing conditions, this lamp circuit is completed only when SW1 is closed by operation of the traverse-control level. The lamp current varies between 1 and 30 mA over a range of signal-input levels covering the white—black contrast of 34 dB; the meter on the control panel reads the lamp current.

Light from the crater lamp is collected by a condenser lens and reflected through two prisms to an objective lens where it is focused accurately on to the drum surface so that the light spot at this point has the correct dimensions. As an initial setting-up procedure the position of the lamp and the height and width of the aperture — between the two prisms — are adjustable. The optics carriage assembly traverses along two tracks to which it is held by spring-loaded rollers; it is driven by a half-nut on the leadscrew. A cam on the mainshaft operates a lever arm to move two spring-loaded driving pawls in a reciprocating action over a toothed wheel on the leadscrew; this causes the leadscrew to rotate progressively in a succession of small arcs and so traverse the optics unit along the drum length. The cam is shaped to give a slow rotation of the leadscrew and a quick return of the driving pawl; movement of the leadscrew which might occur due to the return of the driving pawl is prevented by a phosphor-bronze damping spring which bears upon the side of the toothed wheel. The two spring-loaded pawls are situated

opposite to each other about the toothed wheel on the leadscrew. When the traverse-control lever is in its central position a masking plate ensures that both pawls are lifted clear of the toothed wheel and there is no traverse. The two operated positions of the traverse-control level determine whether only the upper pawl is engaged with the toothed wheel (giving traverse from left to right) or only the lower pawl is engaged (traverse from right to left). When put into either of the traverse positions the control lever

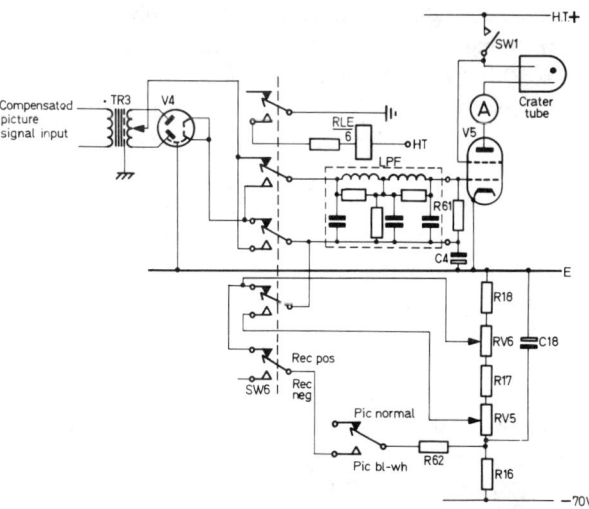

Fig 12.13 Demodulator circuit — phototelegraph receiver

operates a plunger which actuates the shutter mechanism at the back of the drumbox to ensure that the shutter of the light-tight cassette is open only while the drum is being scanned; this lever also operates a microswitch SW1 (Fig 12.13) to complete the circuit of the lamp. At the end of either traverse, the control lever is tripped automatically to its central position, the drumbox shutter closes and the microswitch opens the circuit of the lamp. The position of the scanning spot is indicated on a transparent scale along the front of the receiver. A manual traverse control is also provided.

the local fork oscillator circuit is amplified by V15 and fed to the phase-shift network R78–79–C23, where the voltage produced across R78 is taken to one pair of deflection plates on the cathode-ray tube and the voltage produced across C23 (lagging by 90°) is taken to the other pair of deflection plates. These two voltages have the required difference in phase to produce a circular trace on the screen.

The transmitter fork frequency may be sent either directly, or as a modulation of the carrier frequency if a carrier-type circuit is being used. If sent direct the switch is put to FORK and the trans-

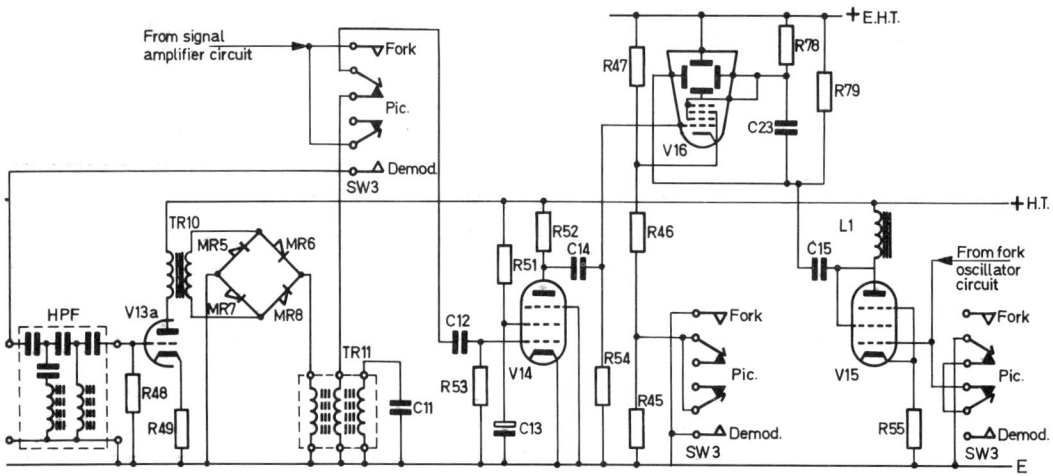

Fig 12.14 Fork-frequency comparison circuit — phototelegraph receiver

Synchronizing. Transmitter and receiver drums must run at the same speed within very close tolerance; both are fed from a 1020-Hz supply. At the receiving station it is usual to compare the local 1020-Hz supply with that sent from the transmitting station for the purpose, unless it can be compared with a local standard of accuracy better than 5 parts in 10^6.

In Fig 12.14 is shown a circuit which enables the transmitter and receiver fork frequencies to be made identical. To compare the two frequencies, a circular trace, with an arc blanked out, is produced on the screen of a cathode-ray tube at the *receiver* fork frequency; the electron beam which describes the trace is interrupted at the *transmitter* fork frequency fed over the line. If the two fork frequencies are identical the cathode-ray tube trace will appear to be stationary; if they differ the trace will appear to rotate in one direction or the other. The output from

mitter fork frequency from the line amplifier is fed to the grid of V14 which is biased so that a square-shaped negative-going pulse is produced for each positive half cycle of the input. These pulses are applied to the control grid of the cathode-ray tube and blank out part of the circular trace.

If the transmitter fork frequency modulates the carrier frequency the switch is put to DEMOD; this applies the incoming amplitude-modulated signal to the grid of V13a via the high-pass filter which rejects any fork-frequency pick-up. The output from this valve is demodulated in the bridge-rectifier network MR5–MR8. The demodulated waveform is the transmitter fork frequency and this is fed via a transformer coupling to V14, the output of which is made to interrupt the electron beam of the cathode-ray tube. If necessary the frequency of the local fork is adjusted until a stationary trace is obtained.

Phasing. The phasing circuit for the receiver is shown in Fig 12.15. When phasing pulses are heard in the monitoring loudspeaker, the switch SW7 is briefly thrown to START and relay RLA operates. **RLA1** holds the relay against the subsequent release of the switch; **RLA2** applies the incoming phasing pulse to the grid of V13b which is normally biased to cut-off by the potentiometer R58–59. When the next phasing pulse arrives V13b momentarily conducts and operates a high-speed relay RLB. At **RLB1** the clutch solenoid RLC is operated to release the twin

ted is from +5 to −26 dBm; with amplitude modulation black level is fixed at 34 dB below white level.

Half-tone or Black-and-white Reception. Between the line circuit and the demodulator circuit the incoming signal passes through a signal amplifier and compensating circuit (Fig 12.16). The functioning of this circuit depends upon the position of the PIC NORMAL/PIC BL-WH key.

With the key in the PIC NORMAL position for

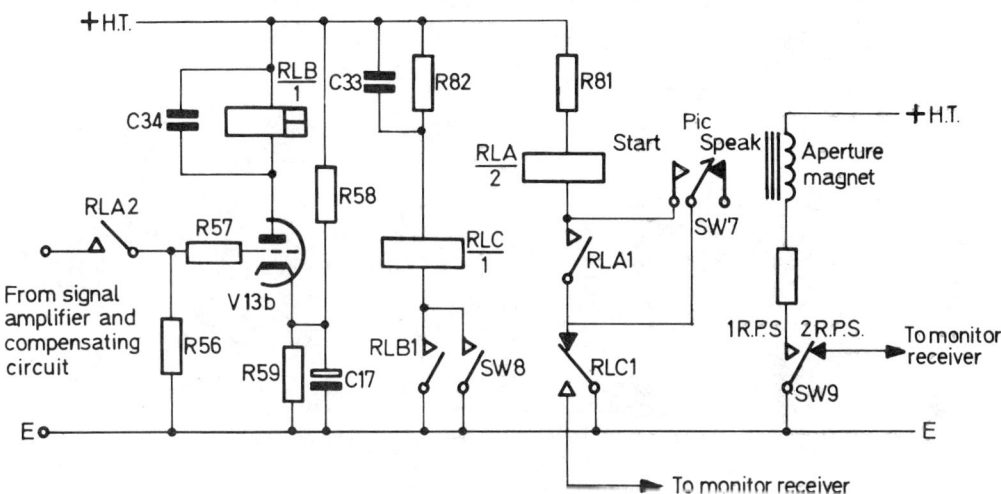

Fig 12.15 Phasing circuit — phototelegraph receiver

pawls of the clutch assembly and the receiver drum commences to rotate in phase with the transmitter drum. The clutch operates a microswitch **RLC1**, releasing relay RLA, and starting up the monitor receiver, if used, in the correct phase.

Line Circuit. The circuit can be switched to operate on a 2-wire or 4-wire termination. A telephone is provided for co-operation with the transmitting station during the setting-up process. A loudspeaker with its amplifier is connected to the incoming line via a high-impedance shunt circuit to monitor all incoming signals except when the telephone is in use. The line signal is fed to the signal amplifier via a 3-position switch LOW-MEDIUM-HIGH associated with attenuator networks and also a SET WHITE LEVEL potentiometer. By these means the total range in white-signal input level which can be accep-

'half-tone' reception, the full range of grey tones between black and white is reproduced, taking into account the characteristics of the recording medium. The incoming-picture signal at the correct level is amplified at V1 whose output feeds two valves V2 and V8 in parallel. Valve V2 feeds the 'black' compensating network while V8 feeds the white-compensating network.

In the black-compensating network a signal corresponding to low-level black or near-black passes from TR1 to TR2 with little attenuation, but a signal tending towards the level whose voltage exceeds the bias voltage across the diode valves V17 and V18, will be attenuated on account of the current which flows via R7, either through V17 and R10 or through V18 and R11, during alternative half-cycles of the carrier.

In the white-compensating network, signals from TR5 are greatly attenuated by current flowing

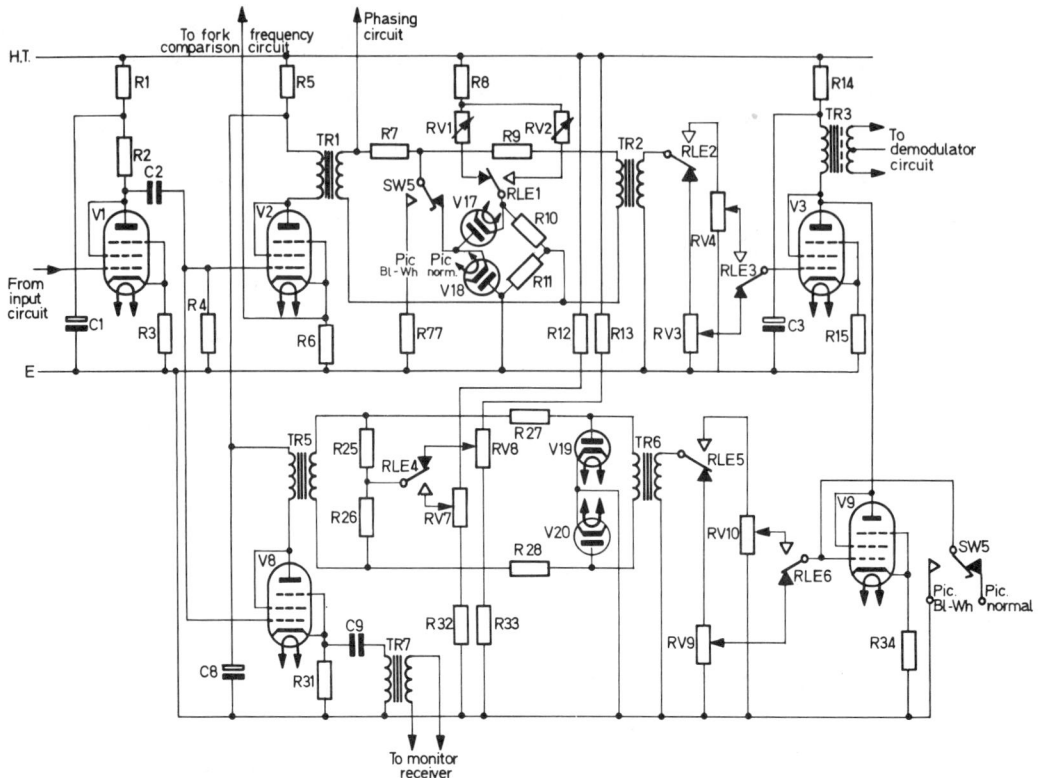

NOTE: RLE shown in REC POS position

Fig 12.16 Signal amplifier and comparator circuit — phototelegraph receiver

through V19 or V20 during alternate half-cycles of the carrier unless they are great enough to overcome the bias voltages of diodes V19 and V20; when the instantaneous value of the signal exceeds this voltage it is not attenuated and the output from TR6 consists primarily of the signal peaks.

The outputs from the two compensating networks are separately fed to the grids of V3 and V9 via pre-set gain controls RV3 and RV9; the amplified outputs from V3 and V9 are combined in the common-anode load TR3 and thence fed to the demodulator and the crater lamp (Fig 12.13).

Alternatively, if the key is moved to the PIC BL-WH position, for reception without regard to the half-tones, the compensating properties of the circuit are eliminated; the output from the white-compensating network is short-circuited to earth and diodes V17 and V18 in the black-compensating network are disconnected; the signal is then amplified by V1 and V3 only. The bias on the lamp amplifier V5 is increased slightly from that for half-tone reception

to reduce the light intensity from the lamp which would otherwise fog the paper. These changes make the near-whites appear pure white whilst mid-greys and near-blacks appear pure black in the received picture. The overall result is that black-and-white material such as line drawings, typescript etc. has greater contrast and effectively better definition.

The incoming signal is also fed to the fork-frequency comparison circuit, the phasing circuit and, if required, to a monitor receiver.

Positive and Negative Reception. If a single print only is required the picture will be received on bromide printing paper (positive reception); if more than one copy is required the picture will be received upon a film (negative reception) so that any number of prints can subsequently be produced from the negative. To change from positive to negative reception (no changes are necessary at the transmitter), three changes are required at the receiver:

(1) For a positive print the transmitter and the receiver must scan the drums in the same direction since the pictures are to be identical. However, for reception on film the picture information has to be recorded on the emulsion surface which in the subsequent photographic printing process will be in contact with the sensitized surface of a sheet of printing paper. The negative and positive copies are therefore mutually inverted, or a mirror image of one another. For this reason, when receiving on film the direction of traverse at the receiving station must be inverted, the direction of rotation of the drum remaining unchanged.

(2) The black- and white-compensating networks have to be changed to suit the photographic characteristics of the film. This is achieved by operation of relay RLE (see Fig 12.13) when the key is thrown from REC POS to REC NEG. Contacts **RLE1** to **RLE6** change the bias voltages for the rectifiers in the black- and white-compensating networks (Fig 12.16) and also bring into circuit different preset gain controls (RV2, RV4, RV7 and RV10 instead of RV1, RV3, RV8 and RV9). The particular functions of these gain controls are seen from Fig 12.11(*b*).

Table 12.2 Muirhead phototelegraph receiver D700 (Technical characteristics)

Index of co-operation	352
Scanning pitch	100 lines/in. (3·9 lines/mm)
Drum diameter	3·52 in. (89·4 mm)
Drum speed	1 and 2 rev/s
Drum length	10 in. (254 mm)
Reception time	17 min @ 1 rev/s
	8·5 min @ 2 rev/s

(3) The black—white signals fed to the lamp must be inverted since, for example, a white signal which must produce white on a positive print must produce black on a negative film. Fig 12.13 shows how the signal inversion is effected; in addition the bias potential for the lamp amplifier V5 is now taken from RV5 (negative reception) instead of RV6 (positive reception).

The power unit provides stabilized voltages for the electronic equipment; a meter with a multipoint switch enables the voltages at important points to be readily checked.

Characteristics of the receiver are given in Table 12.2.

12.7 Phototelegraph operating procedure

Line circuits for phototelegraph traffic are normally used for telephony but nominated for picture traffic, for which purpose they are selected as having suitable characteristics; when required for phototelegraph transmission they are extended as 4-wire circuits to the transmitting and receiving stations, the normal telephone-signalling terminations and any echo suppressors being disconnected at this time.

Amplitude modulation is the normal mode of transmission for line circuits, with white as the greatest and black as the least power level. If radio transmission is used, the same procedure is followed but frequency modulation is used over the radio path.

When the circuit is put through, both stations will have had their equipment in the standby condition for at least five minutes to enable the temperature and performance of it to reach stability. The transmitter will have a picture clipped to the drum; the receiver will have photographic film or paper as required clipped to the drum in the light-tight box on the machine. The medium should lie close to the drum surface to avoid defocusing effects. Unless the information is known from frequent traffic, the transmitting station, using the telephone, advises the receiving station of such characteristics as drum speed, index of co-operation, direction of traverse and whether a half-tone or black-and-white picture is to be sent. The receiving operator puts the DRUM SPEED lever to the desired position (1 or 2 rev/s), puts the PIC NORMAL/PIC BL-WH key to the appropriate position and ensures that the optics carriage is at the extreme and correct end (e.g. left-hand for positive reception) of its traverse. The operator at the transmitting station presses the SET WH button causing the motor to run and the scanning lamp to be illuminated; the position of the optics carriage is adjusted by hand until a white portion of the picture (e.g. the left-hand border) is illuminated by the scanning spot; the OUTPUT control is adjusted until the meter reads the correct white-sending level into a 600-Ω resistor, no signal passing into the line. If the receiving operator requests fork tone, the transmitting operator depresses the '1020-Hz' button (or the '1020-Hz MOD' button if a carrier circuit is being used) and synchronizing tone, modulated by the the carrier frequency if need be, is sent over the line. The receiving operator throws the key to FORK or DEMOD as appropriate and adjusts the fork-frequency control while watching the horse-shoe trace on the cathode-ray screen until it is stationary. When the receiving operator requests a 'White' signal the transmitting operator presses the SEND WH button after ensuring that the white border is still under the scanning lamp and that the output level is correct. The receiving operator then sets the LINE LEVEL key to HIGH and adjusts the SET WHITE control to give a crater lamp-meter reading of 1–5 mA for positive reception (depending upon the make of paper being used) or 27 mA for negative reception. If the incoming-signal level is too low (due to the attenuation of the line being used) and these current figures cannot be atained, the receiving operator puts the LINE LEVEL key to MED or LOW as necessary to cut out attenuators in the line-input circuit and obtain the current figures quoted. When the receiving operator has set white level (the

black level is fixed at 34 dB below the white level) he will ask for a phasing signal, and the transmitting operator presses the PHASE button. Hearing the phasing pulses on the monitoring loudspeaker, the receiving operator flicks the SPK/PIC/START key to START in the period between any two phasing pulses. The receiving drum starts to rotate on receipt of the next phasing pulse, in step with the transmitter drum. When the receiving operator requests 'picture' the transmitting operator ensures that the optics carriage is at the extreme left-hand end of its traverse and then presses the PICTURE button to engage the traverse mechanism. The receiving operator puts the traverse lever to the correct position for positive or negative reception as required. At the receiving station the optics unit traverses parallel to the drum axis; at the end of its traverse the traverse-control lever automatically disengages, the drumbox shutter closes and the lamp current falls to zero. The operator presses the button to stop the motor and removes the drumbox to the dark room for processing. The transmitting operator, on completion of transmission, presses the STANDBY button, returns the optics carriage to the left-hand end of its travel and removes the picture from the drum.

If a receiving monitor set is in use, the receiving operator can ascertain during transmission that reception is satisfactory; otherwise it is necessary for photographic development to have commenced before satisfactory reception can be confirmed. The line circuit can then be released.

12.8 Automatic picture recorder

This machine, illustrated in Fig 12.17, operates automatically, not only for picture reception but for photographic processing also. It not only dispenses with continuous attendance of an operator but also eliminates the need for a dark room. Its applications include reception of press pictures and those from weather satellites. As far as phototelegraph reception is concerned, the design follows the general principles: the machine will operate from any transmitter designed to the appropriate standards.

Operations are controlled from relays and solenoids together with the mechanical devices; pilot lamps denote the stage of operation at any instant. Provision is made for alternative manual control and routine-testing facilities are included. Alarms

Fig 12.17 Automatic picture recorder (K300) (*Courtesy of Muirhead & Co. Ltd*)

are provided to indicate, for example, when paper supply is nearing exhaustion or if the crater lamp should fail.

The picture receiver is left in the standby condition with power switched on. A period of not less than 10 s of unmodulated carrier current provides the starting signal. During a further tone period of 8 s amplifier gain is automatically adjusted to the correct level by a motor-driven potentiometer. Other signals may provide for selecting the drum speed (60 or 120 rev/min) and whether black/white or half-tone material is to be recorded. Phasing, drum rotation and traverse are accomplished automatically following receipt of the phasing signal. Cessation of carrier signal at any stage for a period exceeding 5 s causes the recorder to resume the standby condition.

Photographic paper for finished prints is inserted into the machine in a continuous 300-ft roll, either 8 or 10 in. wide, sufficient for 450 or 325 prints. The paper spool is loaded in daylight, an initial short length being fogged when the paper is threaded into the machine. In operation, a cam-operated mechanism removes the exposed paper from the drum after recording and feeds it into the chemical processor. At the same time a fresh piece of unexposed paper from the roll is fed to the drum in readiness for the next picture which can occur after a delay of only 5 s; this piece is automatically cut by a guillotine to the correct length. A counter shows the number of further prints which can be taken off before re-loading a new paper spool becomes necessary. Recording a new picture can take place while the previous one is being processed; recording takes about 10 min and chemical processing is completed after about 25 s by the ejection into a tray of a damp squeegeed print ready for drying.

To keep processing time to a minimum, a special Ilford rapid-access paper is used with a developing agent contained within the emulsion. After exposed paper has been released from the drum it is drawn through a two-bath processor containing respectively an activator and a stabilizer. The alkaline activator initiates chemical operation of the developing agent while the mildly-acid stabilizer stops the developing process. Finished prints may be stored for several months but if archival-keeping qualities are needed it is necessary to fix, wash and dry in the conventional manner.

The chemical solutions are stored in plastic containers and fed by plastic tubes to the processing trays; circulation by pump takes place during processing. The plastic containers are replaced at intervals of about one week, according to use.

12.9 The Muirhead newspaper-page transmitter (K170)

Newpaper-page phototelegraph equipment is used by the newspaper industry for transmitting a newspaper, a page at a time, to a remote centre; the received films are of sufficiently high quality for printing plates to be made direct from them. High-speed transmission is essential for commercial reasons, resulting in the need for a wideband-transmission circuit — 48 kHz in the system described.

The basic scanning process for transmission and reception follows the principles already described; the main problems here are the great increase in the size and rotational speed of the drum, the higher definition and the reduced scanning time.

The transmitter, illustrated in Fig 12.18, comprises two consoles, a mechanism console and a control console. The former contains the drum assembly with drive mechanism, optics unit and traversing mechanism; the latter houses a control panel with switches, lamps and meters, all electronic units (except the carrier oscillator and modulator) and power supplies. Technical characteristics of this transmitter are given in Table 12.3 which shows a choice from eight

Fig 12.18 Newspaper-page transmitter (K170) (*Courtesy of Muirhead & Co. Ltd*)

Table 12.3 Muirhead newspaper-page transmitter (K170) (Technical characteristics)

Max. copy size	392 x 542 mm
Max. sheet size	416 x 584 mm
Drum diameter	188 mm
Drum length	432 mm
Direction of rotation	Top of drum goes away from observer

Transmission times (min)

Scanning rate (lines/in.)	Drum speed (rev/min)							
	350	400	500	600	650	700	750	1200
300	14·6	12·8	10·2	8·5	7·9	7·4	6·8	4·3
328	15·9	13·9	11·2	9·3	8·6	8·0	7·4	4·7
422	20·5	17·9	14·4	11·9	11·1	10·3	9·6	6·0
600	29·2	25·5	20·4	17·0	15·7	14·6	13·6	8·5
800	38·9	34·0	27·2	22·7	20·9	19·5	18·2	11·3

Carrier frequency, fax,	100 kHz
Carrier frequency, synch.	60·5 kHz
Max. modulating frequency	
Fax. signal	37 kHz
Synch. signal	1 kHz
Bandwidth, total	60–108 kHz
fax.	63–108 kHz
synch.	60·5–62 kHz
Fax. signal-contrast ratio	12 dB
Output level	1 mW into 75 Ω
Phasing signal	Fax. signal (max. level), interrupted once per rev.

Note. The maximum drum speeds for a 48-kHz circuit are to the left of the thick line. A 48-kHz circuit is not suitable for a scanning density of 800 lines/in. at any drum speed.

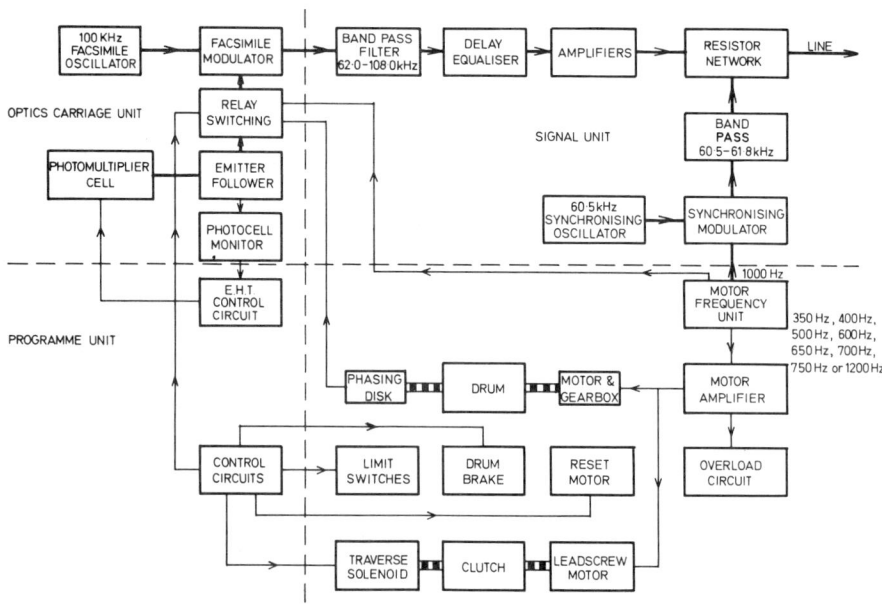

Fig 12.19 Newspaper-page transmitter — block diagram

drum speeds and from scanning rates; the permissible selection will depend largely on the characteristics of the transmission channel. The following description refers to the block diagram (Fig 12.19).

The Optics Carriage Unit. For the optical electrical transducer a photomultiplier cell is used; this is an electronic valve having one cathode with twelve anodes energized at graded potentials, contained within a glass envelope. The minute electron stream released by light striking the cathode is drawn by the positive charge on the nearest anode; the current is increased successively by secondary emission from anode to anode. A useful signal voltage is developed across a resistor in the final anode circuit, the voltage being proportional to the intensity of light reflected from the document on the drum, which strikes the photomultiplier cathode. The principal advantage of the multiplier-type cell is the higher output, which simplifies amplification, reduces amplifier noise and results in a higher signal/noise ratio. At the commencement of each transmission the output from this photocell, focused upon a white page margin, is monitored by an electronic device which includes a comparator circuit and a motor-driven potentiometer to detect any ageing of the cell; the extra-high-tension voltage (250 V) applied to the cell is then automatically raised, if necessary, to compensate for the tendency for reduced emission.

Twin lamps mounted mutually at 90° are used to flood the scanned area with light; this arrangement nullifies the effect of any wrinkling of the wrapped page. In front of the photocell, a multi-aperture disk controlled from a rotary solenoid switch, has five rectangular apertures corresponding to the five possible scanning rates, with a sixth aperture used when adjusting the focus. The correct aperture is selected automatically when setting the scanning rate of the traverse leadscrew. The optics carriage runs along rails, a V-shaped rail holding it at a constant distance from the drum. The carriage is held to the rails by spring-loaded rollers and is driven by a leadscrew between the rails.

The leadscrew is driven, through reduction gears and a 5-speed gearbox, from a hysteresis motor energized by power at one of the eight drum-drive frequencies; the required gear ratio is selected by a rotary solenoid switch. A solenoid-operated dog-clutch couples the leadscrew to its motor at the appropriate time. The optics carriage engages the leadscrew via a nut and traverses from left to right; it is restored by a resetting motor in conjunction with left-hand and right-hand limit switches. The resetting motor can also be controlled for fast traverse either to left or right if it is desired to scan a particular section of the drum.

Picture Drum. The picture drum, which has to rotate at high speed, is supported at each end by air bearings supplied with air at a pressure of 4·6 kg/cm². Any failure of the air supply releases an air-pressure switch to restore the transmitter to the standby condition. A hysteresis motor, energized at the same selected frequency supply as the leadscrew motor, drives the drum through reduction gearing; the drum takes some 12–15 s to reach synchronous speed. During transmission an eddy-current brake, which comprises a copper disk on the drum-shaft running through the field of an electromagnet, applies damping to minimize small fluctuations in drum speed. At the end of transmission the drum is stopped in about 5 s by a powerful electromagnet brake.

Motor-frequency Unit. This unit generates power from a 1-kHz tuning fork-maintained oscillator to provide a means of synchronizing transmitter and receiver; the frequency is accurate to ± 0·5 parts in 10^6. This 1-kHz supply also provides power at eight other frequencies for driving the drum motor and leadscrew motor (both at the same selected speed). Four forks vibrating at frequencies of 1200, 750, 700 and 650 Hz act as efficient acceptance filters to control the frequencies.

Synchronizing. The 1-kHz synchronizing frequency from the motor-frequency unit is used to modulate a carrier frequency of 60·5 kHz generated by a crystal oscillator. The resulting signal (60·5 + 1 = 61·5 kHz) is fed continuously via a band-pass filter which selects the upper side frequency, and injected through a resistive hybrid termination to the transmission circuit where it occupies a position at the lower end of the wideband-frequency spectrum.

Phasing. Phasing pulses generated by a phasing disk with a single radial slit on the motor shaft, in conjunction with the lamp and photo-voltaic cell, are used to modulate the main facsimile carrier supply (100 kHz) for transmission over the circuit at the appropriate stage.

The Facsimile Signal. The facsimile signal from the photomultiplier cell has a maximum value when white is scanned; this signal is fed to a transistor amplifier where, by selecting the output from the collector or the emitter at will, a maximum signal can represent black or white respectively. At this stage the level of the unmodulated facsimile signal can be adjusted by potentiometer. The facsimile signal modulates a 100-kHz carrier generated by a crystal oscillator; the modulated carrier is fed via a band-pass filter and amplifier to the resistive hybrid network where it is combined with the synchronizing signal and injected into the transmission circuit. The amplifier is preceded by a delay-equalizing network which compensates for delay distortion introduced by the band-pass filter. At this point the level of the modulated facsimile signal can be adjusted; power level is measured at the input to the transmission circuit.

Under relay control, the balanced modulator used for the facsimile signal is used alternatively to modulate the phasing signal, the 1-kHz fork tone which initiates traverse at the receiver, or for speech signals for control purposes.

Operation. The sequence of operations is controlled by a programme unit initiated by depressing a START button; lamps indicate the stage reached.

The drum is loaded with copy by a special device which aims at securing close contact between page and drum surface and security when rotating at high speed; drum locks and interlocks are provided. The focus of the optics unit is checked and adjusted if necessary. With the transmitter in the standby condition, switches are operated to select drum speed and scanning rate; the correct aperture is selected automatically for the scanning rate. A switch is thrown to select black or white for maximum line signal.

The transmitter operator ascertains by telephone that the receiver is in readiness, at the same time advising the drum and scanning speeds and whether maximum signal is black or white. At this stage the START button is depressed, causing the drum and leadscrew motors to rotate; lamp indication is given when the drum motor reaches synchronous speed. During this time the comparator circuit operates to set the EHT voltage to the value which enables the photomultiplier cell to give the correct signal output. Phasing pulses are transmitted to enable the receiving operator to phase his drum; he advises the transmitter operator by telephone when this has been accomplished. This done, the transmitter operator depresses the START button a second time: 'traverse start' tone (the 1-kHz fork tone) is then sent for 1 s. Transmitter and receiver commence traversing and the picture is now being transmitted.

On completion of transmission the drum is brought to rest by the electromagnetic drum brake and the carriage unit is restored to the home position. The drum cover can be opened to remove the copy and insert fresh copy for transmission. Full monitoring facilities and precautions against failures are provided.

Fig 12.20 Newspaper-page receiver (K171) (*Courtesy of Muirhead & Co. Ltd*)

12.10 The Muirhead newspaper-page receiver (K171)
The receiver, illustrated in Fig 12.20, comprises separate mechanism and control consoles. Table 12.4 gives the technical characteristics. Reception is on photographic film and can be positive or negative; it

can also be in the form of 'read right' or 'read wrong'. In the latter case the direction of rotation of the receiving motor has to be reversed so that when viewed with the emulsion away from the observer the characters on the received film will be reversed –

focused on to the film drum by an objective lens causing the film to be exposed.

The focus can be checked using a test piece of film fixed to the drum. With the lamp current at the correct value, the drum rotating and the optics

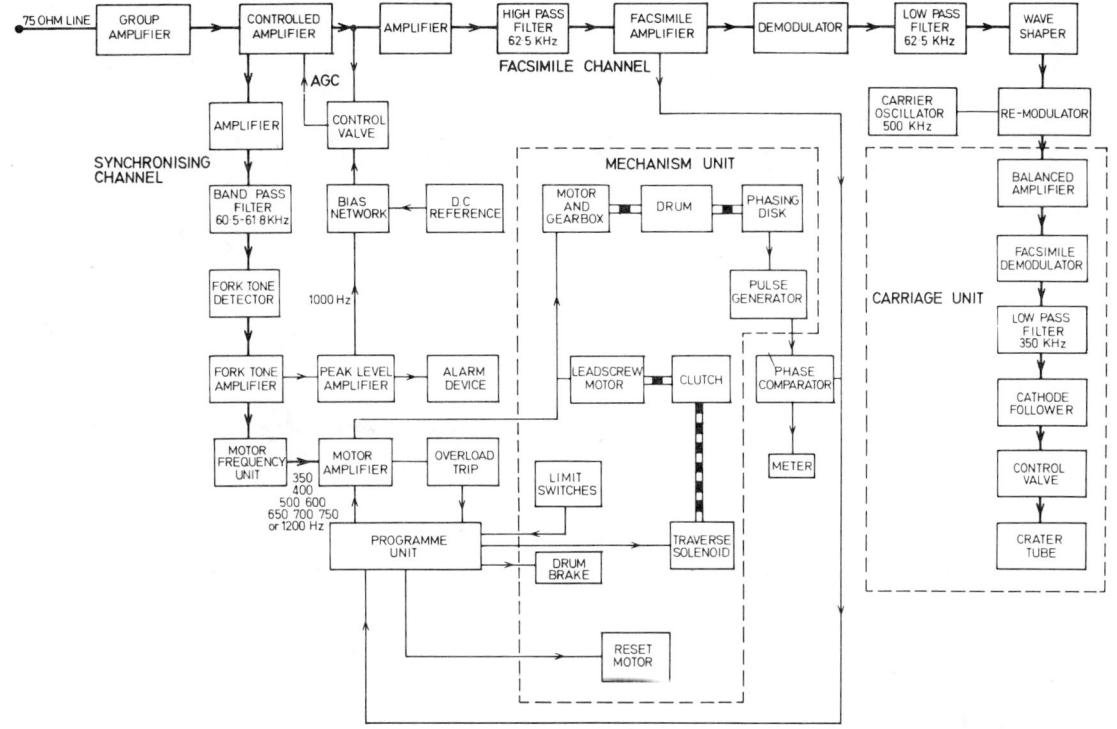

Fig 12.21 Newspaper-page receiver – block diagram

an alternative requirement for preparing printing plates.

The principles of the control panel, drum drive, traverse leadscrew, optics carriage and motor-frequency unit are broadly similar to those of the transmitter; the block diagram of Fig 12.21 shows the general arrangement.

The Optics Carriage Unit. Received facsimile signals are demodulated and passed to a crater lamp which emits light in accordance with the amplitude of the facsimile signal. This light passes through a disk having five apertures; the appropriate aperture is selected by a rotary solenoid switch when the operator sets the scanning-rate switch for the traverse. Light passing through the selected aperture is

carriage traversing, the grain of the film can be viewed through a periscope on the optics carriage and the focus adjusted if necessary.

The Picture Drum. At the receiver the slave drum is driven in synchronism and phase with the master drum at the transmitter. In a dark room, film is wrapped around the drum contained in a detachable light-tight box which is clamped to the receiver and used in daylight. The drum box is fitted with a sliding blind which runs in a channel, between left-hand and right-hand spools, to obscure a narrow horizontal slit running the length of the drum box. The blind carries a shutter which ensures that only the scanned area of the film is exposed to light from the crater tube, and excludes daylight from the slit in the drum box. The

Table 12.4 Muirhead newspaper-page receiver (K171) (technical characteristics)

Max. copy size	
Letterpress	392 x 542 mm
Offset	392 x 521 mm
Film size	
Letterpress	416 x 584 mm
Offset	416 x 558 mm
Drum diameter	
Letterpress	188 mm
Offset	180 mm
Drum length	432 mm
Direction of rotation	Top of drum away from observer for 'read wrong' copy

Reception time (min)

Scanning rate (lines/in.)	Drum speed (rev/min)							
	350	400	500	600	650	700	750	1200
300	14·6	12·8	10·2	8·5	7·9	7·4	6·8	4·3
328	15·9	13·9	11·2	9·3	8·6	8·0	7·4	4·7
422	20·5	17·9	14·4	11·9	11·1	10·3	9·6	6·0
600	29·2	25·5	20·4	17·0	15·7	14·6	13·6	8·5
800	38·9	34·0	27·2	22·7	20·9	19·5	18·2	11·3

Carrier frequency	
Fax. channel	100 kHz
Synch. channel	60·5 kHz
Max. modulating frequency	
Fax. channel	37 kHz
Synch. channel	1 kHz
Synch. signal	1 kHz
Bandwidth	
Total	60—108 kHz
Fax. channel	63—108 kHz
Synch. channel	60·5—62 kHz
Fax. signal contrast ratio	12 dB
Total input level	+5 to −30 dB ref. 1 mW in 75 ohms
Phasing signal	Fax. signal (max. level) interrupted once/rev.

Note. The maximum drum speeds for a 48-kHz circuit are on the left of the thick line.

receiver is fitted with a phasing disk, eddy-current brake and electromagnetic stop brake.

The selected frequency for driving the drum (and leadscrew) is fed through a phase splitter to give a 2-phase supply which is then amplified; operation of the READ RIGHT/WRONG selection switch reverses the polarity of one phase in order to reverse the direction of rotation of the drum motor (the leadscrew-motor supply being unaffected).

Synchronizing. An output from the controlled amplifier is fed after further amplification to a band-pass filter which separates out the synchronizing signal from the combined facsimile plus synchronizing-signal input. The synchronizing signal is demodulated by the fork-tone detector and amplified. The primary purpose of the 1-kHz signal is to feed the motor-frequency unit to lock the frequency of the 1-kHz slave tuning fork; it is also used as a power-level pilot for the controlled amplifier via the bias network and control valve. Locked to the 1-kHz synchronizing frequency, the eight motor-frequency supplies are derived as for the transmitter. The motor amplifier is supplied with one of the eight frequencies; the one selected is used to drive the drum and leadscrew motors in synchronism with the transmitter.

Phasing. An output from the phasing disk on the
drumshaft and pulse generator (a lamp, photo-voltaic
cell and amplifier) is used to indicate when correct
phase relationship with the transmitter is achieved.
The method of shifting the phase of the receiving drum
drum is as follows. The selected frequency which
drives the motor drum is fed to a phase splitter
giving two outputs with a 90°-phase difference; these
two outputs are fed to the stator windings of a phase
resolver. The operator adjusts the phase of the drum
by changing the relative positions of stator and rotor
of the phase resolver. The return signal from the rotor
of the resolver is fed to the motor-frequency amplifier and is used to alter the phase relationship of the
synchronizing signal fed to the drum motor. This has
the effect of changing the angular position of the
drum until it is running in phase with that of the
transmitter.

The phase signals received from the transmitter,
together with the phasing pulses from the receiving-
drum disk, are fed to a bistable trigger which is
switched on and off by the two sets of pulses. When
the drums are in phase, the two sets of pulses are
coincident and the trigger is on or off for a maximum
time, giving a full-scale reading on a meter (or zero,
depending on which pulse is leading) to advise the
receiving operator that further adjustment of the
phase is unnecessary. The two sets of pulses are also
fed to an AND gate which is triggered when they are
coincident; phasing pulses being of very short duration at high drum speeds, the trigger-output pulse is
lengthened sufficiently to operate a relay and light
a lamp indicating that phasing is achieved.

Following phasing, the burst of 1-kHz traverse
tone sent from the transmitter is detected and amplified to fire a cold-cathode tube and operate a relay
which engages the traverse clutch on the leadscrew.

The Facsimile Signal. The received signal is fed via a
group amplifier to the controlled amplifier whose
output is held constant over a wide range of input
level. The received 1-kHz synchronizing tone has a
secondary function as a power-level pilot to control
the amount of feedback in the controlled amplifier.
The constant-level output passes through an amplifier stage followed by a high-pass filter which
prevents the passage of the modulated synchronizing
frequency. The output from this filter passes to the

facsimile amplifier; at this point an output is also fed
to the telephone receiver and loudspeaker unit where
the signal is demodulated for control purposes. The
facsimile signal passes through the demodulator
followed by a low-pass filter which removes any
residual carrier and modulation components; the
facsimile-signal waveform is restored by a wave
shaper.

On the carriage unit the facsimile signal is first
amplified and then demodulated. At this point the
polarity of the facsimile signal can be inverted under
control of the MAX BLACK/WHITE or RECEIVE
POS/NEG switches. The signal is filtered and
passed through a cathode follower to a control
(amplifier) valve and to the lamp. Current in the
lamp and light intensity increase with the applied
signal voltage. At high values of drum speed and
scanning rate, it is important to have an increased
value of lamp current to achieve correct exposure
of the photographic film. For example, at 1200 rev/
min and 800 lines/in. the maximum lamp current
needs to be almost eight times the current required
at 350 rev/min and 300 lines/in. A set of eight
potentiometers, adjusted on first installation, determines the correct maximum lamp current for various
combinations of drum speed and scanning rate; the
appropriate potentiometer is selected by relay control for connection to the control-valve bias circuit
when the DRUM SPEED and SCANNING RATE
switches are operated. Two other preset potentiometers are also selected to bias the control valve for
the correct setting for maximum black or white reception.

A small photodiode, which passes current on
excitation from the residual glow emitted by the
lamp, provides an alarm should the latter fail.

Operation. Receiving operations are carried out by a
programme unit which requires the operator only to
press a START button (twice) and adjust the phase;
lamps indicate each stage of operation.

To receive a picture the drum is loaded with film
in a dark room. With the receiver in standby condition, selection is made of drum speed, scanning rate,
read right/wrong, positive/negative reception and
black/white maximum signal. After advising the
transmitting station of readiness, by telephone, the
receiving operator presses the START button and

the drum motor runs up to synchronous speed in about 12 s (lamp indication). It is then necessary for the receiving operator to adjust the receiving-drum phase to that of the phasing signals received from the transmitter, by adjusting the rotor of the phase resolver until a zero (or full-scale) reading on a meter is obtained and the indicating lamp lights. The operator depresses the START button a second time and advises the transmitting station that the receiver is phased. After a few seconds the receiving carriage commences to traverse, initiated by the traverse signal from the transmitter. Reception proceeds to completion, the drum comes to rest and the carriage restores to the home (left-hand) position. Alarm conditions are provided should the link between transmitter and receiver become broken causing a failure in the synchronizing signal. Provision is made for fast left or right traverse to set the carriage at any specific point and for checking currents, voltages and power levels.

Transmitters and receivers are also available for use on the 60-channel supergroup (248 kHz). Vestigial sideband AM is used with a carrier frequency of 500 kHz, the bandwidth being 240 kHz (312–552 kHz). Drum speeds are 1500, 2000, 2400 and 3000 rev/min; scanning densities are 400, 500, 600, 800 and 1000 lines/in. Depending on the drum speed and scanning rate in use, transmission of a 16 x 24-in. page takes 2·1–10·7 min. Satisfactory transmission requires a channel with differential-group delay not exceeding 4–8 μs over the band.

12.11 Electrosensitive recording paper

It is well over 100 years since first attempts were made to produce a treated paper which would mark a permanent record direct from electric telegraph signals. In the last few decades, such papers have been developed to give good commercial results. Advantages of such a medium are that the recorded message is instantly ready for use without need of photographic or other processing. Facsimile equipment designed on this basis can find a place in any office, the presence of a dark room being unnecessary. These recording papers are capable of excellent results from line drawings, weather charts and printed texts; while they are capable of halftone reproduction, they cannot give the superior-

quality results possible with the phototelegraphic print.

A number of properties are essential in such a paper for commercial use: it must not deteriorate with changes in humidity or temperature, nor be affected by light during storage life; the recorded impression must not fade nor the background discolour; it should neither expand nor contract significantly; it must be capable of recording at sufficiently high speed and should be able to take ink or pencil annotations; if electrolytic, the paper should have high wet strength to avoid tearing when feeding.

Teledeltos Paper. This paper, developed by the Western Union Telegraph Co., consists of a paper stock rendered conductive by uniform impregnation of the pulp with carbon black during manufacture and having an aluminium-backing surface. The front surface is thinly coated with grey opaque material which is sensitive to electric current; such materials as a mixture of lead thiosulphate, which reacts to the current, and titanium oxide to give the light grey colour to provide the 'white' background, have been used. Facsimile signals, either a.c. or d.c. – the latter preferably at a positive potential – are applied to the grey insulating coating of the paper from a wire stylus having a diameter of about 0·010 in. with a pressure of 5–20 g; the circuit is completed through the paper via the aluminium backing to the metallic drum or platen of the recording machine. Tungsten as a stylus material has a high resistance to abrasion from friction and arcing and a longer life than other materials. The method of mounting the stylus is important in order to avoid bounce or fortuitous vibration which would result in excessive sparking and poor definition.

In the recording process, the grey coating is decomposed instantaneously by chemical action from heat produced by the signal currents; to some extent there is dispersion of carbon particles from the paper stock accompanied by arcing at the stylus and charring of the surface to reveal carbon black contained in the paper and record a black impression. The power dissipation in the paper must be regulated by the inclusion of a limiting resistance of a few thousand ohms in the stylus circuit to prevent overheating the paper. Density of marking depends upon the facsimile-signal amplitude, the paper having a limited tone range.

Maximum writing speed depends upon the potential applied to the stylus. For general use with a low-resistance paper, a record speed of 24 in./s will give a distinct marking with a current of 15—30 mA derived from a signal potential of 120 V across the paper; this voltage may result from an initial open-circuit voltage of 220 V required to initiate breakdown of the grey coating. Normal writing speeds lie in the range 10—40 in./s but higher speeds have been obtained under favourable conditions.

Mufax Paper. The action of this paper depends upon an electrolytic process associated with the name of Hogan. It must be used in a moist state and is supplied with the correct moisture content ready for use, in a long roll in a sealed polythene bag; as with photographic materials it has a limited storage life. The paper is of high quality made from resin-bonded sulphite pulp with a high wet strength. Background is dead white, affording maximum contrast with recorded black signals. During manufacture the paper is impregnated with a solution containing correct proportions of chemical materials.

The electrochemical action which takes place in the moist paper during recording is complex and may be regarded as a threefold process. The electrolyte in the paper is rendered electrically conducting by the presence of a suitable chemical substance such as nitrate of potash. A second constituent such as a phenolic compound, normally colourless, forms the black 'colour lake' when electrolytic action occurs with a steel electrode. The third ingredient is an acid, such as oxalic, necessary to neutralize the phenolic compounds to limit spreading the colour lake and also to prevent local action during storage prior to recording.

When current passes between the electrodes, decomposition of the electrolyte takes place at the cathode, commonly in the form of a rotating helix of palladium-silver. At the point of contact of this electrode, an alkali is produced which passes through the paper to the anode — a stainless steel bar known as the *writing bar.* This alkali neutralizes the acid at the point of recording, enabling the iron of the writing edge to be dissolved. Metallic salts so produced combine with the phenolic compound to from an almost black insoluble dye, the colour lake. The quantity of lake formed is, in accordance with the laws of electrolysis, proportional to the signal amplitude and to the time permitted for the action. Spreading of this colour lake is restricted by the presence of unneutralized acid surrounding the recording point. The writing-bar edge is the consumable electrode, necessitating its replacement at intervals, conveniently done at the time of inserting a fresh roll of recording paper. The black impression becomes more intense if the paper is dried at 300°C; alternatively, the paper will dry out naturally in a short time after recording.

Message forms recorded by this process may be kept indefinitely, preferably protected from the action of direct sunlight, with no deterioration except for slight discoloration of the background. The lake-forming compound is volatile and may diffuse through several thicknesses of paper, causing discoloration after a time to any other documents which may be adjacent.

The electrical resistance of the paper falls from 3000Ω at minimum current (white) to about 250Ω at a current of 140 mA. The power required for marking depends upon recording speed and lies between 3—5 W. The threshold of marking voltage is 3 V but for black a value of 45 V is normal; a wide range of half-tone gradations is possible with intermediate values of voltage. Writing speeds of 30 to 40 in./s are normal.

Mufax paper records to a black signal; a photocell reads a white signal. Consequently, white—black inversion must be provided either in transmitter or receiver.

12.12 Facsimile chart transmitter

The transmitter (Fig 12.22) is similar to the phototelegraph transmitter. It is used to transmit weather maps up to 18 in. high by 22 in. wide (457 x 560 mm) and will control any recorder designed to the agreed standards. There is choice of three drum speeds (60, 90 and 120 rev/min), two scanning densities (96 or 48 lines/in.) — hence two indices of co-operation (576 or 288). Signals for start—stop and selection of scanning rate can be sent to give full automatic control, over cable or radio path,

Fig 12.22 Weather chart transmitter (K150) (*Courtesy of Muirhead & Co. Ltd*)

for remote unattended chart receivers. Operation of the transmitter is semi-automatic with a rotary switch operating relays and solenoids; lamp indication is given at all stages.

The Optics Carriage. The scanning area is illuminated by a small lamp energized from a stabilized output of the carrier-current oscillator; the reflected signal from the chart on the drum strikes the cathode of a photomultiplier cell. Lamp current is adjustable to control white-signal level; this adjustment also affects the recording of mid-grey tones. The EHT voltage (about 900 V) for the photocell is derived from the carrier oscillator via an amplifier and voltage doubler; this voltage is stabilized against a d.c. reference voltage using feedback, for any outside variation to the photocell EHT voltage or to the scanning-lamp current would tend to modify the amplitude of the facsimile signal. The carriage traverse is effected by a leadscrew driven from the same motor as the drum; the leadscrew is coupled to the motor drive by a solenoid-operated dog clutch; traverse is provided by a half-nut fixed to the carriage and riding along the leadscrew.

An adjustable end-of-chart stop switch with a pointer enables transmission to be shut down after scanning charts of any width up to 22 in. (560 mm).

The leadscrew can also be driven in either direction and at a higher speed than normal traverse by a reversible resetting motor which is coupled to the leadscrew by a toothed belt. On completion of transmission, the carriage unit is automatically restored to its home position by the reset motor. The carriage can at any time be given a rapid left or right traverse by the reset motor in order to select or omit any part of the chart length during scanning.

The Chart Drum. A 3-position gear lever and gearbox enables one of three drum speeds to be selected. It is unnecessary to transmit a specific signal to indicate the drum speed because the receiving machine assesses it from the frequency of arrival of phasing pulses.

The drum is driven through reduction gearing by a 3000-rev/min hysteresis motor supplied with power at 1 kHz from a tuning-fork amplifier. This amplifier has a high degree of frequency stability and is adjustable within 40 parts in 10^6 by varying the bias on the differential amplifier grids. The fork frequency may be checked occasionally against a standard 1-kHz source; the output is amplified and fed to a phase splitter to produce a $90°$ 2-phase supply, each phase being then amplified to feed the 2-phase synchronous hysteresis motor. A solenoid-operated dog clutch is used to couple the drum to the gear train.

The chart is loaded on to the drum using a transparent wrapper which can be pulled out by the operator against spring tension. It is placed face downwards and when the wrapper is allowed to restore gently under the spring tension the chart becomes wrapped around the drum. Scanning takes place through the wrapper.

Control Signals. The following control signals are generated for transmission to the receiver. They are each applied to the modulator at the appropriate stage under control of the programme switch; start and stop signals can be suppressed under operator control if desired.

(1) *Start signal.* This is a 5-s tone either of 300 Hz to set the receiver for an index of 576, or of 675 Hz for an index of 288. It is generated by a conventional low-frequency oscillator whose tuned anode is adjusted according to the setting of the 2-speed traverse lever.

(2) *Phasing signal.* A set of phasing contacts operated by a cam on the drum close once per revolution. At the required stage a 30-s train of phasing pulses (5% white, 95% black) is derived by applying d.c. potentials to the modulator input, and sent over the transmission channel.

(3) *Start-traverse signal*. Following phasing pulses, carrier current is modulated at 50 Hz for 1 s to send a signal which will operate the traverse clutch at the recorder.

(4) *Blanking signal*. During chart transmission, blanking pulses are generated by a set of contacts similar to the phase contacts to ensure that a white signal is sent while the chart clamping bar is passing the scanned area.

(5) *Stop signal*. The 5-s stop signal at a frequency of 450 Hz is generated by the low-frequency oscillator.

Facsimile Signals. A conventional valve oscillator generates the carrier supply at 1800 Hz or other frequency. The main output from the oscillator is fed to a balanced modulator where it is amplitude modulated by the facsimile signal. Prior to modulation, the upper (black) and lower (white) levels of the facsimile signal are clamped by the use of two diodes such that white signal is at earth potential and black is at a level adjusted by the SET BLACK potentiometer (see Fig 12.29). Modulation depth, the black–white contrast ratio, is 15 dB and controlled by a potentiometer which sets the grid bias of the modulator valves. By designing the output from the photocell to exceed the potential swing required by the modulator grids, a degree of limiting is incorporated which raises the transmitted level for dark grey tones to black level; this is instrumental in assisting the receiver to record darker grey tones of the original chart which might otherwise be recorded as white and lost, causing reduced legibility. The modulated facsimile signals pass through a high-pass filter to suppress unwanted residual facsimile signals and to an amplifer where the power level is set by a potentiometer before passing the modulated signal to the transmission channel.

Operation. The operator selects drum speed by a 3-speed gear lever and scanning density from a 2-speed gear lever, and presses the STANDBY button to apply power to the equipment. A chart is loaded on to the drum and the end-of-chart pointer is set to correspond with the end of the chart to be transmitted. Operation of the TRANSMIT push-button

puts into operation the successive transmission of start and selection signals (unless countermanded), phasing signal, start-traverse signal and facsimile signals. All outgoing signals are monitored on a small loudspeaker; on reaching the end of chart transmission the stop signal is sent, if required, and the carriage restores to the home position; the equipment then returns to the standby condition.

Fig 12.23 Weather-chart recorder (D649) (*Courtesy of Muirhead & Co. Ltd*)

12.13 The mufax chart recorder
The receiver illustrated in Fig 12.23 is the 18-in. chart recorder. It is designed to operate unattended if correct control signals are received from a transmitter; otherwise it can be operated manually. A block diagram is given in Fig 12.24.

Recording Electrodes. The mechanical design of the receiver differs basically from the phototelegraph receiver. Instead of using a drum and optics carriage for scanning, electro-sensitive paper is drawn at constant speed between two electrodes – a stainless steel horizontal writing blade and a rotary helix of palladium-silver – see Fig 12.25. The point of contact with the paper between the writing-blade edge and the helix scans the paper in a horizontal line across the page from left to right for each revolution of the helix. Current proportional to the input signal is passed through the paper at the point of coincidence between the electrodes, causing a grey or

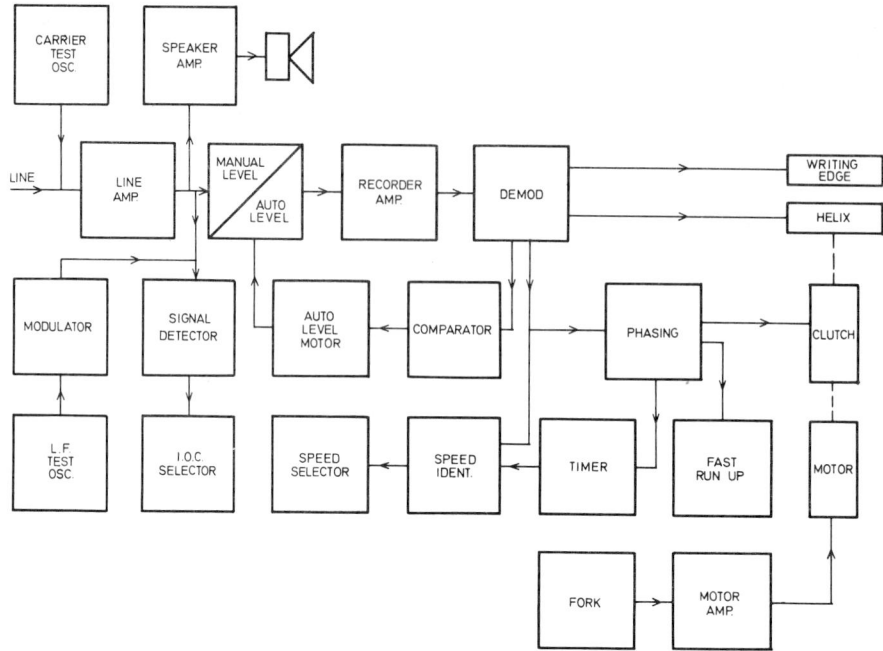

Fig 12.24 Weather-chart recorder — block diagram

black coloration with a density in proportion to the current. The writing edge is held against the paper under spring pressure of 130 g. Marking power is applied to the rotating helix via a carbon brush.

For a helix speed of 120 rev/min the facsimile marking power is applied after amplification direct to the helix and writing-bar electrodes. At the lower speeds of 60 and 90 rev/min a lower power is necessary; series and shunt resistors are switched into circuit by operation of the speed-selector circuit. Maximum recording voltage is held at 40 V, displayed upon a meter, a fairly critical figure to ensure not only a good black-marking potential but also a faithful reproduction of intermediate grey tones.

The writing-bar edge is gradually consumed by electrolytic action; as it has two edges it is usually reversed or renewed with each new roll of paper,

being readily detachable from the recorder. The helix is kept clean by a rotating brush which runs the length of the helix shaft; it is advisable to clean the helix daily by hand, using a small piece of mufax paper.

Facsimile Signals. All received signals are fed to a line amplifier and thence via a potentiometer, which controls the marking power, to a recording amplifier – a push–pull amplifier in which the black–white contrast is adjusted by a potentiometer controlling grid-bias potentials to provide a voltage ratio of 20 : 1 for a 12-dB contrast. The facsimile signal is demodulated by a full-wave bridge rectifier and the d.c. output is fed direct to the recording electrodes.

The marking-power potentiometer is motor-driven and commences from zero on receipt of the start signal and stops when the voltage from the recorder amplifier reaches a pre-determined value when monitored on a comparator circuit.

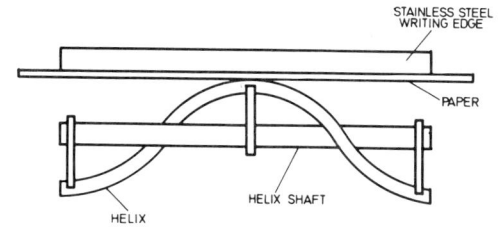

Fig 12.25 Recorder electrodes

Helix Drive. The helix motor, a synchronous hysteresis motor, runs at 3000 rev/min and is coupled by reduction gearing and through a 3-speed gearbox

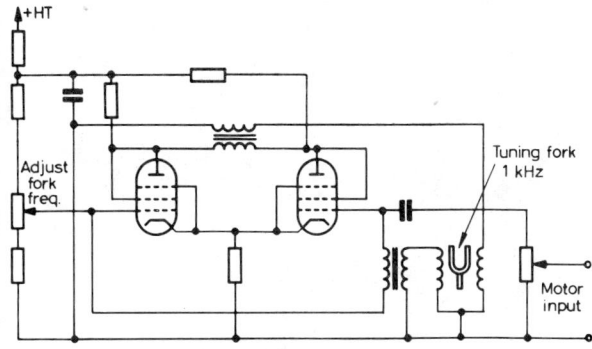

Fig 12.26 Tuning-fork motor supply

and pawl-and-ratchet friction clutch actuated from a solenoid. The motor is energized at 1 kHz from power delivered by a tuning-fork oscillator and amplifier (Fig 12.26). Power from the fork-grid winding is transformer-coupled to a differential amplifier, the output of which is transformer-coupled to the fork-anode winding; a state of oscillation at the fork is maintained. Fork frequency may be varied by 40 parts in 10^6 by the potentiometer controlling grid bias to both valves; when bias is reduced, the valves drive the fork harder and frequency is reduced. The fork is energized in the standby condition, and output is fed via a power amplifier to the motor when required.

There is no synchronizing signal between transmitter and receiver; each machine depends upon stability of frequency supplied by its own fork. With the transmitter regarded as the master frequency, any discrepancy between the two supply frequencies will cause a gradual drift of the recorded chart margin either to left or right; a drift of $\frac{1}{4}$ in. in 12 in. of chart length corresponds to a frequency error of ± 10 parts in 10^6.

Selection of Index of Co-operation. An output from the line amplifier is fed to the signal detector. After further amplification by a valve with automatic gain control and a broadly tuned-anode low-frequency circuit, the signal is demodulated to restore the control frequencies, amplified, and fed to three tuned circuits. These select the appropriate signal which is separately rectified and the d.c. output signal added to a reference voltage on a 'delay' control potentiometer. Each of the control signals, when present,

charges a capacitor which, when sufficiently charged, will fire a cold-cathode valve and operate a relay. This condition — depending upon the receipt of *either* the 300- or 675-Hz start signal — determines the operation of a solenoid which controls the gear ratio in the paper-feed drive. On a helix recorder, it is the paper-feed rate which determines the scanning-traverse rate (and index of co-operation). In addition to selecting the index, the start signal also controls application of power to the helix motor; both these conditions are later countermanded by receipt of the STOP signal. False operations of the signal detector to spurious signals, or during a transmission, is prevented by diodes which provide short discharge times for the signal-storage capacitors and absorb spurious pulses.

Alternatively, the recorder can be started and stopped by receipt and cessation of the carrier signal from transmitters which do not provide the tone-control signals; in this case, automatic selection of scanning rate cannot take place.

Selection of Helix Speed. No specific control signals are received to indicate helix speed; the required information is embodied in the repetition frequency of phasing pulses which occur at 60, 90 or 120 rev/min.

After black level has been set, phasing pulses are fed into a capacitor which charges up during the 95% periods and discharges during the 5% of phasing pulses. The maximum p.d. attained by the capacitor depends upon pulse rate, i.e. it is highest for 60 rev/min and lowest for 120 rev/min. Cold-cathode valves primed to fire at the selected p.d.s. cause selective operation of relays, which establish the correct setting for a rotary solenoid switch to set the position of the speed-change shaft in the gearbox of the helix drive. Speed identification takes place after two phasing pulses and correct speed setting occurs within about 2 s. The receiver can respond to the alternative use of a 50%/50% phasing pulse sent by some transmitters.

Phasing. After helix speed has been selected, the next phasing pulse is directed to a thyratron valve which fires and operates a solenoid to engage a clutch and complete the drive between helix and motor. When the helix commences to revolve, the facsimile signal is extended to the recording electrodes. For manual operation, when the phasing signal is heard in the loudspeaker a PHASING button is pressed until the clutch engages to the next pulse received and drives the helix.

It is possible to bring the recorder into use manually at an intermediate stage when a chart is being received, for example by radio. To bring the received chart into correct phase the operator depresses a PICTURE SHIFT button which brings into the helix-motor circuit a commutator, driven by the helix gearbox, which intermittently reduces the motor power, causing the average speed of the helix to fall. The recorded chart is gradually displaced towards the left of the paper for as long as the button is depressed.

Paper Feeding. A 100-ft roll of paper is contained in a compartment of the receiver, sealed with a rubber gasket to render it airtight. Paper is fed by a feed roller and slipping clutch from the helix-drive motor through a 2-speed gearbox. The paper-feed roller rotates at a speed proportional to that of the helix, the proportion being selected by the gearbox according to scanning rate. Paper feed takes place by friction, the paper being held against the feed roller by a roller at a pressure of 2–3 lb. The paper is held taut by an upper feed roller which is slightly over-driven by a chain drive from the lower feed-roller gear, but a slipping clutch prevents excessive tension in the damp paper. A tear-off blade enables a completed chart to be removed from the machine.

If the recorder has stood for any appreciable time between recordings, exposed paper near the writing bar will have dried out and become unsuitable for recording. A separate paper-feed motor is energized for a few seconds at commencement of transmission to feed off the 2–3 in. of dried paper.

A small perspex viewing window in the paper chamber shows the amount remaining on the roll at any time; when the roll diameter coincides with two lines scribed on the window, 20 ft of paper remain unused. As a reminder, a marginal red line appears for 72 in. of the last 82 in. of the roll. Finally, a pair of contacts close if paper becomes exhausted and brings about alarm conditions.

Fig 12.27 Business-document transmitter (D901) (*Courtesy of Muirhead & Co. Ltd*)

12.14 The Mufax business-document transmitter (Fig 12.27)

This machine is designed for use in a commercial office to transmit the contents of any document up to $14 \times 8\frac{5}{8}$ in. (355×220 mm) to a receiver situated in a remote office. Communication takes place over the public telephone system, the connection being first established by dialling; when the correspondent answers, both parties switch from telephone to facsimile machine. Alternatively, the equipment is suitable for use over a leased telephone circuit with or without alternative speech. The equipment is designed to be used by office staff without specialized skill. The transmitter operates automatically from the moment of pressing the START button. A buzzer draws attention when the machine shuts down automatically on completion of transmission. A lamp signal acknowledges satisfactory reception by the receiving station. The machine is of similar design to the chart transmitter; a block diagram is shown in Fig 12.28.

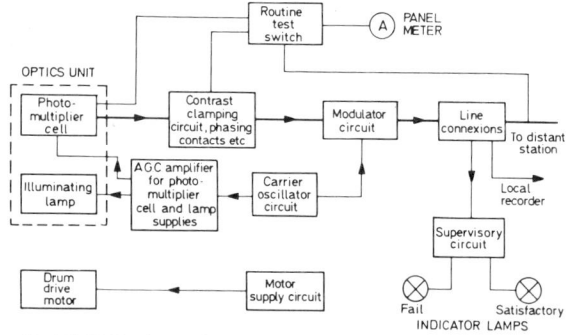

Fig 12.28 Business-document transmitter — block diagram

The Message Drum. The message is placed within a spring-loaded transparent wrapper. If not too large, several documents can be loaded side-by-side on the drum. If a document with a dark background does not completely cover the surface of the drum, a backing sheet of the same colour must be used to maintain the correct average background colour. An end-of-message pointer corresponding to the right-hand limit of the document on the drum is set by the operator.

For inland communication where the electric-power system is synchronized throughout the network (as in the United Kingdom), the 50-Hz mains supply drives the motor and provides synchronism; otherwise a 1-kHz tuning fork or crystal is provided. When the START button is pressed, the 50-Hz hysteresis motor drives the drum through reduction gearing at a single fixed speed, dependent upon the transmission characteristics of the line, e.g. 240 rev/min for the public exchange system and 360 rev/min for a leased circuit.

Optics Carriage. The scanning lamp is energized from the carrier-current supply; the latter is also used to provide the EHT voltage (900 V) via a voltage doubler for the photomultiplier cell which responds to light reflected from the document. Scanning lamp and EHT supplies are stabilized to render them independent of mains-voltage variations. Lamp current can be adjusted by the operator within narrow limits for setting white-signal level according to a marked sector of a meter scale; some latitude is provided if the background colour of the form is other than white in order that it shall appear white on the recorded copy. Scanning density is 90 lines/in. ($M = 264$).

The leadscrew is driven from the drum motor; the drive is extended via a dog clutch when it is engaged by the traverse solenoid after the phasing-pulse train has been transmitted. A captive nut on the carriage unit is traversed from left to right by the rotation of the leadscrew. When the carriage reaches the end-of-message pointer a microswitch causes the traverse solenoid to disengage the dog clutch; a reset motor is energized to drive the leadscrew via gearing in the reverse direction to restore the optics carriage to the left-hand (home) position, where a second microswitch stops the reset motor.

Phasing. The drum motor quickly reaches synchronous speed and black signal interrupted by white-phasing pulses is transmitted for about 5 s — measured by an RC circuit and a valve-operated relay TR (Fig 12.29). These pulses are generated from a contact PH1 interrupted by a cam on the drumshaft. The grids of the modulator are given a slight positive potential corresponding to black signal; each time phase contacts PH1 close, the modulator grids are earthed — equivalent to white signal.

On conclusion of the 5-s period relay TR operates; traverse is started by **TR1** which controls engagement of the traverse clutch while **TR2** extends the photocell output to the modulator. The phasing contacts continue to operate and provide 'blanking' — during the period when the image of the metal edges of the transparent wrapper falls on the scanning aperture; a white pulse is transmitted instead of the near-black pulse which would otherwise have been transmitted.

The Facsimile Signal. Carrier current is generated at 1500 Hz by a tuned-anode oscillator having a short-term frequency stability better than $\pm 0.1\%$ and an amplitude stability of $\pm 2\frac{1}{2}\%$ for mains-voltage variations of $\pm 10\%$. Carrier current is fed to a balanced modulator (Fig 12.29) where it is amplitude-modulated by the facsimile signal and passed to the line transformer.

The optical contrast between black and white in a business document is 25–30 dB. Contrast in the transmitted signal is deliberately reduced to 12 dB (i.e. carrier amplitude for white is 12 dB below black level) since it is known that delay distortion becomes less harmful as contrast ratio is reduced. The facsimile signal voltage developed across load resistor R in the anode circuit of the photocell is fed via capacitor C but clamped by MR1 and MR2 which fix the upper (black) and lower (white) limits before passing it into the modulator. Assuming that the drum is stationary, the modulator is biased to black by a small positive voltage nearly at earth potential. When the drum rotates, an area of black passes the scanning aperture and output voltage from R rises to about earth potential. This represents a change of a few volts in the positive direction which, although it is coupled by C to the modulator input, is restricted to approximately 1 V by MR2 and the potential on the slider of RV.

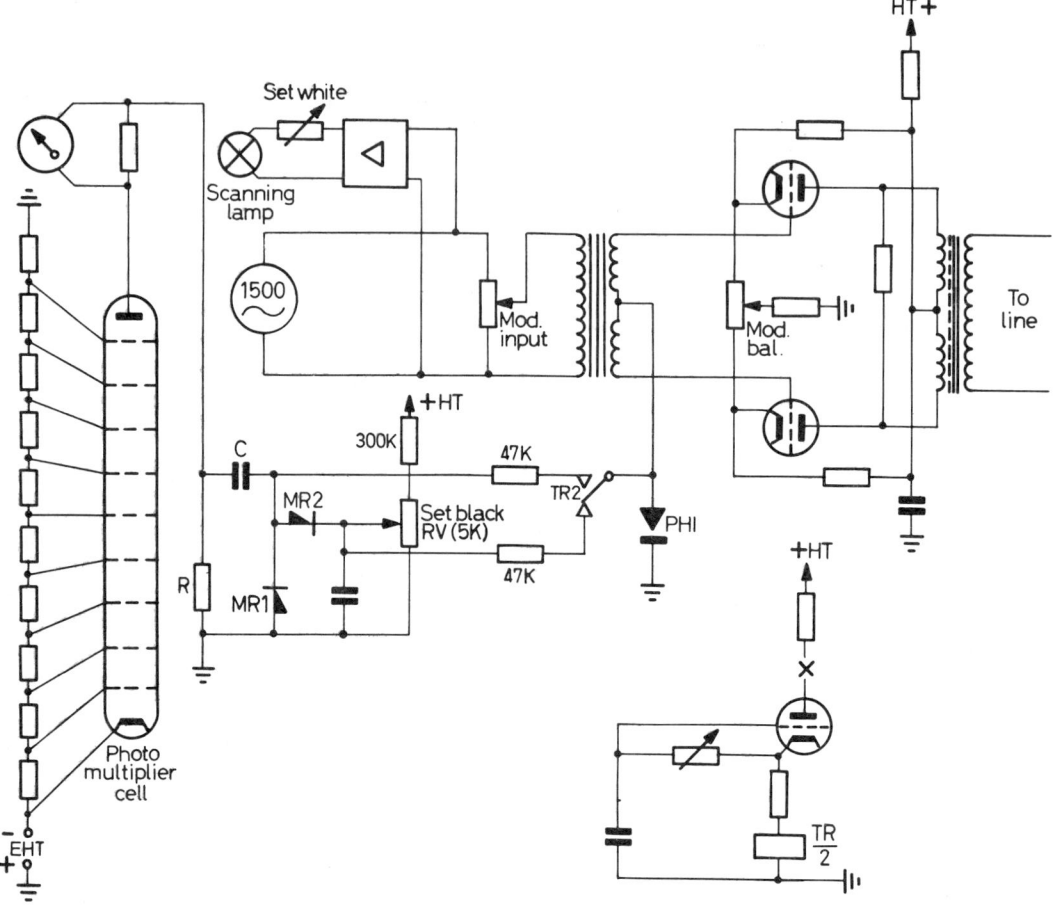

Fig 12.29 Facsimile balanced modulator

This process continues throughout transmission, signals being restricted to earth potential for white and to the positive potential on the slider of RV for black. During long periods of black signal the capacitance of C is sufficient to hold the modulator grids at the correct positive voltage in spite of the tendency for them to return to near-earth potential.

The excursion of output-potential voltage from the photcell exceeds the permitted excursion of 1 V for the modulator input. Excess voltage is used to offset any slight variations in background whiteness, resulting in an improvement of appearance of the received copy.

The above description relates to the AM model. More recent models use FM with 1500 Hz for white, 2100 Hz for black, and with 300 and 450 Hz for start and stop signals respectively.

12.15 Business-document receiver (Fig 12.30–31)

This receiver operates with the transmitter of Fig 12.27. The recorder operates from electro-sensitive paper scanned by a helix and writing bar. The machine is for unattended operation, apart from the need to switch an incoming telephone call to the facsimile machine if over the public system. Synchronism is dependent on the 50-Hz supply mains, if synchronized, otherwise a tuning fork or crystal is used. Time intervals required for controls are generated from RC circuits with cold-cathode or thyratron valves, and relays. The electronics and power unit (not shown) can be used as a base on which to stand the recorder.

A 200-ft roll of electro-sensitive paper, supplied in a sealed polythene bag which contains also a replacement writing bar, is placed in an airtight com-

Fig 12.30 Business-document receiver (D900) (*Courtesy of Muirhead & Co. Ltd*)

partment sealed with a rubber gasket. An alarm indicates when the paper roll is almost exhausted (30 in.).

Input signals are fed to the amplifier via a high-pass filter which cuts off any spurious signals below 400 Hz. When the transmitter operator presses the START button, reception of carrier frequency in the tuned-signal detector operates relays which prepare the receiver by starting the helix and leadscrew motor and also the fast-paper-feed motor. A few inches of unrecorded paper are fed out rapidly to ensure that moist paper lies between helix and writing bar; this fast paper feed can be overruled by the operator if several documents are to be recorded in succession.

On public-exchange calls, line attenuations vary widely over successive connections to different points on the network; the receiver must accommodate attenuations in the range 0 to −40 dB. A comparator circuit measures the level of rectified input signal against a preset reference voltage; a power-driven potentiometer increases input voltage to the marking amplifier until a balance is struck. The next phasing pulse to arrive is applied to a thyratron valve which fires and operates the clutch to couple up the motor with helix and leadscrew. This clutch is of the double-ratchet and pawl type acting with a pair of gear wheels with staggered teeth to reduce pick-up time. At the same time a solenoid operates to press the writing bar against the paper, from which it has hitherto been withheld to avoid the paper being marked by phasing and other signals; the heater bar is also energized at this stage. Facsimile signals are fed to the recording electrodes via the amplifier and demodulator; the receiving amplifier is given a non-linear characteristic which expands the contrast to give an output-voltage ratio of at least 20 : 1 (26 dB). If FM is used, the limiter will cope with a 40-dB input range and the motor-driven potentiometer is unnecessary.

The recorded document is dried out by an electric-heater bar — adjustable for off, low, normal or high power — before being fed out. If the transmitted document has a white, off-white or lightly-coloured background it will be reproduced as white; deeper-coloured backgrounds (if not too dark) are reproduced as off-white.

On completion of transmission, carrier signal ceases and the receiver shuts down, except for (1) opera-

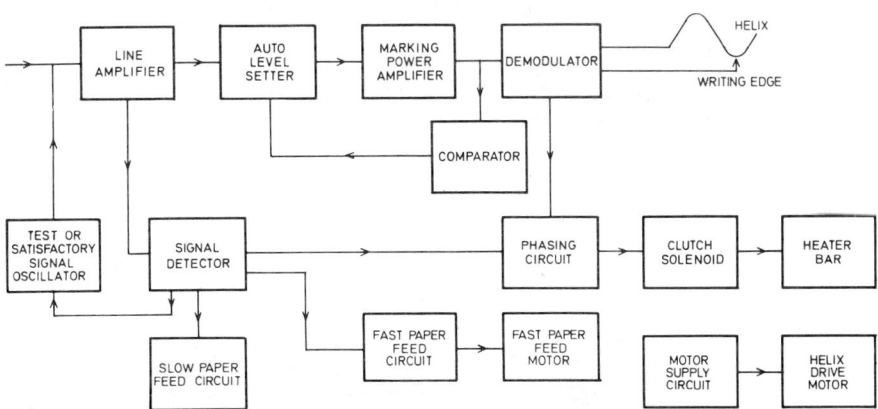

Fig 12.31 Business document receiver — block diagram

ting a tuned anode-signal oscillator (1 kHz) to send 1-s signal to the transmitter to acknowledge reception; (2) continuing to run the helix-motor drive for two minutes to feed the end of the document over the heater bar, after which the motor drive is stopped and heater power cut down; (3) operating a buzzer to call attention to the waiting message.

Facilities are provided to couple a transmitter and receiver at the same station for alternative exchange of messages in either direction.

12.16 'Courier' document transmitter and receiver (K400)

This machine, which uses Mufax electro-sensitive direct-recording paper, is for use over the general switched telephone network. Frequency modulation is used, with black = 1300 Hz and white = 2100 Hz. Other VF tones, 1150 and 950 Hz, denote satisfactory or unsatisfactory reception respectively. The machines are driven from 50-Hz (or 60-Hz) mains supply, which also services for synchronization between transmitter and receiver on integrated mains networks. (A crystal-controlled power supply is available when required.)

The Transmitter. After the leading edge of the document is inserted in the machine, transmission takes place automatically. When scanning is completed the transmitter reverts to standby condition.

A strip of the document is brightly illuminated from a pair of fluorescent lamps. An image of this illuminated strip is focused on to an optical slit through a folded optical system using a mirror and lens. The optical slit allows light from a small strip of the image to pass to a rotating drum fitted with a second optical slit in the form of a helix. The combination of fixed slit and rotating helical slit produces a moving aperture which scans the image of the illuminated strip of document. A light guide inside the rotating drum directs the light from an elemental area of the image on to a photomultiplier cell. Variations in the brightness of the image cause the photomultiplier-cell current to vary, and this in turn varies the frequency of the signal fed from the transmitter.

Small sections of approximately 0·010 in. thick are projected by the optical system on to the photo multiplier. The output from this is proportional to the intensity of light falling on it, so that a d.c. voltage analogue of the tone, or greyness, is obtained. The voltage is applied to a voltage-controlled oscillator so that the signal which is transmitted consists of a frequency which varies with the light tone. White paper will produce a frequency of 2100 Hz and black a frequency of 1300 Hz, with intermediate tones between these two extremes. To ensure that documents printed on coloured paper are not printed by the receiver with a grey background, the lightest colour seen by the photomultiplier is interpreted by the AGC circuit as being white, ensuring that maximum contrast is achieved on the print.

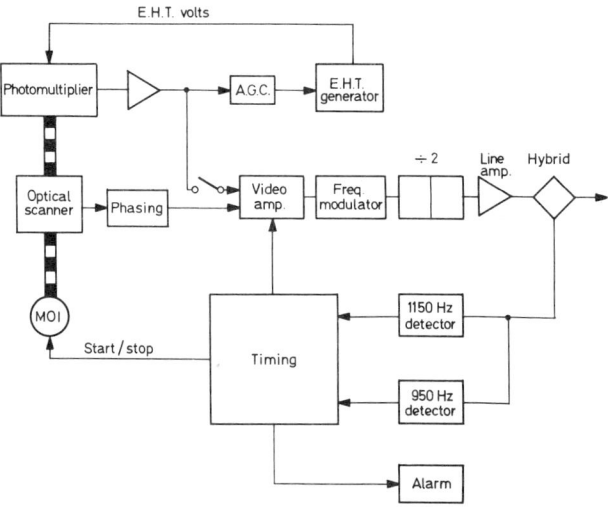

Fig 12.32 'Courier' document transmitter — block diagram

The block diagram (Fig 12.32) illustrates in a simplified manner the functions of the transmitter circuits. The optical scanner and feed mechanism project areas of the image sequentially on to the photomultiplier. Output from the photomultiplier is amplified and passed to the AGC circuit in the EHT generator and to the FM modulator and line-amplifier circuit. Current from the photomultiplier is approximately 20 μA for white and zero for black. This is converted, at the output of the amplifier, to −8 V for white and 0 V for black. The output of the photomultiplier is controlled by the AGC loop in the EHT generator; it is necessary to maintain this output at −8 V for the lightest tone on the copy. By varying the EHT voltage to the photomultiplier, the sensitivity is varied to compensate for a coloured background.

The tone output signal is fed to the FM modulator circuit where it is processed and used to control the frequency of a multivibrator; the frequency of the multivibrator is a function of colour tone. This frequency is divided by a bistable stage and passed, via a low-pass filter and a hybrid transformer, to the line for transmission. The hybrid prevents the transmitted signal from being passed to the input of the frequency-sensitivity detectors. In the same way, when a SATISFACTORY or NOT SATISFACTORY signal tone is received back from the receiver, the hybrid prevents this from being fed to the transmitter-signal circuits.

Switching various transmitter functions is controlled by timing circuits. These consist of a series of d.c. amplifiers with RC timing networks which drive relays to initiate appropriate functions at correct times. The insertion of a document trips fingers on a feeler bar attached to a microswitch and initiates the starting sequence. Similarly, when the document passes through the machine and releases the microswitch, this initiates the sequence to finish scanning and switch off.

With the transmitter switched ON, if the receiver is in correct working order and the facsimile signal is satisfactory, a tone at 1150 Hz is sent back to the transmitter and passed, via a hybrid transformer, to the 1150-Hz detector. Output from this detector is fed to timing circuits where it completes the starting sequence. On the other hand, if the receiver is inoperative, no signal will be received and a warning buzzer will sound.

The Receiver. Once the receiver has been switched ON it will function automatically, its operation being controlled by the signal from the distant transmitter.

The received signal, after processing, is detected in a discriminator circuit; this produces a d.c. output which is a voltage analogue of the received frequency. This is used to drive a power circuit which produces the marking current passed through the electro-

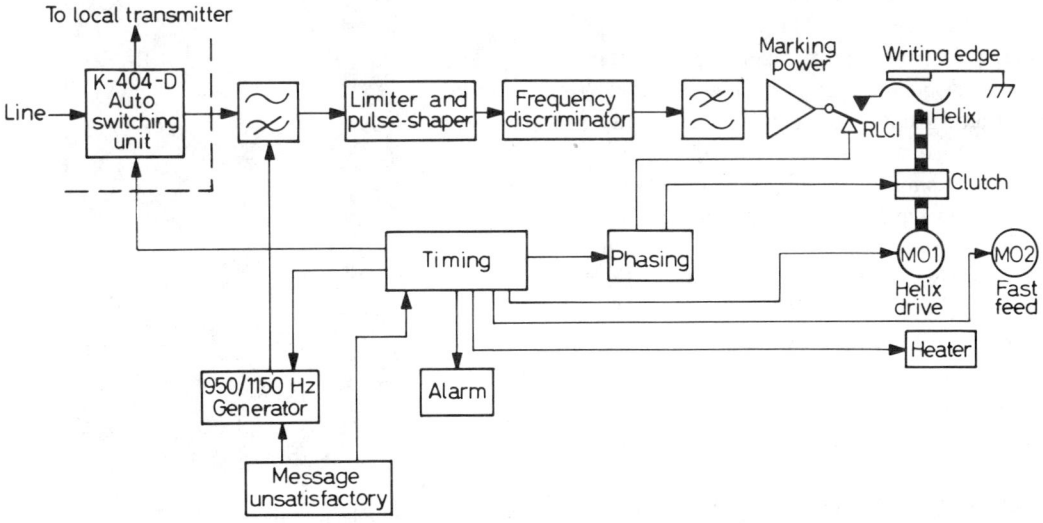

Fig 12.33 'Courier' document receiver — block diagram

sensitive paper. This paper passes through the helix and writing-edge scanning electrodes so that it is scanned in a series of close and parallel lines, synchronously with the scanning of the original document at the transmitter.

The block diagram (Fig 12.33) shows the essential functions of the receiver. In standby condition the receiver monitors the line circuit. When a signal is received from the remote transmitter, the receiver produces an output which is fed to the automatic switching unit to switch the telephone out of circuit and connect only the facsimile receiver to the line. Provided that the receiver is in a condition to accept the signal, a 1150-Hz SATISFACTORY signal tone is sent back to the transmitter to initiate the transmission sequence. If the receiver is loaded with insufficient paper, transmission is inhibited and the warning buzzer operates.

Before starting to record, the receiver runs paper through to ensure that only damp paper is between the helix and the writing edge. Synchronization with scanning at the transmitter is achieved by the phasing circuit, which detects phasing information from the transmitter, to ensure that the start of a line of printed copy coincides with the start of a line of transmitted copy. Once the receiver has completed this sequence the printed copy will be a faithful reproduction of transmitted information.

In the received signal the frequency represents the tone of each area of transmitted copy. This frequency is detected in the discriminator circuit in the receiver and used to produce a d.c. signal proportional to the tone received. This current is passed through the electro-chemical paper to print a tone which varies through the greys from white to black.

When transmission of the copy is completed, the receiver sends a further SATISFACTORY signal tone to the transmitter before closing down and operating the buzzer. The receiver continues to run for a period to ensure that all the paper containing copy is passed over the heater bar. Should the copy be unsatisfactory, the operator can press the UNSATISFACTORY button at any time. This transmits a 950-Hz signal to initiate the warning buzzer at the transmitter, causing it to shut down; it then completes the closing-down sequence for the receiver also.

The index $M = 264$; scanning rate is 180 lines/min; scanning density 90 lines/in., the maximum

document width is 9 in. The phasing signal is black interrupted by white for 8% of a scanning line.

12.17 Monitor recorder

When a phototelegraphic picture is being received, the quality of the finish is not known until photographic processing is almost completed; the result could have been marred by a faulty line or noisy radio circuit. The monitor set uses direct-recording electro-sensitive paper and quality of reception can be observed at all times. Another use is at an art editor's desk to enable him to anticipate what pictures are being received. For some purposes the quality of the half-tone picture received may itself be suitable for making a printing block.

The monitor follows the general principles of helix writing-bar recorders. It is designed to start and to phase automatically and to operate at drum speeds and scanning rates of equipment being monitored.

12.18 Facsimile tape recorder (Fig 12.34)

At a large picture-telegraph station, a picture may arrive for onward transmission at a time when a free

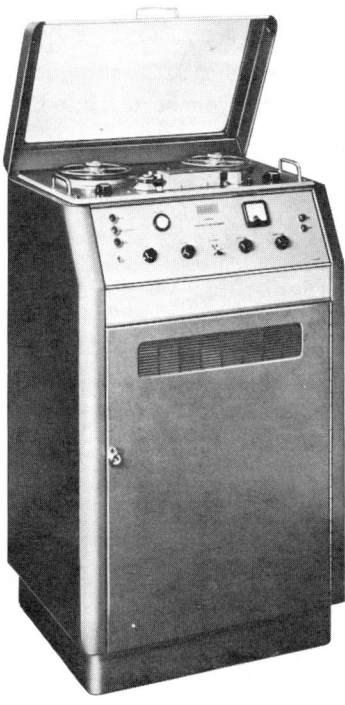

Fig 12.34 Facsimile tape recorder (*Courtesy of Muirhead & Co. Ltd*)

outgoing circuit is not available or perhaps a radio circuit is temporarily unserviceable. Pictures can be relayed by retransmitting a received phototelegraph print, but even though the second scanning takes place at right angles to the direction of the original there is bound to be some loss in definition.

The tape recorder meets this need, the main technical problem being uniformity of tape speed to the capstan-head motor on the tape deck. Despite any tape stretch or slip that may occur, the facsimile signal is reproduced in the same continuous synchronism as that of the originally-transmitted signal. Signals on the tape may be recorded and replayed using either AM or FM. The retransmitted picture can be monitored on either a photo-telegraph receiver or a direct-recording monitor.

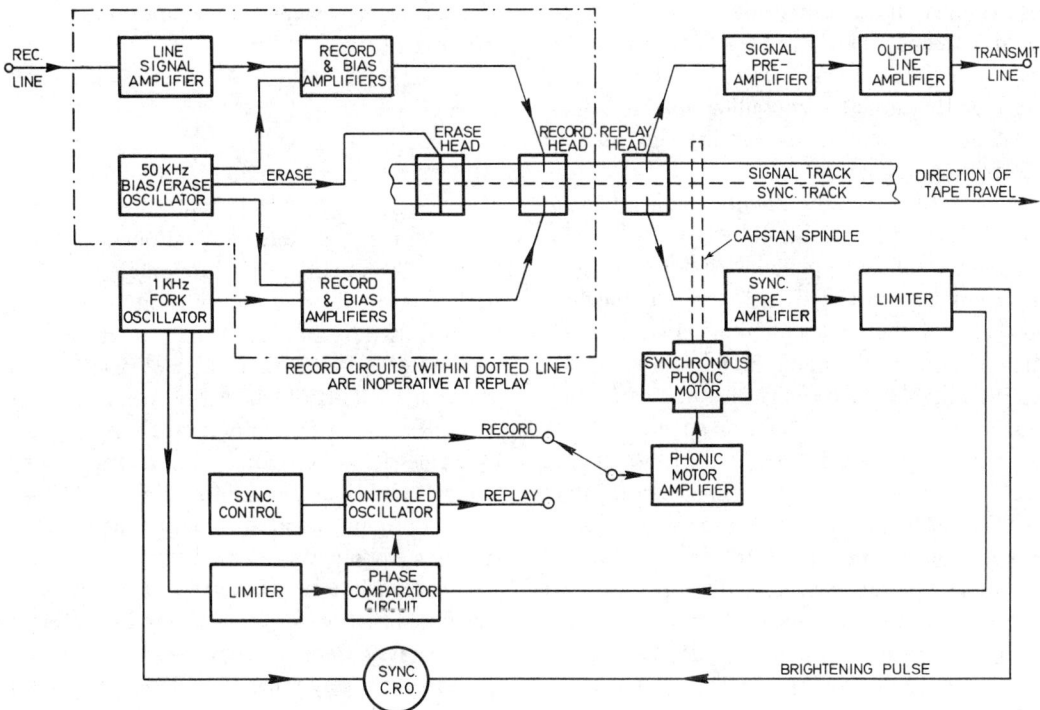

Fig 12.35 Facsimile tape recorder — block diagram

usual high-grade limits of the conventional drum transmitter.

A tuning fork adjustable over a range of 20 parts in 10^6 is used as the frequency source for synchronism. The recorder uses standard $\frac{1}{4}$-in. wide magnetic tape; the general arrangement of the recorder is shown in Fig 12.35. Facsimile signals for storage are recorded on the upper track, while on the lower track standard frequency from the fork is recorded. When the tape is 'replayed' for retransmission of the stored picture, the reproduced standard-frequency signal from the other tape track is compared with the 'live' standard frequency direct from the fork; any frequency difference is used in a synchronizing circuit to control the speed of the

Tape speed is 11·8 in./s for picture telegraphy and 7·5 in./s for document and weather-chart recording. The capstan spindle is driven by a phonic motor from the standard fork source. In addition to the normal facilities of a tape deck, provision is also made for setting black and white levels, a monitoring loudspeaker and a cathode-ray oscillograph which provides an indication of tape-speed variations while recording; on replay, the synchronizing control is used to set tape speed at midpoint of the synchronizing range.

When recording, the tape is drawn past the erase, record and replay heads by the synchronous phonic motor which is supplied with power at the fork frequency of 1 kHz. Existing signals on both tracks

of the tape are first erased by the erase head at
50 kHz. The facsimile signal and the 1-kHz standard
fork-frequency signal are then simultaneously
recorded on the upper and lower tape tracks respectively. As the tape passes the replay head the facsimile signal from the upper tape track passes through
the normal replay circuit; it is possible to monitor
signals on the tape almost as soon as they are
recorded (100 ms later at 11·8 in./s). As the tape
passes the replay head during a recording, the
standard frequency signal from the lower track
passes through the pre-amplifier and limiter circuits;
the trace on the cathode-ray oscillograph indicates
the small phase variations between live and replayed
standard-frequency signals.

On replay, the recording circuit is inoperative and
the tape is drawn past the heads by the synchronous
phonic motor which is supplied with power at the
frequency of the controlled oscillator – nominally
1 kHz. Facsimile signals from the upper track pass
through the pre-amplifier and the line-output amplifier to the TRANSMIT terminals for retransmission.
Simultaneously, the standard-frequency signal from
the lower track and the live standard fork-frequency
signal are fed, after limiting, to the phase-comparator
circuit. Output from the phase-comparator circuit
varies the frequency of the controlled oscillator,
which in turn controls the speed of the phonic motor
to keep phase difference between live and replayed
synchronizing signals to a minimum. Picture signals
are retransmitted in the same continuous synchronism with the ultimate receiver as that of the original
signal.

The induction starting motor for running the
phonic motor up to synchronous speed, the tape-feed
and take-up motors are mains driven. The two latter
are connected in series and their torques are in
opposition; for rewinding a spool, the power feed to
either motor can be progressively short-circuited
leaving the other in operation. During recording or
replaying, only the take-up motor (and phonic
motor) operates.

For high-quality results, special attention must be
given to avoidance of dust and other foreign material
on the tape or magnetic heads; tape-cleaning pads
and a capstan-cleaning ribbon are embodied in the
equipment.

12.19 AM/FM conversion

The design of a phototelegraph equipment usually
provides for amplitude modulation and demodulation. On radio circuits it is preferable to use
frequency-modulated transmission to avoid impairment caused by amplitude changes when fading
occurs. At the same time, it is not practicable to use
direct modulation of the radio carrier on account of
the difficulty of maintaining, at these very high
(decametric) radio frequencies, the very close frequency stability demanded for facsimile operation.
Instead, frequency modulation of the picture signal
is applied at audio frequencies using a nominal
centre frequency of 1900 Hz with a ± 400-Hz frequency deviation for the white and black signals;
intermediate-frequency deviations correspond to
intermediate grey tones. The frequency-modulated
signals can be transmitted over any telephone circuit
connecting the phototelegraph station with a radio
station. This type of signal is referred to as a sub-
carrier frequency-modulated (SCFM) signal since,
at the radio-transmitting station, it will be further
modulated with a radio-frequency carrier for transmission by the single-sideband (SSB) or independent-
sideband (ISB) method, for which either AM or FM
could in principle be used.

To provide the SCFM signals, converters are used
to convert from AM to FM for transmission and to
AM from FM for reception. It is important to maintain the related characteristics of the AM system
with the 34- or 12-dB white–black contrast ratio
when converting to the 800-Hz frequency deviation
range of the FM system to reproduce the half-tone
facsimile range faithfully. The graph of Fig 12.36
provides a guide to the manner in which frequency

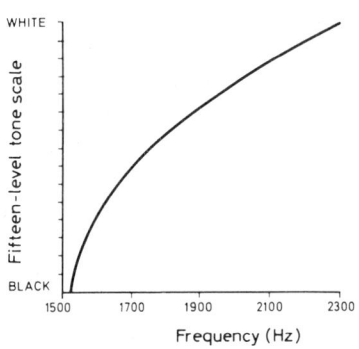

Fig 12.36 AM/FM phototelegraphy response curve

range is related to fifteen levels of black—white tone scale (see Fig 12.40).

To convert the AM output from a facsimile transmitter into an FM signal, the input signal is fed via an input-level potentiometer to an amplifier followed by a demodulator. This output is filtered and applied to a multivibrator which is free-running at 9200 Hz. The facsimile signal-voltage variations drive the multivibrator frequency between the limits 9200 and 6000 Hz corresponding to black and white limits. The multivibrator output passes through two 'divide-by-two' stages which reduce the black—white limits to 2300 and 1500 Hz; this signal is fed to line at the appropriate power level via an amplifier. The two limit frequencies can be checked against a reference signal provided by a crystal oscillator.

For converting an incoming SCFM signal to AM for a phototelegraph receiver, the input signal is applied to a band-pass filter to suppress signals at extraneous frequencies; it then passes to a limiter stage which cuts off variations of the input signal and gives an output of constant amplitude. This FM signal is passed to a discriminator having two tuned circuits in its anode, with resonant frequencies slightly above and below the 2300- and 1500-Hz limits. The outputs from the two tuned circuits are rectified giving d.c. output signals of opposite polarity for black and white, together with intermediate values (i.e. zero output at 1900 Hz). These facsimile signals pass through a low-pass filter to remove the products of demodulation; they are fed to a remodulator — a balanced amplitude modulator supplied with a carrier frequency of 1900 Hz (or other required frequency). A CONTRAST potentiometer controls the facsimile-signal input to the modulator to provide the required black—white contrast (e.g. 34 dB) in the output signal fed to a phototelegraph receiver.

12.20 Siemens Hellfax facsimile equipment

These machines record facsimile signals using a percussion system developed by Dr R. Hell in which signals are marked on ordinary paper by a small inking wheel controlled from an electromagnet energized by the facsimile signals.

One model, the KF108, is designed as a transceiver for transmitting from a document or receiving upon a sheet of paper 8·27 x 5·23 in.(211 x 132 mm).

The drum needs to be loaded with paper each time a message is to be received, and the recording head must be lowered on to the drum. The machine is capable of black-and-white recording without intermediate half-tones. The transmitter uses photoelectric scanning and both drum and leadscrew are driven by a synchronous motor dependent upon mains frequency for synchronization.

Using the same recording principle, continuous-roll receivers for 18-in. weather charts and for 8·5-in wide documents are also available. The principle of operation of the Siemens Hellfax Model BS133 recorder for documents is shown in Fig 12·37. The drive motor is controlled from electric-mains supply or from a fork; operation is automatic once a start signal is received from the transmitter.

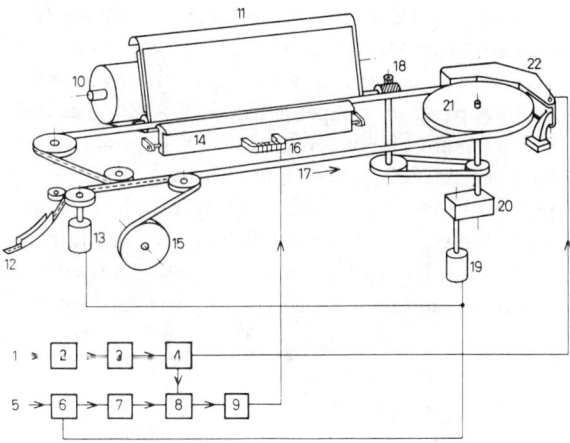

Fig 12.37 Siemens Hellfax recorder (BS133)

1 Signal input
2 Signal output
3 Automatic-gain control
4 Demodulator
5 Power network
6 Power supply
7 Frequency divider
8 Automatic-phasing circuit
9 Printing-bar amplifier
10 Paper-supply roll
11 Paper guide
12 Erasing-belt exit
13 Erasing-belt motor
14 Printing bar
15 Erasing-belt roll
16 Printing magnet
17 Recording belt
18 Paper-feed gearing
19 Main-drive motor
20 Motor gearbox
21 Recording drum
22 Recording system

The start pulse from the remote facsimile transmitter passes through an automatic gain-control amplifier (2) and a demodulator (3) to the output stage (4) which starts the motors (13 and 19) by means of a relay. At the same time an automatic-phasing circuit (8) detects proper phase, and after a few seconds the recording process starts. The modulated facsimile signals are held at a preset level by automatic gain control (2). The demodulator (3) rectifies signals and eliminates carrier. Signals are then reshaped to form square pulses; after the pulse lengths have been corrected they are amplified in the output stage (4) and passed to the recording system (22). They are finally used to control an electromagnet such that each pulse causes the magnet armature to press a small inked recording wheel against a continuously-moving endless plastic belt (17). An exact copy of each line scanned by the facsimile transmitter is transferred to the recording belt. The belt is subsequently passed in front of the printing paper on a paper guide (11). As soon as the starting point of a line has reached a determined spot near the left paper edge, a printing bar (14) presses the belt against the paper and the entire line is printed at once. Control pulses for the printing-bar magnet are derived from power mains via a frequency divider (7) and they are checked by the automatic-

phasing circuit (8). The pulse rate corresponds to the line-recording rate (180 or 120 lines/min). After printing a line, an erasing belt (12, 13, 15) moving in the opposite direction removes residual ink from the recording belt, which is now ready for the next line.

A companion transmitter, the Model FA123, uses flat-bed scanning in place of the more traditional drum. The document for transmission is fed into a slot in front of the machine; this automatically starts transmission and on completion the document is ejected through an exit slot. The machine is driven and synchronized by electric-supply mains; alternatively, a fork may be needed for synchronism. A schematic diagram of the transmitter is given in Fig 12.38.

With the power switched on, insertion of the original document causes a switch to energize the lamp and the motor; at the same time the transmitter is connected to line. The document on paper guide (1) is advanced steadily by transport rollers (2) which pass it on under mask (3) and it emerges from exit slot (4). Mask (3) covers all but the required scanning line (5) which is illuminated by a fluorescent lamp (6) supplying chopped light of about 8 kHz. Light reflected from this strip is deflected by the stationary mirror (8) to rotary mirror (9) which, depending upon its momentary posi-

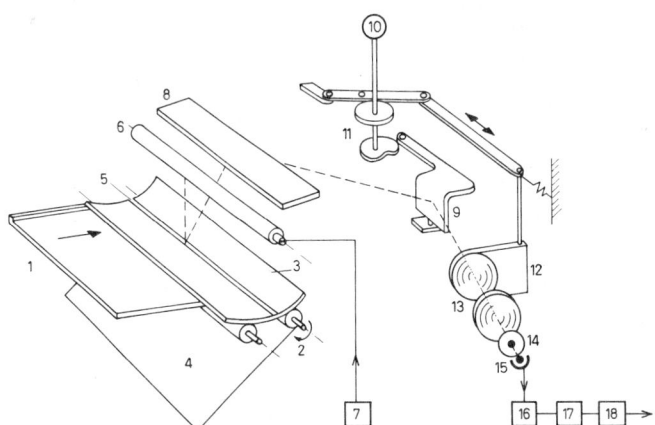

Fig 12.38 Siemens Hellfax transmitter (FA123)

1 Paper guide
2 Transport rollers
3 Mask
4 Exit slot
5 Scanning-line window
6 Fluorescent lamp
7 Lamp-voltage generator
8 Stationary mirror
9 Rotary mirror

10 Drive motor
11 Cams
12 Correcting lens
13 Optical system
14 Aperture
15 Photomultiplier
16 Automatic gain control
17 Demodulator
18 Modulator

tion, reflects the image of a certain small portion of the line to the photomultiplier cell (15) via optical system (13). The synchronous motor (10) drives cam (11), which in turn rotates mirror (9) in such a way that each line is scanned continuously from left to right with uniform speed; after completion of one scan the scanning spot rapidly returns to the starting position for the next line. In this way all parts of each line are successively presented to the photomultiplier cell. The unavoidable distortion of perspective accompanying the scanning of a line is compensated by a moving correcting lens (12), which follows a path controlled by a second cam (11). An aperture (14) placed immediately in front of the photocell limits the image to a neatly-defined spot. The photocell converts the chopped light, modulated by the varying brightness of the document scanned, into a proportionally modulated 8-kHz voltage. Using an amplifer with automatic gain control (16) the amplitude corresponding to the background colour of the original document is set to the white level. The modulated signal is rectified and the 8-kHz carrier eliminated in the demodulator (17). The resulting d.c. voltage can have all possible values between the white maximum and the black minimum, according to the varying brightness of the original. A trigger circuit set for a given threshold value separates the intermediate values into white (above the threshold) or black (below the threshold); an output of square pulses with equal amplitude is obtained only for voltages within this range. After correction of pulse duration and rounding of the edges, a modulator (18) amplitude-modulates a carrier with these pulses – either 1900 Hz for the scanning rate of 180 lines/min or 1500 Hz for the 120 lines/min model; the signals are then transmitted to line via the line transformer.

The index of co-operation is 288; the line-scanning rate (i.e. the equivalent of drum speed) is either 120 or 180 lines/min; and the scanning density 4·2 lines/mm (approximately 106 lines/in.) These machines are used by subscribers to the telephone service or over leased circuits.

12.21 Black/white facsimile systems
Three new facsimile equipments have recently been introduced for use over the public-telephone network. Their main characteristics are shown in Table 12.5, together with those of the *Courier* equipment. As no two of these machines are compatible they cannot interwork.

The need to reduce transmission time for this type of message equipment has long been appreciated as an economic necessity; with most types of document the transfer efficiency is very low since long periods of white transmission usually occur. There are three possible ways by which improved efficiency could be achieved, though these have not yet been embodied in practical equipment.

Typical of these is a system in which alternate high- and low-frequency signals represent black, and an intermediate frequency represents white. The transmission of a white, black, white, black, white sequence results in the transmission of intermediate, high, intermediate, low and intermediate frequencies, giving a fundamental modulating signal of half that which would be obtained if black and white were represented by single frequencies. This system, which is akin to the alternate-mark-inversion technique employed on some pulse-code modulation systems, usually precludes the transmission of half-tones and requires a transmission system having a high signal-to-noise ratio.

Most documents to be transmitted have areas which contain no information, such as inter-typescript-line and inter-paragraph gaps. Where the scanning line runs parallel to the typescript lines, a considerable saving of time should be possible. Each line is scanned and if it contains information it can be rescanned or the information can be held in storage before being transmitted. If no information is present, a skip signal is sent to line, and time is allowed for the scanning and writing devices to advance one line.

In most documents transmitted over a facsimile link, white predominates, probably by a factor of 10 : 1 or more. There is considerable scope for saving time by speeding the transmission of the predominating shade. Still further economy can be achieved by speeding the transmission of either shade and pausing only where a transition is to be transmitted. Various methods can be employed for this purpose, but it is necessary to provide either variable-velocity scanning, or storage of one line of information (approximately 800 bits), at both transmitter and receiver if the flow of information to the modulator is to be kept near to the maximum that the circuit can convey.

An alternative method is to inspect blocks of line

Table 12.5 Black-white facsimile telegraph-machine characteristics

Machine designation	Country of manufacture	Type of machine	Recording process	Index of co-operation	Scanning pitch (mm)	Scanning rate (lines/min)	Useable portion of whole scan line (per cent)	Modulation system	Frequencies used (Hz)
E.M.I. *Emifax HF 146-4*	Germany	Drum transceiver	Inked stylus percussive on ordinary paper	264	0·265	180	90	FM	2100 (white) 1300 (black)
Muirhead, *Courier K-400-D & K-401-D*	U.K.	Separate flat-bed transmitter and receiver	Helix and writing edge on damp electrolytic paper	264	0·28	180	92	FM	2100 (white) 1300 (black)
Plessey, *Remotecopier KD 111*	Japan	Flat-bed transmitter and receiver mounted in same cover	Electrostatic charge, deposited by stylus on electrostatic sensitive paper	296 or 198 (operator selectable)	0·265 or 0·395	180	88	FM	2100 (black) 1500 (white)
Rank-Xerox *400 Telecopier*	U.K.	Drum transceiver	Electro-sensitive paper, marked by current flow from stylus	264 or 176 (operator selectable)	0·265 or 0·395	180	94·9	FM	2100 (black) 1300 (white)

information in, say, 10-element blocks, and if all elements are black or white a simple code is transmitted, otherwise the information is transmitted as scanned. This method also requires storage or variable-velocity scanning.

Other methods include signalling the length of any block of one shade, or the intervals between shade transitions.

All these methods of information compression introduce considerable complexity into the equipment; this is reflected in cost and reliability, and most imply black-and-white transmission only.

12.22 Xerography

The process of xerography (dry writing) enables facsimile messages to be recorded directly upon ordinary paper without need for further processing; the same principle is used for office duplicating machines and in a high-speed line printer for computer output. A visible image is created from an electrostatic-charge pattern on a surface; this pattern is used to obtain a visible image which can be transferred to a sheet of paper where it can be fixed. The method does not involve chemicals and it need use only dry processes.

There are two main methods of producing an electrostatic image — by a direct discharge from a shaped metal electrode or by optical illumination of a photoconductor. The most useful photoconductor is selenium, which for this purpose is evaporated as a thin layer about 50 μm in thickness upon a substrate of aluminium, the latter being a plate or drum. The selenium is charged positively to 600 or 800 V — equivalent to sensitizing a film. Selenium has the property that in the dark it is an insulator but when exposed to light it becomes conductive. If the selenium surface is exposed to an optical image, the photoconductor becomes conducting with the result that the positive charge on the plate leaks away to the base. The potential of the surface decreases according to the intensity of illumination, the time of exposure and the spectral distribution of the light. An electrostatic image is formed corresponding in intensity and shape to the original optical image. This image can be made to develop a visible image by depositing a powder which is attracted either to the image or non-image portion of the plate.

Small insulating particles, about $\frac{1}{3}$ mm in diameter, are mixed with very fine insulating particles a few microns in diameter, the materials being such that they charge one another oppositely. The large particles are called the carrier and the smaller the toner; the combination is called the developer. There is an attractive force between carrier and toner. For printing, the visible image on the selenium is transferred by a discharge method on to a piece of plain paper. The positively-charged toner adheres to the image area. The paper makes contact with the xerographic plate and a negative discharge is played on to the upper surface. The positively-charged toner is pulled off the plate on to the paper by the attraction of negative charges, and comes away with the paper when the latter is peeled off. Nearly all the powder is so transferred from selenium to paper and the visible toner image is fixed to the paper in one of several ways, e.g. fusing by heat. High-quality images can be produced in this way with a resolution up to 50 lines/mm or more.

A practical embodiment of the xerographic

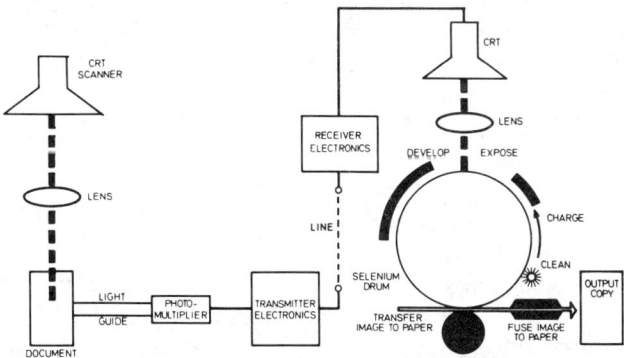

Fig 12.39 Xerox high-speed facsimile system

process for facsimile telegraphy is illustrated in Fig 12.39. At both transmitter and receiver, scanning is performed by a moving spot of light from a cathode-ray tube. At the transmitter the document is fed forward while the light spot makes rapid line-scanning sweeps; light reflected from the document passes through a light guide to a photomultiplier cell; the signal is modulated before passing to the line circuit. At the receiver, the signal is demodulated and fed to the cathode-ray tube to control the intensity of the moving light spot which is projected on to the selenium-coated drum.

Such a high-speed process is naturally best adapted to high-speed line (or radio) transmission. Scanning rates of 135 and 190 lines/in. are used; the time required to transmit a document measuring 11 x 8·5 in. (280 x 216 mm) is of the order of one minute. This transmission requires a wideband circuit of 48 kHz.

12.23 Telewriter

This is an analogue device in which a pen or pencil is connected to the arms of a pantograph arranged to provide two distinct electric signals (d.c. or a.c.) on a 2-co-ordinate position basis when the pen writes a message for transmission. In conception very old, more sophisticated versions of the apparatus have appeared in recent years using capillary-type pens embodying a mercury standby switch.

In the model known as the *Electrowriter*, the vertical and horizontal signals at the transmitter are derived from position transducers connected to the pen pantograph; the range of pen travel along either ordinate is used to vary the X and Y frequencies. This frequency variation is effected by the relationship between the stator and rotor positions of a rotary transformer which forms part of the tuned circuits of the transmitter oscillators. The motion of the pen towards or away from the paper is used to modulate one of the frequencies at 120 Hz.

In the receiver, each co-ordinate signal is used in a local servo loop to control motion of the receiving pen by the aggregate motion of two shafts. The shafts are controlled by d'Arsonval movements with a high-permeability rotating core producing a high torque and having a maximum deflection of $\pm 30°$. The displacements of the motor shafts connected to pantograph linkages are determined from the

instantaneous frequencies received over the line, applied to a discriminator and amplified. Paper feed is initiated by moving the transmitting stylus to a marked position at the upper left-hand edge of the page. In this corresponding position, the pen linkage at the receiver operates a switch which actuates the paper feed. These devices are of limited application and the quality of the received copy is not strictly a distortionless facsimile of the original.

12.24 Test chart

If doubt occurs about the fidelity of a picture received by phototelegraphy, this cannot be resolved without a comparison between the transmitted and received prints; such a comparison can rarely be made short of posting one of the prints to the remote station. For this reason, agencies operating a photo-telegraph service have found it beneficial to use a chart containing critical photographic detail which can be transmitted for test purposes; the receiving station can then compare the received print with a copy of the test chart (held at the receiving station) which is identical with that used for transmission. These charts are also valuable for laboratory investigations and in development of new machines.

Hitherto many different test charts have been in use, but the CCITT has met a need by developing a standardized test chart suitable for both phototelegraphy and black-and-white direct recorders. This chart, illustrated in Fig 12.40, is divided into 13 numbered sections. At the transmitter the chart should, if necessary, be cut to fill the drum circumference, cutting along (or near) the same numbered lines in the two sections (10). The white and black transmitted levels must correspond to the steps numbered 1 and 15 in the tone scale of section (1).

Sections 1 and 2. These are the tone scale (tone wedge) having 15 density steps varying from white to black and vice versa. These scales will show up deficiencies in half-tone reproduction.

Section 3. A group of black hyperbolic lines, their thickness and spacing diminishing from left to right from 1 to $\frac{1}{6}$ mm according to the adjoining scale. This will show up deficiencies and limits to definition.

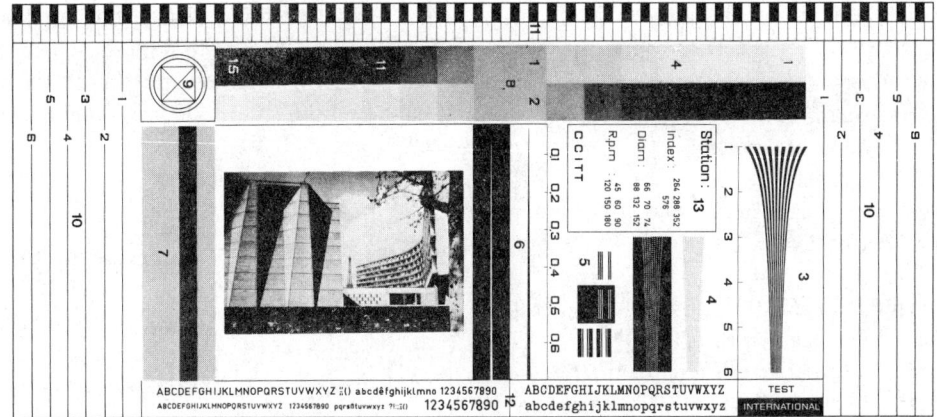

Fig 12.40 CCITT facsimile test chart (1st edition)

Section 4. Similar to section 3 but limited to scanning densities lying between points 3 and 6.

Section 5. Three black–white patterns to specified dimensions. This will show up any bias in restoration of black after white, or vice versa.

Section 6. Black and white lines tapering from 0·7 mm according to the scale. This will show up the repetitions caused by echo.

Section 7. Tones of step 11 against a background of step 5. Variations in level, noise or disturbing frequencies will be apparent.

Section 8. A half-tone photograph (Unesco House, Paris). (In the second edition of CCITT test chart, this photograph of the UNESCO building in Paris is superseded by a portrait of the head of a child. Other details of the chart remain basically unchanged. The two charts are compatible.)

Section 9. Two concentric circles differing in radii by 1 mm enclosing a square with diagonals. Any disagreement between the index of co-operation at transmitter and receiver will cause the circle to be produced as an ellipse and the square as a rectangle. The ratio of the sides of the rectangle will reveal the ratio between the indices. The axis of the ellipse will show which index is the greater. The square will become a parallelogram if there is a small speed difference.

Section 10. Adjustment lines to accommodate different drum diameters.

Section 11. White rectangles with alternate black rectangles. The black and white rectangles will be deformed due to speed differences. If there is a wide difference (e.g. 1 and 2 rev/s) the chart will be repeated (e.g. twice).

Section 12. Typographical characters in various styles and sizes.

Section 13. A space for the station code; the index of co-operation, drum diameter and speed can be indicated by ringing around the appropriate value (separate charts would be required for different machines at a station).

12.25 CCITT recommendations

In no branch of telecommunications is the need for compatibility between transmitter and receiver more vital than in facsimile telegraphy. The CCITT has studied these problems and issued a number of technical recommendations in some detail. The more important characteristics are summarized here.

T1. Phototelegraph Apparatus

Scanning: 'Negative' direction at transmitter. At receiver, 'negative' direction for positive (print), but 'positive' direction for negative (film) reception.

('Negative' and 'positive' directions are not defined.)

Scanning density: F = 4 lines/mm approx. (according to values of M and D).

Index of co-operation: M = 352 (or 264).

Co-operation factor: ($C = \pi M$, for flat-bed scanning) = 1105 (or 829).

Dimensions. Drum diameter D = 66, 70 or 88 mm.

Drum factor, transmitter: $\leqslant 2\cdot4$

   receiver: $\geqslant 2\cdot4$

Drum speed (rev/min): 60, 90, 120 or 150 (according to conditions); tolerance ±10 parts in 10^6.

Synchronization: better than 5 parts in 10^6.

Phasing: 3% of scanning line used. Phasing signal 95% black, 5% white.

Modulation: AM: white/black power ratio = 30dB (maximum for white); carrier frequency f_c = 1300 Hz (audio circuits), f_c = 1900 Hz (carrier circuits). FM: white = 1500 Hz; black = 2300 Hz; mean = 1900 Hz.

T2. Black/white Apparatus (Not applicable to Meteorological purposes, for which WMO issues Recommendations)

Scanning: Positive direction.

Scanning density: 3·85 lines/mm.

Scanning-line frequency: 120 (or 180) per min

Index of co-operation: M = 264 (C = 829).

Dimensions: Documents up to 210 x 297 mm

Drum Diameter = 68·5 mm

Drum factor: transmitter: $\leqslant 4\cdot5$

   receiver: $\geqslant 4\cdot5$

Flat-bed scanner, total scanning line = 215 mm (15 mm for margins).

Continuous recorders: Paper width = 210 mm.

Modulating frequency (max.): $f_m = \pi M n/120$ Hz (n = scanning lines/min).

Phasing signal: Black = 95%, white = 5%.

Modulation: AM (leased circuits): f_c between 1300 and 1900 Hz. FM (leased or switched circuits): black = 1300 Hz, white = 2100 Hz (f = 1700 ± 400 Hz).

Power output: AM: between −7 and 0 dBm for black; white level is 15 dB below black level. FM: between −10 and 0 dBm.

Power input: Between 0 and −40 dBm (refers to black if AM).

T4. Remote Control of Black/white Apparatus. A procedure (too lengthy to reproduce here) based upon supervisory signals consisting of black/white reversals at 100 bauds, 200 bauds and 5% white/95% black.

T10. Black and White Transmission on Leased Circuits

AM: f_c = 1300 Hz (audio circuits) or 1900 Hz (carrier circuits).

FM: Black = f_0 − 400 Hz; white = f_0 + 400 Hz (f_0 = 1300 or 1900 Hz).

Power (at the trunk circuit): −10 dBmO (simplex); −13 dBmO (duplex).

T10A. Black/white Transmissions on General Switched Telephone Network

FM: black, f_B = 1700 − 400 Hz; white, f_W = 1700 + 400 Hz.

Power output: 10 dBmO.

Power input: −40 dBmO (minimum).

T11. Phototelegraph Transmissions on Telephone-type Circuits

Sent power: white = 0 dBmO (AM); −10 dBmO (FM). Black: 30 dB below white level.

Attenuation distortion: AM, 8·7 dB.

Phase distortion: $\Delta t \leqslant 1/2 f_p$ (f_p = maximum modulating frequency).

T12. Range of Phototelegraph Transmissions on Telephone-type Circuits. The permissible difference (for a world-wide chain of 12 circuits) between the minimum group delay throughout the frequency band transmitted and that at the upper and lower limits of this band are:

	Frequency-band limits (ms)	
	Lower	Upper
International chain	30	15
Each of the national 4-wire extensions	15	7·5
On the whole 4-wire chain	60	30

T15. Phototelegraph Transmissions over Combined Radio and Cable Circuits. (Corresponds to CCIR Recommendation 344–1.)

SCFM: Centre frequency = 1900 Hz. White and phasing f_W = 1500 Hz. Black f_B = 2300 Hz.

Limits for frequency stability are also given.

T16. Transmission of Meteorological Charts over Radio Circuits. (Corresponds to CCIR Recommendation 343–1.)

SCFM: Centre frequency: f_0 = 1900 Hz. Black f_B = 1500 Hz. White f_W = 2300 Hz.

T20. The Test Chart.

Questions under study by the CCITT include: Use of vestigial sideband modulation; Correction of phase distortion (use of modern submarine cables); High-speed facsimile telegraphy on wideband circuits; Black/white facsimile telegraphy by numerical coding to indicate the positions of black elements; Phototelegraphy over satellite circuits; Phototelegraph colour transmissions; Terms and definitions.

References

1 Mertz, P. and Grey, F., 'A Theory of Scanning and its Relation to the Characteristics of the Transmitted Signal in Telephotography and Television', *Bell System Technical Journal*, **13**, p. 464 (1934).
2 Cole, A. W. and Smale, J. A., 'The Transmission of Pictures by Radio', *Proc. IEE*, **99**, p. 325 (1952).
3 Carter, R. O. and Wheeler, L. K., 'A Phototelegraph Transmitter-Receiver Utilizing Sub-carrier Frequency-modulation', ibid. p. 335.
4 Bell, J., Davidson, J. A. B. and Phillips, E. T. A., 'Some Recent Developments in Phototelegraphy and Facsimile Transmission', ibid. p. 344.
5 Everett, R., 'A Picture Transmitter Employing a New Light-sensitive Device', *Muirhead Technique*, **19**, p. 26 (1965).
6 Everett, R., 'The Automatic Recording and Processing of News Pictures', ibid. p. 23.
7 Starr, A. T., 'Xerography', *Proc. IEE* 8, p. 529 (1962).
8 Westaway, H. E., 'A Fresh Look at Facsimile for Document Transmission', *POEE Journal*, **66**, p. 18 (1973).

Appendix A
Abbreviations Used

A	The passive binary state = space or 0
ABS	Out of service (absent)
a.c.	Alternating current
AGC	Automatic gain control
AM	Amplitude modulation
ARQ	Automatic request for repetition
ASCII	American standard code for information interchange
BQ	Response to RQ
BS	British standard
BSI	British Standards Institution
B/W	Bothway
CCE	Common control equipment
CCIR	International Consultative Committee for Radio
CCITT	International Consultative Committee for Telegraphy and Telephony
CDC	Characteristic distortion compensation
d.c.	Direct current
DER	Faulty (dérangé)
DSB	Double sideband
ECB	Engineering control board
EHF	Extra-high frequency
EMF	Electromotive force
FDM	Frequency-division multiplex
FM	Frequency modulation
FRXD	Fully-automatic reperforator transmitter-distributor
FSM	Frequency-shift modulation
GDF	Group distribution frame
GRQ	Gated RQ
GTX	Gentex
HF	High frequency
HRDF	High-frequency repeater distribution frame
I/C	Incoming
IDF	Intermediate distribution frame
IF	Intermediate frequency
ISB	Independent sideband
ISO	International Standards Organization
ITS	Inland telegraph service
ITU	International Telecommunication Union
LF	Low frequency
LSB	Lower sideband
MCVF	Multi-circuit voice frequency
MDF	Main distribution frame
MF	Medium frequency
MOM	Wait (moment)
MUF	Maximum usable frequency
NC	No circuits
NP	Spare line or level
OCC	Engaged (occupé)
O/G	Outgoing
OPM	Operations/min
OTO	Overseas telegraph office
PBX	Private-branch exchange
PCM	Pulse-code modulation
PD	Potential difference
PM	Phase modulation
PRM	Printergrams
PTS	Proceed-to-select
RF	Radio frequency; receive filter
RIS	Radio-interference suppressor
RQ	Request for repetition
S & T	Speak and test
SCFM	Sub-carrier frequency modulation
SF	Send filter
SHF	Super-high frequency
SLC	Station line circuit
SSB	Single sideband
TAF	Test-access frame
TAS	Teleprinter automatic switching
TCHR	Trunk circuit hold and retest
TDF	Trunks distribution frame
TDM	Time division multiplex
TDMS	Telegraph distortion measuring set
TEF	Time-efficiency factor
TLX	Telex
TMS	Transmission measuring set
TRC	Tested repetition cycle
TRQ	Tested RQ

UHF	Ultra-high frequency	VHF	Very-high frequency
USB	Upper sideband	VLF	Very-low frequency
USR	Uniform spectrum random (noise)	VSB	Vestigial sideband
VF	Voice frequency	WMO	World Meteorological Organization
VFT	Voice-frequency telegraph	WRU	Who are you?

Appendix B
Standard Logic Circuits

The design of complex electronic digital circuits has been greatly assisted by the use of a relatively small number of basic electronic logic-circuit units, each designed to provide a specific type of function. Their input and output circuits are designed for ready interconnection among themselves and between other units. The term *logic*, applied to represent the operation of binary devices used in digital equipment, derives from Boolean algebra, originated by George Boole in 1847 as a systematic mathematical treatment of logical relations. A logical circuit element is one which combines two or more inputs to produce an output, this output being related to the inputs in a manner defined by the logic function performed by the circuit element.

Digital logic circuits were developed originally using thermionic-valve circuits, but the transistor has advantages of economy in power and space requirements. It also gives a close approach to an ideal switch; in the conducting state (saturated or *bottomed*) the voltage drop across the transistor is of the order $0.1-0.5$ V; in the non-conducting or cut-off state the impedance may be as high as 10 MΩ.

The more simple of these circuit units are described briefly in this appendix. The detail given should be regarded as illustrative only, since the presentation may differ according to the designer and manufacturer. These circuits will operate at speeds up to 20 kilo-operations/s — in some cases up to 100 ko/s.

Fig B.1 (*a*)—(*h*) shows typical circuit configurations. The preferred method of setting a trigger circuit comprising p-n-p transistors to a required condition is to use a positive-going wavefront or pulse to turn off a conducting transistor at the base. The coupling to the activating circuit is capacitative so that either a pulse or a step waveform which is differentiated by a capacitor is equally effective. To discriminate against small interfering pulses the capacitor

is connected to the base of the transistor through a back-biased diode, a suitable bias being about 1.75 V derived from a potential divider to the negative supply. Clamping diodes are used to maintain the output voltage of a cut-off transistor at about $V_C = -6$ V.

(a) Bistable trigger. Using very little energy this device provides indefinite storage of binary-signal states, remaining (without further outside control) in either the 0 or 1 state to which it is triggered. The trigger consists of two interconnected transistors, each of which can produce an output, and it may be in either of two states. The circuit is triggered from one stable state to the other by application of a suitable drive pulse to the base of either transistor.

If VT1 is conducting and the potential at control 2 is zero, a positive-going step input applied at input 2 will be differentiated by the input circuit and applied to the base of VT1. This transistor will be momentarily cut off and its collector potential will approach that of the negative supply. Due to the interconnection of the two transistors, the base of VT2 will become negative and VT2 will conduct. As a result, its collector will approach earth potential and the base of VT1 will become positive, holding this transistor non-conducting.

If the potential at control 2 is sufficiently negative, no pulse will be passed by the associated diode and the state of the bistable circuit will remain unchanged.

A corresponding sequence of events takes place if VT2 is conducting: a potential is applied at control 1 and a positive-going step voltage is applied at input 1.

The bistable trigger, or *toggle*, can be used for a variety of purposes such as storage, pulse frequency dividing chains, counting chains and shift registers.

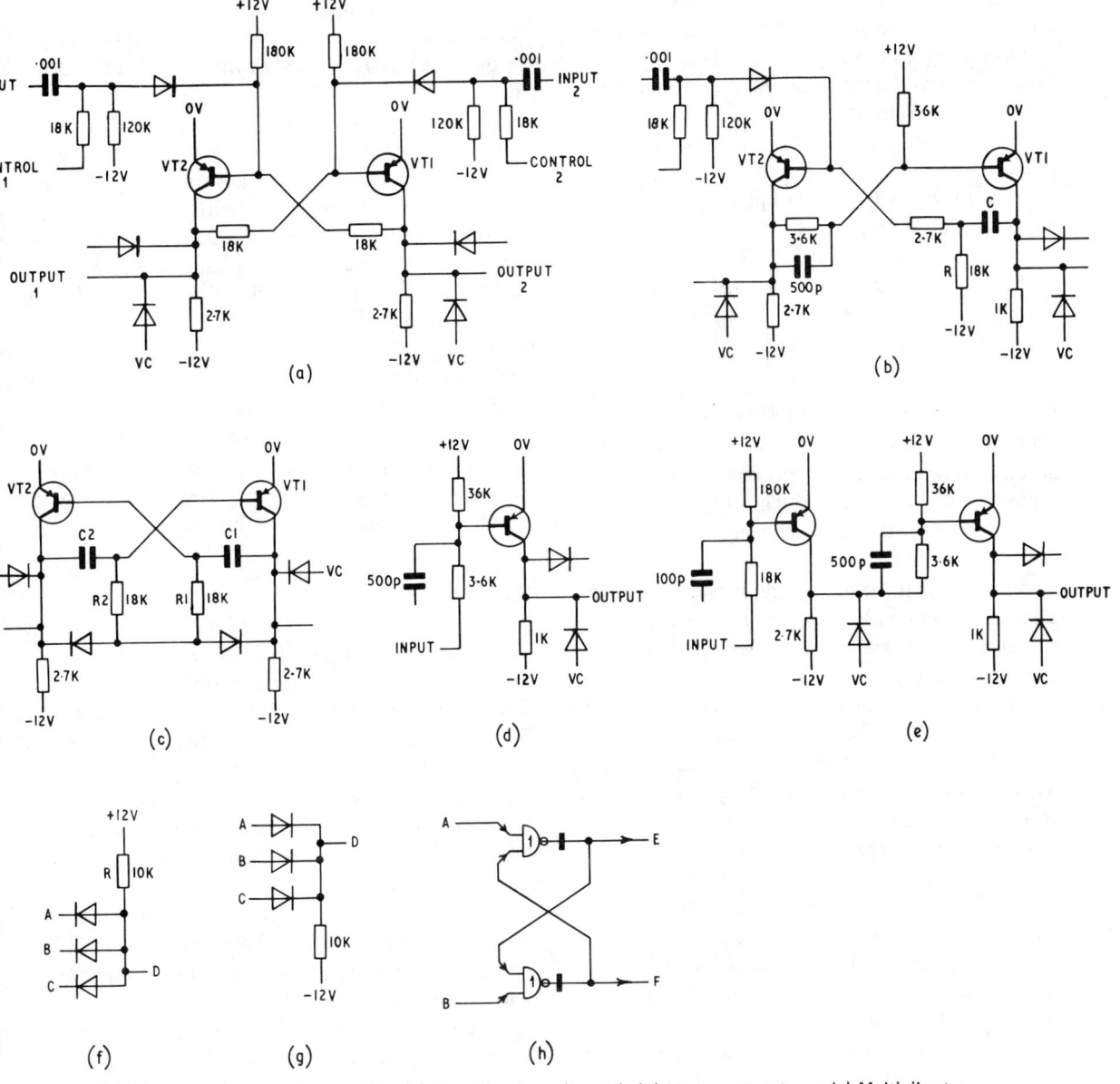

Fig B.1. Transistor-switching circuits: (a) Bistable trigger (toggle); (b) Monostable trigger; (c) Multivibrator; (d) Inverting amplifier; (e) Erect amplifier; (f) AND gate; (g) OR gate; (h) Negatory condition.

(b) *Monostable trigger.* This circuit provides an output pulse of specified duration (depending upon the value of C) in response to a pulse or step input signal. It is triggered from its quiescent state to the operated state by the input signal, returning automatically to the quiescent state when the designed period expires.

This trigger comprises two transistors with resistance coupling between VT1 base and VT2 collector, and RC coupling between VT1 collector and VT2 base. The circuit has one stable position in which VT2 is conducting. When the circuit is triggered a quasi-stable condition is set up during which VT1 conducts for a period which is determined by the time constant RC.

In the stable condition, VT2 is conducting and its collector approaches earth potential. Due to the interconnection this potential holds VT1 cut off.

When the circuit is triggered, by a positive pulse applied to the base of VT2, this transistor is cut off and its collector becomes negative. This negative potential is transferred to the base of VT1 which conducts, the collector becoming negative. Capacitor C is charged, holding the VT2 base at positive potential, and so cut off. A quasi-stable condition is now set up with VT2 cut off and VT1 conducting. The charge on C decays exponentially until the potential is reached at which VT2 again conducts in the stable state and VT1 is cut off. During the period while C is charged a positive pulse is delivered at the output from VT1 collector.

(c) Multivibrator. This oscillating device provides a source of square-wave signals where precise timing accuracy is not required. Operation to either state is unstable, the device switching automatically from one state to the other at a frequency determined by the circuit components.

Assuming that the multivibrator is running and VT1 has just commenced conducting, the collector potential of VT1 will have changed rapidly from that of the supply to earth potential. This positive-going change of potential is transferred via C1 to the base of VT2, reverse biasing the base-emitter junction and holding the transistor well below cut off. Due to R1, the capacitor C1 slowly discharges and the potential at the base of VT2 approaches that of its emitter. When the base of VT2 becomes slightly negative with respect to the emitter, the transistor starts to conduct and its collector potential approaches earth potential. This positive-going change of potential is transferred via C2 to the base of VT1 and reduces the collector current of that transistor, resulting in the potential at the collector of VT1 becoming more negative. This change of potential is transferred via C1 to the base of VT2, increasing the forward base-emitter bias and thus the collector current. As a result of the closed loop, this action takes place very rapidly; VT2 is switched quickly from cut off to conduction, and VT1 from conduction to cut off. Since the collector-base circuits of VT1 and VT2 are symmetrical, a similar sequence of events takes place when VT2 becomes conducting. The period between any two instants at which one of the transistors conducts is largely determined by the time constants of R1C1 and R2C2.

(d) Inverting amplifier. This provides an output signal which is inverse in form from the input signal. It provides a NOT function with a high loading output.

If a positive voltage (state 1) is applied to the input, the base of the transistor will be positive with respect to the emitter. The transistor is cut off, the collector being at negative potential (state 0).

Alternatively, if a negative voltage (state 0) is applied to the input, sufficient base current will flow to saturate the transistor. The collector voltage drops virtually to zero (state 1).

(e) Erect amplifier. This amplifier provides an output signal substantially identical to its input, but with a higher load rating. It uses two transistor stages, the second of which neutralizes the phase change introduced by the first stage, so that overall there is no phase change between input and output.

Gates. An electronic gate is a device with an input, an output and one or more controls; the control and input may be indistinguishable from one another. When a certain combination of potentials is impressed on the controls, an output appears at the output terminals and the gate is said to be open or conducting. Under all other combinations of input and control potentials, the gate remains closed, or non-conducting.

(f) AND gate. The diagram shows three inputs A, B and C (though a greater number could be applied) and a single output D.

If any one or more of the three input voltages are low (0 V) the associated diode will conduct and the p.d. across R will be approximately V volts. The output at D is therefore $0 V$.

If *all* the input voltages are at $+V$ volts, no diode conducts; there is no current in, nor p.d. across R and the output remains at $+V$ volts. Hence there is an output pulse only when *all* the inputs are at $+V$ volts.

When using positive-going pulses, $+V$ volts corresponds to state 1 and $0 V$ to the binary state 0. The combinations of input signal (for A and B) and the resultant output signals are shown in the function Table B.1, commonly referred to as a *truth table.*

(g) OR gate. The diagram shows three inputs, A, B and C (though a larger number could be applied) and a single output D. The circuit conditions are reversed from those of the AND gate.

With 0 V applied to all three inputs, none of the diodes conduct, there is no current in, nor p.d. across R, and 0 V appears at the output D.

If any one or more of the inputs is at $+V$ volts, one or more of the diodes conducts and $+V$ volts appears at the output D. There is an output pulse if any one of the inputs is at $+V$ volts. Table B.1 shows the output conditions for all combinations of input signals (at A and B).

Magnetic-Core Storage. Binary signals can be stored on small, ring-shaped magnetic cores of ferromagnetic material (e.g. ferrite) having a hysteresis curve which is approximately rectangular (Fig B.2). For such purposes, magnetic cores have the advantages of very low power consumption, high reliability and relative immunity from interference surges. The switching time is of the order of a few microseconds only so that sharing of logic and switching circuits is possible.

The magnetized core can have either of two saturated states, attained by applying a magnetizing force of at least $\pm H_C$ AT/m, the resulting magnetic-flux density being $\pm B_{SAT}$ Wbs/m^2. On removal of

Table B.1 Truth table for simple logical operations

Inputs		Output for					
A	B	AND	OR	NAND	NOR	Equivalence operation	Non-equivalence operation
0	0	0	0	1	1	1	0
0	1	0	1	1	0	0	1
1	0	0	1	1	0	0	1
1	1	1	1	0	0	1	0

(h) NAND and NOR gates. The operation of negation changes an 0 to a 1 and vice versa, and is equivalent to NOT, since if a binary signal is not 1 it must be 0. The NOT-AND or NAND operation, and the NOT-OR or NOR operation are shown at *(h)*; they are equivalent to the AND and OR respectively, followed by a negator.

A NAND gate is a logic configuration designed to give an output 0 state, only when all the inputs are in state 1.

A NOR gate is a logic circuit designed to give an output 1 state, only when all the inputs are in the 0 state.

(i) Equivalence gates. The other two operations shown in Table B.1 are defined for two inputs only. In the EQUIVALENCE operation the output signal represents 1, only when both the input signals are alike; and in the NON-EQUIVALENCE operation the circuit output signal represents 1, only when the input signals differ from one another. The latter is often called the EXCLUSIVE-OR operation, while the OR operation above may be called the INCLUSIVE-OR operation.

the magnetizing force ($H = 0$), the core remains indefinitely in the approximately saturated state to which it was set, without further stimulus, due to the property of remanence. The flux density now has the value B_R which is very slightly less than B_{SAT}, the hysteresis loop not being strictly rectangular. The hysteresis loop provides two possible

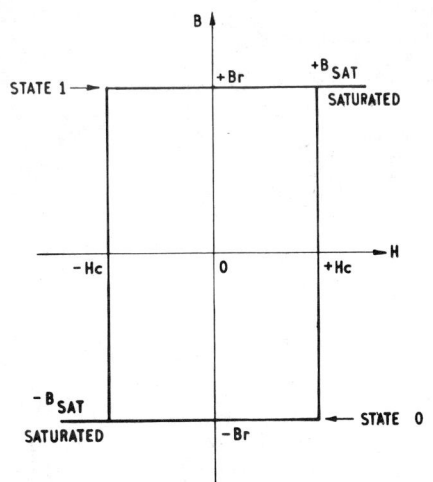

Fig B.2 Rectangular *B/H* hysteresis loop

discrete magnetic states designated for identification as *switched, set* or 1; and *normal, reset* or 0. Switching from one state to the other will occur only if the applied magnetizing force exceeds the critical value $\pm H_C$ (the coercivity).

Magnetizing forces are applied by passing current pulses through toroidal windings on the core, or on wires simply threaded through the core. According to the direction of the magnetizing current, either the 1 state or the 0 state can be applied.

If a second current pulse (of saturation value) is applied to the core (either through the same or another winding on the same core) either of two conditions will arise, depending upon the relative current directions.

If the second pulse is in the *same* direction as the first (assumed positive), the magnetizing force increases from $H = 0$, rises beyond $+H_C$ to the saturation value, and falls again to $H = 0$. In the ideal case there is no change in resulting flux density, but in practice this rises and falls very slightly (due to the non-rectangularity of the loop), the 1 state being retained.

If the second pulse is *opposite* in direction to the first, the magnetizing force changes from $H = 0$, exceeds the value $-H_C$, and restores to $H = 0$. As a result the flux density changes from $+B_R$ to $-B_R$. The state of the core is consequently changed from 1 to 0; the new state 0 is stored, the previous state 1 being suppressed.

An additional winding on the core can be used to detect when a change of flux takes place. Any flux change will induce an EMF (in any winding on the core). This EMF will be negligible, proportional to the small flux change if the successive pulses are in the same sense. For the large and rapid flux change which results from successive dissimilar pulses, an EMF of about 0·5 V turn will be generated.

One method of utilizing the storage property of the core is to have two series of pulses, designated A and B in Fig B.3. The A pulses contain the information to be stored and so can occur in various patterns. The B pulses are used for interrogating the cores and usually occur at regular intervals. An A pulse sets the core and is always followed by a B pulse in the opposite direction which resets it, causing a large EMF output. Should the core be already in the reset state when a B pulse arrives, the flux is changed very little and the EMF generated is negligible. When a B pulse is followed by an A pulse, a large EMF is again induced, but now with opposite polarity, permitting ready distinction.

Stored information is destroyed when the core is interrogated. After reading, the core is always reset to the 0 state; if the information read out was a 1 bit, this must be written back (using a rewrite cycle) if it is required to be available for a subsequent interrogation. It follows that in storing information it is necessary to write in only the 1 bits, since a store which has been interrogated or cleared will already be in the 0 state.

Core Matrices. Batches of cores with similar functions are combined into matrices in which common write and read circuits can serve many cores.

A simple 2 x 2 matrix is shown in Fig B.4. The vertical and horizontal thin lines represent wires threading the cores, while the thick, short diagonal lines represent an end view of the cores, one core being at each crossing point of the horizontal and vertical wires. It is possible to write information into any core by passing a current corresponding to half the switching magnetizing force through each of any two wires threading it. These currents are usually known as *half-write* currents, or $I_S/2$.

Consider the storage of two words, each consisting of two bits – the first word 10 and the second 11. The writing-in sequence is as follows, assuming that all cores are initially in the reset condition.

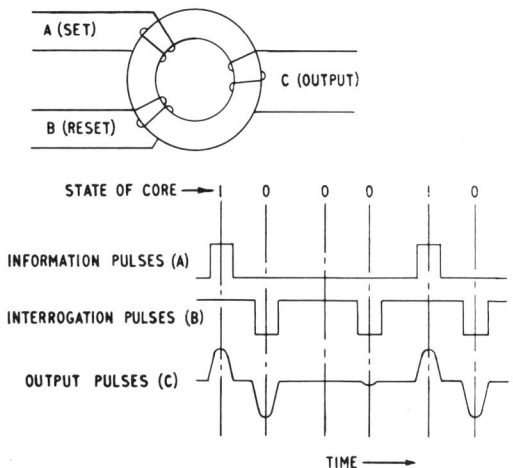

Fig B.3 Magnetic-core storage

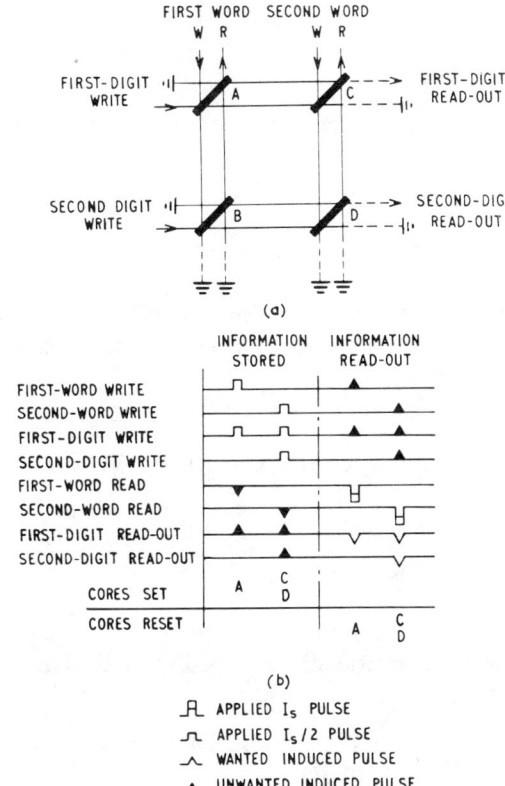

FIRST WORD   SECOND WORD
W   R   W   R

FIRST-DIGIT WRITE — A — — C — FIRST-DIGIT READ-OUT

SECOND DIGIT WRITE — B — — D — SECOND-DIGIT READ-OUT

(a)

	INFORMATION STORED	INFORMATION READ-OUT
FIRST-WORD WRITE	⊓	▲
SECOND-WORD WRITE	⊓	▲
FIRST-DIGIT WRITE	⊓	▲
SECOND-DIGIT WRITE	⊓	▲
FIRST-WORD READ	⊔	⊔
SECOND-WORD READ	⊔	⊔
FIRST-DIGIT READ-OUT	▲ ▲	⌄
SECOND-DIGIT READ-OUT	▲ ▲	⌄
CORES SET	A C / D	
CORES RESET		A C / D

(b)

⊓ APPLIED I_S PULSE
⊓ APPLIED $I_S/2$ PULSE
⌃ WANTED INDUCED PULSE
▲ UNWANTED INDUCED PULSE

Fig B.4 2 x 2 core matrix: (a) formation of matrix;
(b) pulses on matrix leads

For the first word a positive $I_S/2$ pulse is applied to the first-word write wire (W) at the same time as a positive $I_S/2$ pulse is applied to the first-digit write wire, but no signal is applied to the second-digit write wire. Core A will be set and core B will remain reset.

For the second word a positive $I_S/2$ pulse is applied to the second-word write wire (W) at the same time as positive $I_S/2$ pulses are applied to the first-digit and second-digit write wires. Cores C and D will both be set.

The information can be read out, one word at a time, at any subsequent moment by the application of a full negative I_S pulse to the appropriate word read wire — e.g. for the first word a negative I_S pulse is applied to the first-word read wire (R), and this resets core A but hardly affects core B, causing a large output to appear on the first-digit read-out wire and a very small output on the second-digit read-out wire. These pulses are of insufficient magnitude to affect cores C and D, so the information in the second word is undisturbed and can be read out later by application of a negative I_S pulse to the second-word read wire.

It can be seen from Fig B.4(b) that quite a number of unwanted pulses exist besides those required for reading out the stored information. Discrimination between wanted and unwanted pulses can be achieved by means of diodes (for polarity), limiters (for amplitude) and strobes (for time discrimination).

Possible economies in the matrix can be obtained by combining the functions of the read and write wires so that each core is threaded by only two wires. This means that the input and output pulses have to be switched by external circuits.

Magnetic cores can be readily used for other logic functions such as gates, counter circuits, shift-registers, etc.

References

1 French, J. A. T. et al. 'Outline of Non-linear Transistor Characteristics and Applications', *POEE Journal*, **56**, p. 268 (1964).
2 Morton, W. D., 'Semiconductor Device Developments: Integrated Circuits: Standard Logic Circuits', ibid. **60**, p. 110 (1967).
3 Banham, H. and Hillen, C. F. J., 'The Use of Magnetic Cores in Logical and Memory Circuits', ibid. **56**, p. 46 (1963).

Appendix C
A.C. Power Levels

A telecommunication transmission circuit, in part or whole, can be represented by the element shown in Fig C.1. Power is applied at a source E and transmitted into a receiver load Z_0. The power values at source and load are usually, though not necessarily, different.

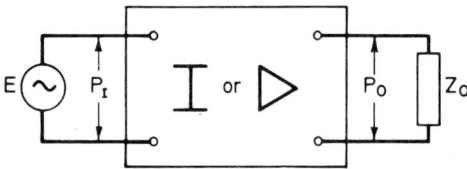

Fig C.1 Basic transmission system

It is convenient to compare or measure the power values at points along a telecommunication circuit using logarithmic units. This is because, in a chain of connections, the power output at any point always bears some *ratio* to the input power. Calculations of power ratios by arithmetic processes are cumbersome. When logarithmic units are used, power gains and losses of the links in the chain, expressed in such units, can be simply added algebraically.

The transmission unit is the decibel (dB) equal to one-tenth of the bel (B), which is simply the common logarithm (to the base 10) of the power ratio. If the power at input and output terminals of a line or equipment is written as P_I and P_O respectively, the power ratio is P_I/P_O. The common logarithm of this ratio is $\log_{10} P_I/P_O$. The power loss ($P_I > P_O$) or gain ($P_O > P_I$) in dB is:

$10 \log_{10} P_I/P_O$ dB (loss), or $10 \log_{10} P_O/P_I$ dB (gain)

If, instead of powers, the voltages V_I and V_O or the currents I_I and I_O are measured, then *provided that the two voltages or currents are measured in*

relation to the same impedance (usually the characteristic impedance Z_O) the power gain or loss is:

$20 \log_{10} V_I/V_O$ dB (loss), or $20 \log_{10} V_O/V_I$ (gain), or

$20 \log_{10} I_I/I_O$ dB (loss), or $20 \log_{10} I_O/I_I$ dB (gain)

since power $P = VI = V^2/Z_0 = I^2 Z_0$, so that the power ratio is $(V_I/V_O)^2$ or $(I_I/I_O)^2$.

For example, if the power is halved due to the insertion of an attenuation network, i.e. the power ratio is 2/1,

loss = $10 \log_{10} 2 = 10 \times 0.30103 = 3$ dB approx. ($\log_{10} 2 = 0.30103$)

Similarly, the power ratios 10, 100, 1000, . . . (logarithms = 1, 2, 3, . . .) are expressed as 10 dB, 20 dB, 30 dB, . . .

A value quoted in dB is not a power but a ratio of powers — more accurately the logarithm of a power ratio. A value in dB has no dimension, it is simply a number.

By way of example, suppose that a certain telephone circuit consists of lines with a total attenuation of 45 dB and includes two amplifiers giving transmission gains of 12 and 14 dB respectively; in addition, an average loss of 1 dB occurs at each of six switching points and 2 dB is allowed for reflection loss. The total attenuation amounts to 45 (lines) + 6 (switching points) + 2 (reflection) = 53 dB. The total gain is 12 + 14 (amplifiers) = 26 dB. The net loss is 53 − 26 = 27 dB.

For testing and measuring lines and equipment it is customary to use a power of 1 mW across a load of 600 Ω. Measurements made at points (Z_0 = 600 Ω) in a chain of networks with respect to the fixed power value $P_I = 1$ mW are referred to as the transmission *level* with respect to 1 mW sent power. For convenience, the value is written as x dBm, the m indicating that the value referred to is in relation to a sent power of 1 mW into 600 Ω. For example, with respect to 1 mW (= 10^{-3} W), a power of 1 μW (= 10^{-6} W) has a power ratio $10^{-3}/10^{-6} = 10^3$ equivalent to 30 dB; hence 1 μW may be expressed as −30 dBm. Similarly, 1 pW (= 10^{-12} W) has, with respect to 1 mW, a power ratio of $10^{-3}/10^{-12}$, equivalent to 90 dB, and so may be written as 1 pW = −90 dBm. The signs + and − are used to express power levels which are respectively above and below 1 mW, e.g. +30 dBm = 1 W, −30 dBm = 1 μW. A power of 1 mW may be expressed as 0 dBm since the power ratio 1 mW/1 mW = 1 and $\log_{10} 1 = 0$.

Several similar abbreviations are used for convenience. In the symbol dBr, the r relates to a *reference* point which is specified. The most commonly-used reference point in line transmission is the point at which the test power of 1 mW is applied (across 600 Ω) – the 0 dBm point. Power levels with respect to *this* testing point are expressed in dBm0.

In a 4-wire transmission circuit the *zero relative point* is the 2-wire point of a 2-wire/4-wire termination; the power at the 'go' channel terminals of the 2-wire/4-wire termination is expressed as −4 dBm0, the loss (in that direction) through the hybrid terminating transformer being approximately 4 dB.

The term dBa refers to F1A weighted-noise measurement above a reference noise level of −85 dBm, i.e. 0 dBa = −85 dBm (at 1 kc/s).

As stated above, an expression such as x dBm implies an actual value of power P_O since $P_I = 1$ mW and x is known. For convenience the values of power corresponding to various power levels (dBm) are given in Table C.1.

Table C1 Values of power corresponding to power levels (dBm)

dBm	Power	dBm	Power
+10	10 mW	−13	50 μW
+9	8 mW	−14	40 μW
+7	5 mW	−17	20 μW
+6	4 mW	−20	10 μW
+3	2 mW	−21	8 μW
0	1 mW	−23	5 μW
−1	800 μW	−24	4 μW
−3	500 μW	−27	2 μW
−4	400 μW	−30	1 μW = 10^6 pW
−7	200 μW	−40	0·1 μW = 10^5 pW
−10	100 μW	−50	0·01 μW = 10^4 pW
−11	80 μW	−90	1 pW

Administrations in Continental Europe use a transmission unit called the *neper* (N) and its submultiple the decineper (dN). This is also a logarithmic unit based upon a ratio but differs from the bel and decibel in two respects: (1) it is based on natural or *naperian* logarithms to the base e (e = 2·71828); and (2) it is based on the current or voltage ratio (in equal impedances, usually Z_0 = 600 Ω). Hence:

$$\text{loss (or gain)} = 10 \log_e V_I/V_O \text{ dN, or } 10 \log_e V_O/V_I \text{ dN}$$

$$= 10 \log_e I_I/I_O \text{ dN, or } 10 \log_e I_O/I_I \text{ dN}$$

$$= 5 \log_e P_I/P_O \text{ dN, or } 5 \log_e P_O/P_I \text{ dN}$$

In the last expression, since $I \alpha V \alpha \sqrt{P}$, a value $\frac{1}{2} \log_e (P)$ is involved.

From the relation ($\log_{10} e$ = 0·4343), the following conversion factors are obtained:

1 dB = 1·151 dN and, conversely, 1 dN = 0·8686 dB.

Index